Towards Nearly Zero Energy

URBAN SETTINGS IN THE MEDITERRANEAN CLIMATE

Towards Nearly Zero Energy

URBAN SETTINGS IN THE MEDITERRANEAN CLIMATE

ANNARITA FERRANTE
Department of Architecture
University of Bologna

ELSEVIER

Amsterdam • Boston • Heidelberg • London
New York • Oxford • Paris • San Diego
San Francisco • Singapore • Sydney • Tokyo
Butterworth Heinemann is an imprint of Elsevier

BH

Butterworth Heinemann is an imprint of Elsevier
The Boulevard, Langford Lane, Kidlington, Oxford OX5 1GB, UK
50 Hampshire Street, 5th Floor, Cambridge, MA 02139, USA

Cover Image Athens 2015, Courtesy of Maria Beatrice Bettazzi

British Library Cataloguing-in-Publication Data
A catalogue record for this book is available from the British Library

Library of Congress Cataloging-in-Publication Data
A catalog record for this book is available from the Library of Congress

ISBN: 978-0-08-100735-8

For information on all Butterworth Heinemann publications visit our website at http://store.elsevier.com/

CONTENTS

FOREWORD

Annarita Ferrante's book, *Towards Nearly Zero Energy: Urban Settings in the Mediterranean Climate* is a remarkable scientific contribution. It is a complete and scientifically advanced work that successfully investigates the synergies, the relations, and the trade-offs between buildings and the surrounding urban built environment.

The tremendous increase of the urban population, combined with an unprecedented increase of the boundaries of our cities, has created significant energy, environmental, economic, and social problems that mandate an immediate solution. Particularly in Southern Europe, these problems have become very significant and have increased the stress level in our societies.

Through specific case studies, this book identifies problems and presents their major characteristics, but it mainly tries to offer practical solutions for immediate implementation. It presents very well organized studies that identify the mitigation and adaptation potential of major technologies for buildings, and all are presented in a very practical and accessible way.

The idea of zero-energy buildings in zero-energy settlements that Ferrante promotes is a very attractive one and without a doubt her ideas will be a major theme for the future of architecture. Furthermore, very few books and articles of such a practical nature are available in the international literature, which is something that makes this book unique.

Mat Santamouris

CHAPTER 1

The Framework of Urban Built Environments

1.1 INTRODUCTION

In some ways, nowadays the concept of green buildings belongs to the "history of architecture," as the first prototype solar houses and the first attempts to achieve zero heating date back to the 1950s (Hernandez and Kenny, 2010). Over the past several decades, energy-oriented innovations in building technology have emerged in many areas of the building construction sector (Brown and Vergragt, 2008).The latest projects are aimed at setting to zero the energy demand and the carbon emissions of new housing developments and even of a whole city. Working toward this goal is the well-known urban village BedZED, the Beddington Zero Emission Development, winner of the prestigious Energy Awards in Linz, Housing and Building category, Austria, 2002 (Dunster et al., 2009). Another well-known example is the first zero waste/zero carbon emission city designed by Norman Foster, in Abu Dhabi, Masdar City, where the urban development is designed as a huge, positive energy building complex, resulting in a self-sustaining, car-free, urban environment. Finally, a pilot city plan to set to zero the carbon emissions of the entire city of Copenhagen (The Climate Plan, City of Copenhagen, 2009) shows how to make this city the world's first carbon-neutral capital by 2025, by means of using biomass in power stations, erecting windmill parks, increasing reliance on geothermal power, and renovating the district heating network.

We may assume that at present, energy-saving technologies have been successfully developed, that energy-efficient building designs have been experienced and extensively publicized, and that we own a large set of best available technologies (BATs). In other words, we now possess the technical knowledge to design and construct zero-energy buildings and city as well. "Moreover, extensive monitoring of local, national, and international building stocks means we know more than ever before about the precise potential for improved energy performance" (Guy, 2006).

Towards Nearly Zero Energy
http://dx.doi.org/10.1016/B978-0-08-100735-8.00001-5

Today, the concept of nearly zero energy building (nZEB) has gained great international attention. It represents the principal target of the next generation of buildings; in fact, building design construction and practices need to respond quickly to the mandatory nearly zero energy requirements for new construction within a few years (EU Directive on Energy Performance of Buildings, 2011; EU Communication, 2011a, 2011b).[1] The increasing interest in nZEBs, the recent European and national directives on energy performance of b (EPB), and more easily accessible BATs and renewable energy sources (RES) all seem to point to further exploitation of BATs and better penetration of RES into new building construction. However, current construction practices in the new housing sector show limited signs of change (Ferrante and Semprini, 2011). Furthermore, despite growing investments in RES technology (Bürer and Wüstenhagen, 2009), feed-in tariffs and, in general, policy incentives (Bulkeley, 2010), additional investments are needed to reduce carbon emissions and fossil fuel consumption.[2] "Needless to say, this is particularly challenging in a context of global economic slowdown such as the one the world is currently experiencing" (Masini and Menichetti, 2010).

This concern is particularly true within the Mediterranean areas, where mild winters and hot summer conditions, together with different – and more critical – social and economic factors determine a chronic resistance to the widespread penetration of nZEB principles in building construction practices. Here the observed growing interest in nZEB usually results in prototype models and experimental pilot studies whose real impact on building practices and construction is very limited. Furthermore, commercial and industrial interests by specific categories of landlords or private owners in green buildings has also caused the production of highly visible, fashion-oriented and contemporary buildings that are often located in the

[1] In the frame of the legislative plane, recently the European Parliament (Directive 2010/31/EU on the EPB), amending the previous 2002 EPB Directive, has approved a recast, proposing that by 31 December, 2020 all new buildings shall be nearly zero-energy consumption and will have to produce as much energy as they consume on-site. See also Task 40/Toward Zero Energy Solar Buildings, IEA SHC /ECBCS Project, Annex 52.

[2] The treaty issued by several NGOs calls for a doubling of market investments by 2012 and quadrupling by 2020 to attain the proposed carbon emission reduction targets (Meyer 2010). As reported by Guy (2006), according to the United Nations Environmental Programme (UNEP) there is still an "urgent need for the incorporation of EE issues to be included in urban planning and construction."

framework of superimposed, universal, and/or incremental master plans and radical urban operations (Charlseworth, 2006), whose real impact on current behavior and smaller building practices is, again, negligible (Brown and Vergragt, 2008).

In general, the majority of pilot models and constructions cited so far refer to newly conceived buildings and new development plans. Although new nZEB concepts and experiments might have been the first priority in the previous decade, in more recent years it has become widely acknowledged that renovating dwellings is the most significant opportunity to reduce global energy consumption and greenhouse emissions. In fact, while there was detailed research on and the implementation of new buildings designed to perform in a zero-energy world, it is important to remember that even if, going forward, we built all new zero-energy buildings, we would only make a very small contribution toward the reduction of energy consumption and emissions due to the energy-consuming tendencies of the existing building stock as a whole (Ravetz, 2008; Kelly, 2009). This is especially true in the European Union (EU), due to three main geo-political, concurrent factors:

First, the consolidated and historical characterization of the EU continent, where refurbishment represents the major need: three-quarters of the buildings standing today, including the residential stock, are expected to remain in use in 2050 and the demolition rate is about 0.1% per year. (Highly energy efficient new-build rates are approximately 1% additional per year [JWG, 2013]).

Second, the EU is facing a challenging economic crisis, with the construction and real estate sectors representing the worst affected market sectors and where energy efficient (EE) refurbishment is often indicated and supported as one of the potential solutions for the real estate/building construction crisis (Kaklauskas et al., 2011).

Finally, the EU faces uncertainties on energy imports. Since buildings represent 30% of the total energy consumption, the consequent gas and oil reduction in this sector could substantially contribute to reducing the EU's dependency on energy imports up to a hypothetical scenario of energy self-sufficiency for all the European countries.

In particular, the need to focus on saving energy in the building sector of the Mediterranean area is illustrated by a vast array of political, economic, and social factors. Globally, governments are called upon to reduce greenhouse gas (GHG) emissions. Worldwide, the Mediterranean region has the third largest growth in carbon emissions, especially in those countries that export oil, which make up 74% of the total emissions of the region.[3]

It is evident that the current challenge is to widen the nZEB technical knowledge in existing built environments, shifting the methodological and technical achievements on energy-efficiency from newly conceived buildings toward the rehabilitation of existing building stock (Ferrante, 2014), since the large amount of existing stock represents the biggest challenge and the larger potential in the reduction of energy demand and carbon emissions. This is urgent, and trial in fragile sectors like the social residential areas of the Mediterranean cities is imperative (Santamouris et al., 2007).

A systematic approach to retrofit buildings has been reported and discussed by Zhenjun et al. (2012) and Ferreira et al. (2013), and a specific set of tools for energy retrofit including economic cost-benefit models have been developed by Rysanek and Choudhary (2013) and by Kumbaroglu and Madlener (2012). Furthermore, studies, design proposals (Ferrante and Semprini, 2011), and measurements (Morelli et al., 2012) on building energy retrofitting have been investigated and discussed to achieve energy-saving tools and techniques toward nZEBs in existing buildings.

Despite the recent attention received by the topic from researchers and building market actors and notwithstanding the stress on the goal of reaching nZEBs within 2020, so far, only about 1.2% of Europe's existing buildings are renovated every year. In fact, buildings and the existing urban compound of modern cities still face a number of hurdles and bottlenecks to retrofit uptake:

First, there are high up-front costs, long payback times of retrofitting interventions, and a lack of access to available and affordable finance

[3] These countries include Algeria, Iran, Iraq, and Saudi Arabia. A growing awareness about the fact that fossil fuel resources are limited is emerging in these countries. In recent years, some of the previously net exporting countries became net importing countries. Also the (still) oil exporting countries will be facing with future problems to maintain high growth rates of energy supply. While in the past long-term power development plans have been focused on supplying more energy from more fossil-fuel burning power plants it is high time to redirect those investments to the supply of Green Energy and, more important, to increasing EE measures, thus increasing energy saving, and in parallel decrease rate of energy consumption. Relocating investments into sustainable energy production and consumption cannot only provide a cleaner and safer future but can actually reduce investments in energy infrastructure (MED-ENEC, 2015).

(economic barriers and financial barriers), all of which represent a major obstacle.[4]

Second, insufficient incentives and rigidity of current regulations, neither of which allow significant improvements in existing buildings nor give priority to deep energy renovation (legislative barriers), is another significant hurdle.[5]

Third, the delay of states and regions in producing strategies for building energy renovation threatens the adoption and application of measures

[4] The economic and financial barrier is indicated as the major obstacle also in MED-ENEC (2015). In this report are listed the following main subcauses in the framework of economics gaps limiting the energy retrofit uptake. They are: energy tariffs subsidized that distort the market and make investments look less attractive/not financially feasible; the absence of Green Finance guideline to incentivize commercial banks to lend to energy efficient investments (in the Mediterranean regions not all recommended technologies and the best available technologies (BAT) seem to be economical and financially feasible); the lack of public funds, public sector funds that very often are not properly targeted at the end-user; Public sector funds that, if and when available, often fail to leverage commercial lending; loans provided with no or little accompanying measures such as technical assistance; reluctant commercial banks in lending to EE because they do not understand the EE savings potential, so they do not assume EE investments considered to be high risks (with the consequent increase of interest rates) resulting in EE investments even less attractive; lack in staff capacity; lack of internal procedures, lack of cash-flow based lending procedures; high transaction costs also connected with the small size of investments; EE investments are considered to have high up-front costs, making pay-back periods long, and increase risk for investor. Finally, the report indicates that generally banks require high collateral, which can limit future investments.

[5] The obstacles in the legislative framework are also considered in the list of major barriers toward retrofit uptake (MED-ENEC, 2015). The following main sub-causes in the framework of legislative gaps limiting the energy retrofit uptake are listed: EE policies that do not formulate clear objectives and the connected weak enforcement of the policies; the policies that do not provide for clear priorities (no benchmarks such as standard labeling programs, EE building codes, certification systems to establish transparency. No standard energy auditing procedure, no standard energy auditor trainings, no certified graduation processes. No reporting standards on EE, no sufficient metering to document savings. Government support programs are developed but information not disseminated, leading to little or no demand); the absence of development of convincing implementing programs; the absence of government budget to kick start identified EE priorities; the absence or scarcity of reporting mechanism deployed; the weak capacity in the relevant authorities. One crucial observation in the MENA region is that the decision-making processes are very centralized, making a swift and locally adapted implementation difficult. The absence of a holistic approach is also indicated, for example there is a clear omission of cooperation between ministries and public authorities, while energy efficiency actions require different Ministries and authorities to cooperate between them and major stakeholders. In this context, the private fund and banking sector stakeholders are left out of consideration of energy efficiency although they are undoubtedly among the major facilitators of EE investments.

for EE as foreseen by article 4 of the Directive 29/2012/EU (Memberstate dependent legislative barrier).

Finally, other major bottlenecks for retrofit uptake are due to a general lack of awareness among final users (urban dwellers, tenants, owners, and house managers) and small enterprises operating in the building sector.[6]

By building on this backdrop, this book aims to contribute to the current debate on nZEB in the retrofitting of the built environment, pointing out the potential for major renovation in existing urban settings that has not been discussed so far. Using in-depth case study investigations in selected representative urban areas, the book seeks to answer the following questions:

- Is it technically feasible to achieve nZEB targets in existing urban buildings and compounds in Mediterranean cities?
- Which are the main barriers and constraints hindering the practical feasibility of retrofitting existing buildings toward nZEBs?
- How might we overcome those barriers?

1.2 URBAN ENVIRONMENTS AND ENERGY RELATED ISSUES

Cities are the "guilty victims" of energy demand pressures and climate change. On the one hand, cities are responsible for around 75% of energy consumption and 80% of greenhouse emissions, despite accounting for only

[6] The deployment of best practices, such as pilot projects or development of champions, is very often missing. Technical understanding of EE and its implication in practice is very limited among population in the Mediterranean regions; in the context of building construction/refurbishment, there are little skills to install, maintain, and monitor EE equipment and materials. There are not or not sufficient local equipment providers or production of EE material; no understanding of traditional (EE) materials and building techniques. Furthermore, in the frame of final users and inhabitants a number of barriers can be numbered. Among these barriers, difficulties are encountered in the quantification and appreciation of energy savings in individual equipment, and these are not easy to measure; lack in consumer awareness is another hurdle since benefits of saving are often not perceived as big, because each of the measures/appliances, when taken singularly, consume only a small amount of energy. This problem is exacerbated, as the information on EE equipment/material and its benefits are not readily available, which makes the information process burdensome and time-consuming. Finally, split incentives in decision-making processes represent a major barrier in the frame of home tenure: developers very often do not want to invest, because they are not directly the ones to save on energy, while tenants are not involved in the construction/renovation processes but are the ones that have to pay the energy bill at the end.

about half of the world's population (While, 2008). Cities have an extremely great impact on the environment, as they concentrate most of the population of the globe and most economic activity in sectors of building, transportation, and industry. Urban areas (especially in the warmer regions) often offer a very low climatic quality and therefore use large amounts of energy for air conditioning in summer. In general urban zones use even more electricity for lighting (Bitan, 1992).

Throughout the world, human beings are moving into cities (EU Report, 2010) and today more than two-thirds of the European population lives in urban areas (EU Report, 2011). With their concentration of population, cities consume resources disproportionately, contributing to the production of climate change (May et al., 2013). Urban growth has reached such a peak that bypasses, reversals, and new ways of development are needed.[7]

Conversely, cities are the victims of climate change. Phenomena such as the urban heat island (UHI) effect and its health consequences on urban dwellers, and the exposure of coastal areas and riverside cities to flooding and natural disasters represents the main threats in the Mediterranean region. Besides their energy and environmental needs, many cities in this specific region are now facing new and persistent problems of unemployment, poverty, and social exclusion. As a combined outcome of the labor market crisis on the one hand, and the functional or "Ford" city of the twentieth century on the other hand, much of the urban space is devalued[8] and an

[7] Some significant and alarming figures: the world population has grown from 2 to 6 billion, and soon will reach 7 billion, while the percentage of human beings living in cities has increased from 3% in 1800 to 14% in 1900 and is estimated to rise from the current 50% to 75% in 2050. The figure for Europe is still higher: 83% of the population are expected to live in cities by 2050 (EU Report, 2010). The average temperature on the Earth's surface has suffered an increase of +0.6% and is estimated to reach 1.5% by 2030. The progressive increase of global warming will specifically raise urban temperatures and heat island effect. Forecasts prepared by the Intergovernmental Panel for Climate Change indicate that, under reasonable assumption about future human activities (its mid-range "business-as-usual scenario"), concentrations of greenhouse gases will rise sufficiently over the next century to cause a world-wide rise in average temperature of between 1.5°C and 4°C.

[8] The urban space of the contemporary cities is often deteriorated, starting, for example, from the street level, where pedestrian paths and accessibility of buildings are often destroyed, either by closing all potential social/public attractiveness on ground level or by the infill of garages, causing a tunnel effect, which makes the streets unattractive and unsafe for pedestrians to pass.

increasing societal need for human recognition within depersonalized urban environments is emerging.[9]

The environmental question, as reflected in the issues of climate change, resource depletion, and loss of biodiversity, is nowadays arguably the foremost worldwide public concern, and urban planning, monitoring, and mitigation are essential parts of environmental and energy policy. The growing urbanization of the world population makes cities central players in the search for new approaches to environmental management.[10]

Thus cities, which use most of the environmental resources, are called upon to adopt programs in accordance with the concept of sustainability. The large, complex series of environmental dysfunctions produced by the urban-related use of resources is multifaceted: it produces effects at the local level, such as high energy consumption in the built environment, and air

[9] Redesigning energy technologies in the urban areas is certainly a major scientific challenge. However, succeeding in this endeavor requires more than getting the engineering right (Webler and Tuler, 2010). Thus, energy efficiency in urban settings is more than a technical problem. Recent studies have suggested that more focus should be placed on the social aspects at community level and that energy users should be engaged in their role of citizens. In fact, developing more sustainable consumption and production systems depend upon consumers' willingness to engage in "greener" and more collective behaviors (Peattie, 2010). In this frame, local urban communities have inimitable advantages in providing infrastructure for more sustainable consumption environment; in fact, different types of low-carbon communities as a context to reduce carbon intensity are emerging at different scales (Heiskanen et al., 2010). Existing literature (Mulugetta et al., 2010; Guy, 2006) stresses the need for a clear transfer from a technical/economically based urban theory to a human based and socio-technical urban vision to achieve a greener behavior in urban environment. Timid attempts in the sustainable "redesign" of urban places by locally based communities are arising in the spatial sphere of the urban street environments. In this context, it is important to recall that some of Europe's leading innovation nations have included user-driven or user-centered innovation as a way of providing innovative products and services that better correspond to user needs and therefore are more competitive. User-driven innovation (EU Communication, 2009) is closely associated with design, and involves tools and methodologies developed and used by designers. The re-designing of existing buildings by engaging final users is also occurring in different contexts of EU and worldwide.

[10] The need for rapid and concerted action on the global scale has been officially recognized by the international community in events like the United Nations Conference on Environment and Development, held in Rio de Janeiro in 1992. Agenda XXI, one of the agreements emanating from this conference, addressed the specific issue of climate change. This is a complex and controversial phenomenon related to the increase concentrations of naturally occurring greenhouse gases in the atmosphere due to human activities. These gases, which include carbon dioxide and methane, trap solar heat in the atmosphere, so warming up the surface of the planet. The main source of greenhouse gases is burning of fossil fuels, such as coal, oil, and gas for energy, very often in a rather inefficient way.

and water pollution at the urban microclimate, but it also produces regional and global effects, such as global warming.

1.2.1 Heat Island in Mediterranean Cities

It is largely demonstrated that the progressive increase of global warming will cause urban temperatures to rise, thus producing the heat island effect. This will have a tremendous impact. For example, after the Messina, Italy earthquake of 1908 (which caused about 83,000 deaths), the hot summer of 2003 with ~ 70,000 deaths, mostly in the cities, was the second heaviest natural disaster of the last 100 years in Europe.[11]

As reported by Santamouris (2001), countries in the Mediterranean region have recorded a huge increase in summer cooling demand, especially evident in urban areas. The effects of global warming are of concern for both the environment and for humans in the area.[12] Here, the average yearly air temperatures are expected to increase between 2.2 and 5.1°C

[11] As reported by Santamouris (2006), during the summer period, high ambient temperatures and heat waves cause dramatic problems to vulnerable populations living in overheated households. In France, the estimated death toll of the 2003 heat wave was about 15,000 deaths. According to the EuroSurveillance (2005), an estimated 22,080 excess deaths occurred in England and Wales, France, Italy, and Portugal during and immediately after the heatwaves of the summer of 2003. Additionally 6,595–8,648 excess deaths have been registered in Spain, of which approximately 54% occurred in Aug., and 1,400–2,200 even occurred in the Netherlands, of which an estimated 500 occurred during the heatwave of Jul. 31 to Aug. 13. In parallel, it is reported that approximately 1,250 heat-related deaths occurred in Belgium during the summer of 2003, almost 975 excess deaths during Jun.–Aug. in Switzerland, and 1,410 during the period Aug. 1–24 in Baden-Württemberg, Germany. Studies in Europe and the United States (Michelozzi et al., 1999; Michelozzi et al., 2005), show that the greatest excess in mortality was registered for people in low socioeconomic status, leaving in buildings with improper heat protection and ventilation.

[12] The Mediterranean climate is the less extensive of mesothermal climates, according to the 20th century geographical classification developed by German climatologist Wladimir Köppen (1846–1940), which continues to be the authoritative map of the world climates in use today. Currently, the upgraded version of Köppen classification uses six letters to divide the world into six major climate regions, based on average annual precipitation, average monthly precipitation, and average monthly temperature. As also reported, Köppen defined the Mediterranean climate (or Etesian climate) as the area where: (1) the mean temperature of the coldest month is between 3 and 18°C; (2) the summer season is generally dry and the rainfall amount of the wettest month is at least three times greater than that of the driest month; (3) the mean temperature of the warmest month is above 22°C; and (4) the mean annual rainfall amount (in mm) is higher than 20 times the mean annual temperature (in °Celsius). The first three conditions also refer to semiarid and arid regions adjacent to Mediterranean climate zones. Thus, the crucial difference between the Mediterranean and

(*Continued*)

(in summer between 2.7 and 6.4°C). On the northern side of the Mediterranean basin, the increase (to about 38°C) is likely to be most pronounced in winter. This is forecast to occur by 2100, although more recent research indicates that the time span may be even shorter (Hansen et al., 2007). The overall 2°C air temperature increase represents a critical limit beyond which dangerous climate changes could occur by 2030.[13] All climate forecasts also converge onto similar considerations for extreme events like the increase of the frequency, intensity, and duration of heat waves, and a clear drop in the number of rainfall days with a consequent increase in the length of the longest rainfall-free periods.

As a consequence of the heat balance, air temperatures in densely built urban areas are higher than the temperatures of the surrounding rural zones, which is known as the "heat island" (HI) phenomenon. The HI effect is due to many factors (Santamouris, 2001; Yamashita, 1996; Akbari, and Konopacki, 2005): the canyon geometry of cities, the thermal properties of certain materials that increase storage of sensible heat in the fabric of the city, the anthropogenic heat, and the urban greenhouse. Research studies on this subject usually refer to the "urban HI intensity", which is the maximum temperature difference between the city and the surrounding area (Santamouris, 2001).

Zinzi (2010) reports interesting data about the Mediterranean region, defining it as a "geographical complex entity" which consists of 23 seaside states with about 600 cities, 46,000 km of coastline, and more than 450 million inhabitants. This constitutes 7.2% of the world's population, 9% of

adjacent arid climate zones is the mean annual rainfall. The Mediterranean Sea contributes to the temperate warm climate, retaining heat in summer and releasing it in winter. The majority of the regions with Mediterranean climates have relatively mild winters and hot summers. However, winter and summer temperatures may vary greatly between different regions. In the case of winters, for instance, Lisbon experiences very mild temperatures in the winter, with frost and snow practically unknown, whereas Thessaloniki has colder winters with annual frosts and snowfall. As a further example, inland France is very different from the southern borders of Egypt or Libya. Nonetheless, many similarities can be found for a wide portion of countries bordering the basin: in almost all the coastline cities, the minimum yearly average temperature is within 5–10°C and the maximum is within 27–34°C, with the highest values being recorded in the Turkish coastline and Cyprus.

[13] These climatic conditions in the Mediterranean area would result in one additional month of summer and in a longer duration of tropical nights (namely temperatures always above 20°C). Another consequence is a decrease in rainfall of around 30% with respect to the actual average standards.

total energy supply, 10% of electricity consumption, and 8% of CO_2 emissions. According to Zinzi (2010), in this complex area, a first simple partition can be assessed on a socioeconomic basis: the north rim European states and the south and east rim states, with their transition economies.

The Mediterranean is widely recognized as the area of the world in which the call for sustainable development encompasses all main issues because: (1) it represents a fragile eco-region where development is already set back by environmental damage, and (2) it is an example of a common contrast between the developed northern part and the developing southern part.

The demographic trend in the Mediterranean region is quite alarming: the north rim population increased by 14% from 1975 to 2005, when the maximum expansion occurred and when it reached 190 million inhabitants. The south rim population has almost doubled in the same period, accounting now for more than 258 million people (Zinzi, 2010). The south rim trend appears to be critical in terms of environmental impact because all the issues are strictly related to the massive urban sprawl.

The urban population increased from 42% of the total to 62% in 2005. Another consequence of urban sprawl in several countries is that more than 20% of the population moved into major cities; the percentage increased to 26% in Turkey and more than 35% in countries such as Greece and Portugal. Percentages of more than 60% were recorded in smaller states where the population is concentrated in a few urban areas, such as Israel or Jordan. According to the trend, it is expected that another 70 million people will live in metropolitan areas by 2025, with about 90 million expected to dwell in the coastal urban settlements. Thus, the future of Mediterranean countries relies on the implementation of new development models based on a conscious integration of environmental, social, and economic issues.

As also reported by Zinzi (2010), a strong impact of the urban heat island is on energy use in buildings.[14] According to OECD/IEA (2007), the building sector, including both residential and nonresidential, is experiencing

[14] The outdoor air temperature increase has several implications at the level of energy consumption of buildings: (1) energy consumption increase for cooling the building; (2) increase of peak cooling demand and, as a consequence, chillers' size increase; (3) energy price increase; (4) less "granted" energy supply especially in crucial periods, as in summer heat waves; and (5) reduction in the efficacy of bio-climatic and passive cooling strategies, often based on night natural ventilation techniques, when the outdoor air temperature decreases and shows several degrees lower with respect to the indoor air.

an incremental trend in energy consumption, accounting for one-third of global end uses (the total rises to more than 50% when considering the sole electricity consumption). The relative electricity consumption increase was 2.3% in 2005 with respect to the year before, but the value for the building sector was 3.7%. A decisive role in this context is played by room air conditioners and cooling systems mainly used in dwellings, whose market is under a continuous positive trend (Zinzi, 2010), also considering additional increasing electricity consumption in the next few years, because of the growing energy demand of transition economies.

Global warming and urban heat island increase the ambient temperature and intensify the energy consumption for cooling purposes of buildings (Santamouris et al., 2015). In fact, the major impact of ambient temperature rise is related to the possible increase of the peak electricity demand that obliges power utilities to construct additional power plants and probably increase the cost of electricity supply. Furthermore, as illustrated by Santamouris and colleagues, the potential increase of electric energy consumption puts both the consumers and the electricity networks under stress. The review illustrates the results of 15 studies examining the impact of ambient temperature on total electricity consumption, showing how the penalty of ambient overheating is quite high and depends mainly on the characteristics of the building stock, the climate zone, the urban form, and the type of energy services provided. The potential increase of the electricity demand per degree of temperature rise varies between 0.5% and 8.5%. In particular, for the case study of Athens, a percentage increase of the base electricity load per degree of temperature increase equal to 4.1% was found.[15]

1.2.2 The Case Study of the Athens Metropolitan Area (AMA)

The city of Athens represents a highly significant pilot study, both in the context of energy issues and for the ever-emerging problems of degraded urban environments.

The urban history of Athens in the last 170 years (1830–2000) is marked by expansion – its population grew from 9,000 inhabitants

[15] The potential increase of the peak electricity demand triggered by the ambient warming is also high (Santamouris et al., 2015). Existing studies show that the peak electricity demand increases from 0.45 to 4.6% per degree of ambient temperature rise. This corresponds to a penalty of about 21 (±10.4) W per degree of temperature rise per person.

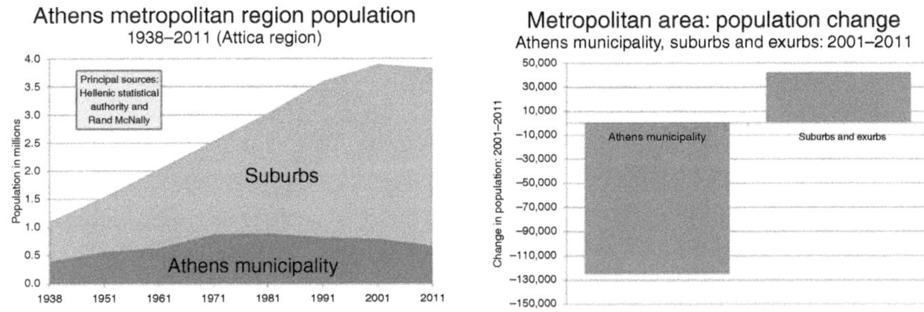

Figure 1.1 ***Athens growth population.*** *(Source: Wendell Cox,* The Evolving Urban Form: Athens; *Principal Source: Hellenic Statistical Authority & Rand McNally).*

in 1824[16] to 3.75 million in 2014 (considering the Athens Metropolitan Area, in the Attica region, with its 58 municipalities).[17]

The Athens city area is a metropolitan area (AMA) located at the south end of the Attica peninsula, on the Aegean Sea. Since 1951, suburban and exurban Athens has accounted for 95% of the growth in the metropolitan region, adding 2.2 million new residents, compared to approximately 100,000 for the Athens municipality (Wendell, 2013). Since 1971, all of the population growth has been in the suburbs and exurbs (Fig. 1.1).

Even with its current slow and (and sometimes negative) growth, the Athens urban area remains among the densest in the developed world (Figs. 1.1 and 1.2). This place, Athens is slightly ahead of London (15,300 per square mile or 5,900 per square kilometer), about double the density of Toronto or Los Angeles, and more than four times that of Portland.

From the administrative and urban point of view, the territory of Athens can be represented by a synergy of conurbations involving the historical town, the central area of the city, and the other 34 municipalities, for a total of 35 municipalities forming the prefecture of Athens. The monumental

[16] During the years of the Greek Revolution, Athens was a small city (in 1824 has 9000 inhabitants and occupies extension 116 hectares) destroyed and depopulated by war events. Nevertheless played an important ideological role in the symbolism of the birthplace of democracy and the source of Greek culture and civilization – values that have been adopted by the European Enlightenment and the Greek Revolution. Therefore, Athens as the capital would have the ideological authority to incarnate the center of the scattered Greek ethnicity.

[17] The 2013 edition of Demographia World Urban Areas indicates that the Athens urban area has a population of 3.5 million, living in a land area of 225 mi^2 (580 km^2), for a density of 15,600 per mi^2 (6000 per km^2).

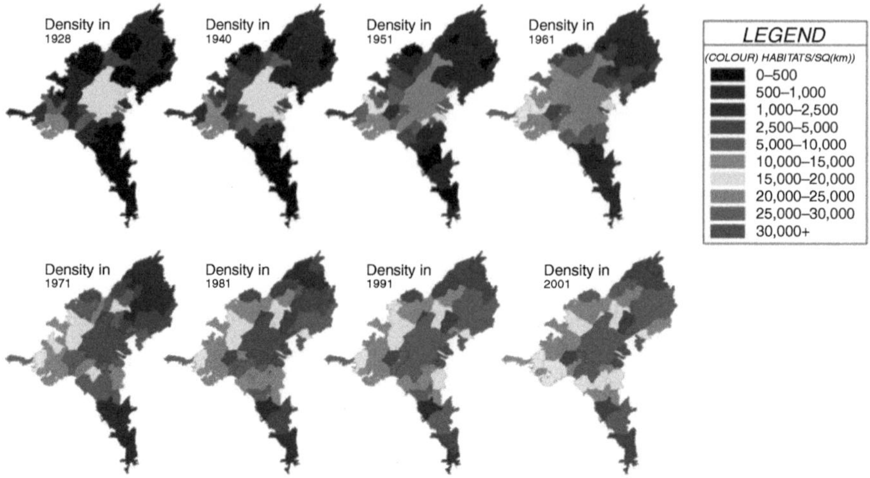

Figure 1.2 ***The evolution of urban density (in terms of inhabitants per square meters) in Athens by districts from 1971 to 2001.*** *(Source: http://en.wikipedia.org/wiki/File:Population_Density_in_Athens.PNG).*

urban structure of Athens dates back to the Otto period (1833–1864)[18], as the logical consequence of selecting Athens as the new capital. The plan is a clear response to the ideological role that a capital city plays. The monumental and hierarchical structure has, as a main reference, the Acropolis, a

[18] Otto and the Bavarian power installed in Nafplion in January 1833. That period it is developed a very keen competition and skepticism among various cities for the establishment of the capital of the state. Since 1831, the architects Kleanthis and Schaubert, old students of Schinkel, begin to prepare a plan for a new city capital in the region of Athens. The ideological acceptance of the state as it is necessary the existence of the project and new standards. The design of architects Kleanthis and Schaubert, one the most advanced for its time, proposed a combination of rectangular grid and radial connections of central spots and large open spaces and marked with its basic choices the center of Athens. In terms of technical design, it is one of the best examples of the era. The area of the project occupies a total of 300 hectares. The plan was submitted to King Otto in late 1832 and adopted in Jul. 1833, while at the same time, the decision was made to transfer the seat of government from Nafplio to Athens. The decision on the establishment of the capital in Athens ended a long-term uncertainty. However, it marked the beginning of a literally chaotic situation on the city plan, because, from the start and beyond any doubt, it appeared that the financial and technical resources to implement the plan were minimal. The successive modifications of the plan, however, changes the projected position of public buildings and especially the palace, the large decrease (to full elimination in practice) of the area reserved for excavations can be attributed in interests on speculation in land, of Athenians and foreign owners of land. These interests were the ones that intertwined with the aims of political parties and the disputes between the state and the municipality and the lack of resources finally led to inactivity on the project and the majority of the original structure, except for the layout of the main axes of the historic triangle – Panepistimiou Avenue and Pireos streets, Ermou and Athena streets – (May, 1929; Biris 1995) that remains in the urban form of the city today.

symbol of the ideological prestige of Athens, and the Old Royal Palace. The position of the ministries surrounding the palace expresses the nature of the new administrative and political power.

As in many other cities, in Athens the trend toward eradicating the suburbs started early, in the period between 1864 and 1909,[19] and was fully eradicated later in the 1970s. The plans for the suburbs from the beginning of the last century are anarchy plots, with no emphasis on urban design or organized public space. Thus, the expansion of the city through suburbs occurred since 1900 (Sarigiannis, 2000); at that time Athens had already acquired one of the urban characteristics that distinguishes the city even now – mismatched large areas in relation to its population size, with extreme (from low to high) density variation and a considerable urban sprawl, where the interests of small and medium property holders almost[20] determine the entire processes of expansion (Asteriadi, 2011).

A nonsolid, but very rapid growth of the manufacturing sector and the consequent rapid expansion of the city occurred again in the period

[19] During this period, until the rise to power of Venizelos and World War I, the growth in industry formalized the role of Athens and Piraeus as the dominant economic pole of the country. Athens gets stronger, with the construction of railways and roads that connect the new capital with Peloponnese and Thessaly and the opening of the Isthmus of Corinth, promoting the capital as the new economic center of the country. Athens' industry was concentrated mainly west of Keramikos area, where the gas and the silk factory are established, from which respective regions took their names (Gazi and Metaksourgio area) and in the north of the port of Piraeus. During that time, there was not a proper study for organizing the overall structure of the city. The pre-existing plan for the center of Athens, after the constant changes and slight extensions to Klenze's plan, had more than 173 amendments, which concern mostly roads narrowing or elimination of public free areas (Biris, 1995). Typical design proposals of that period are the extensions of the existing urban design, with the suburbs been designed without connection to the city, approved ad hoc. It is indicative the situation of the suburb Kato Patissia, closing the Avenue September 3rd, despite the fact that this was planned to be one of the main arteries of the town (this "closing" remained a major obstacle to the urban tissue of Athens until now). In 1907, the population of Athens exceeds the 260,000 inhabitants (Biris, 1995; Mantouvalou, 1988).

[20] However, it is important also to remember another important aspect of the period, which refers to the modernization and role of the state: to restore the capital's connections with the territory, the urban transportation and infrastructure networks (train, gas, electricity, water) is developed, in parallel to the western metropolises' similar construction. But again, it is indicative that even within this context, because of the discussions and oppositions of the main stakeholders of the time with respect to construct roads able to solve the congestion of the central part of the city brought to the result that such infrastructures were never performed.

between the two World Wars,[21] contributing once more to the fragmentation and real estate development of large areas of suburban land without a plan and with scarce forecasting of possible integration with the existing fabric of the city.[22]

The same tendencies continued after the Second World War and even until the present time. However, a dramatic element seems to define the evolution of the urban design space in the Athens Metropolitan Area: a contradictory relationship with respect to the historical time, characterized by the fact that while there is an ideal fascination of and desire to continue an unbroken continuity of Hellenism, the actual urban path is marked by consecutive shocks and discontinuities. The construction of buildings of the 1950s and 1960s, built by entrepreneurs and manufacturers, with the method of payment, sometimes illegally constructed without even an architect's study, turned Athens into an unidentifiable European capital.

The densification trend sped up in the 1980s, and continued at the periphery. Throughout this period, many urban plans were drafted for

[21] The large international financial crisis was combined with the installation of 1,220,000 refugees in Greece. The war period (Balkan, World War I, the Asia Minor campaign), intensified the severity of the problem in Athens compared with other cities of the country, not only in terms of the population but also from an economic and political point of view. The crisis of housing in Athens is acute, while the economic crisis and high inflation block the possibility of housing construction. Rent controls that were imposed in 1916 resulted in the total suspension of all the construction activity. Thus, the addition of 246,000 refugees in Athens comes at a time when there is already a very acute housing shortage (Travlos and Kokkos, 1977; Vergopoulos, 1978).

[22] Illegal construction intensified with the tolerance of the state in the hope that it would help to defuse the housing demand. During this period of illegal widespread construction, further legislative tools were developed for economic cooperatives. In this context, it is worth mentioning that there are some cases where local plans were studied and applied. It is the case of some of the social housing program for the accommodation of refugees. The houses that were built were very few in relation to the number needed. A very large percentage of the refugees were living in huts, constructed by themselves or by people that were interested when there was an available plot from the state. The construction of some urban compounds in different parts of Athens took place (building block settlements for refugees in places like Kaisariani, Dourgouti, Kokkinia, Aghios Dimitrios, Agia Varvara, Peristeri, etc.). These were erected and positioned in urban plots according to the experimental modernist principles implemented in Germany, in particular by Walter Gropius. Besides the suburbs mentioned before, there are some suburbs for the middle class that have been developed as private businesses (Psychiko, Ekali Filothei, Ilioupoli) and are designed as model garden cities, after similar suburbs in Europe (central, round, large open spaces, radial arrangement, gardens).

the overall organization of the city. Alongside these projects, a new urban legislation was revised, often going hand in hand with the urban development of the western countries of Europe.

The city has expanded without following a harmonic overall plan and the organization is considered problematic and very far from the desired European patterns. Typical modern houses characterize the current aspect of the capital. Apartment buildings built both in the center and the periphery took the place of the architectural valuable neoclassical houses (Travlos and Kokkos, 1977).

Nonetheless, Athens appears to retain potential in terms of the ability to rethink its development and rebound.

Surveying the urban planning history of Athens, we encounter the traces of multiple forces; these forces have sometimes amplified and sometimes canceled out one another. The attempts at organized construction, the production of surplus value, the institutional and political shifts, the social reception of issues relating to the city's cultural heritage and environment, the natural disasters, and the influxes of immigrant and refugee populations have all fed into the ever-changing urban landscape.

Ultimately, the cityscape is the face of Athenian society, which is why the history of the formation of Athens' urban fabric has so much to tell us about Greek society as a whole.

The 20th century brought separation and dispersal of buildings to an extent unparalleled in city history. Aerial photos and ground observation confirm this unambiguously – the sharp contrast of built form between the old "city" and its newer additions is inescapable (Grammenos, 2011). However, the 20th century also ushered a new form of agglutinated settlement, the vertical block, which can equal several earlier horizontal blocks in habitable space and thus dramatically increase the potential for the concentration of people.

To classify the different densities encountered in the Athens metropolitan area (AMA), a preliminary survey of the different urban districts has been conducted by visiting, observing, and analyzing the different areas and suburbs in the Attica region. The ethnographic observation has shown that different densities and different urban textures and morphologies shape the urban contexts of Attica. Obviously, they are also characterized by different climatic data, orientation, and so forth at the urban scale, but the territorial morphological asset of the natural boundary conditions around the Athens metropolitan area is the main cause of the different climatic conditions within the different parts of the city, as it will be shown in the next section.

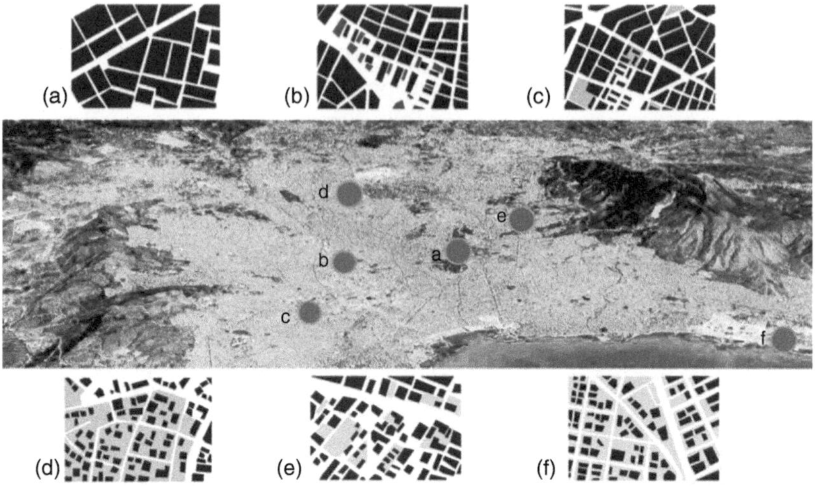

Figure 1.3 ***Main urban textures in the Athens metropolitan area.***

In this preliminary classification, six main areas have been selected to explicate different urban textures and morphologies in the north, south, center, east, and west areas of the AMA. Fig. 1.3 depicts a scheme of the typical urban textures in the AMA, ranging from very dense (a,b) down to suburban (c,d) and peripheral (e,f).

1.2.2.1 The Urban Heat Island in the AMA

In the context of heat island research, the city of Athens represents a highly significant pilot study. Here, increasing urbanization and deficiencies in development control in the urban environment have important consequences on the thermal degradation of urban climate and the environmental efficiency of buildings in the AMA (Cartalis et al., 2001; Hassid et al., 2000). As a consequence of heat balance, air temperatures in densely built urban areas of Athens are higher than the temperatures of the surrounding rural zones.

In Athens, the UHI is the foremost threatening issue affecting the energy demand and the comfort conditions in the built environment. But the Athens climate also presents a significant heating demand in the winter period, as will be illustrated in Chapter 2, when we will consider the total energy demand of selected representative areas. In this context, it is important to recognize that although Athens has winter temperatures that are milder than the freezing winters of Northern Europe, it still has a significant heating demand in its winter season.[23] January is, on average, the coldest month, with usual temperatures around 10°C (50°F), while the average low temperature is 7°C (45°F), and the average high temperature is 13°C (55°F). The lowest ever recorded temperature in Athens in January is −2°C (28°F), while the highest ever recorded temperature is 21°C (70°F). As a combined result of the recent economic crises and energy demands in winter, much of the low-income population has been found to be unable to cover their housing energy needs. Hence, these people live in temperature conditions that are well outside established comfort limits[24] (Santamouris et al., 2014). Therefore, Athens must be considered as a highly representative case study for energy demand conditions, ranging from significant heating needs and high cooling demand in the winter and summer seasons, respectively.

Data compiled by Ferrante (1998) and surveys performed in Athens on the HI intensity, involving more than 30 urban stations, show that during hot summer seasons, urban stations present temperatures that are significantly higher than the ones recorded in the comparable suburban stations (the gap varies from 5 to 15°C). As a consequence, the cooling load of reference buildings in the city center is about twice the value of equivalent buildings in rural areas. Previous research developed within the frame of the research project POLIS in Athens (Ferrante et al., 1998) has shown some appropriate procedures to design the use of natural components – such as green roofs and pedestrian permeable surfaces – within urban canyons.[25] The design of outdoor spaces – even if reduced to the envelope of the buildings because of existing urban constraints within thickly-built urban areas – as well as the use of natural components have been regarded as key means to improving

[23] Athens has a subtropical Mediterranean climate (according to the Köppen classification) since it receives just enough annual precipitation to avoid Köppen's semiarid climate classification. The dominant feature of Athens' climate is alternation between prolonged hot and dry summers and mild winters with moderate rainfall. In fact, rainfall occurs largely between the months of October and April (average of 414.1 mm of yearly precipitation has been registered. July and August are the driest months, where thunderstorms occur sparsely, about once or twice a month. Winters are mild and rainy, with a January average of 8.9°C (48.0°F). Snowstorms in Athens are generally infrequent (though more frequent in the northern suburbs of the city) and can cause disruption when they occur.

[24] For the whole period of the study (Santamouris et al., 2014), climatic data on the outdoor environment were provided by the National Observatory of Athens. The minimum temperatures were recorded during January 2013, (0.9°C), while the corresponding minimum for December was 5.6°C. The average temperatures for December and January were 11.1°C and 10.5°C, respectively.

[25] They are briefly illustrated in Chapter 2 of this book (Mitigations on Urban Canyons in Athens).

urban conditions in relation to both microclimate and reduction of pollutants. By "making-up" the building's surfaces and elevation facades with green components or shading devices, four different scenarios have been proposed in four different urban canyons in downtown Athens. Experimental software research models have been used to quantify the positive effects of these selected passive techniques. Obtained results clearly indicated that outer surfaces' alternative design acts as prior microclimate modifier and deeply improves outdoor air climate and quality (up to two-thirds °C reduction in ambient temperature) (Santamouris, 2001). In Chapter 2, results of these studies are reported with more detailed information.

Other significant physical factors in the thermal performance of urban environments are wind flows and air circulation (Santamouris et al., 1999), (Ricciardelli and Polimeno, 2006) as well as air stratification within urban canyons and open areas. It is clear that the heat island effect and the microclimatic conditions typical of open urban canyons appear to be strongly influenced by thermal properties of the materials and components used in the buildings and on the streets (Bitan, 1992; Buttstädt et al., 2010). The comparative research carried out by GRBES, the Research Group on Buildings Energy and Environmental Studies at the University of Athens, demonstrated that the use of cool-colored materials (Synnefa et al., 2007) and thermochromic building coatings can contribute to energy savings in buildings, providing a thermally comfortable indoor environment and improved urban microclimatic conditions (Karlessi et al., 2009).

Thus, morphological and spatial geometry of the urban textures, as well as the thermal properties of surface coatings and green surfaces have a strong potential on the energy performance and cooling demand reduction in urban settings.

Athens is subject to a strong and predominant "heat island" effect with seasonal temperatures rising 5–7°C, especially concentrated in the higher densely populated areas of the city center. Here, the moderate sized buildings of the suburbs have been progressively substituted by prominent structures, both considerably high and massive, generally squat and heavy, with a reciprocal proximity that alters the winds' flow and further increases soil sealing and heat accumulation. While mitigating the winter cold season, the morphological configurations of the central Athens area lead to a drastic climate change during the hot summer period, generating extreme high temperatures and consequently hazardous health conditions. Of course, this effect also undermines the energy use; in particular, it negatively affects the cooling need and the same performance of the cooling systems since in the maximum temperatures, the air conditioning systems' performance may decrease up to a quarter compared to the average seasonal efficiency.

As reported by Founda (2011), the urbanization of Athens and its effect on the temporal variation of air temperature have been stressed by many researchers, even in very early studies concerning climatic changes of the city. Most studies carried out up until the late 1980s identified the UHI effect on the minimum rather than on the maximum temperature (Katsoulis and Theoharatos, 1985; Katsoulis, 1987). The UHI is reported to be more pronounced in winter and is related to the heat produced by human and anthropogenic activities, especially in the evening hours, whereas mean and maximum temperatures are supposed to be less affected by the urban effect due to the influence of the sea breeze (Katsoulis, 1987). In more recent studies, however, it has been reported that the UHI effect is manifested mainly in the daily maximum temperature, which has increased significantly during the summer since the mid-1970s (eg, Philandras et al., 1999; Founda et al., 2009; Mihalakakou et al., 2004).

Climatic measurements from almost 30 urban and suburban stations as well as specific measurements performed in 10 urban canyons in Athens, Greece, have been used to assess the impact of the urban climate on the energy consumption of buildings (Santamouris, 2001). It is found that for the city of Athens, where the UHI intensity exceeds 10°C, the cooling load of urban buildings may be doubled and the peak electricity load for cooling purposes may be tripled, especially for higher set-point temperatures, while the minimum coefficient of performance (COP) value of air conditioners may be decreased up to 25% because of the higher ambient temperatures. Regarding the potential of natural ventilation techniques when applied to buildings located in urban canyons, it is found that, mainly during the day, this is seriously reduced because of the important decrease of the wind speed inside the canyon. Air flow reduction may be up to 10 times the flow that corresponds to undisturbed ambient wind conditions. Thus the UHI in Athens was found to have a stronger effect on the daily maximum summer temperature (up to 10–15°C for urban and rural/suburban regions); (Santamouris, 2001; Santamouris, 2007, Livada et al., 2002).

Using an ensemble of regional climatic scenarios, Founda and Giannakopoulos (2009) have illustrated that, by the end of the century, maximum and minimum summer air temperatures in Athens are expected to increase by approximately 4°C with respect to 1961–1990. It is also reported that the exceptionally hot summer of 2007 is likely to be the "normal summer" of the period 2071–2100.

Many are the studies, the measurements, and calculations in the AMA that confirmed the existence of a strong HI phenomenon (Mihalakakou et al., 2002; Mihalakakou et al., 2004; Santamouris et al., 1999). The association of the HI with synoptic climatic conditions has been identified (Livada et al., 2002) and the influence of the surface temperature and wind conditions have been analyzed in literature (Papanikolaou et al., 2008; Stathopoulou et al., 2009). In parallel, the impact of various mitigation techniques involving cool and reflective materials has been identified (Doulos et al., 2001; Karlessi et al., 2009; Santamouris, 2001, 2007, 2010). As already stated, all performed research studies on this subject – referring to the "urban HI intensity" as the maximum temperature difference between the city and the surrounding area – demonstrate that the AMA represents a highly significant pilot study: during hot summer seasons (corresponding to the HI upper limit), urban stations present temperatures significantly higher than the ones recorded in the comparable suburban stations (the gap varies from 5 to 15°C). Furthermore, a detailed statistical analysis of the HI characteristics and distribution in the greater Athens area has been carried out (Giannopoulou et al., 2011) using temperature data from 25 stations distributed on the city. Fig. 1.4 reports the location of the meteorological stations used this study.

The presence of the mountains Egaleo at the western part and Parnitha at the northern part and also of the mountains Penteli and Hymettus at the northern and eastern parts, respectively, act as the major natural obstacles at the territorial scale, since it is a barrier against the north winds which dominate during the summer period at the Athens area (Etesian winds). In the analysis, it is concluded that the higher air temperature values were found firstly at the western part, mainly due to the combined effect of industrialization and poor ventilation given by the territorial natural

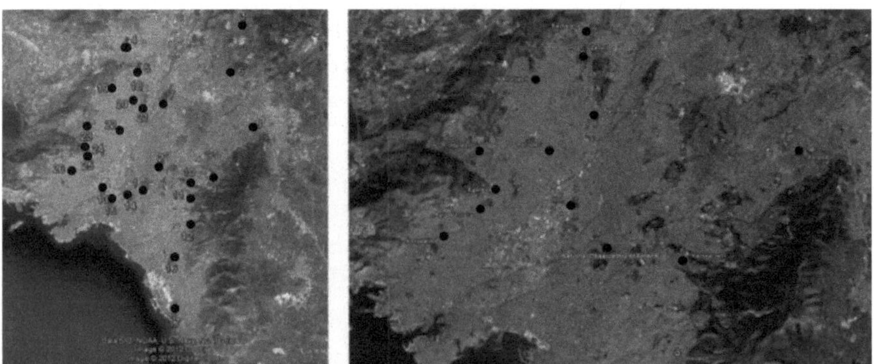

Figure 1.4 ***Location of meteorological stations.***

obstacles mentioned previously. Secondly, high temperature values have been found at the center of Athens, due to the traffic and anthropogenic heat as well as to the presence of continuous impermeable surfaces and densely built areas. Conversely, lower values were found at the northern and the eastern parts of the greater Athens area (Tables 1.1 and 1.2). According

Table 1.1 Values of monthly maximum air temperature for all monitoring stations and mean diurnal and nocturnal temperature differences from the reference station located in the center of the city during June, July, and August

Area	Station	June	July	August
Center of the city	Athens Univ.	33.9 ± 2.65	35.1 ± 2.95	31.9 ± 196
Northern part	N. Erythrea	30.8 ± 2.79	32.4 ± 3.16	31.4 ± 2.63
	Ano Liossia	31.6 ± 2.97	33.8 ± 2.92	32.0 ± 2.10
	Kamatero	33.6 ± 2.61	35.3 ± 3.10	32.1 ± 2.17
	N. Philadelphia	30.0 + 2.35	32.6 ± 2.39	31.1 ± 1.52
	Maroussi	29.6 ± 2.47	32.0 ± 2.19	30.2 ± 1.48
	Mean	31.12 ± 2.980	33.23 ± 2.994	31.36 ± 2.114
Eastern part	Zografou	30.6 ± 2.54	32.4 ± 2.55	30.6 ± 1.78
	Kessariani	32.4 ± 2.99	33.8 ± 2.90	31.5 ± 2.33
	Ilioupooli	32.0 ± 2.47	34.0 ± 2.69	31.5 ± 1.73
	Byronas	32.7 ± 2.79	35.1 ± 2.71	33.5 ± 2.05
	Agia Paraskevi	28.1 ± 1.84	31.0 ± 2.03	30.7 ± 1.47
	Mean	31.17 ± 3.027	33.25 ± 2.930	31.57 ± 2.146
Southern part	Glyfada	29.5 ± 2.67	32.8 ± 2.11	32.6 ± 1.29
	Renti	31.7 + 3.82	35.9 ± 2.90	36.2 ± 1.82
	Elliniko	30.2 ± 3.57	34.6 ± 2.70	35.9 ± 1.58
	Kallithea	32.0 ± 3.61	35.0 ± 2.12	33.6 ± 1.63
	Mochato	32.9 ± 3.38	35.9 ± 1.85	34.5 ± 1.31
	Mean	31.27 ± 3.608	34.84 ± 2.606	34.57 ± 2.503
Western part	Korydallos	30.4 ± 2.41	33.1 ± 2.12	31.2 ± 1.58
	Agia Varvara	32.0 ± 2.76	34.1 ± 2.31	31.8 ± 1.79
	Haidari	33.1 ± 2.98	34.1 ± 2.55	33.0 ± 2.46
	Egaleo	31.1 ± 2.98	33.8 ± 2.35	32.5 ± 1.55
	Petroupoli	33.3 ± 3.25	36.9 ± 3.22	36.2 ± 1.89
	Peristeri	32.8 + 2 62	40.0 ± 4.36	40.6 ± 4.37
	Ilion	32.3 ± 3.02	34.5 ± 2.50	34.2 ± 2.93
	Agioi Anargyroi	32.5 ± 3.23	36.1 ± 2.03	34.7 ± 2.67
	Zefyri	31.6 ± 3.38	35.3 ± 2.82	35.0 ± 1.37
	Mean	32.13 ± 3.065	35.33 ± 3.391	34.37 ± 3.618

Source: Giannopoulou et al. (2011).

Table 1.2 Number of nighttime and daytime hours in which the air temperature exceeds 30°C

Area	Station	June (Day)	June (Night)	July (Day)	July (Night)	August (Day)	August (Night)
Center of the city	Athens Univ.	159	6	271	20	164	3
Northern part	N. Erythrea	93	0	161	1	99	0
	Ano Liossia	151	1	260	9	167	0
	Kamatero	216	1	294	13	195	1
	N. Philadelphia	86	2	227	13	154	0
	Maroussi	102	1	254	7	107	0
	Mean ± s.d.	129.4	1.0	239.2	8.6	144.4	0.2
Eastern part	Zografou	167	2	218	3	117	0
	Kessariani	147	2	265	9	142	0
	Ilioupooli	136	5	260	9	155	0
	Byronas	188	11	323	50	282	13
	Agia Paraskevi	31	1	173	15	130	1
	Mean ± s.d.	133.8	4.2	247.8	17.2	165.2	2.8
Southern part	Glyfada	66	0	239	13	213	3
	Renti	126	5	299	18	298	4
	Elliniko	92	0	277	3	287	0
	Kallithea	140	42	290	23	248	4
	Moschato	115	11	339	44	297	13
	Mean ± s.d.	107.8	11.6	288.8	20.2	268.6	4.8
Western part	Korydallos	118	5	305	28	177	4
	AgiaVarvara	183	6	327	38	197	6
	Haidari	167	8	295	19	198	3
	Egaleo	106	7	265	26	223	4
	Petroupoli	168	5	299	27	245	6
	Peristeri	210	6	349	24	322	0
	Ilion	174	1	311	13	259	0
	Agii Anargyri	168	1	334	17	301	0
	Zefyri	149	1	285	8	278	0
	Mean ± s.d.	160.3	4.4	307.8	21.2	244.4	2.6

Source: Giannopoulou et al. (2011).

with those results, it can be concluded that HI intensity presents its maximum concentration in the center and the western part of Athens area, with up to 5°C with respect to the minimum values.

Concerning the maximum daily air temperature values, it was found that the differences between the eastern and northern stations during June and July and also between the center of Athens and the western stations during July were not statistically significant. As observed, the northern and the eastern parts of the Athens area present a similar temperature regime as a result of the high percentage of green areas and the absence of an industrial zone. On the contrary, the center and the western part of Athens present a similar temperature regime due to the lack of green spaces, the densely built areas, and traffic.

The study of the mean maximum air temperature values for each month and for all the stations has concluded that during June, higher temperatures are recorded at the center of Athens and also at the western part, while during July and August, the higher temperatures are found at the southern and the western part of the greater Athens area. In this study, the distribution of the value of Humidex (H) has been estimated by means of statistical methods. Humidex values have been linked to the probability distribution of the conditions of discomfort ($H > 30$ $H > 40$), and the number of hours that the difficult conditions persist (see Table 1.3). From the study of those cases where $H > 40$, considering the air temperature, relative humidity, and wind speed, it appeared that great discomfort or dangerous discomfort conditions occur, with either very high air temperatures or lower air temperature values, in combination with high relative humidity values (Giannopoulou et al., 2014).

Four significant deductions have been drawn from performed measurements and consequent comparative results:

1. The geographical position of the Athens area and the morphological territorial assets of its boundary conditions characterized by the presence of the surrounding height of the mountains (which exceed 1000 m) contribute to the development of high summer air temperatures in the whole AMA.
2. High-air temperatures are also reinforced by the increased urbanization, industrialization, anthropogenic heat, and the lack of vegetation.
3. In particular, during July and August, the mean and maximum air temperatures at the city center and at the western part of the city are much higher than the corresponding values for the northern and northeastern part of the AMA.

Table 1.3 Monthly air temperatures and relative humidity in five geographical areas of Athens

	June		July		August	
Places	**T (°C)**	**RH (%)**	**T (°C)**	**RH (%)**	**T (°C)**	**RH (%)**
Center of the city	27.0 ± 3.7	40.6 ± 14.5	29.3 ± 3.4	43.8 ± 12.8	28.1 ± 2.9	42.3 ± 9.4
Northern part	26.4 ± 2.1	41.3 ± 12.9	28.5 ± 2.1	44.1 ± 14.1	27.3 ± 1.4	38.7 ± 11.6
Eastern part	26.8 ± 2.1	41.9 ± 11.2	29.0 ± 2.0	42.6 ± 11.6	27.7 ± 1.6	40.6 ± 8.5
Southern part	26.9 ± 2.5	48.0 ± 13.4	29.7 ± 1.7	44.4 ± 13.8	29.0 ± 1.3	39.2 ± 9.8
Western part	27.2 ± 2.3	41.6 ± 13.1	29.8 ± 2.0	40.9 ± 12.5	28.7 ± 1.5	39.2 ± 8.9

Source: Giannopoulou et al. (2014).

4. Furthermore, from the analysis of the mean diurnal and nocturnal air temperatures in all stations and from the difference between them and the reference station located at the center of Athens, it is possible to conclude that HI during the night period is mainly observed at the western part of the city.

Thus, the western Athens metropolitan area, together with central Athens, presents the highest HI intensity of the whole AMA (see Table 1.3).

1.2.3 Policy Background and Nearly Zero Energy Studies

Building on this background, it appears to be extremely difficult to show potential achievement of deep renovation up to nearly zero energy targets in the Athens area. The main question we introduced at the beginning of this chapter "Is it really feasible to technically achieve nZEB targets in existing urban buildings and compounds of Mediterranean cities?" sounds particularly challenging given the climatic and urban conditions described in the previous paragraph.

Before proceeding any further, it is essential to point out what nZEB is assumed to be in the current legislative framework, and briefly report some notes on its definition. In the following paragraph, some case studies – past and recent – are also briefly discussed to enable understanding about which lessons can be recovered. Finally, we present a brief discussion on how to widen the technical knowledge in nZEBs and achieve a deeper penetration in building practice.

1.2.3.1 On the Definition of Nearly Zero Energy Buildings

As introduced briefly at the beginning of this chapter, increased attention on nZEBs has led to the European Commission's proposal in November 2008 for an update of the 2002 *Energy Performance of Buildings Directive* (EPBD). Subsequently, a recast of the same directive was adopted by the European parliament and the Council of the European Union (European Council for an Energy Efficient Economy, updated February 2011), in order to strengthen the energy performance requirements in order to clarify and streamline some of its provisions. Among the highlights of the recast is a strengthening of the energy performance requirements of new buildings across the European Union. For new buildings, the recast fixes the year 2020 as the deadline for all new buildings to be "nearly zero energy" (sooner for public buildings – by the end of 2018) while, for existing buildings, member states are required to draw up national plans to increase the number of nZEBs, though no specific targets have been yet set. Article 2(1a) gives a qualitative definition of what a "nearly zero energy building" is: "A 'nearly zero energy building' is a building that has a very high energy performance. The nearly zero or very low amount of energy required should be covered to a very significant extent by energy from renewable sources, including energy from renewable sources produced on-site or nearby." From this qualitative definition, a wide spectrum of additional specifications have been derived, as illustrated by various authors in literature. In particular, the definition has originated a widespread debate about the issues pertaining to terminology and definitions of buildings that consume very low or zero energy (or carbon), including those with net energy production or are, in other words, "energy positive."[26]

A list of related documents of support measures in accordance with Article 10 Directive, 2010/31/EU, have been developed by different countries (Denmark, France, Greece, Lithuania, Romania, etc.).[27] The subject matter of the directive is defined (Article 1) as the "Performance of buildings within the Union, taking into account outdoor climatic and local conditions, as well as indoor climate requirements and cost-effectiveness." In particular, the directive lays down requirements as regards: (1) the common general framework for a methodology for calculating the integrated energy performance of buildings and building units; (2) the application of minimum requirements to the energy performance of new buildings and new building units; (3) the application of minimum requirements to the energy performance of: (a) existing buildings, building units and building elements that are subject to major renovation; (b) building elements that form part of the building envelope and that have a significant impact on the energy performance of the building envelope when they are retrofitted or replaced; and c) technical building systems whenever they are installed, replaced or upgraded; (4) national plans for increasing the number of nearly zero-energy buildings; (5) energy certification of buildings or building units; (6) regular inspection of heating and air-conditioning systems in buildings; and (7) independent control systems for energy performance certificates and inspection reports.

[26] Comprehensive review studies on the definitions of Zero Energy Buildings (ZEBs), including life cycle assessment (LCA) are reported by many authors (Hernandez and Kenny, 2010; Marszal et al., 2011). A list of related documents of support measures in accordance with Article 10 Directive 2010/31/EU, have been developed by different countries (Denmark, France, Greece, Lithuania, Romania, etc.).

[27] Some countries, like the United Kingdom, have already established comparable targets for all new housing, which will see "net-zero" achieved by 2016.

The directive also states that its requirements are minimum requirements and shall not prevent any member state from maintaining or introducing more stringent measures. These measures shall be compatible with the treaty on the functioning of the European Union and the commission must be notified about them.

As previously noticed, currently a wide range of terms and descriptions are used in discussions on low- and zero-energy buildings. The definition of an nZEB used by the Industry Committee during the negotiations on the recast has been replaced by the less stringent requirement "nearly zero energy building" as opposed to the original "net zero energy building"[28] in the final document adopted. In parallel with the variation of these definitio.ns with respect to the different challenging level of energy requirements, there is not a fully harmonized understanding about what a "net zero energy building" really is. An international study was started in 2008 within the framework of the International Energy Agency "Toward Net Zero Solar Energy Buildings" (Task40/Annex 52, 2008) to attempt to enhance the overall understanding of the concept; thus the objective of this work is to study current net zero, near net zero, and very low energy buildings and to develop a shared, common, and harmonized international framework (Sartori et al., 2010), including design tools, innovative solutions, and guidelines. This includes new requirements on comfort and energy performances (Sartori et al., 2010) grid interactions, and an exhaustive benchmarking of the existing nZEBs already built around the world.

Four well-documented definitions – net zero site energy, net zero source energy, net-zero energy costs, and net-zero energy emissions – have been reported by Torcellini et al. (2006), where pluses and minuses of each condition are discussed.[29] In summary, Torcellini et al. (2006) identified the following main definitions of zero-energy buildings (ZEB):

1. *Net zero site energy*: A site ZEB produces at least as much energy as it uses in a year, when accounted for at the site.
2. *Net zero source energy*: A source ZEB produces at least as much energy as it uses in a year, when accounted for at the source. Source energy refers

[28] "A net zero energy building is where, as a result of the very high level of energy efficiency of the building, the overall annual primary energy consumption is equal to or less than the energy production from renewable energy sources on site."

[29] These definitions have been applied and validated to a set of low-energy buildings for which extensive energy data were available; the study shows the design impacts of the definition used for ZEB and the large difference between definitions.

to the primary energy used to generate and deliver the energy to the site (this definition is equivalent to the Industry Committee's definition).

3. *Net zero energy costs*: In a cost ZEB, the amount of money the utility pays the building owner for the energy; the building exports to the grid are at least equal to the amount the owner pays the utility for the energy services and energy used over the course of the year.
4. *Net zero energy emissions*: A net zero emissions building produces at least as much emissions-free renewable energy as it uses from emissions-producing energy sources (this is the case of the "Zero Carbon Building," or the ZCB).

Table 1.4 summarizes the nZEB definitions and the corelated pluses and minuses.

1.2.3.2 Energy Performance in the Mediterranean Areas: Examples and Lessons from the Past

To understand the possible strategies to be adopted for low- or zero-energy buildings, it is also important to recall the basic bioclimatic principles of ancient and traditional architecture, which can be either used to make a proposal for the recovery of vernacular constructions with peculiar bioclimatic strategies and/or translated in the present modern buildings (Cañas and Martìn, 2004).

The importance of vernacular or traditional construction as a frame of reference model of bioclimatic architecture is reported by many authors in the literature (Cañas and Martìn, 2004; Coch, 1998). All over the world, houses, small towns, and villages of the past collectively contain some of the best-preserved climatic conscious and aesthetically enjoyable traditional architectural types and techniques. Furthermore, current passive techniques used in reducing the cooling demand of buildings are based on or derive from systems and components used in the traditional and vernacular architecture.[30]

Different research studies in literature on interesting examples of vernacular and traditional architecture have been selected. They are briefly illustrated in the following part of this paragraph.

[30] With respect to the representative or even monumental architecture, the popular architecture has been structured by people as a direct and basic response to their own needs and values (Coch, 1998). These buildings show a greater respect for the existing environment. They use local materials and techniques as far as possible, repeating and slowly modifying the course of history models and take fully into account the constraints imposed by the climate. In spite of the plurality of solutions within the different geographical contexts, it is also interesting to observe how practically identical architectural models are developed in similar climates with highly different cultures and very distant locations.

Table 1.4 ZEB definitions summary

Definition	Pluses	Minuses	Other
Site ZEB	• Easy to implement. • Verifiable through on-site measurements. • Conservative approach to achieving ZEB. • No externalities affect performance, track success over time. • Easy for the building community to understand and communicate. • Encourages energy-efficient building designs.	• Requires more PV export to offset natural gas. • Does not consider all utility costs (can have a low load factor). • Not able to equate fuel types. • Does not account for non-energy differences between fuel types (supply availability, pollution).	
Source ZEB	• Able to equate energy value of fuel types used at the site. • Better model for impact on national energy system. • Easier ZEB to reach.	• Does not account for nonenergy differences between fuel types (supply availability, pollution). • Source calculations too broad (do not account for regional or daily variations in electricity generation heat rates). • Source energy use accounting and fuel switching can have a larger impact than efficiency technologies. • Does not consider all energy costs (can have a low load factor).	• Need to develop site-to-source conversion factors, which require significant amounts of information to define.

(Continued)

Table 1.4 ZEB definitions summary (*cont.*)

Definition	Pluses	Minuses	Other
Cost ZEB	• Easy to implement and measure. • Market forces result in a good balance between fuel types. • Allows for demand-responsive control. • Verifiable from utility bills.	• May not reflect impact to national grid for demand, as extra PV generation can be more valuable for reducing demand with on-site storage than exporting to the grid. • Requires net-metering agreements such that exported electricity can offset energy and non-energy charges. • Highly volatile energy rates make for difficult tracking over time.	• Offsetting monthly service and infrastructure charges require going beyond ZEB. • Net metering is not well established; often with capacity limits and at buyback rates lower than retail rates.
Emissions ZEB	• Better model for green power. • Accounts for non-energy differences between fuel types (pollution, greenhouse gases). • Easier ZEB to reach.	• Need appropriate emission factors.	

Source: Torcellini et al. (2006).

Analyzing and reporting 212 cases of old buildings in Spain, significant points have been highlighted: (1) The bioclimatic strategies used in popular constructions correctly respond to the conditions imposed by the climate; (2) Selected strategies and technological tools for the protection against the solar radiation may be found in buildings located in the southern middle part of Spain, where the solar radiation received is very high, while, opposingly, strategies for the use of solar radiation appear in the northern middle part of Spain where the solar radiation is lower; (3) Strategies for the protection against cold temperatures agree with the regions of Spain where the temperatures are minimum; (4) Generally, technological tools for the protection against rain are located mainly in areas where the averages of precipitations are high. In particular, passive regional strategies are: (a) Andalusian "patios" whose function is to accumulate fresh air; (b) Light color of the façade as a mechanism for the protection against solar radiation, since light colors reject the solar rays reducing solar heating; (c) Use of vegetation for the shading of the housing; (d) Orientation of the building for the collection of higher amount of solar radiation in winter than in summer.

Other energy efficient design strategies have been investigated by Manioglu and Yılmaz (2008) in Mardin, a town situated in the hot, dry area of the southeastern Turkey. This study examined building types and bioclimatic cooling techniques by evaluating them in terms of design criteria such as building types and forms, selection of the area, distance between buildings, orientation, and building envelopes. Typical features of this traditional settlement are: (1) orientation (southeast); (2) close distance between buildings; (3) plan type (ie, the courtyard houses); and (4) compact forms to minimize the area affected by the solar radiation. In particular, in the traditional architecture of Mardin, the arranging of close form types in courtyards produced shaded areas. In the courtyards, with the help of water and plants for evaporative cooling, the floor temperature is further minimized by natural elements for cooling (channels and water poured out from the pool).[31] Massive envelope details of these houses are very convenient for continental climates, where the summers are severe with high variations in daily temperature. The thermal mass slows down the heat transfer through the envelope, thus producing lower indoor daytime temperatures. The high heat capacity of the opaque component provides a high time lag for the transmission of the

[31] "Eyvan" and "revak," typical spatial elements of the traditional houses, are used to create shady and cool living spaces during the day.

outside temperature to the internal area, while the low transparency ratio minimizes the direct solar radiation gained through the windows. In this climate, other precautions against the solar radiation are: (1) minimization of the area and the number of windows; (2) construction of a window at a high level to block the floor radiation; (3) minimization of the absorptivity of the facades by white or light colors; (4) providing natural ventilation especially at night; (5) constructing a part of the house underground, which is always cooler than the outer ambient temperature. The research study also attempts to compare the energy performance between a traditional house and a modern house (built after 1990): a simplified thermal evaluation and comparison of a traditional house with a contemporary house have been given by using data derived from measurements of physical data and questionnaires, which have been carried out for 100 buildings. The modern building was constructed by contemporary techniques in compliance with the instructions of Energy Conservation Standard of Turkey. For the measurements, rooms with similar features in modern and traditional houses were chosen. (The main façade of these two rooms are southeast oriented; they present a flat roof; they are situated in the upper floor of both houses; no shading device was used on the facades of both houses).[32] After comparing the obtained results, it has been observed that air temperature and surface wall temperature are much lower in the traditional building than in the modern. In particular research result shows, quite stable and low indoor air temperature values during all the reference day in the traditional building.[33] Similar studies and similar results comparing modern and traditional architecture have been performed in other hot and humid parts of the world.

Architectural structure and environmental performance have been reported also for the case study of the traditional buildings in Florina, in northwestern Greece (Oikonomou and Bougiatioti, 2011). This study was based on the documentation and the analysis of the architectural and bioclimatic aspects of a sample of forty remaining houses of the 19th and the beginning of the 20th centuries. The analysis has been performed considering

[32] The thickness of the walls is 0.8 m in traditional house and 0.25 m in modern house. Thermal properties of the building envelope materials were stone wall for the traditional house 2.33 λ (W/mK) and brick for modern house 0.46 λ (W/mK). Single glazed wooden window framed present a U value of 4.5 W/m^2K in traditional house while double-glazed, PVC window framed realize a U value of 2.6 W/m^2K in modern house.

[33] Satisfyingly, the measurements also meet the user's perception of indoor temperature in summer, which has been obtained by a questionnaire carried out with 100 occupants in the selected modern and traditional buildings at the same time with the measurements.

the building types and form, the materials and the construction techniques, whereas the analysis of bioclimatic aspects involves the thermal behavior of the building shell, as well as the thermal and the visual comfort conditions. Traditional architecture of Florina is characterized by proper southern or eastern orientation of the buildings and by the exploitation of the prevailing winds. To a large extent, the buildings are oriented to achieve the best possible exploitation of solar radiation for passive solar heating. Natural materials are used with great efficiency and according to their physical properties and thermal characteristics (density, heat capacity, time-lag), as well as their durability, improving the inter-seasonal thermal behavior of the various spaces and enforcing the bioclimatic function of the buildings. In this way, the winter living spaces, which are situated on the ground floor, are characterized by increased thermal mass and reduced thermal losses due to the increased wall thickness and the reduced number of openings, whereas the summer living spaces on the upper floors are characterized by reduced thermal mass and enhanced ventilation due to the light-weight walls and the increased number of windows. As a whole, the traditional buildings evaluated in this study resulted in highly energy efficient buildings, both in the summer and in the cold winter season.[34]

The energy and microclimatic performance of traditional buildings also have been observed in the well-known case of "Sassi," historical hypogenous buildings in southern Italy (Cardinale et al., 2010). The "Sassi" district of Matera, a UNESCO World Heritage since 1993, is described as an exceptional example of traditional bioclimatic Mediterranean architecture. The authors found that both hypogenous units and half-hypogeum – where half of the building is constructed from stone bricks out of the ground level – optimally perform during both summer and winter conditions. They analyzed the energy performance of the hypogenous building structures during one calendar year, finding, again, that the huge thermal mass of the walls ensured constant and regular microclimate indoor conditions throughout the seasons, without relevant differences in the daily thermal oscillation. In spite of some

[34] Computer-aided thermal analysis have shown that for the coldest day (26th January), the main winter living spaces had slight diurnal variation and range (around 0°C with outside temperatures below −12°C. For the hottest day (25th August), while the exterior temperature ranged from 22°C (early in the morning) to 34°C (around noon), the summer living spaces resulted warmer than the outdoor air temperatures in the morning and in the night (25°C), but significantly fresher around noon (29°C). When natural ventilation is inserted as an additional parameter (passive cooling strategy) in the computer analysis, the inside temperature ranges from 23.5 to 27.5°C.

air change need (thermal-hygrometric comfort values in deep hypogenous units are not fully reached), these structures present a null thermal balance during mid-season, while in the summer, the floor loses heat, thereby cooling the environment. The opposite occurs in winter.[35]

As a general rule, it can be stated that incorporation of traditional or traditional-based building techniques into newly designed building may help the design process to use low-cost and achievable, locally based construction practices toward the achievement of zero-energy balance and zero onsite CO_2 emission within the Mediterranean climate (Ferrante and Cascella, 2011). In the reference case study (a newly designed housing development for a site located in the south of Italy), the concurrence of different building components aims to achieve multipurpose objectives within the same building frame.

These components mainly derive from the re-elaboration of traditional forms and techniques (building type – a courtyard house presenting a good balance among natural ventilation and building compactness, high mass envelope features and construction from the local practices, selected passive tools for energy saving). The same components are integrated with systems for energy micro-generation from renewable energy sources (RES), combining existing technologies from RES into traditional building types and consolidated construction practices.

1.2.3.3 Main Passive Technological Tools for Building Cooling

The responsible sources of cooling demand or thermal discomfort in buildings can be listed in the following order: solar radiation, relative humidity, and air temperature (Coch and Serra, 1996). Because of this, to optimize a building's energy performance, rational design and use is crucial. Solar radiation control is the most important strategy and, although much has been written about the efficacy of transparent shading systems, it is important to assess the impact of opaque envelope solar control as well.

[35] The monitoring campaign has confirmed all results obtained through the dynamic analysis. By quantifying the total energy balance of the hypogenous structures, the authors found the following: (1) during the winter season, the heat flow loss from the walls was balanced by the positive energy gain through the floor, which stabilized the temperature that remains constant at about 12–13°C; (2) during mid-season, exchanged heat fluxes are essentially null, resulting in a constant evolution of temperature with values of 15–16°C; and (3) in summer, the heat flow dispersed from the floor counterbalances the incoming heat flow for transmission and ventilation, reducing high summer temperatures. During the summer season, the indoor temperature is in the range of 18–21°C (especially in deep hypogea). It can be concluded that these buildings were built as "zero energy" houses and they can be used today, after conservative and light-method based restoration processes, with limited or null use of technology energy systems.

As shown by Santamouris (2006), passive cooling relies on the use of solar and heat control techniques, heat amortization, and heat dissipation techniques. Intensive research studies carried out on the topic have permitted to develop advanced and low cost systems and techniques that, when applied, can highly contribute to decrease the cooling needs of buildings and improve indoor environmental quality. In particular, as reported by Santamouris (2006), the development of high reflectivity coatings for the building envelope can considerably decrease the solar input to the buildings, while new developments on ventilation technology permits successful dissipation of the excess heat into the ambient air, improving indoor comfort and decreasing the concentration of indoor pollutants.

Various passive cooling systems and their applicability to different climates and building types, as well as strategies for minimizing cooling needs by building design (eg, ventilation cooling, radiant cooling, evaporative cooling systems, ground cooling, cooling of outdoor spaces adjacent the building, etc.) have been comprehensively investigated by Givoni, 1994, 1998a, and 1998b.

More recently, Zinzi (2010) has highlighted the need to fruitfully apply passive building technologies to mitigate the cooling demand increase, reducing the energy consumption in cooled buildings and improving the thermal comfort in not-cooled buildings. The author suggests cool materials as possible strategy options for reducing building cooling loads, since they are material building components that stay cool under the sun because of high solar reflectivity and thermal emittance properties. In particular, because of the high horizontal solar radiation at Mediterranean latitudes during the cooling season, roofing systems may be considered key and strategic envelope components for the solar control (Zinzi, 2010). The case-study-based research reported by Zinzi (2010) demonstrates the positive impact of the technology in terms of cooling and total energy savings as well as on the indoor thermal conditions in Mediterranean buildings. In particular, the author states that the effect of cool roofs has an important impact on the following: (1) Energy performance of insulated buildings – built in the Mediterranean north rim – with the optimization of heating and cooling performance; (2) Energy performance of existing noninsulated buildings below 40 degree latitude and for all the existing buildings to be renovated with energy measures such as roof insulation; (3) Temperature profiles in noninsulated buildings and with high external surface-to-volume ratio. These configurations ensure enhanced performance of cool roof technology, with a strong decrease of discomfort hours. More insulated and compact structures still benefit from cool roof applications, but the advantages may be amplified by the contextual application of other passive techniques, such as night ventilation, more radical window shading strategies, and increased thermal mass; (4) Positive impact in terms of energy performance backs up other critical issues related to cool roofs: the environmental impact in terms of mitigation of the UHI and an effective answer to global warming, which is critical in large urban areas.

Performances of various passive techniques in hot and arid regions by using roof cooling were also performed by other studies (Nahar et al., 2003). Here, the evaporative cooling has been found to be the best solution for the conventional roof, but it requires high water consumption. Therefore, pieces of white glazed tiles stuck over the roof have been selected as a valid alternative tool to reduce heat load from the roof and decrease the indoor air temperatures, among a range of other passive techniques.[36]

One of the best performing cooling strategies via the roof-cooling technique is the use of vegetated green roofs (Niachou et al., 2001). This technology will be diffusely mentioned as a retrofitting strategy in the future development of this work (see Chapters 2 and 3).

Integrated water and energy management in the urban context is one major route toward environmental sustainability and the reduction of carbon dioxide emissions into the atmosphere. This passive cooling system has been investigated, in terms of expected benefits for the building and the urban environment, by many authors. A review of related worldwide experiences is reported for comparison purposes by Fioretti et al. (2010). This investigation is related to the specific climatic context of the Mediterranean region. Full-scale experimental results are provided from two case studies, located in northwest and central Italy, consisting of two fully monitored green roofs on top of public buildings. The attenuation of solar radiation through the vegetation layer is evaluated as well as the thermal insulation performance of the green roof structure. The daily heat flowing through the roof surface was quantified; results showed that the green roof outperforms the reference roof, therefore it significantly reduces the daily energy demand

[36] A large set of possible tools were used over the roof of different experimental units for cooling the environment inside test structures: (1) painting of roof with white cement; (2) thermal insulation over the roof; (3) nocturnal cooling, that is, a shallow pond with a 100 mm water column 40-mm thick and movable; (4) thermal insulation over the roof; (5) evaporative cooling, that is, roof is provided with gunny bags soaked with water with the help of a storage; (6) tank and dripper for controlling water without any pump, (7) broken white glazed tile pieces stuck over the roof, (8) air void insulation, that is, inverted earthen pots having 100-mm diameter and 125-mm height were provided over the roof, and (9) roof covered with Sania, a local insulating material used over the huts in the arid regions and the controlled unit without any treatment.

of the building. In fact, the most important effects created by a green roof consist of the reduction of the surface temperature and the attenuation of temperature fluctuations. In a traditional roof, made of highly absorptive materials, the horizontal external surface can reach very high temperatures due to the combination of the air temperature and the solar irradiation absorption, which determines a strong thermal load. By this analysis, it was possible to evaluate the time delay effects due to the green roof rather than the reduction in temperature alone. The green roof has a lower temperature level due to the plant shading, insulation, and evaporative transpiration of the foils apparatus. As for water management, it is confirmed that green roofs significantly mitigate storm water runoff generation – which may occur even in a Mediterranean climate – in terms of runoff volume reduction, peak attenuation, and increase of concentration time.

Green roofs, high-reflection roofs, wall planting, green parks, roadside trees, and the like are all desirable from the viewpoint of building and urban planning. Improvement in the surface of buildings and structures that have been covered with cement and asphalt concrete are some of the measures to mitigate the urban heat island phenomenon (Ferrante and Mihalakakou, 2001; Julia et al., 2009; Sfakianaki et al., 2009). The use and effects of highly reflective painting for cooling roofs and road pavements has been examined by many authors (Santamouris, 2001; Spala et al., 2008; Shashua-Bar et al., 2010). Among others, Akbari et al. (1997, 2009), from the heat island group of the Lawrence Berkeley National Laboratory, have studied different solutions and techniques for reducing UHI's effects and energy consumption by different building fabrics and urban surface types.

Surface heat budget on green roofs and high-reflection roofs for mitigation of the UHI have been studied by Takebayashi and Moriyama (2007).[37]

[37] In this study, surface temperature, net radiation, and water content ratio of green roofs and high reflection roofs have been observed, comparing the heat and water budget to each other. In the daytime, the temperature of the cement concrete surface, the surface with highly reflective grey paint, bare soil surface, green surface, and the surface with highly reflective white paint have been observed to be in descending order. On a surface with highly reflective white paint, the sensible heat flux was observed to be small because of the low net radiation due to high solar reflectance. On the green surface, the sensible heat flux was small, too, because of the large latent heat flux by evaporation, although the net radiation was large. On the cement concrete surface and the surface with a highly reflective grey paint, the sensible heat fluxes have been observed to have almost the same values because of similar solar reflectance. Methods to estimate the quantity of evaporation, evaporative efficiency, heat conductivity, and thermal capacity are explained, and the observation data is applied to these methods.

Green roofs and high reflection roofs have performed significantly in successfully mitigating UHI and keeping buildings cooler.

Ventilation is another strategic passive technique for building cooling. Proper ventilation of urban buildings can contribute significantly to decreasing the energy consumption for air conditioning, improving indoor comfort, reducing the concentration of indoor pollutants, and protecting public health (Santamouris, 2006). Design and integration of efficient natural ventilation systems and components like wind and solar towers improve indoor thermal comfort. Natural ventilation can contribute to improving thermal comfort, decreasing the need for air conditioning, and improving indoor air quality in the developed world. Experimental studies have shown that effective night ventilation in office buildings, for example, may reduce the overheating hours by half and reduce the cooling load by at least 55%.

Important research studies have been carried out on appropriate and advanced ventilation techniques (Santamouris and Wouters, 2006). The main achievements of these studies are:(1) better understanding of the air flow phenomena and of the expected comfort benefits, in particular in the dense urban environment, and development of efficient and practical procedures to design natural and hybrid ventilation systems and configurations (Allard and Ghiaus, 2006); and (2) technological developments mainly on the field of hybrid and mechanical ventilation that greatly contribute to a more comfortable and healthy indoor environment (De Gids, 2006).

The appropriate design of windows permits an increase in air speed in households and improves comfort by cooling down the human body through the mechanisms of convection, radiation, and perspiration. Cross-ventilation and vertical ventilation increase airflow rates and lower internal temperatures. Although the appropriate levels of air velocity to achieve comfort is a widely discussed topic within the scientific community (Tanabe and Kimura, 1994; Arens et al., 1984), studies performed in hot and humid climates confirm that increased air speed, especially at higher temperatures, enhances thermal comfort (Hien and Tanamas, 2002).

Vertical air extraction and wind catchers are also very efficient techniques. In particular, wind-catcher systems have proven to deliver effective natural ventilation. Once again, these systems derive from past technologies, employed in buildings in the Middle East for more than 3000 years. In the modern wind catchers, the principles of wind effect and passive stack effect are considered in the design of the stack, which is often divided into two halves or four quadrants/segments with the division running the full length of the stack. Although they have different names in different parts of the southeastern Mediterranean region, the basic physical principles may be investigated according to similar geometrical forms (Bahadori, 1994; Elmualim et al., 2001; McCarthy, 1999).

An increasing number of wind-catcher and cool-tower systems (Etzion et al., 1997; Hurdle, 2001; Swainson, 1997) have been applied into commercial buildings and residential buildings (ie, the Queen's Buildings at Demonfort University and the Ben-Gurion University of the Negev – or the International Centre for Desert Studies in Negev, Israel) and their performance is still evaluated by many authors in literature. Among many others, Li and Mak (2007) performed the evaluation of a wind-catcher system using computational fluid dynamics, demonstrating that wind-catcher cooling potential is very effective, although a precise design study has to be planned in relation to the wind conditions of the specific site, since its behavior is greatly influenced by the external wind speed and direction with respect to the wind-catcher quadrants.

Furthermore, as illustrated in the examples from traditional architecture, the significance of mass is of crucial importance for the reduction of cooling load in buildings. This is an important property of the building envelope for energy conservation, since excess heat is stored in it and dissipated at a later stage when needed. In this way, the indoor temperature fluctuations are regulated and overheating is avoided. These temperature fluctuations could be due to: (1) diurnal fluctuations of outdoor temperatures; (2) internal gains; or (3) incident solar radiation especially in rooms with south glazing surfaces. The aspects relating to mass are of particular significance for countries with large diurnal fluctuations for the displacement and dissipating potential of mass. In these contexts, ensuring high values of superficial mass and adequate levels of the damping and phase-displacement of the thermal wave is essential. Thus, many authors (Santamouris, 2001, 2006; Givoni, 1994, 1998a; Cañas and Martìn, 2004; Ferrante and Cascella, 2011), reported research and applied studies on the effectiveness of thermal mass in the bioclimatic design for building cooling in hot or Mediterranean areas.

1.2.3.4 On the Combination of Passive and Active Techniques for Building Cooling

In the search for further energy exploitation within the building sector, researchers are attempting to discover, measure, and evaluate the energy-saving potential achievable from the combination of active and passive techniques. In this context, automatic control and implementing an active system to prevent overheating in summer represent crucial strategies. In a recent study, Ochoa et al. (2008) have explored the influence of incorporating intelligence in buildings in hot climates, to achieve glare and solar radiation control. Through the perspective of energy consumption and user comfort, with an emphasis on lighting, the study shows how decisions made in the early design stages can affect those of later ones.[38] Finally, the combination of optimized active features (fan ventilation elements and automatic controls) with passive design strategies offers very promising energy savings and glare control at the same time.[39]

Furthermore, current studies are investigating the potential of combining both active energy generation systems from renewable energy sources (RES) and passive techniques. By coupling active features and correct passive design strategies, consistent savings may be achieved. In particular, recent studies exploiting the radiative cooling capacity of active energy producers have led to the production of a new photovoltaic-thermal (PVT) system (Eicker and Dalibard, 2011), which has been developed with the goal of providing both electrical and cooling energy for buildings. Experimental studies of PVT collectors were carried out in Stuttgart to validate a simulation model, which calculates the night radiative heat exchange with the sky. Larger PVT frameless module surfaces were then

[38] By means of computer energy modeling, a prototype office unit is used to evaluate energy performance and visual comfort in three parametric series. The first one is the result of the incorporation of active features alone, the second one is guided by intelligent passive design strategies, and the third one is the combination of both approaches. Here, to analyze the influence of active features and passive design, an office module will be studied in the city of Haifa, Israel (32.51N, 351E), with a regional climate usually classified as Csa (warm summers and mild winters) in the Köppen scale (also "meso-thermal/humid continental," due to its location in the Mediterranean coastal plain). The building is a free-standing structure, with similar glazed facades in all four main orientations, with the view portions made of clear double glazing. Results show that energy savings in the "passive design strategies only" (sun shading elements, daylight redirection elements, glazing elements, etc.) were among the largest – from 60 kWh/m^2 per year (base-case) to around 24 kWh/m^2 per year.

[39] Combining active features and correct passive design strategies gives consistent savings of around 50–55%, compared to a conventional situation. By these options, intelligent or smart buildings offer flexibility and convenience, such as rapid temporary changes, options to open individual windows or operate specific blinds.

implemented in a residential ZEB and tested under the climatic conditions of Madrid.[40]

1.2.3.5 Current Research Studies and Projects on Zero-Energy Building in the Mediterranean Climate

Other studies have attempted to evaluate the energy and comfort behavior of building models by comparing their predicted and/or measured values in different climate conditions. Simulation is employed as the main research method for these studies. By simulating a standard passive house (a house with heating energy demand equal to 10–16 kWh/m^2 y) in different locations throughout the Mediterranean and European regions, Grove-Smith (2010) illustrated how the general and basic principles of passive houses seem to work in Mediterranean warm climates; this finding also has been considered generally valid, although for southern locations like Palermo and Seville, passive systems conceived for northern climates would require de-humidification measures, as can be observed by obtained sensible and latent values in Table 1.5.

Table 1.5 Simulated results of a same passive building in different locations

	Mannheim	Torino	Madrid	Seville	Palermo
Insulation wall [cm]	25	20	10	8	6
Insulation roof [cm]	35	25	25	20	20
Insulation basement [cm]	20	15	6	0	0
U window frames [W/(m^2K)]	0.72	0.72	0.72	1.6	1.6
U-value glazing [W/(m^2K)]	0.7	1.2	1.2	1.2	1.2
Humidity control for cooling	No	Yes	No	No	Yes
Heating demand [kWh/(m^2 y)]	15.6	14.8	12.7	4.6	3.1
Sensible cooling [kWh/(m^2 y)]	0	0.8	0.4	4.2	7.2
Latent cooling [kWh/(m^2 y)]	0	2.3	0	0	7.2

Source: Grove-Smith (2010).

[40] Measured cooling power levels were between 60 and 65 W m^{-2}, when the PVT collector was used to cool a warm storage tank and 40–45 W m^{-2}, when the energy was directly used to cool a ceiling. The ratio of cooling energy to electrical energy required for pumping water through the PVT collector at night was excellent with values between 17 and 30. The simulated summer cooling energy production per square meter of PVT collector in the Madrid/Spain climatic conditions is 51 kWh m^{-2} a^{-1}. In addition to the thermal cooling gain, 205 kWh m^{-2} a^{-1} of AC electricity is produced under Spanish conditions.

Therefore, it seems that some of the winter requirements (sufficient insulation of roofs, walls, double low-e glazing, south-facing windows, some thermal mass) may be useful for energy saving in the summertime if solar shading/control, light control, and appropriate ventilation techniques are also applied.

Among the recent research studies on low- and zero-energy housing is the experience of a regional project financed by the European Union (Wenzel, Deutsche Gesellschaft für Technische Zusammenarbeit mbH, GTZ, German Agency for Technical Cooperation). This project supports the design, construction, and monitoring of ten low-energy demonstration buildings in ten southern and eastern Mediterranean countries (MED-ENEC www.med-enec.com). In the report, some important issues on energy efficiency in the Mediterranean are discussed. In particular, it states that: (1) High energy savings can be achieved in buildings with a variety of partly mature and partly innovative technologies; (2) Low- and even zero-energy houses and green buildings are technically feasible in the Mediterranean region; (3) Economic considerations limit the broad application of some of these technologies and thus reduce the technical potential for energy savings in a large-scale dissemination; (4) The most cost-efficient technology mix, according to the type of building, the climate zone, energy prices, and the availability of know-how and technologies needs to be identified;(5) High transaction costs such as substantial initial learning and search costs put the profitability of low-energy buildings in the region at risk and constrain the development of respective markets; (6) When using a cost-efficient technology mix and if mitigating transaction costs, low-energy buildings become attractive with energy savings of 20–60% in most of the countries, and incremental costs of 10–15%, as well as short pay back periods; (7) Government support and incentives are necessary for overcoming the initial high transaction costs and market failures and for boosting energy efficiency in buildings; and, finally, (8) Subsidies on energy are the most important single constraint for broad dissemination in some of the countries in the Mediterranean region.

Results of these pilot projects show that high primary energy savings of up to 100% – compared to conventional buildings in the same country – are technically possible. These energy savings correspond to anywhere from 3 to 307 tons of avoided CO_2 annually, according to the different sizes of the buildings and the chosen energy efficiency concept. The average primary energy savings achieved is 57%, compared to a conventional building of the same size and comfort (concerning heating and cooling).[41]

The cost-effective results have been analyzed to deliver an overview of the economic performance indicators of the pilot projects. In this context, the payback periods have been considered as the more important indicators. The simple payback calculation method has been used by dividing the value of the annual energy costs savings (assuming a 5% annual increase in energy prices) by the incremental costs of the building, compared to a conventional building. When analyzing the economic performance figures, major differences among the pilot projects become obvious. On the one hand, the Lebanese Project is the most attractive with payback occurring in roughly one year. The paybacks for five other projects seem to be moderate with a payback period of around 10 years. However, four projects are clearly unattractive with paybacks of 18 to nearly 70 years. On average[42] the new buildings needed incremental investments of about 30%, and show a payback period of about 20 years. If, however, broad dissemination of low-energy buildings in the region is the first priority, economic considerations, that is, the relation of energy savings to additional cost, limit the use of available technologies to the most cost-efficient "smart technology mix." The feasible technology mix may be different according to the type of building, the climate zone, the national energy prices, and the availability of know-how and technologies. The integrated and mixed use of different energy forms to be

[41] Finally, the behavior of the inhabitant is crucial for realizing the theoretical savings potential. In fact, when shading devices are not used properly, when windows are left open while heating or cooling, energy-efficient technologies may not have the desired effects. In some MED-ENEC pilot projects (Jordan and Israel), users of the building receive written guidelines and explanations for the handling the installed equipment and for energy-efficient procedures and behavior. In addition, a part of the theoretical savings potential may be "lost" in comfort increase, that is, the rooms are a bit warmer in winter and slightly colder in summer ("rebound effect"). This is particularly true in countries where thermal comfort is rather poor. It should be noted that significant additional energy savings potential exists on the urban planning level.

[42] Except the Tunisian project, which is a special building with rather expensive unconventional features. In this case, attracting "green" clients by demonstrating the "state of the art" and maximizing the energy savings was a major objective. Therefore, a mix of mature and innovative but rather expensive technologies such as large photovoltaic electricity generation or solar cooling was chosen. This approach makes the pilot project an interesting place to visit and learn from, but reduces potential for dissemination.

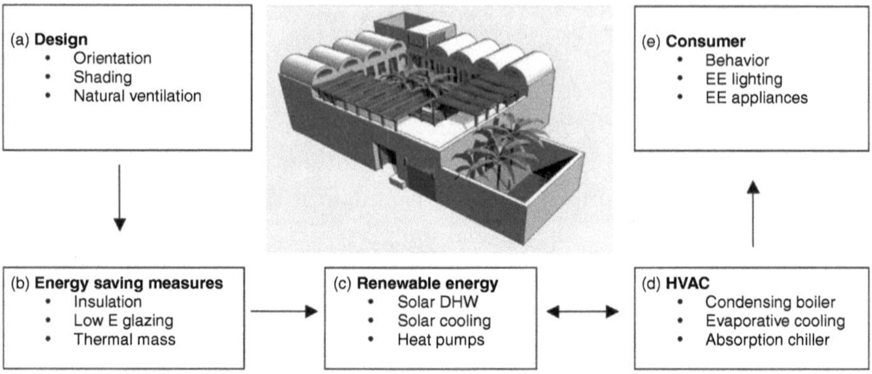

Figure 1.5 ***Integrated EE-approach for buildings.*** *(Source: MED-ENEC, Carsten Petersdorff, ecofys).*

considered in passive and low-energy building within the hot regions is summed up in Fig. 1.5.

This integrated approach helps reduce incremental costs and improve the payback period. Experiences from pilot projects like the MED-ENEC pilot projects lead us to move toward more specific and regionally related assessment, considering the ranking of technologies according to their cost-effectiveness in the specific country or region. While some technologies, such as photovoltaic capacity, are sensible to national framework conditions (eg, they may be cost-effective in Israel, where a feed-in tariff exists, or in African countries, where grid connection is significantly lower than in the Mediterranean region), some others may seem less attractive. Thus, passive design features, such as orientation of the building, shading, natural ventilation, and the use of daylight are free or very cheap in the design phase, and are therefore always cost-efficient for new buildings. Insulation of roofs and walls as well as efficient lighting and solar water heaters prove to be equally in the top ranking.

The MED-ENEC pilot project developers have encountered major constraints, especially high and often uncompetitive costs. These effects are not specific for the MEDA region (MGI 05/2007); other groups have reported similar conditions. One of the conclusions that may be derived from the MED-ENEC pilot project is that "government must lead." This is true not only for the Mediterranean region but for all regions and countries. Setting regulations, and providing standards and enforcing them, are necessary but not sufficient conditions for success. As the value chain in the building sector is long and diversified with plenty of stakeholders acting on individual perceptions of risks and benefits, it is of the highest importance that economic signals such as energy tariffs, incentives, and sanctions orient the actors in the market. However, resistance to change, initial market failures, and high-transaction costs usually slow down or even impede development of the market for energy-efficient products and services. An integrated package encompassing regulations and standards (including control and enforcement), financial support and incentives, information, training, and education, as well as research and development activities, has to be part of improved economic framework conditions. Further research to identify the most effective (and cost-effective) policy mix must be developed, since finding a "best" policy instrument is probably impossible.[43] The lessons learned from the MED-ENEC project seem to be applicable for most emerging and developing countries. An appropriate policy mix according to different framework conditions is always necessary. Countries with very low energy prices need strong political will through enforced regulations and standards and significant subsidies for energy-efficient products and services while reducing non-targeted energy subsidies and, at the same time, protecting the vulnerable part of the population. Countries where framework conditions are more favorable and where the private sector is strong should focus instead on policy packages with information, training, quality control, technology transfer, and credit program components.

Recent progress has been made on low-energy and passive buildings by the research and projects of the Passivehaus Institute. In the last ten years, northern and central Europe (and in particular, Germany) has seen increasing interest in the Passivhaus Construction Standard. Homes built according to this standard are buildings that provide a comfortable indoor climate in winter without the need for a conventional heating system.[44] The success of the Passivhaus Institute in Darmstadt,

[43] Each country needs to work out a comprehensive analysis of individual framework conditions and barriers and has to design an adapted policy package. In this process, stakeholder participation, definition, and monitoring of quantitative objectives and the synergetic combination of instruments are major general success factors (IEA 03/2008, UNEP 09/2007, ECOFYS 03/2007, WEC 2008, GTZ/Wuppertal Institute/UNEP 2007).

[44] To achieve this standard, it is essential that the building's space heat load does not exceed 10 W/m^2 living area in order to be able to use a simple air pre-heater. Simulations and measurements have shown that for the typical German climate such a design leads to an annual demand for space heating of 15 $kWh/(m^2\ y)$. Passive houses therefore require roughly 85% less energy for heating than a house built to existing German building regulations. The total primary energy demand, including household electricity, is limited to 120 $kWh/(m^2\ y)$. In 1991, Wolfgang Feist and Bo Adamson applied the passive design approach to a house in Darmstadt, with the objective of providing a showcase low-energy

(*Continued*)

home for the German climate at a reasonable cost. By 1995, based on the successful first experiment (both in terms of energy consumption and comfort), the same passive systems were applied again in a second construction in 1995 in Groß-Umstadt. From these experiences, Feist had codified the passive design of the Darmstadt and Groß-Umstadt homes, into the Passivhaus standard. The standard fundamentally consists of three elements: (1) an energy limit (heating and cooling); (2) a quality requirement (thermal comfort); and (3) a defined set of preferred passive systems that allow the energy limit and quality requirement to be met cost effectively. The "Passive-On" consortium has therefore formulated a revised proposal for the application of the Passivhaus standard in warm European climates, taking into account the aforementioned climatic issues and contents. According to Passive-On, the Passive house standard for the warm European climate is defined in six points: 1. *Heating criterion:* Useful energy demand for space heating not exceeding 15 kWh per m^2 of net habitable floor area per year.[2] of net habitable floor area per year. 2. *Cooling criterion:* Useful, sensible energy demand for space cooling not exceeding 15 kWh per m^2 net habitable floor area per year. 3. *Primary energy criterion:* The primary energy demand for all energy services, including heating, domestic hot water, auxiliary, and household electricity, does not exceed 120 kWh per m^2 net habitable floor area per annum. 4. *Air-tightness:* If good indoor air quality and high thermal comfort are achieved by means of a mechanical ventilation system, the building envelope should have a pressurization test (50 Pa) result according to EN 13829 of no more than 0.6 ach-1. For locations with winter design ambient temperatures above 0°C, a pressurization test result of 1.0 h-1 is usually sufficient to achieve the heating criterion. 5. *Comfort criterion room temperature, winter:* The operative room temperatures can be kept above 20°C in winter, using the aforementioned amount of energy. 6. *Comfort criterion room temperature, summer:* In warm and hot seasons, operative room temperatures remain within the comfort range defined in EN 15251.

To achieve the Passivhaus standard, it is necessary that indoor summer temperatures, more specifically operative temperatures, remain lower than the maximum temperatures defined by the EN 15251 standard. According to EN 15251 standard, acceptable comfort temperatures depend on the type of system used to provide summer comfort. If cooling is provided by an active system, then indoor temperatures must respect those defined by the comfort model originally proposed by Fanger or the predicted mean vote (PMV) model (Fanger, 1967, 1994; Fanger et al., 1974, 1975, 1988). Instead, if summer comfort is provided by passive cooling strategies, the higher temperature limit is set by the adaptive model, which is the model taking into account the ability of occupants of buildings to adapt to the prevailing climate (adaptive comfort model). Compared to the Fanger Model, the adaptive model considers a wider range of temperatures as "comfortable" and therefore allows for easier integration of passive cooling technologies. For example, applying the adaptive algorithm defined in the EN 15251 standard to typical annual weather data predicts maximum summer neutral temperatures (in correspondence with a sequence of hot days) for Frankfurt, Milan, Lisbon, and Seville as, respectively, 26.1°C, 27.2°C, 26.7°C, and 28.7°C. As a comparison, a building cooled by an active air conditioning system will work to a fixed set point chosen between 23 and 26°C. In more recent years, some international standards (e.g., the US norm ASHRAE 55 2004 and the European norm EN 15251) have proposed adaptive comfort models based on in-field comfort surveys. These have thus replaced the previous Fanger-based temperature standards with "adaptive" ones for indoor temperature in naturally ventilated buildings.

Germany naturally led to the question of whether this program might be applicable in other countries, with particular reference to the Mediterranean region. This question is central to some recent research dissemination projects funded under the IEE program by the European Commission, such as the "Passive-On" project (http://www.passive-on.org/en/), which primarily addresses the question of the applicability of Passivhaus standards in southern Europe. In fact, although in central Europe (eg, Germany, Austria, northern Italy, etc.) passive design is increasingly associated with the Passivhaus standard, this is not necessarily the case in southern Europe (eg, Spain, Italy, Portugal, and Greece). The project examined how to take the Passivhaus concept forward, especially in southern Europe. In these regions, the problem of household energy use is not only to provide warm houses in winter but also, and in some cases more importantly, to provide cool houses in summer. The Passive-On project has lead to three major outcomes: (1) For architects and building designers, the project has developed design guidelines and enhanced the PHPP software design tool for developing cost-effective, all-season passive houses in both heating load and cooling load climates; (2) For policy makers, the project has provided a set of policy proposals that examine the barriers and the solutions that EU, national, and local governments can adopt to promote the more widespread development of passive houses.

The Passivhaus standards have been applied in the five partner countries of Passive-On (France, Spain, Portugal, Italy, and the United Kingdom). They have been performed as national "exercises," formulated according to the standard typology of a semidetached three-bedroom house. The exercises revealed that heating loads are relatively low in many southern European countries and generally stay below the 15 kWh/m^2 mark. Comparatively, however, they are marginal to other household energy requirements such as water heating, lighting, and appliances. Results showed that in many cases, there are cooling loads to take into account but that often these can be met by passive strategies alone.

For the Portugal case, for example, the starting point for the Passivhaus proposal was a single floor, two-bedroom house, complying with the Portuguese national building regulations for 2006. The current proposal takes into account the local climate (case study for Lisbon), the construction standards, and the technical and economic framework.[45] The three main aspects explored in the proposed house are: relation with the sun, ventilation for cooling, and high thermal mass to control temperature swings. The house combines the ability to collect solar heat (large south windows) and the capacity to regulate inside temperature with its high thermal inertia. Since solar availability is quite high in Portugal, even during the heating season, a key factor in this house is the relation with the solar radiation, captured both directly (windows) and indirectly (thermal solar system). Large windows are mainly oriented south, which increases the useful solar gains during winter. Smaller areas are oriented east and west and minimal areas are oriented to the north. Solar protection is chosen according to the orientation: overhangs to the south windows, thus reducing the solar incidence during summer, and exterior Venetian blinds in all windows. Strategies adopted for the design of a Passivhaus for the heating and

[45] Here the level of insulation in walls and roofs goes well beyond the national standards and the air infiltration is controlled (about 0.8 ach at 50 Pa). A very important feature of the proposal is the use of a thermal solar system. The current proposal extends the solar installation of thermal systems from the compulsory limit use of Portuguese thermal regulation for buildings (solar system for domestic hot water) to also cover a significant portion of the heating demand, by increasing the solar panel area and using a low temperature hydraulic heat distribution (radiant floor). As proposed for the Passivhaus standard, the active heating and cooling capacity is limited to 10 W/m^2. The extra costs of the proposed Passivhaus for Portugal were 57 €/m^2 with a payback period of 12 years. To further reduce heat losses and gains, 150 mm and 100 mm of insulation are proposed for the roof and exterior walls, with U-values of 0.23 W/m^{2}K and 0.32 W/m^{2}K, respectively. Windows facing south correspond to around 60% of the total glazed area; about 20% of glazed area faces east and another 20% west. Low-emission double glazing can be very effective in colder climates of Portugal, but in most situations, standard double glazing is more cost-effective (U-values of 2.9 W/m^{2}K for standard double glazing and 1.9 W/m^{2}K for low-emission glazing are considered). Performance: energy and comfort. The annual heating energy demand of the Passivhaus proposed for Portugal has been estimated as 16.9 kWh/m^2, of which 11 kWh/m^2 are supplied by the solar system (in this analysis, priority of the solar system is given to heating and the solar fraction for domestic hot water is 48%). The annual cooling energy demand is 3.7 kWh/m^2. The sum of net heating and cooling demand is 9.6 kWh/m2 year, against limits of heating and cooling by regulatory tools in Lisbon, are 73.5 and 32 kWh/m^2 year, respectively. The current house, with an active cooling, has a Fanger comfort index of 811 (the house is penalized by the influence of the radiant temperature of the high glazed area). If no active cooling is present, the adaptive comfort index (AI2) applies (ASHRAE 55). For the proposed Passivhaus for Portugal, the AI2 was 16. For this house, the resultant temperature is kept below 25°C for 71% of the occupied time, and below 28 °C during 98% of the occupied time. In winter, the low power heating system of 10 W/m2 is in use, resulting in only 8% of time with a resultant temperature below 19.5°C (lower resultant temperature achieved is 18°C).

cooling climate of Lisbon can be successful, both regarding the energy demand limits and the comfort level requirements.

1.3 DISCUSSION

Since today the reserves of conventional energy resources are drastically depleted, it becomes imperative to look at the reduction of energy demand as the primary option for decreasing energy consumption and to achieve natural cooling within our buildings. To reach this objective, it is important to look into the past for suggestions and possible solutions. In no way is it being suggested that we should copy traditional building types or abandon the available technology to return to construction practices of the past. However, there are basic principles that can be derived from traditional architecture, for example: correct orientation and placement of buildings and their components, openings to promote day lighting and natural ventilation during the cooling period, distribution of thermal mass, the use of local materials, and so forth. These practices may be considered as the first (and lower cost) bioclimatic options to design or redesign a low-energy building. In fact, the architectural design process usually moves from the conceptual features (definition of massing, orientation, form) to the specific requirements (thermal mass, solar and lighting control, mechanical ventilation type).

It should be taken into account that, as the design process advances, earlier and often free of costs- decisions, which could have an enormous influence on the building performance, are costly and difficult – when not impossible – to change (Ferrante, 2012).

Examples from the analyses of pilot case studies (MED-ENEC project, Passivhaus) have shown that passive design features, such as orientation of the building, shading, natural ventilation, and the use of daylight are free or very cheap in the design phase, and are therefore always cost-efficient for new buildings. Insulation of roofs and walls as well as efficient lighting and solar water heaters proved to be equally in the top ranking of cost-effective solutions, in most cases.

However, although climatic building design strategies are important, it cannot be stated that any one of them has a higher priority than any other; the large number of elements existing in the market today, together with constraints limiting their applicability, make it necessary for designers to have tools that help them to identify the best combinations for any specific situation. In this context, the cost-effective analysis, together with the life cycle evaluation of solutions, may represent useful components of a strategic framework plan aimed at the incorporation of zero-energy concepts in building construction.

Under the assumption that sustainability in construction should be first and foremost achieved through the passive building conception, to achieve nZEB buildings we should make use of the major advantage from building type and the potential energy subsystems building components (area of external surfaces/volume ratio[46] walls, orientation, ventilation strategies, etc.) and that of its surroundings (landscape, shape, and orientation). For example, as it can be observed from the analysis of the traditional bioclimatic examples in architecture, in warm and hot climates and/or during the summer season, a permeable form of the building type can be advantageously designed to achieve shadowed areas and improved natural ventilation. Inversely, for the winter period, the same passive principles should lead to a solution providing a more compact form and reduced values of surface-volume (S/V) ratio, which is the most beneficial strategy during the winter, especially for inland locations with high seasonal thermal range and fluctuations. Thus, taking into account the need for ventilation (cross-ventilation and vertical air extraction provided by internal courtyard and vertical ducts) the S/V ratios in Mediterranean climate should consider a reasonable balance between natural ventilation requirements in the hot summer period and thermal outflow in the colder winter season. A good compromise could be found in the use of variable or movable systems and components (ie, glazing or opaque sliding systems) to close off parts of the building, thus resulting in a more compact, less dissipative space, to reduce the external surface areas in the cold winter season; variable components can be removed or kept open in the summer season, so as to increase the surface area of heat exchange and ensure ventilation (especially at night, when air temperature decreases). The search for such an adaptive strategy for buildings brings to mind the concept of adaptive comfort adopted by humans and also the similar adaptation capacity of many other living organisms. Unfortunately, this strategy is often connected with higher investment and maintenance costs.

[46] The surface-area-to-volume ratio is also defined the surface-to-volume ratio and it is variously denoted Sa/Vol or SA:V. It represents the amount of surface area per unit volume of an object or collection of objects. For a given volume, the object with the smallest surface area (and therefore with the smaller surface-area-to-volume ratio) is the sphere. By contrast, objects with tiny spikes will have very large surface area for a given volume.

As a general rule, when conceiving and designing options for buildings in the Mediterranean zone, we should always remember that northern European design decisions sometimes have been adopted in southern latitudes with counterproductive results (eg, large windows, or solar greenhouses, high compactness, high-insulation performing materials, etc., may produce summer discomfort in the Mediterranean region). Generally, the building envelope design should be focused on technological solutions able to guarantee an adequate balance between the thermal insulation requirements in the winter season and the thermal mass in the hot summer season. In this framework, as an example, constructive typology in masonry brick walls ensures high values of superficial mass and adequate levels of the damping and phase-displacement of the thermal wave (Ferrante and Cascella, 2011).

With reference to legislative measures in the frame of ZEBs, as pointed out by Torcellini et al (2006) "A good ZEB definition should first encourage energy efficiency, and then use renewable energy sources available on site. A building that buys all its energy from a wind farm or other central location has little incentive to reduce building loads, which is why we refer to this as an off-site ZEB." Thus, taking into account that "it is almost always easier to save energy than to produce energy" it is necessary to:

- Set specific energy supply options and establish a hierarchy of them in relation to the specificity of sites, energy prices, and cultural and social aspects.
- Consider ZEB as an additional step with respect to the Passivhaus principles, not as an alternative to them. Since the energy savings from passive cooling and heating is the first and foremost criteria to be considered in ZEB, it would be particularly important to fix an "internal limit" of passive requirements within ZEBs. In this way, ZEB can be considered not different from but integrated with the passive house concepts. A new standard should be framed in order to encourage low-energy buildings and not only high-energy performing buildings.
- Suggest new forms for future standards, which are more in keeping with the objectives of the EPBD on zero-energy building (European Council for an Energy Efficient Economy, February 2011).

Of course, setting regulations and standards and enforcing them is a necessary but not sufficient condition for success. As the value chain in the building sector is complex, with plenty of stakeholders acting on individual perceptions of risks and benefits, it is of highest importance that economic signals such as energy tariffs, incentives, and sanctions orient the actors in the market. Furthermore, integrated packages of regulations/standards and financial support/incentives should be combined with information, training, and education as well as research and development activities.

Thus, to achieve the correct approach toward ZEBs, it is necessary to start from the radical reduction of energy demand for cooling, heating, lighting, and ventilation; the use of RES to cover the remaining energy demand should be considered as the secondary and consequent step.

Reported examples and studies have shown the technical feasibility of ZEBs in the context of European and Mediterranean regions.

Over the years, there has been a growing interest among stakeholders, architects, and entrepreneurs to include energy-efficient tools and techniques in buildings as a way to achieve buildings that comply with stringent energy codes and national goals of reducing dangerous emissions, together with improving corporate image. Reported case and research studies demonstrate the technical feasibility of nZEBs both in colder and warmer areas of the Mediterranean region.

Despite growing interest in nZEB buildings, given the recognized increasing problems related to global overheating and fuel poverty, there is a crucial risk of underestimating the growing energy demand for cooling and heating purposes in the Mediterranean region. This is mainly due to the following, often intertwined and concurrent factors:

- More attention paid to nZEBs in the northern and colder climate areas compared to the southern regions. Consequently, a relatively new approach of passive and low-energy buildings within the nZEB and current European policies directive needs to be further investigated with more specific reference to cooling demand and summer energy performance.
- Use of light, low-inertia building components in modern architecture and culture, with subsequent limitation of the capacity to control building thermal performance with respect to mass and dense buildings materials, ultimately resulting in further high energy consumption
- Lack of knowledge and understanding of the potential to use available construction systems and techniques (which can be also derived from traditional passive cooling systems)
- Less favorable social and economical conditions in the Mediterranean countries, compared to the northern and central Europe
- Technical, economic, and even social barriers with respect to the integration of passive techniques and RES to achieve nZEBs in the context of existing building stock and urban settings.

1.4 FURTHER STEPS

As introduced in the first part of this chapter, the majority of these studies refer to newly conceived buildings and new development plans; but, obviously, when approaching a refurbishment project, the freedom degrees are extremely reduced when compared to the design of a new building. Thus, the further challenge would be to widen technical ZEB knowledge in existing built environments, shifting our technical understanding on energy efficiency from new developments and newly conceived buildings into the reality of existing urban settings. In the Mediterranean areas, there is a special need for efficient schemes where zero-energy requirements, energy construction quality, and technological flexibility can fit with the different economic and social possibilities, within an extended building stock, often in degraded or abandoned conditions. These existing urban settings represent the biggest challenges both in carbon terms – because of the large amount of existing building stock – and for the social impact they may generate on the local economy. In this case, the integration of passive technologies and renewable energy sources are the real challenge for the architectural and engineering projects of future generations. Yet, these actual built environments, with their special concentration of technical and socioeconomic problems, are ripe for innovative responses to energy-related issues and climate change.

Thus, to deal with the main issues arising from the state-of-the-art in low- and zero-energy building within the warm and hot climate regions in the Mediterranean area, we need further research studies and design implementation. We do need to take the chance to intervene in the real material parts of the actual cities, in a project-based, but sorted and ranked manner. In these hypothetical scenarios, according the different degrees of possible transformation of existing built environments, the project is not the scope, but becomes the main the tool of our research study, the effective device to verify the techno-economic feasibility of systemic retrofitting.

First and foremost, these studies should be indented to turn future designers' and public stakeholders' priorities away from new pilot zero-energy experiments and toward demonstration buildings to spread interventions in the existing buildings of actual urban environments; we should then analyze, share, and present this opportunity as a creative and sustainable challenge for future generations of architects, planners, industrial designers, and researchers. In fact, there is thus a lack of comprehensive information on the possible intersections between the BATs and a concrete inclusive approach to users' needs. This is because close collaboration between physicians, engineers, architects, energy companies, stakeholders, and urban dwellers, amongst others, is missing, which is probably why progress in this area is still limited to date.

In this context, recent studies, design proposals, and simulations of building energy retrofitting scenarios (Ferrante and Cascella, 2011) has already proved that huge energy savings may be achieved in winter by adding different coatings to existing buildings. In particular, the combination of new building coatings such as sun-spaces or buffer zones and RES, drastically reduce the energy performance indexes of the buildings up to the target of an nZEB. These studies, performed during the winter season so far, need to be further investigated in order to quantify the cooling demand reduction of the proposed scenarios during hot summer conditions.

Finally, real urban environments, or RUEs (streets, squares, courtyards, and connected residential buildings) are here indicated as the core of the search for new intersections between urban dwellers and energy-related issues. An effective action should explore the technical mechanisms that can promote the concrete synergies between economic constraints, users' expectations, and EE systems and production, in renewed forms of urban retrofitting.

The complexity of the urban environments in the Mediterranean areas has highlighted an increasing need for:

- Further exploration of urban geometry for achieving better understanding of current energy demand within the different residential areas of the city. This further exploration will be observed and quantified with reference to the building blocks in the urban textures considered as the "solid" part of the city and to the open-street environments, considering the buildings and the related open spaces as the core focus of energy investigation and trying to outline the possible cumulative effects of all the buildings and open areas of the city.
- Energy retrofitting actions in existing building stocks within localized, ground-based urban environments
- Feasibility of small-scale and distributed RES in local urban environments

Thus, the next steps to be undertaken may be summed up as follows:

1. In order to evaluate the energy demand and potential of an urban environment, it is possible to select reference case studies, representative of the different urban conditions of a whole discretized city model based on different urban units, corresponding to the building blocks and the related street/open area in RUEs.

2. In these "units," retrofitting actions toward zero-energy RUEs can be hypothesized and validated exploring both passive physical components (green and EE techniques for surfaces and coating materials) and energy micro-generating technologies by RES – solar, photovoltaic (PV), Aeolian, and so forth.
3. The economical feasibility of these actions in RUEs can more easily lead to economically and socially competitive actions and lead the drive toward a zero-energy future.

The goals of next chapters are to study, evaluate, and demonstrate the technical feasibility of nZEBs throughout representative RUEs of the AMA. To achieve this purpose, the energy performance of selected typical residential buildings in the urban Athens area will been investigated in detail in order to properly identify the buildings' energy requirements in winter and summer conditions and, finally, to propose a targeted retrofitting hypothesis toward the goal of increasing the number of nZEBs.

REFERENCES

Akbari, H., Konopacki, S., 2005. Calculating energy-saving potentials of heat island reduction strategies. Energy Policy 33 (6), 721–756.

Akbari, H., Kurn, D.M., Bretz, S.E., Hanford, J.W., 1997. Peak power and cooling energy savings of shade trees. Energy Build. 25, 139–148.

Akbari, H., Menon, S., Rosenfeld, A., 2009. Global cooling: increasing world-wide urban albedos to offset CO2. Climatic Change 94, 275–286.

Allard, F., Ghiaus, C., 2006. Natural ventilation in the urban environment. In: Santamouris, M., Wouters, P. (Eds.), Ventilation for Buildings – The State of the Art. Earthscan Publishers, London.

Arens, E.A., Blyholder, A.G., Schiller, G.E., 1984. Predicting thermal comfort of people in naturally ventilated buildings. ASHRAE Trans. 90, 272–279, 1B.

Asteriadi, S., 2011. Liminal words: Transit migration in Athens. MSc City Design and Social Science, London School of Economic and Political Science. www.sarcha.gr/Resources.aspx?file...Asteriadi.pdf.

Bahadori, M.N., 1994. Viability of wind towers in achieving summer comfort in the hot arid regions of the Middle-East. Renew. Energ. 5, 879–892.

Biris C., 1995. Athènes du XIXe au XXe siècle, Melisa edition, Athens: 1995, first edition Athens: 1966. p. 21.

Bitan, A., 1992. The high climatic quality of the city of future. Atmos. Environ. 26B, 313–329.

Brown, H.S., Vergragt, P.J., 2008. Bounded socio-technical experiments as agents of systemic change: the case of a zero-energy residential building. Technol. Forecast. Soc. 75, 107–130.

Bulkeley, H., 2010. Cities and the Governing of Climate Change. Ann. Rev. Environ. Res. 35, 229–253.

Bürer, M.J., Wüstenhagen, R., 2009. Which renewable energy policy is a venture capitalist's best friend? Empirical evidence from a survey of international cleantech investors. Energy Policy 37, 4997–5006.

Buttstädt, M., Sachsen, T., Ketzler, G., Merbitz, H., Schneider, C., 2010. Urban temperature distribution and detection of influencing factors in urban structure, ISUF; International Seminar on Urban Form, Hamburg and Lubeck 2010.

Cañas, I., Martìn, S., 2004. Recovery of Spanish vernacular construction as a model of bioclimatic architecture. Build. Environ. 39, 1477–1495.

Cardinale, N., Rospi, G., Stazi, A., 2010. Energy and microclimatic performance of restored hypogeous buildings in south Italy: the "Sassi" district of Matera. Build. Environ. 45, 94–106.

Cartalis, C., Synodinou, A., Proedrou, M., Tsangrasoulis, A., Santamouris, M., 2001. Modifications in energy demand in urban areas as a result of climate changes: an assessment for the Southeast Mediterranean region. Energ. Convers. Manage. 42 (14), 1647–1656.

Charlseworth, E., 2006. Architects without Frontiers: War, Construction and Design Responsibility. Architectural Press, Oxford.

City of Copenhagen, Copenhagen Climate Plan, 2009. The technical and environmental administration city hall, 1599 CopenhagenV, 2009.

Coch, H., 1998. Bioclimatism in vernacular architecture, Renewable and Sustainable Energy Reviews, 67–87.

Coch, H., Serra, R., 1996. Summer comfort solutions in Mediterranean areas. In: Proceedings of WREC – World Renewable, Energy Conference, Denver, CO.

De Gids, W., 2006. Advanced components for ventilation. In: Santamouris, M., Wouters, P. (Eds.), Ventilation for Buildings – The State of the Art. Earthscan Publishers, London.

Doulos, M., Santamouris, Livada, M., 2001. Passive cooling of outdoor urban spaces: the role of materials. Sol. Energy 77, 231–249.

Dunster, B., Simmons, C., Gilbert, B., 2009. The ZEDbook: Solutions for a Shrinking World. Taylor and Francis, New York.

Eicker, U., Dalibard, A., 2011. Photovoltaic–thermal collectors for night radiative cooling of buildings. Sol. Energy 85, 1322–1335.

Elmualim, A.A., Awib, H.B., Teekaram, A.J., Brown, R.G., 2001. Evaluating the performance of a windcatcher system using wind tunnel testing. In: Proceedings of the Twenty Second Annual Aivc Conference, 11–14, September 2001, Bath, UK. Belgium: Air Infiltration and Ventilation Centre, 1, INIVEEIG, 2001. 29.1–29.12.

Etzion, Y., Pearlmutter, D., Erell, E., Meir, I.A., 1997. Adaptive architecture: integrating low-energy technologies for climate control in the desert. Automat. Constr. 6, 417–425.

EU Communication, 2011a. Energy Efficiency Plan 2011, Comm. from the Commission to the European Parliament, the Council, the EU Economic and Social Committee and the Committee of the Regions. Brussels, 8.3.2011 COM (2011) 109 final.

EU Communication, 2011b. A resource-efficient Europe – Flagship initiative under the Europe 2020 Strategy. Comm. from the Commission to the EU Parliament, the Council, the EU Economic and Social Committee and the Committee of the Regions. Brussels, 26.1.2011 COM (2011) 21.

EU Parliament, Report on the proposal for a Directive of the European Parliament and of the Council on the Energy Performance of Buildings (recast) (COM [2008]0780-C6-0413/2008-2008/0223 [COD]), 2009.

EU Parliament, 2010. Directive 2010/31/EU of the EU P of 19 May 2010 on the energy performance of buildings.

EU Report, 2010. World and European Sustainable Cities. Insights from EU research, Directorate-General for Research. Directorate L, Science, economy and society, Unit L.2. Research in the Economic, Social sciences and Humanities. European Commission, Bureau SDME 07/34, B-1049 Brussels, 2010. Socio-economic Sciences and Humanities; EUR 24353 EN.

EU report, Cities of Tomorrow: Challenges, Visions, Ways Forward, 2011. European Commission – Directorate General for Regional Policy, Luxembourg: Publications Office of the European Union, 2011, 112 pp.

Eurosurveillance, Volume 10, Issue 7, 01 July 2005. Editorial THE 2003 EUROPEAN HEAT WAVES, accessed at: http://www.eurosurveillance.org/ViewArticle.aspx?ArticleId=552.

Fanger, P.O., 1967. Calculation of thermal comfort: introduction of a basic comfort equation. ASHRAE Trans. 73 (2), III4.1–III4.20.
Fanger, P.O., 1994. How to apply models predicting thermal sensation and discomfort in practice. In: Oseland, N.A., Humphreys, M.A. (Eds.), Thermal Comfort: Past, Present and Future. Building Research Establishment, Garston, UK, pp. 11–17.
Fanger, P.O., Langklide, G., 1975. Interindividual differences in ambient temperatures preferred by seated persons. ASHRAE Trans. 81, 140–147.
Fanger, P.O., Melikov, H., Hanzawa, H., Ring, J., 1988. Air turbulence and sensation of draught. Energy Buildings 12, 21–39.
Fanger, P.O., Ostergaard, J., Olesen, S., Lund-Madsen, T.H., 1974. The effect of man's comfort of a uniform air flow from different directions. ASHRAE Trans. 80 (2), 142–147.
Ferrante, A., 2012. Zero- and low-energy housing for the Mediterranean climate. Adv. Build. Energ. Res. 6 (1), 81–118.
Ferrante, A., 2014. Energy retrofit to nearly zero and socio-oriented urban environments in the Mediterranean climate. J. Sustain. Cities Soc. 13, 237–253.
Ferrante, A., Cascella, M.T., 2011. Zero energy balance and zero on-site CO_2 emission housing development in the Mediterranean climate. Energ. Build. 43, 2002–2010.
Ferrante, A., Mihalakakou, G., 2001. The influence of water, green and selected passive techniques on the rehabilitation of historical industrial buildings in urban areas. J. Sol. Energy 70 (3), 245–253.
Ferrante, A., Semprini, G., 2011. Building energy retrofitting in urban areas. Procedia Eng. 21, 968–975.
Ferrante, A., Santamouris, M., Koronaki, I., Mihalakakou, G., Papanikolau N., 1998. The Design Parameters' Contribution to Improve Urban Microclimate: An Extensive Analysis within the Frame of POLIS Research Project in Athens, The Joint Meeting of the Second EU Conference on Energy Performance and Indoor Climate in Buildings and Third International Conference IAQ, Ventilation and Energy Conservation. Lyon, France.
Ferreira, J., Pinheiro, M.D., de Brito, J., 2013. Refurbishment decision support tools review: energy and lifecycle as key aspects to sustainable refurbishment projects. Energy Policy 62, 1453–1460.
Fioretti, R., Palla, A., Lanza, L.G., Principi, P., 2010. Green roof energy and water related performance in the Mediterranean climate. Build. Environ. 45, 1890–1904.
Founda, D., 2011. Evolution of the air temperature in Athens and evidence of climatic change: a review. Adv. Build. Energy Res. B 5 (B), 7–41.
Founda, D., Giannakopoulos, C., 2009. The exceptionally hot summer of 2007 in Athens, Greece – atypical summer in the future climate? Global Planet. Change 67, 227–236.
Founda, D., Giannakopoulos, C., Sarantopoulos, A., Petrakis, M., Zerefos, C, 2008. Estimating present and future fire risk in Greece: links with the destructive fires of summer 2007. Geophys. Res. Abstracts 10.
Giannopoulou, K., Livada, I., Santamouris, M., Saliari, M., Assimakopoulos, M., Caouris, Y.G., 2011. On the characteristics of the summer urban heat island in Athens, Greece. J. Sustain. Cities Soc. 1, 16–28.
Givoni, B., 1994. The Passive Cooling of Buildings. John Wiley & Sons Inc., New York.
Givoni, B., 1998a. Effectiveness of mass and night ventilation in lowering the indoor air temperatures. Energ. Build. 28, 25–32.
Givoni, B., 1998b. Climate Considerations in Building and Urban Design. John Wiley & Sons Inc., New York.
Grammenos, F., 2011. European Urbanism: Lessons from a City without Suburbs. http://www.planetizen.com/node/48065.
Grove-Smith, J., 2010. Passive Houses in the Mediterranean Climate, Passivhaus Institut, Darmstadt, Germany- www.passiv.de; IG Passivhaus – www.ig-passiv.de; Passivhaus Dienstleistung GmbH – www.passivhaus-info.de.

Guy, S., 2006. Designing urban knowledge: competing perspectives on energy and buildings. Environ. Plann. C Gov. Policy 24, 645–659.

Hansen, J., Sato, M., Ruedy, R., Kharecha, P., Lacis, A., Miller, R., Nazarenko, L., Lo, K., Schmidt, G.A., Russell, G., Aleinov, I., Bauer, S., Baum, E., Cairns, B., Canuto, V., Chandler, M., Cheng, Y., Cohen, A., Del Genio, A., Faluvegi, G., Fleming, E., Friend, A., Hall, T., Jackman, C., Jonas, J., Kelley, M., Kiang, N.Y., Koch, D., Labow, G., Lerner, J., Menon, S., Novakov, T., Oinas, V., Perlwitz, Ja., Perlwitz, Ju., Rind, D., Romanou, A., Schmunk, R., Shindell, D., Stone, P., Sun, S., Streets, D., Tausnev, N., Thresher, D., Unger, N., Yao, M., Zhang, S., 2007. Dangerous human-made interference with climate: a GISS model E study. Atmos. Chem. Phys. 7, 2287–2312.

Hassid, S., Santamouris, M., Papanikolaou, M., Linardi, A., Klitsikas, N., Georgakis, C., et al., 2000. The effect of the Athens heat island on air conditioning load. Energ. Build. 32, 131–141.

Heiskanen, E., Johnson, M., Robinson, S., Vadovics, E., Saastamoinen, M., 2010. Low-carbon communities as a context for individual behavioural change. Energy Policy 38, 7586–7595.

Hernandez, P., Kenny, P., 2010. From net energy to zero energy buildings: defining life cycle zero energy buildings (LC-ZEB). Energ. Buildings 42, 815–821.

Hien, W.N., Tanamas, J., 2002. The effect of wind on thermal comfort in the tropical environment, in: Proceedings of the International Symposium on Building Research and the Sustainability of Built Environment in the Tropics, Jakarta, Indonesia.

Hurdle, A., 2001. Is natural ventilation the new wind of change? Build. Serv.Environ. Eng. 24 (7), 26–27.

IEA, 2008. Energy, efficiency requirements in building codes, energy, efficiency policies for new buildings (IEA Information Paper). Jens Laustsen: OECD/IEA, International Energy Agency.

Julia, E., Santamouris, M., Dimoudi, A., 2009. Monitoring the effect of urban green areas on the heat island in Athens. Environ. Monitoring Assess. 156 (1–4), 275–292.

JWG, 2013. Towards assisting EU Member States on developing long term strategies for mobilising investment in building energy renovation (EU EED Article 4), Composite Document of the Joint WG of CA EED, CA EPBD and CA RES. http://www.ca-eed.eu/reports/art-4-guidance-document/eed-article-4-assistance-document.

Kaklauskas, A., Kelpsiene, L., Zavadskas, E., Bardauskiene, D., Kaklauskas, D., Urbonas, M., Sorakas, V., 2011. Crisis management in construction and real estate: Conceptual modeling at the micro-, meso- and macro-levels. Land Use Policy 28, 280–293.

Karlessi, T., Santamouris, M., Apostolakis, K., Synnefa, A., Livada, I., 2009. Development and testing of thermo chromic coatings for buildings and urban structures. Sol. Energ. 83 (4), 538–551.

Katsoulis, B.D., Theoharatos, G.A., 1985. Indications of the urban heat island in Athens, Greece. J. Climate Appl. Meteorol. 24, 1296–1302.

Katsoulis, B., 1987. Indications of change of climate from the analysis of air temperature time series in Athens, Greece. Climat. Change 10, 67–79.

Kelly, M.J., 2009. Retrofitting the existing UK building stock. Build. Res. Info. 37 (2), 196–200.

Kumbaroglu, G., Madlener, R., 2012. Evaluation of economically optimal retrofit investment options for energy savings in buildings. Energ. Build. 49, 327–334.

Li, Liu, Mak, C.M., 2007. The assessment of the performance of a wind-catcher system using computational fluid dynamics. Build. Environ. 42, 1135–1141.

Livada, I., Santamouris, M., Niachou, K., Papanikolaou, N., Mihalakakou, G., 2002. Determination of places in the great Athens area where the heat island effect is observed. Theor. Appl. Climatol. 71, 219–230.

Manioglu, G., Yılmaz, Z., 2008. Energy efficient design strategies in the hot dry area of Turkey. Build. Environ. 43, 1301–1309.

Mantouvalou, M., 1988. Apo tin akropoli tis Athinas sto limani tou Pirea. Schedia anaplasis astikon periochon, {From Athena's acropolis to Piraeus port. Drawings of regeneration of urban areas}, NTUA– Politechnico di Milano, 1988 (in Greek).
Marszal, A.J., Heilselberg, P., Bourrelle, J.S., Musall, E., Voss, K., Sartori, I., Napolitano, A., 2011. Zero energy building – a review of definitions and calculation methodologies. Energy Buildings 43, 971–979.
Masini, A., Menichetti, E., 2010. The impact of behavioural factors in the renewable energy investment decision making process: conceptual framework and empirical findings. Energy Policy, 1–10.
May, E., 1929. Die Frankfurter Wohnungspolitik, Publikation 2, Internationaler Verband fürn Wohnungswesen.
May, T., Hodson, M., Marvin, S., Perry, B., 2013. Achieving "sistemic" urban retrofit: a framework for actions. In: Swan, W., Brown, P. (Eds.), Retrofitting the Built Environment. First ed. John Wiley & Sons, Inc. Oxford, OX4 2DQ, United Kingdom.
McCarthy, B., 1999. Wind towers: Details in Building, Academy Ed. West. John Wiley & Sons, Inc., Sussex, UK.
MED-ENEC, 2015. Financing Energy Efficiency in the Building Sector in the MENA Region. Energy Efficiency in the Construction Sector in the Mediterranean, www.med-enec.eu.
Meyer, A., 2010. A Copenhagen Climate Treaty, Version 1.0, 2010 Karlsruhe.
Michelozzi, P., Perucci, C.A., Forastieri, F., Fusco, D., Ancona, C., Dell'Orco, V., 1999. Inequality in health: socio-economic differentials in mortality in Rome, 1990–95. J. Epidemiol. Community Health 53 (11), 87–93, 6.
Michelozzi, F., De' Donato, L., Bisanti, A., Russo, E., Cadum, M., De Maria, M., D'Ovidio, G., Costa, A., Perucci, C.A., 2005. The impact of the summer 2003 heat waves on mortality in four Italian cities. Euro. Surveill. 10 (7), 161–165.
Mihalakakou, P., Flocas, H.A., Santamouris, M., Helmis, C.G., 2002. Application of neural networks to the simulation of the heat island over Athens, Greece, using synoptic types as a predictor. J. Appl. Meteorol. 41 (5), 519–527.
Mihalakakou, P., Santamouris, M., Papanikolaou, N., Cartalis, C., Tsangrassoulis, A., 2004. Simulation of the urban heat island phenomenon in Mediterranean climates. J. Pure Appl. Geophys. 161, 429–451.
Mithraratne, N., 2009. Roof-top wind turbines for microgeneration in urban houses in New Zealand. Energ. Build. 41, 1013–1018.
Morelli, M., Rønby, L., Mikkelsen, S.E., Minzari, M.G., Kildemoes, T., Tommerup, H.M., 2012. Energy retrofitting of a typical old Danish multi-family building to a "nearly-zero" energy building based on experiences from a test apartment. Energ. Build. 54, 395–406.
Mulugetta, Y., Jackson, T., van der Horst, D., 2010. Carbon reduction at community scale. Energy Policy 38, 7541–7545.
Nahar, N.M., Sharma, P., Purohit, M.M., 2003. Performance of different passive techniques for cooling of buildings in arid regions. Building Environ. 38, 109–116.
Niachou, K., Papakonstantinou, K., Santamouris, M., Tsangrassoulis, A., Mihalakakou, G., 2001. Analysis of the green roof thermal properties and investigation of its energy performance. Energ. Build. 33, 719–729.
Ochoa, C.E., Capeluto, I.G., 2008. Strategic decision-making for intelligent buildings: comparative impact of passive design strategies and active features in a hot climate. Building Environ. 43, 1829–1839.
Oikonomou, A., Bougiatioti, F., 2011. Architectural structure and environmental performance of the traditional buildings in Florina, NW Greece. Build. Environ. 46, 669–689.
Papanikolaou, N., Livada, I., Santamouris, M., Niachou, K., 2008. The influence of wind speed on heat island phenomenon in Athens, Greece. Intl. J. Ventilation 6 (4).
Peattie, K., 2010. Green consumption: behaviour and norms. Ann. Rev. Environ. Res. 35, 195–228.

Philandras, C.M., Metaxas, D., Nastos, P., 1999. Climate variability and urbanization in Athens. Theor. Appl. Climatol. 63, 65–72.

Ravetz, J., 2008. State of the stock: what do we know about existing buildings and their future prospects? Energy Policy 36, 4462–4470.

Ricciardelli, F., Polimeno, S., 2006. Some characteristics of the wind flow in the lower urban boundary layer. J. Wind Eng. Indust. Aerodynam. 94, 815–832.

Santamouris, M. (Ed.), 2001. Energy and Climate in the Urban Built Environment. James and James Science Publishers, London.

Santamouris, M., 2006. Special issue of the solar energy program devoted to natural ventilation in urban areas (editorial). Sol. Energy 80, 369–370.

Santamouris, M., 2007. Heat island research in Europe: the state of the art. In: Santamouris, M. (Ed.), Advances in Building Energy Research. Earthscan, London, p. 2007.

Santamouris, M., 2014. Cooling the cities – A review of reflective and green roof mitigation technologies to fight heat island and improve comfort in urban environments. J. Sol. Energy 103, 682–703.

Santamouris, M., Wouters, P., 2006. Ventilation for Buildings – The State of the Art. Earthscan Publishers, London.

Santamouris, M., Mihalakakou, G., Papanikolaou, N., Assimakopoulos, D.N., 1999a. A neural network approach for modelling the heat island phenomenon in urban areas during the summer period. Geophys. Res. Lett. 26 (3), 337–340.

Santamouris, M., Papanikolau, N., Koronakis, I., Livada, I., Asimakopoulos, D., 1999b. Thermal and air flow characteristics in a deep pedestrian canyon under hot weather conditions. J. Atmospheric Environ. 33, 4503–4521.

Santamouris, M., Papanikolaou, N., Livada, I., Koronakis, I., Georgakis, C., Argiriou, A., et al., 2001. On the impact of urban climate to the energy consumption of buildings. Sol. Energy 70 (3), 201–216.

Santamouris, M., Kapsis, K., Korres, D., Livada, I., Pavlou, C., Assimakopoulos, M.N., 2007. On the relation between the energy and social characteristics of the residential sector. Energ. Build. 39, 893–905.

Santamouris, M., Sfakianaki, A., Pavlou, K., 2010. On the efficiency of night ventilation techniques applied to residential buildings. Energ. Build. 42, 1309–1313.

Santamouris, M., Alevizos, S.M., Aslanoglou, L., Mantzios, D., Milonas, P., Sarelli, I., Karatasou, S., Cartalis, K., Paravantis, J.A., 2014. Freezing the poor: indoor environmental quality in low and very low income households during the winter period in Athens. J. Energ. Build. 70, 61–70.

Santamouris, M., Cartalis, C., Synnefa, A., Kolokotsa, D., 2015. On the impact of urban heat island and global warming on the power demand and electricity consumption of buildings – A review. Energ. Build., 119–124.

Sarigiannis, G., 2000, Athens 1830–2000: Evolution, Planning, Transportation (in Greek language).

Sartori, I., Napolitano, A., Marszal, A., Pless, S., Torcellini, P., Voss, K., 2010. Criteria for definition of net zero energy buildings. Proceedings of EuroSun 2010. Graz, A.T., Smith, M., Whiteleg, J., Williams, N., 1998. Greening the Built Environment. Earthscan Publications Ltd, London, UK.

Sfakianaki, A., Pagalou, E., Pavlou, K., Santamouris, M., Assimakopoulos, M.N., 2009. Theoretical and experimental analysis of the thermal behaviour of a green roof system installed in two residential buildings in Athens, Greece. Intl. J. Energ. Res. 33 (12), 1059–1069.

Shashua-Bar, L., Tsiros, I.X., Hoffman, M., 2010. A modeling study for evaluating passive cooling scenarios in urban streets with trees. Case study: Athens, Greece. Build. Environ. 45, 2798–2807.

Spala, A., Bagiorgas, H.S., Assimakopoulos, M.N., Kalavrouziotis, J., Matthopoulos, D., Mihalakakou, G., 2008. On the green roof system. Selection, state of the art and energy potential investigation of a system installed in an office building in Athens, Greece. Renewable Energy 33, 173–177.
Stathopoulou, M., Synnefa, A., Cartalis, C., Santamouris, M., Karlessi, T., Akbari, H., 2009. A surface heat island study of Athens using high-resolution satellite imagery and measurements of the optical and thermal properties of commonly used building and paving materials. Intl. J. Sust. Energ. 28 (1), 59–76.
Swainson, M.J., 1997. Evaluation of the potential of solar chimneys to drive natural ventilation in non-domestic buildings, PhD Dissertation, Cranfield University, UK.
Synnefa, A., Santamouris, M., Apostolakis, K., 2007. On the development, optical properties and thermal performance of cool coloured coatings for the urban environment. Sol. Energ. 81, 488–497.
Takebayashi, H., Moriyama, M., 2007. Surface heat budget on green roof and high reflection roof for mitigation of urban heat island. Build. Environ. 42, 2971–2979.
Tanabe, S., Kimura, K., 1994. Effects of temperature, humidity and air movement on thermal comfort under hot and humid conditions. ASHRAE Trans. 100 (2), 953–960.
Torcellini, P., Pless, S., Deru, M, 2006. Zero energy buildings: a critical look at the definition. National Renewable Energy Laboratory D. Crawley U.S. Department of Energy to be presented at ACEEE Summer Study Pacific Grove, California. Available from: http://www.nrel.gov/docs/fy06osti/39833.pdf
Travlos I., Kokkos A., 1977. Poleodomia kai architektoniki {Urbanism and architecture}, History of Greek Nation, Vol. XIII, Athens: 1977 (in Greek), pp. 512–528.
Vergopoulos, C., 1978. I Elliniki oikonomia apo to 1926 eos to 1935 {The Greek economy from 1926 until 1935} History of Greek Nation, Vol. XV, Athens: 1978. (Original book in Greek), pp. 327-342.
Webler, T., Tuler, S.P., 2010. Getting the engineering right is not always enough: researching the human dimensions of the new energy technologies. Energy Policy 38, 2690–2691.
Wendell, C., 2013. The Evolving Urban Form: Athens. http://www.newgeography.com.
While, A., 2008. Climate change and planning: carbon control and spatial regulation. Viewpoint in town planning review 78 (6), vii-xiii.
Yamashita, S., 1996. Detailed structure of heat island phenomena from moving observations from electric tram-cars in metropolitan Tokyo. Atmos. Environ. 33 (3), 429–435.
Zhenjun, Ma, Cooper, P., Daly, D., Ledo, L., 2012. Existing building retrofits: methodology and state-of-the-art. Energ. Build. 55, 889–902.
Zinzi, M., 2010. Cool materials and cool roofs: potentialities in mediterranean buildings. Adv. Build. Energ. Res. Vol. B, 201–266.

INTERNET LINKS

http://www.eie.gr/archaeologia/gr/chapter_more_9.aspx#_17
http://revistas.ucm.es/index.php/CHCO/article/viewFile/40066/38495
http://users.clas.ufl.edu/kapparis/aoc/athens.htm
Keep cool: https://ec.europa.eu/energy/intelligent/projects/en/projects/keep-cool-ii

CHAPTER 2

Nearly Zero Energy Urban Settings (ZEUS)

2.1 SEARCHING FOR HOMOGENEOUS URBAN AREAS AND URBAN UNITS TO DEPICT A CITY MODEL FOR ATHENS

Urban and geographical studies of urban textures' discontinuity within and around the city have been highly and deeply investigated by the discipline of urban morphology (Whitehand, 2009; Conzen M.R.G, 1988, 2004; Conzen, P.P, 2009). In particular, these studies are concentrated on the different urban structures derived from the tendency of the expanding metropolis toward the homogeneous conglomeration of the flat cityscape around the city core, which is the foremost dominant mark of the so-called dispersed city (Segal and Verbakel, 2008).[1]

Processes of suburbanization have made it difficult to distinguish the growth of the city from its suburbs in many different urban contexts (Phelps et al., 2010). Phelps and colleagues also reported a large set of terminology innovations describing outer cities or edge cities, which are variously termed "techno-burbs" (Fishman, 1987), "edge cities" (Garreau, 1991), "edgeless cities" (Lang, 2003), or even long-standing processes of sub-urbanization (Bruegmann, 2005; Hayden, 2003; Berger, 2014).

Thus, the general outcome of the dispersed city seems to be a society where the values of the rural and the urban blend, resulting in more uncertain places, indifferent to both the city and the countryside (Donadieu, 1998). As a result, the urban-rural areas, conquered by gray industrial fields, expanding transportation networks, fragmented built areas, and social desolation, urgently call for new ideas. Correspondingly, the loss of human recognition within the city is even more striking as we move from the city center outwards.[2]

[1] One of the consequences of this process is that the farming community around the cities is sharply declining, uncertain about the images that the future has in store for it, since its spaces are continuously invaded by the expanding city: on the one hand the city is increasingly suburban and the rural areas are correspondingly becoming increasingly urbanized (Donadieu, 1998).
[2] Generally, the more recent expansions in periurban contexts, compared to historic areas, present different features of legibility of urban form, in the specific correlation with pathways, public and open spaces. The strong corelation among closed and open spaces has been

(Continued)

Towards Nearly Zero Energy
http://dx.doi.org/10.1016/B978-0-08-100735-8.00002-7

The study of this discontinuity does present deep co-relations to the interpretation of urban landscape as the geographical mosaic of urban units (Whitehand, 2009) and the Conzenian theory of fringe belts formation (Conzen, 1988, 2004) whose definition, conceptualization, theory, and applied validation on several case studies have been reported by a large set of contributions in urban morphology (Conzen, 2009; Curdes, 2010). As reported by Whitehand (2009), there is a need for much greater clarity in the methods of characterizing and delimiting the urban units and their aggregation in the mosaic of the urban landscape and for a wider appreciation of their role in planning. The methodological approach toward the knowledge of urban context can also be seen as the typological process of the buildings, at the different scalar correlations among building types, pathways, and textures forming the overall urban structure (Caniggia and Maffei, 1984).

As shown in the previous chapter, so far the metropolitan Athens area has expanded according to the same process. Furthermore, due to additional peculiar gaps derived from the process of substitution and transformation of buildings involving the historical parts of the city, a consistent suburban character appears to be diffusely present in the overall context of the city, embracing the central and peripheral areas as well. The Athens cityscape reveals a contradictory, dense, and disordered but fascinating structure as a whole (Fig. 2.1).

Here, the dramatic city scene seems to define a sort of a conflicting relationship with respect to the historical time: buildings of the 1970s and 1980s overlap the historical urban path, contrasting the old, residual, often abandoned, but still lasting, historical houses of the last century and marking the urban landscape with consecutive shocks and discontinuities.

As described in Chapter 1, the strong densification trend sped up in the 1980s in the central areas of the cities, destroying and substituting many historical buildings while continuing its spread at the periphery. Nowadays, the higher and massive buildings of the 1950s, 1960s, up to the 1980s, coexist with lower, historical buildings, creating a series of gaps among the volumetric structure of the city, with a variable series of "blind" façades

reported by Hillier (1996), which also refers to previous studies when declaring that we cannot create a space inside without also making a space outside. Thus it's possible to point out a clear conflict between the "calibrated" structure of historical villages developed along natural or anthropic boundaries and the absence of these relations in recent expansions. In fact, moving outward from the city center, generally plots become more fragmented, less tied to fixation lines (Conzen, 2009).

Figure 2.1 ***Discontinuities, historic, and new buildings in the cityscape of Athens.*** Bettazzi Maria Beatrice, Athens, 2015.

facing public urban spots and open areas, thus infusing them with an unidentifiable urban character (Fig. 2.2). Overall, gaps between buildings do exist in many areas of the city and between different building types (Figs. 2.3–2.5).

In this complex scenario, it is extremely ambitious to aim at identifying urban compounds to be assumed as different and repeatable units of a possible discretized city model. Nonetheless, to evaluate the energy demand and the energy-saving potential of urban environments, it is necessary to select reference case studies representative of the different urban conditions and corresponding to the building blocks and the related street/open area in real urban environments, the aforementioned RUEs.

In an attempt to achieve a categorization of typical and significant urban environments, it is necessary to introduce the concept of urban density, since it represents the most crucial aspect to be studied and applied in the planning and redesign of new and existing urban contexts.

2.1.1 Notes on the Concept of Urban Density

Urban density, in its various and different definitions, is the basic tool of the urban zoning; it is conceived as the most diffuse and necessary tool to shape cities, imposing an order on the urban development, and regulating future growth. In a city as large and as eclectic as Athens, zoning regulations are crucial to preserve the existing built environment and, at the same time, foster potential innovative new developments. In large cities and metropolitan areas, zoning provides specifications on end-use and size of buildings,

Figure 2.2 ***A nine-story massive building of the 1980s (on right and in the background) coexists with a two-floor historical building.*** A.F., Evripidou central Athens area, 2014.

contributing to the characterization of diversity of the many neighborhoods that constitute the city.[3]

As discussed, Athens has been characterized by an uncontrolled growth of massive and quite tall buildings in the early 20th century, due to the

[3] Different classes can be based on the discrete values (strong, medium, weak) of three parameters in urban planning: building continuity (continuous, medium, discontinuous), surface density (strong, medium, weak), building height (low, medium, high). The surface density is the ratio between the surface of the buildings at the ground level and the surface of the concerned area. A strong building continuity indicates that buildings predominantly structure

(*Continued*)

Figure 2.3 ***A new building, two-levels high, in the corner between two five-floor buildings (on the right and in the background).*** A.F., Central Athens area, 2015.

urban spaces. Adjacent buildings, streets and artificial pavements cover more than 80% of the area. Vegetation is scarcely used, except in linear arrangement. A building discontinuity indicates that buildings and artificial pavements cover with discontinuity large spaces. Vegetation surfaces coexist with mineral spaces. The use of discrete values implies to get threshold values. Main used parameters are: Building continuity (Discontinuous: isolated buildings; Means: some group of adjacent buildings; Continuous: linear or block arrangement); Surface density (Weak: less than 15%; Means: from 15 to 30%; Strong: more than 30%); Building height (Low: from one to four floors; Means: from five to ten floors; High: more than ten floors). Threshold values may be adjusted or modified to take into account various urban contexts. To limit the study of this work and to fit the correspondence with the Athens area case study, we will mainly deal with downtown urban fabrics characterized by medium and strong density, continuous, medium-high height.

Figure 2.4 ***A new three-floor building between two nine-story buildings.*** A.F., Evripidou central Athens area, 2014.

improvement of building techniques, like the use of reinforced concrete structures. Consequently the sudden enlargement of housing demand, has resulted from the continuous increase of the population.[4]

[4] Half a century later, a Modern emphasis had grown over public plazas and open spaces in general. Moreover the new car-based society required new spaces for parking. Therefore the next release of the zoning resolution included car park requirements and bulk regulations and provided extra floor bonus incentives for the creation of public plazas. Nowadays the zoning resolution is a planning tool constantly updated by large cities, with some main focuses like promoting mixed uses, keeping the city streets vibrant, and generating zoning for special districts, to enhance the character of special neighborhoods.

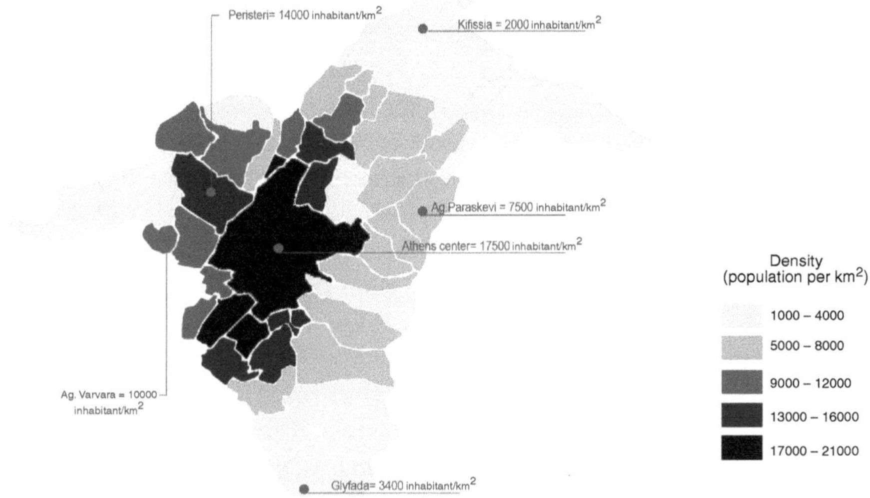

Figure 2.5 ***Density of population in six representative districts of Athens.***

The urban density in a residential area can be calculated in different ways, but there are essentially three that are most usual: (1) the ratio between the number of residential units and the urban horizontal surface; (2) the ratio between the number of inhabitants and the urban horizontal surface; and (3) the ratio between the residential surfaces and the urban horizontal surface. Although the first method is the most commonly used by the urban planner in new developments,[5] it has the disadvantage that it is strictly dependent from the actual destination and use of the specific buildings. The second ratio index (2) is used by landscape designers, geographers, and infrastructure engineers, since it is extremely useful in the calculation of the environmental urban load in a specific built context and the consequent dimensioning of its infrastructural network systems. The third ratio index is the floor area ratio (FAR), and it is indicated as a measurement to identify the density of a particular region (Reale, 2008). This index can be calculated by dividing the sum of all the built surfaces by the square meters of the total urban open area. FAR is one of the more commonly used indexes by urban planners and designers. In fact, in the zoning resolution, the floor area is defined as the sum of the gross areas of several floors of a building or buildings, measured from exterior faces of exterior walls or from the center lines of walls separating two buildings. In Figs 2.5 and 2.7 are mapped the density of population and the FAR in six representative districts of Athens. Even if it does not provide information about the functional use of the urban surfaces (public

[5] It is especially used from Spanish urban planners and architects (Mozas and Fernandez, 2004).

spaces, services, real occupied dwellings, etc.), it represents the most effective ratio to express the numerical incidence of the existing volumetric forms in the built environment. In particular, a FAR below 1 represents a low density, 1–2 represents a medium density, and above 2 indicates a high/very high urban density[6]; a FAR around 3 or more is considered extremely high.

In other terms, the FAR is the total floor area on a zoning lot, divided by the lot area of the same zoning lot. FAR is also sometimes called floor space ratio (FSR), floor space index (FSI), site ratio, or plot ratio; it is the ratio of a building's total floor area (gross floor area) to the size of the piece of land upon which it is built. This value is often used as a zoning regulatory tool for different zoning districts and it can also refer to limits imposed on such a ratio. The floor area ratio can be used in zoning to limit the amount of construction in a specific area. For example, if the relevant zoning ordinance permits construction on a parcel, and if construction must adhere to a 0.10 FAR, then the total area of all floors in all buildings constructed on the parcel must be no more than one-tenth the area of the parcel itself.

The corresponding formula is:

$$FAR = \frac{\text{Total area of all the buildings' floor } \textit{or total gross floor area}}{\text{Area of the urban plot}}$$

The floor area of a building usually does not include cellar spaces, elevator or stair bulkheads, accessory water tanks or cooling towers, attic space providing structural headroom, floor space in open or roofed terraces, bridges, breezeways or porches, floor space used for mechanical equipment, the lowest story of a residential building, floor space in exterior not enclosed balconies, and the like.

If the area of the plot is 100 m^2, then 100 m^2 of gross floor area has been built on the plot. In this case, four floors of 25 m^2 each are built on a site of 100 m^2, with a resulting FAR of 1.0. Thus, the same FAR could have been obtained in the different following options:

A one-story building on 100% of the site ($1 \times 1 = 1.0$)
A two-story building on 50% of the site ($2 \times 0.50 = 1.0$)
A four-story building on 25% of the site ($4 \times 0.25 = 1.0$)

[6] A FAR of 1.5 is quite high, although this density is not unusual in historical city centers like Florence, Venice or central Paris, and is considerably exceeded in most of Manhattan. It requires four-story buildings and narrow streets with limited interior courtyards. Higher buildings would leave more room for streets and gardens, but they are hardly present in the old historical centers of the cities.

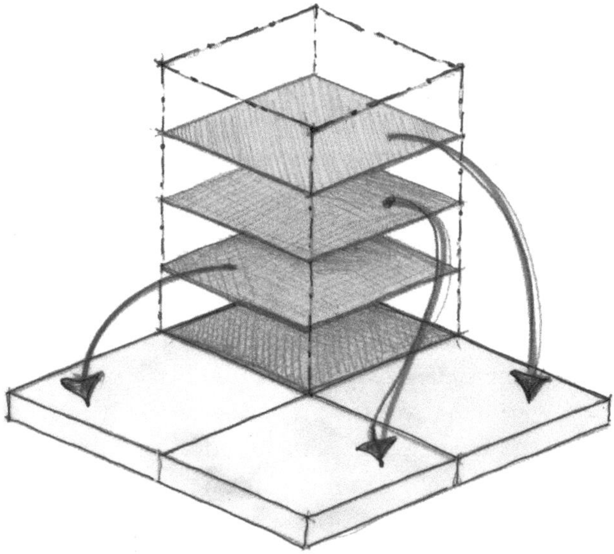

Figure 2.6 ***Floor area ratio (FAR) in the case of a four-story building covering one-quarter of the site.***

The example reported shows a FAR of 1.0 in the case of a four-story building covering one-quarter of the site (Fig. 2.6).

Thus, the same FAR can express either a single-story building consuming the entire allowable area in one floor, or a multistory building that rises

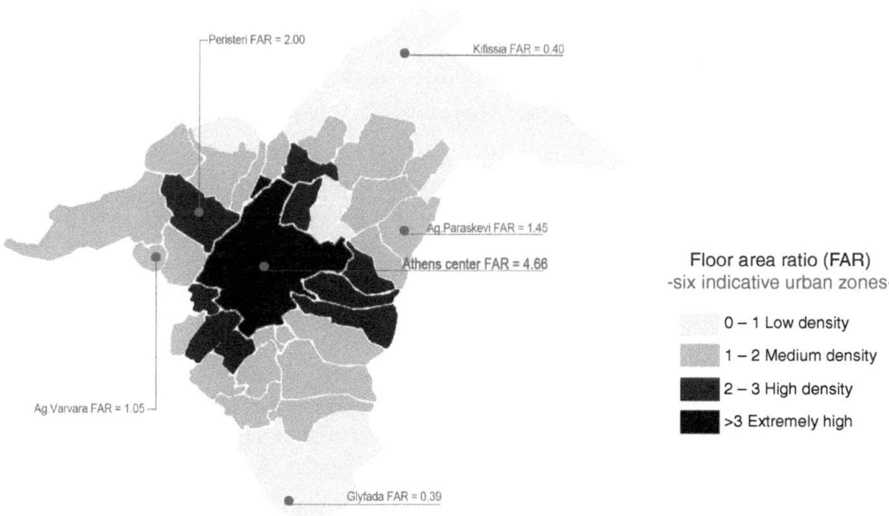

Figure 2.7 ***FAR in the six selected contexts of Athens.***

higher above the plane of the land. In this second case, the building will consequently result in a smaller footprint with respect to a single-story building of the same total floor area. Notwithstanding the manifest degree of approximation of the FAR, which disregards important other parameters, like height, width, or length, it is important to notice that the floor area ratio well correlates with other factors relevant to zoning regulation, such as the total parking area that would be required for an office building, the total number of units that might be available for residential use, total load on municipal infrastructural services, and so forth. In fact, the amounts of these other important urban parameters are nearly constant for a given total floor area, regardless of how total gross floor area is distributed in parallel and/or perpendicularly with respect to the ground level.

Nevertheless, the form is the "genetic code" for the relationship between the building and the urban microclimate and disregarding the shape of urban fabrics, the way they "confine" with the outdoor spaces, would imply a macroscopic misstep.

As a matter of fact, two basic parameters

1. the compactness of buildings and building blocks,
2. and the ability of structures to exchange air

are fully related with the form of the building envelope with respect to its surrounding open areas and the way they border one another. To account for this, the factor of compactness, expressed by the envelope area to volume ratio (Ae/V) has to be considered. The compactness is measured by the ratio of the total surfaces of external envelopes (Ae) and the total volume (V) of the buildings or consecutive building blocks; the surface includes the floor space at the ground level, while deducting wall-to-wall surfaces between adjacent buildings. Thus, at the urban level, outer surfaces connected with a similar neighboring house will not be included (no heat exchange between thermal zones assumed at the same indoor temperature).[7]

A cross-section of building blocks and different densely built urban areas in the major parts of Athens has been chosen. Overall, we compared in the main different urban contexts characterized by different population per km^2

[7] The surface-area-to-volume ratio has physical dimension L−1 (inverse length) and is therefore expressed in units of inverse distance. As an example, a cube with sides of length 1 cm will have a surface area of 6 cm^2 and a volume of 1 cm^3. The surface to volume ratio for this cube is thus given by the following formula: SA:V= 6 cm^2/1 cm^3 = 6 cm^{-1}. For a given shape, SA:V is inversely proportional to size. A cube 2 cm on a side has a ratio of 3 cm^{-1}, half that of a cube 1 cm on a side. The basic geometric forms have different ratios: the ball is 4.83 cm^{-1}, the cube, as shown is 6 cm^{-1}, a pyramid 7.21 cm^{-1}.

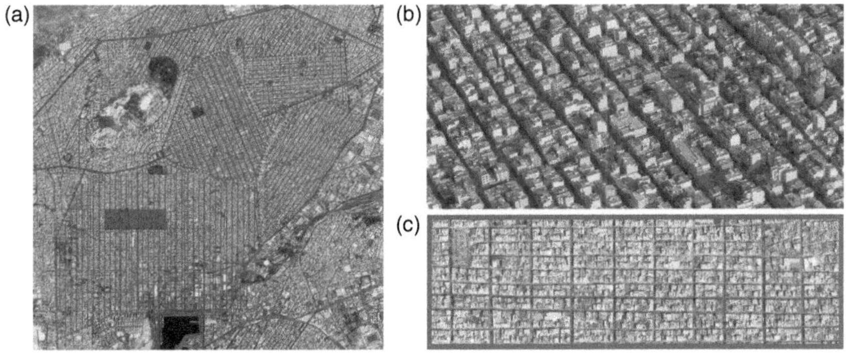

Figure 2.8 (a) An example of "homogeneous areas" at the large urban scale within the Athens central area; (b, c) Aerial view and zenith view of the urban section (a) within the homogeneous district.

Figure 2.9 ***A detailed 3D view of the sample reported in Fig. 2.8.***

as reported in Fig. 2.5. Within these main urban areas, a further subdivision has been identified to extrapolate areas with homogeneous textures, as reported in Figs. 2.8 and 2.9.

By means of 3D reconstruction of these "apparently homogeneous" districts (Fig. 2.10), we compared the Ae/V ratio for different urban block types. The Ae/V ratio of the encountered blocks and urban structures includes all annexes behind the buildings.

Results are shown in Table 2.1.

As observed, being the ratio between the envelope surfaces and the volume a building type depending index, a large range of Ae/V ratios'

Figure 2.10 ***An example of the 3D models used to calculate Ae/V ratio for different urban block types in the homogeneous districts.***

Table 2.1 Different densities in the selected urban districts

AMA's selected districts	Number of inhabitants/km²	FAR	(Ae/V)
Kifissia	2,000	0.40	0.28–0.68
Glyfada	3,200	0.39	0.33–0.63
Aghia Paraskevi	7,500	1.45	0.38–0.59
Aghia Varvara	10,000	1.05	0.28–0.43
Peristeri (social housing urban complex)	14,000	1.99	0.27–0.45
Center (Evripidou courtyard)	17,500	4.66	0.27

variations has been recorded in similar homogeneous districts. As an example, in the elegant district of Kifissia, north of Athens, both detached houses, villas, and residential apartment buildings' blocks are present, thus resulting in the coexistence of houses and urban blocks form very porous to highly compact. On the other side, a similar variation has been encountered in the opposite case of the central Athens area. In fact, given the typical consecutive shocks and discontinuities in the urban landscape of Athens, a large set of envelope area to volume ratio (Ae/V) has been recorded in the same homogeneous district of downtown Athens. Here, if all classical blocks from the last century have a ratio lower than 0,5, the juxtaposition of higher building blocks with lower buildings – either historical or new – increases the amount of external envelopes' surfaces. The ratio of the irregular morphologies ranges from 0,27 in highly dense and homogeneous urban courtyards up to 0,68, reaching values, which are comparable to low-dense urban districts like Kifissia and Glyfada.

Once again, the articulated geometry and the urban morphology depicted in the urban environments of the Athens metropolitan area has highlighted the need for a more detailed, case-study based exploration of selected urban geometry in localized real urban environments. In the following section, additional considerations are discussed to finally identify three potential case studies.

2.2 CASE STUDIES' SELECTION AS POTENTIAL REITERATIVE UNITS AT THE DISTRICT CITY SCALE

The selection of representative case studies should respond to:

1. The need of comprising several urban textures including both dense urban contexts gathered in the typical courtyards of the historical parts of the city and the peripheral urban areas where the stand-alone building types represent the large majority with respect to other buildings.
2. The necessity of properly comparing both low-dense, medium, and high-rise building types and connected urban areas in order to develop a comprehensive study considering very different urban environments in a complex and articulated area like the metropolitan area of Athens.[8]
3. The ambitious goal of estimating the feasibility of nearly zero energy settings in the whole city, including urban settings affected by poorest climatic conditions in terms of heat island effect, higher temperatures, lowest rate of natural ventilation, urban degradation, etc.[9]

[8] The comparison of different aspects in different city contexts is at the basis of the strategic aim of achieving nearly zero urban settings throughout a whole city and to understand the potential integration of mutual intersections among the different physical components (green, surfaces, coating materials, new buildings envelopes) and their effects on urban climate.

[9] In this context it is necessary to note that we need to deal with easily accessible geometrical and climatic measured data. We referred here to developed and on-going research and correlated studies of GRBES (Group Building Environmental Studies), Department of Physics at the National and Kapodistrian University of Athens, coordinated by Professor Mattheos Santamouris on specific case study of the Athens metropolitan area. Measurements have been performed by GRBES in many areas of downtown city and on selected representative areas in the periphery (Municipalities of Aghia Varvara, Peristeri, Kifissia, Glyfada, etc.).

With reference to this last point, we recall here the analyses reported in Chapter 1. They highlighted that, during July and August, the city center and at the western part of the city present much higher mean and maximum air temperatures than the corresponding values in the northern and northeastern part of the AMA.[10] Being that the western areas, together with the central Athens zone, are affected by the highest heat island (HI) intensity of the whole AMA, they are assumed to be the ideal candidates due to their poorest climatic conditions. These aspects, together with the need for considering the different density requirements, dense urban contexts gathered in the typical courtyards and peripheral urban areas with stand-alone building types, has led to the selection of three representative urban compounds, with three very different urban textures and building types, located in the critical areas of the AMA with respect to climatic conditions and urban degradation: Aghia Varvara, Peristeri, and Evripidou.

In the western urban districts of Aghia Varvara and Peristeri, we selected two urban settings occupied by social housing of the 1960s, with typical structures built by reinforced concrete and infill walls. Conversely, in the Evripidou central Athens areas, the selected case study consists of a very dense urban setting made up of adjacent buildings with different dimensions grouped in the distinctive courtyard of the historical parts of the city. The selection of these three real urban settings is representative because of their morphological, climatic, urban, and building type aspects. In fact, these three urban areas also represent three differently dense urban compounds (Fig. 2.11). As far as it regards the energy demand investigation, the three case studies may be assumed as potential reiterative units at the district city scale.

2.3 THREE REPRESENTATIVE URBAN ENVIRONMENTS IN THE AMA

Thus, coherently with the critical issues previously discussed, the research study has been conducted on case studies specifically selected for their representativeness, both in terms of geographical location and climatic

[10] As shown, this is due to the natural territorial morphology of the Athens Area and to increased urbanization, impermeable areas, industrialization, anthropogenic heat and the lack of vegetation. Furthermore, from the analysis of the mean diurnal and nocturnal air temperatures in all stations and from the difference between them and the reference station located at the center of Athens, it is possible to conclude that HI during the night period is mainly observed at the western part of the city (Aghia Varvara and Peristeri).

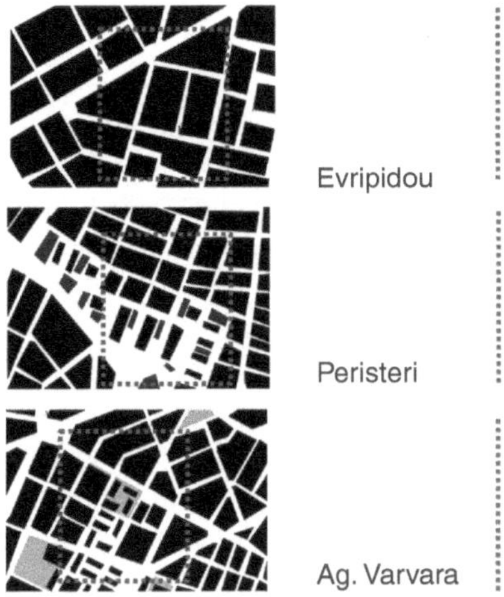

Figure 2.11 ***A schematic representation of the three selected areas in Athens.***

Figure 2.12 ***Aghia Varvara compound presents similar building blocks with different orientation and layouts.***

conditions (the western and central part of the AMA) and for the constructive and building types in each specific urban context.

The two western compounds in the Athens' peripheral area both consist of stand-alone block buildings, although there are important differences and variations in the building types and consistency of these two RUEs: while the Aghia Varvara compound presents similar building blocks with different orientation and layouts, the Peristeri Workers House urban setting is featured with different building types, namely the double block buildings, the tower, and the block buildings (Figs. 2.12 and 2.13).

From the constructive point of view, all the building types in the two RUEs consist of a series of block buildings with a structure made of reinforced concrete and infill walls. This construction type is massively present

Figure 2.13 ***Peristeri workers house urban setting is featured with different building types (the double block buildings, the tower, and the block buildings).***

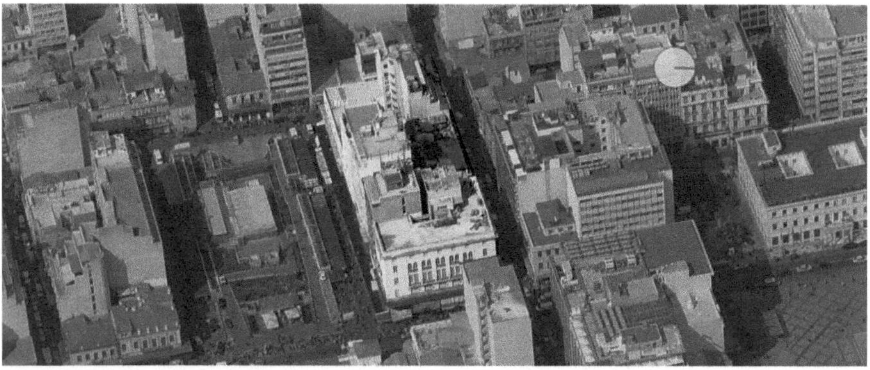

Figure 2.14 ***Evripidou central Athens courtyard with the typical aggregation of the historical center of the city.***

throughout AMA suburbs and it is also typically connected with similar building blocks all over Europe (Fig. 2.14) (Table 2.2).[11]

For the case study of Aghia Varvara, two building types recur, with different orientations, throughout the whole social housing urban compound: they have been named as building "Type A" and "Type B" (Figs. 2.15 and 2.16 and correspondent facades in Figs. 2.17–2.20).

The base case building used in the simulation is a four-story residential building with a roof area of 337 m^2, total floor area of 1093 m^2 and 15 m

[11] Generally, modern buildings constructed after World War II represent the larger majority (about the 60%) of the existing building stock in EU. This percentage even increases (about the 70–75%) if we confine the analysis within the boundaries of the Mediterranean European countries (Greece, Cyprus, Spain, Portugal, etc.) and slightly increases for Italy, as well (about 65%) (Ferrante, 2014).

Table 2.2 Different densities in the selected case studies

Selected case studies	Number of inhabitants/ plot (estimated)	FAR	(Ae/V)
Aghia Varvara	10,000	1.05	0.42*
Peristeri	14,000	1.99	0.28*
Evripidou	17,500	4.66	0.95

* In the case of Peristeri and Aghia Varvara Ae/V values refer to the single building types (tower, block, double block).

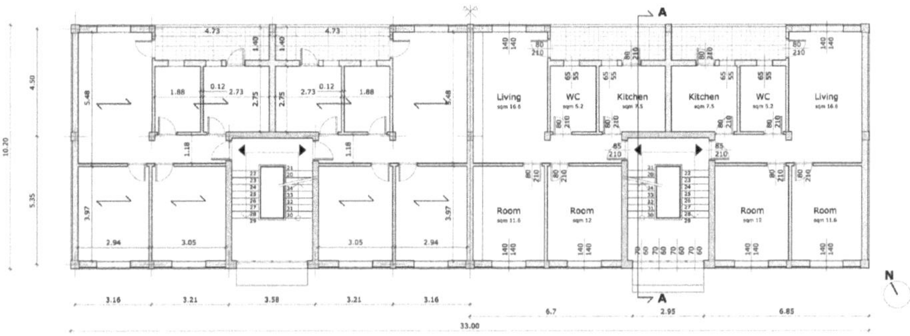

Figure 2.15 ***Block building Type A in Aghia Varvara (type plan).***

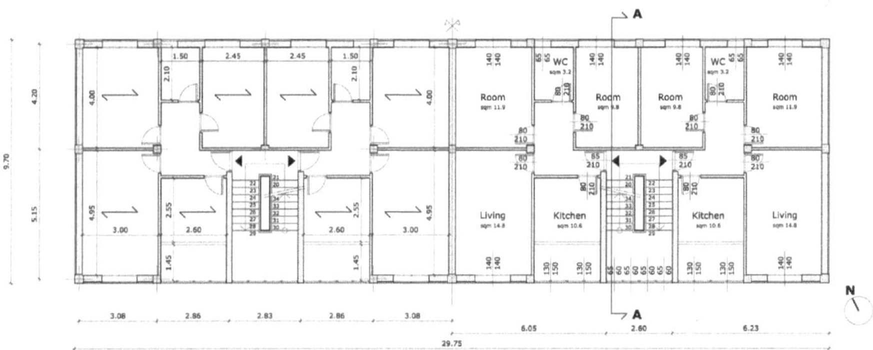

Figure 2.16 ***Block building Type B in Aghia Varvara (type plan).***

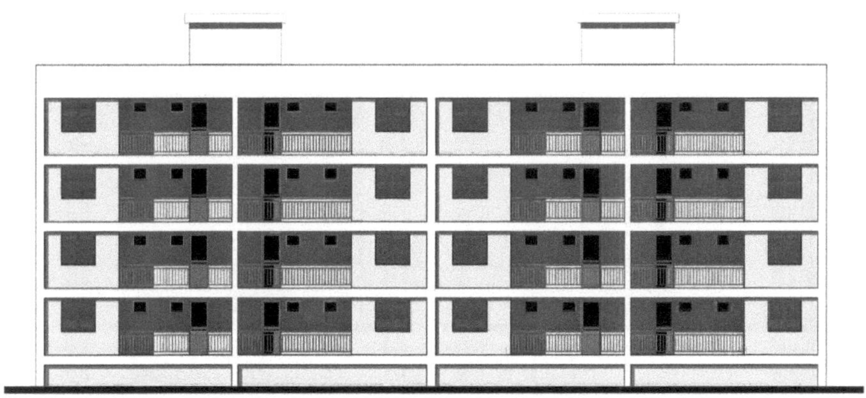

Figure 2.17 ***North front block building Type A.***

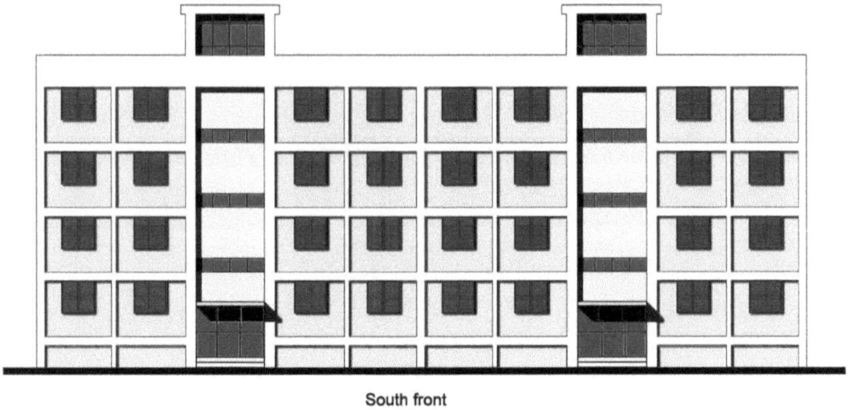

Figure 2.18 ***South front block building Type A.***

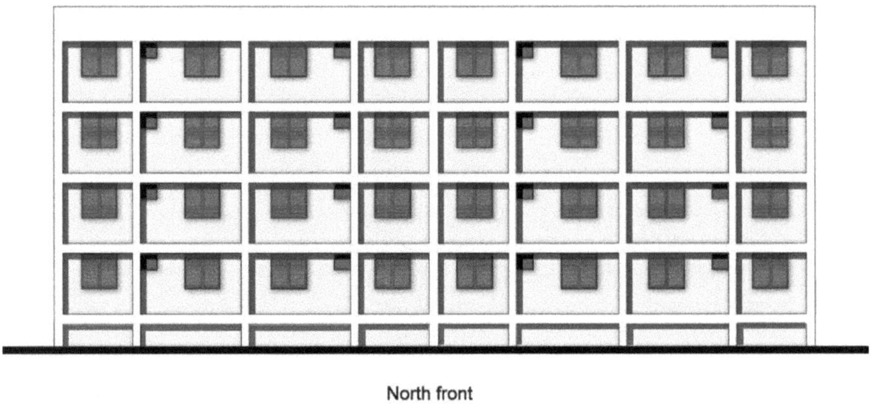

Figure 2.19 ***North front block building Type B.***

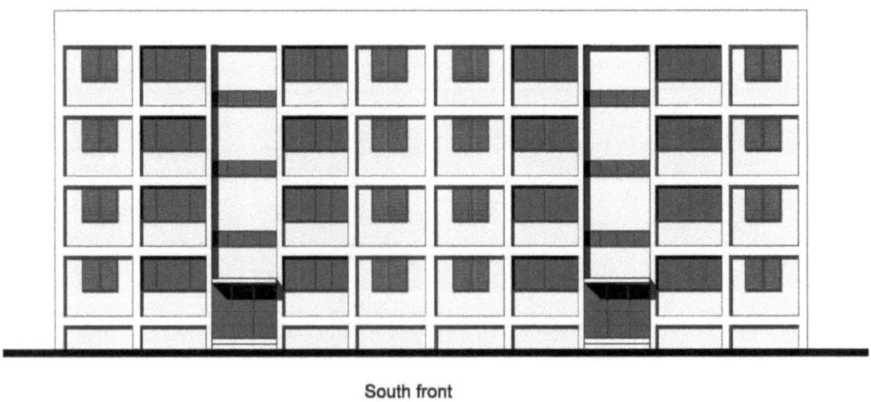

Figure 2.20 ***South front block building Type B.***

height. Indeed, these two building types slightly differ for internal distribution and partitions and consequently they could be defined as two variants of the same building type (Fig. 2.21).

Conversely, the urban compound in Peristeri is characterized by the presence of massive volumes with different building types and building

Figure 2.21 ***Aghia Varvara urban compound formed by recurring building types presenting different orientation and layouts.***

geometry.[12] In particular, the Peristeri urban settings consists of 12 stand-alone buildings segmented in four main building types:

1. Three "tower buildings" (T11)
2. Three double building blocks north-south oriented (A7)
3. Four building blocks east-west oriented (T7)
4. Two building blocks north-south oriented (B6)

All these buildings are residential with the sole exception of the B6 building type, which houses small businesses, shops, offices, and retail at the ground floor level on the south side.

For Peristeri urban settings, real facades' reconstruction of the different building types has been used (Figs. 2.22–2.28) in order to understand current modifications/appropriations of the building spaces indicating thermal and comfort needs of the inhabitants. This analysis has also been developed as a first step toward the ethnographic analysis in the whole urban context.[13]

The ethnographic observation of end-users appropriations that are visible on the buildings' envelopes can be summarized in the following four types of modifications introduced by urban dwellers in the specific Peristeri urban compound case study:

1. The addition of shading devices
2. HWAC and heat pump components hanging on the facades
3. Modifications in window size
4. Space appropriation of balconies and loggias to create additional internal spaces

[12] The urban compound extension is 37,820 m^2, of which 25,713 (68%) is occupied by building construction and impermeable surfaces (parking areas and streets); the remaining 32% is a green open area. The build area represents the 29% (7,504 m^2) of the total impermeable surfaces.

[13] These analyses have been developed as a first step to depict and predict urban dwellers' behaviors and expectations and use them as an in-put in the redesign process of the buildings. Thanks to the complementary research developed at the School of Psychology, Trinity College Dublin, TCD, in the Aerospace Psychology Research Group (APRG), the study has been implemented with the use of ethnographic research to inform the technical aspects of design with a proper consideration of the end user needs and expectations. To inform this study with the criteria of user-centered design and user requirements specifications a social survey and meetings with the inhabitants have been performed in collaboration with the Municipality of Peristeri. In order to support the development of EE solutions in socio-oriented RUEs in the use of appropriate social engagement methods and user-driven design technologies, it is necessary to involve users, build competences, create/implement a real demand for EE solutions. The social engagement methods include: (1) functional analysis of socio technical systems; (2) human and organizational factors best practices, (3) user requirements elicitation and ethnographic research. The use of ethnographic research and the real representation of the buildings through photos is a key passage to inform the technical aspects of design with a proper consideration of the end user needs and expectations.

Tower building (T11)

- 1 staircase
- 4 apartments/floor
- 10 floors
- 40 total apartments

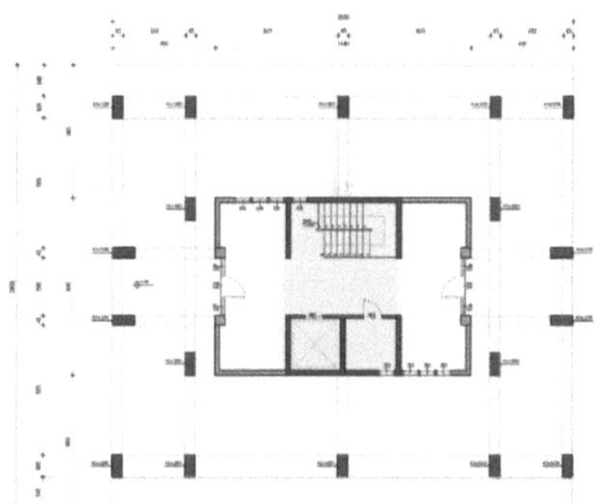

Figure 2.22 ***Ground-level plan and type plan of the tower buildings type in the urban compound of Peristeri (T11).*** On the left, location of the tower building type in the urban setting of Peristeri.

The ethnographic survey has revealed a generally higher presence of modification by users and inhabitants in the south, west, and east facades of the buildings.[14]

Conversely, the urban courtyard selected as the reference case study of Evripidou is representative of the dense built historical area in the core

[14] These modifications have been used to drive the redesign process related to the energy retrofitting of the buildings, tackling them by means of an interdisciplinary, both sociotechnical and physical-engineering approach. Research and design have been focused on the following important incremental/alternative aspects: (1) Study of the energy demand/potential of indoor building types; this research addresses the indoor energy performance of existing building types in the searching of a higher target (zero energy/emissions RUEs), as well as the potential of mutual synergies between outdoor and indoor microclimate improvements. (2) Feasibility studies of possible new additional envelopes and spaces/volumes to be tackled both as passive

(Continued)

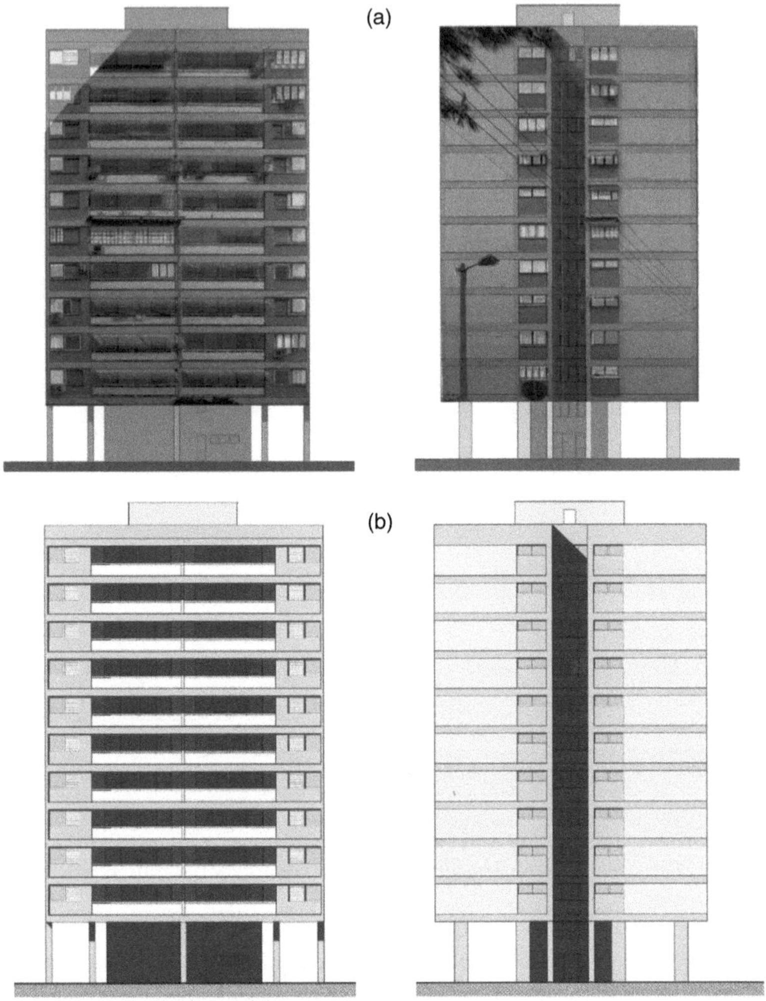

Figure 2.23 South and north facades (a); east and west facades (b) with photographic reconstruction of the tower building block in Peristeri.

tools (incrementing thermal mass, insulation, building compactness) and as possible forms of responding to urban dwellers' requirements. As it will be shown in Chapter 1, the potential adds on can be also considered as a result of combined ethnographic and end user needs research aimed at delivering forms of customized and variable components of self-expression in urban environment with respect to urban dwellers and users' expectations. For the case of Peristeri, in order to meet social aspects related to the possible retrofitting options, a social survey has been developed in collaboration with the School of Psychology, Trinity College Dublin, in the Aerospace Psychology Research Group (APRG). Thanks to the close and effective collaboration with the Municipality of Peristeri, the developed questionnaire (translated into Greek language) has been delivered to sample building blocks (the towers and the block building T7).

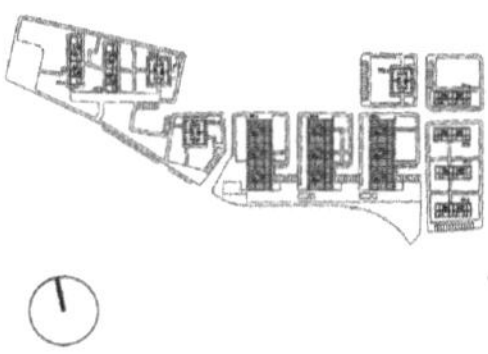

Double building block type (B6)

- 3 stairs
- 15 apartments/floor
- 6 floors
- 90 total apartments

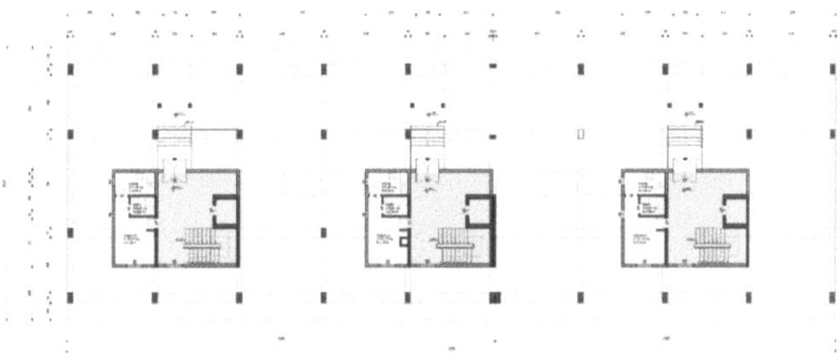

Figure 2.24 ***Type floor (up) and ground floor (down) of one of the three double building blocks (B6).*** On the left, location of the double building block type in the urban setting of Peristeri.

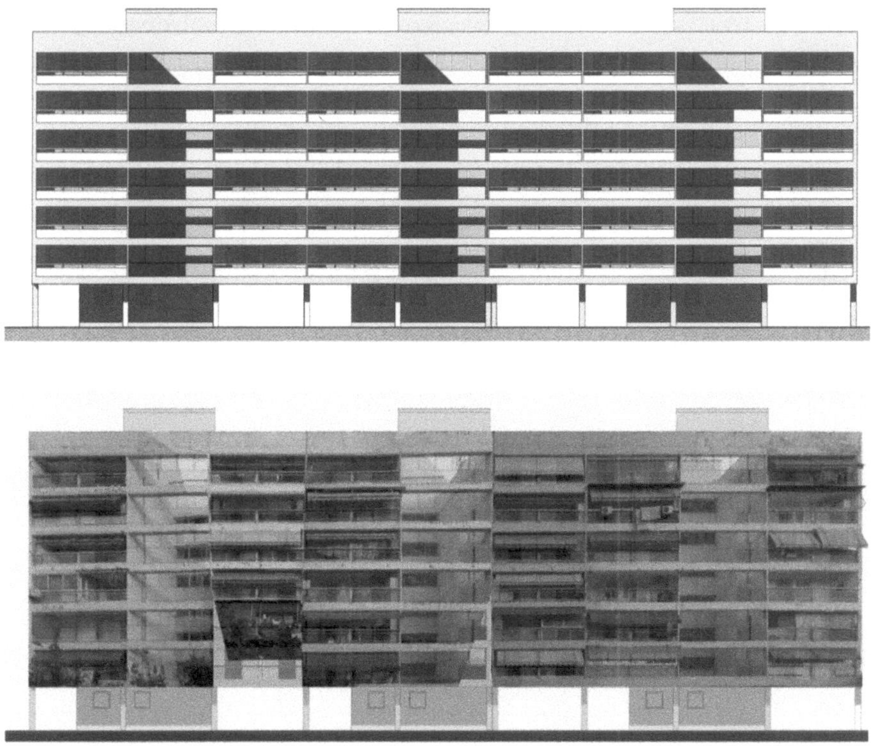

Figure 2.25 ***East facade and photographic reconstruction in the double building blocks in Peristeri.***

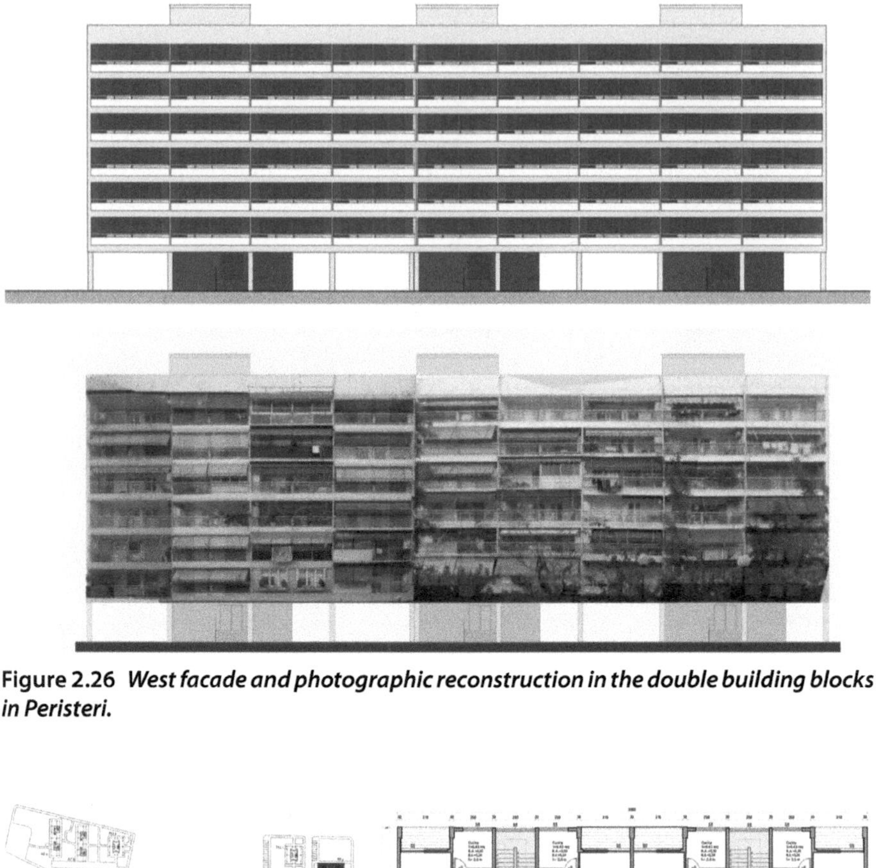

Figure 2.26 ***West facade and photographic reconstruction in the double building blocks in Peristeri.***

Building block type (T7)

- 2 stairs
- 4 apartments/floor
- 7 floors
- 28 total apartments

Figure 2.27 ***Ground and type floor of one of the four building blocks (T7).*** On the left, location of the double building block type in the urban setting of Peristeri.

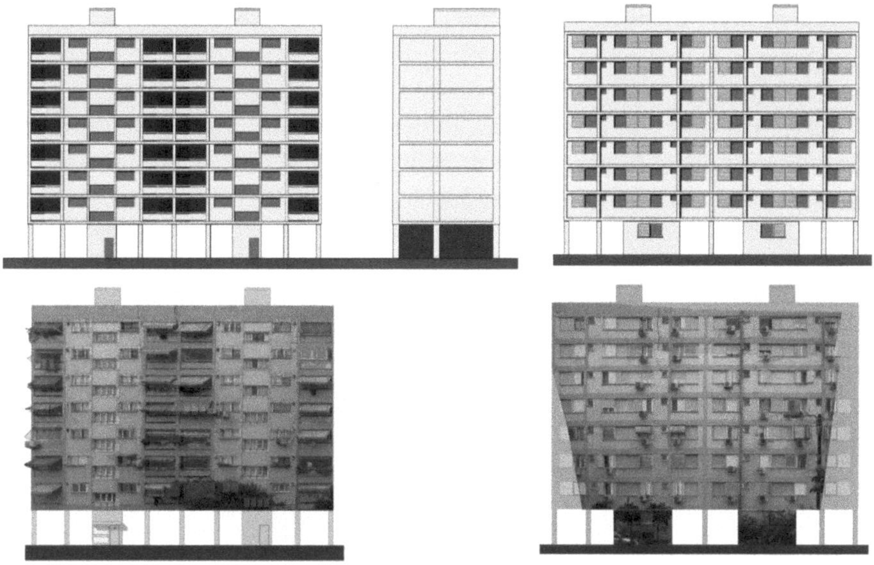

Figure 2.28 ***North and south facades and photographic reconstruction in the building block in Peristeri.***

center of the city; more specifically, the selected area is also enhanced by the presence of many public services like the municipal hall, several civic and administrative offices as well as the central Athens market.[15] A single urban compartment between via Sofokleous and Athinas Street formed by 11 adjacent building blocks – mainly residential housing – has been selected for further investigation. The majority of these buildings' blocks present one floor for retail at the ground level. The type of construction dates back to between 1965 and 1970 for almost all buildings, with the exception of the older buildings 4 and 6, which date back to the 1940s.

As observed, among the main differences between this case study and the previously examined urban compounds is the population density, clearly increasing from Aghia Varvara (11,000 inhabitant/km^2), through Peristeri (14,000 inhabitant/km^2) up to the center of Athens (18,000 inhabitant/km^2). Nonetheless, the most significant data to be considered in this case is the FAR, which reaches an extremely high value. In fact, while the

[15] The considered area of Psiri is surrounded by the streets Peiraios, Athinas, Evripidou, and Tsaldari. The total surface of the area is 105.116 m^2, of which 99.607 m^2 is occupied by buildings and infrastructure, while the remaining 5509 m^2 present small green spots and courtyards as fragments between the buildings. The analyzed area occupies a surface of 2.600 m^2, 2291 m^2 of buildings and 309 m^2 of interior courtyards. All buildings have a gross floor area of 11.249 m^2 (see Fig. 2.29).

calculated FAR for the Psiri district ranges from 2.5 to 3.5, the correspondent value of the selected area, considering it strictly within the borders of the surrounding streets, grows up to 4.33 (Fig. 2.29).

As for the previous case studies, all the supporting structures consist of a frame of reinforced concrete beams and pillars and opaque components of semi-perforated (hollow) masonry blocks (Figs. 2.30–2.32).

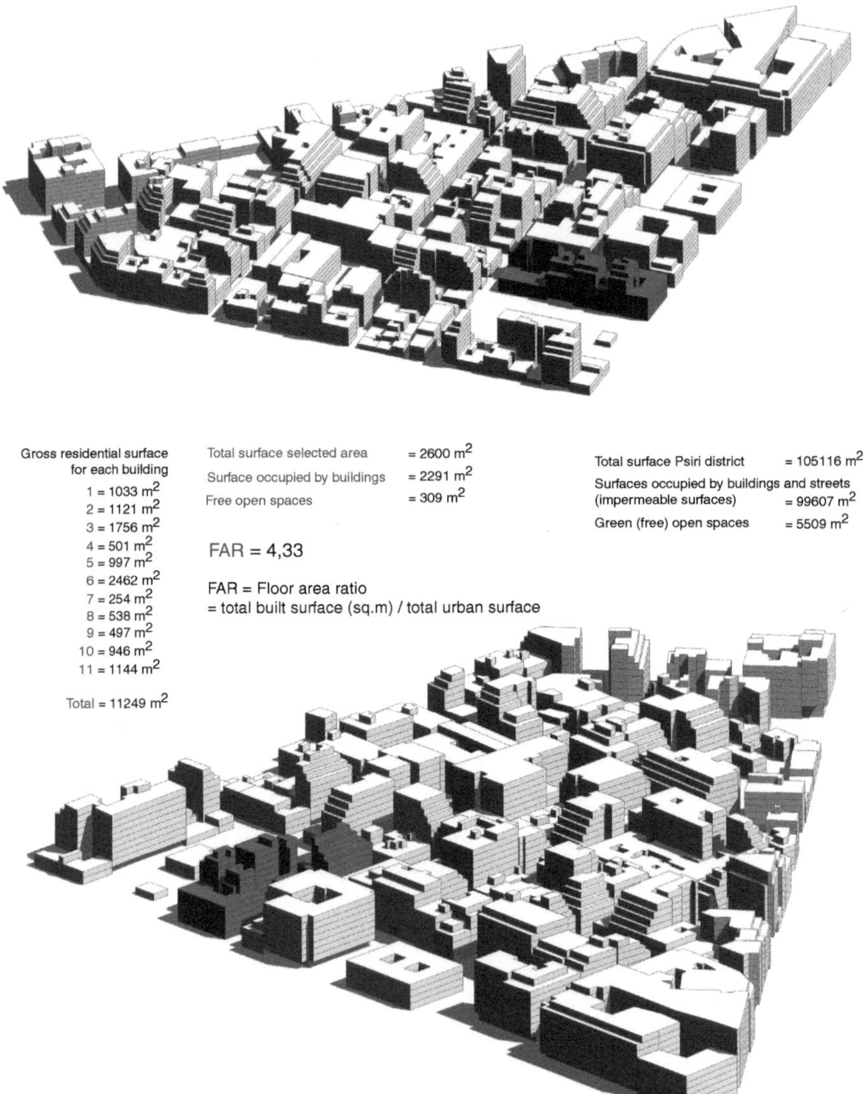

Figure 2.29 ***3D reconstruction of Psiri district and calculation of correspondent density expressed as FAR for the selected urban courtyard (darker).***

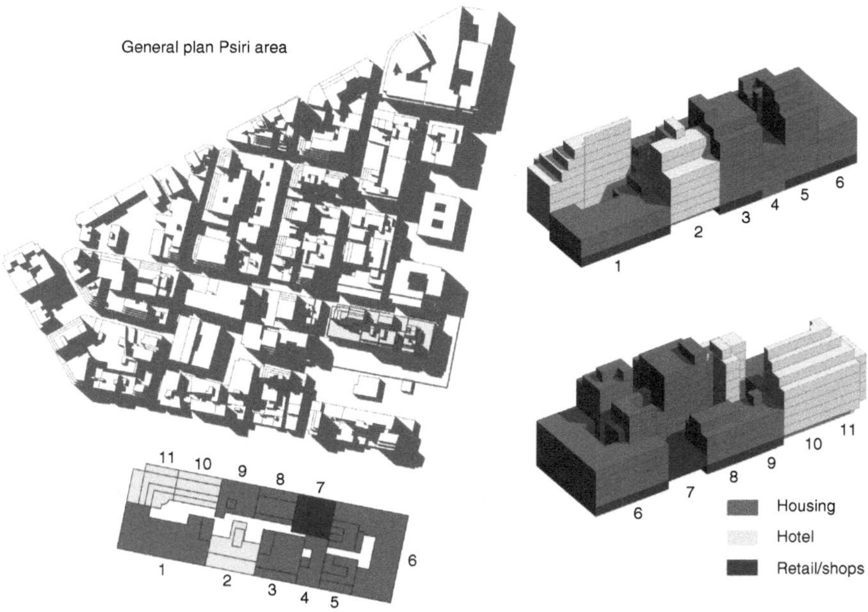

Figure 2.30 ***Site map and use destination of the selected buildings (darker, 7) in the urban courtyard of Evripidou.***

Taking into account the overall complexity that has emerged in the urban morphology of Athens, its unidentifiable urban character, the diffuse presence of gaps between buildings and the diverse building types, the selected representative urban settings can be assumed as different but replicaple units in the representation of the residential areas in the whole a city.

2.3.1 The Energy Demand in the Selected Urban Environments

The evaluation of the energy performance of existing buildings is necessary; it is the preliminary step toward defining the most appropriate solutions to reducing buildings' energy requirements and examining whether and how they can be turned into nearly zero-energy buildings. Thus, to study and evaluate the technical feasibility of nZEBs, the aforementioned real case studies have been analyzed by using simulation tools[16] to assess the energy

[16] All simulations have been performed by using Design Builder, a fully featured interface using EnergyPlus platform to detect energy, carbon, lighting, and comfort performance of buildings. EnergyPlus is the US Departments of Energy's third generation dynamic building energy simulation engine for modeling building, heating, cooling, lighting, ventilating, and other energy flows. It has been validated under the comparative Standard Method of Test for the Evaluation of Building Energy Analysis Computer Programs BESTEST/ASHARE STD 140.

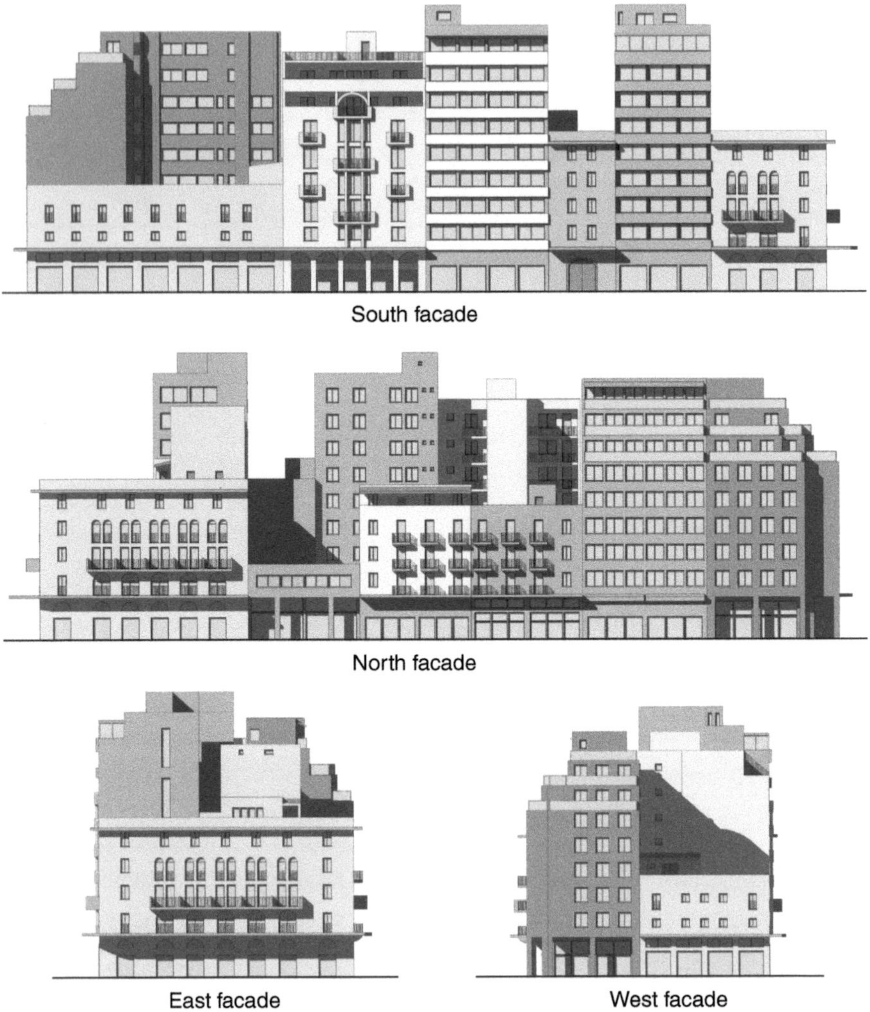

Figure 2.31 ***Buildings' elevations – south, north, east, and west oriented.***

performance in the as-built scenario. To run energy simulations in buildings, it is necessary to define the internal thermal zones of a built environment,[17] Thus, separate volumes for each building or set of buildings have to be identified as correspondent thermal zones,[18] In order to avoid an excessive

[17] A list of levels and sublevels from the building up to the division of opaque and transparent parts is generally created to run accurate dynamic simulations.

[18] 3D models for each case study have been developed to redesign the exact positioning of all opaque and transparent surfaces and climate data settings.

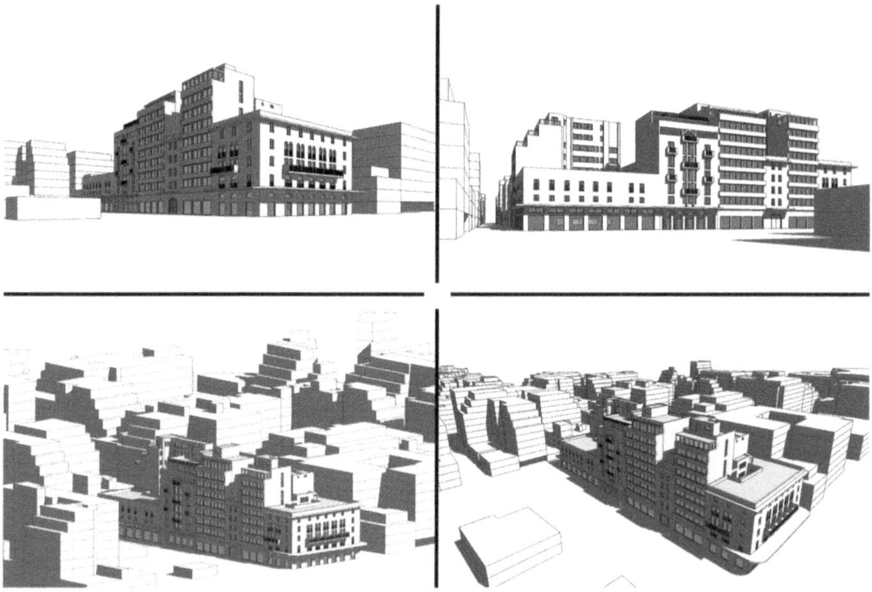

Figure 2.32 ***Reconstruction of the main 3D perspective views in the Evripidou urban courtyard.***

fragmentation of data and achieve comparable results among the case studies, a number of common input data has been assumed and estimated.[19] The operating programs of the systems follow the schedule indicated in Table 2.3.

[19] Among the common characters assumed for the different cases is the plant systems' equipment. Each building is assumed equipped with a central heating system using gasoil as fossil fuel. In all the visited building of Peristeri and Aghia Varvara cast iron radiators are installed in the rooms of the apartments, therefore they are assumed as terminal heating plant components in all case studies. In all the different urban settings, the ethnographic observation has showed how the majority of residential units (about 95%) are equipped with air conditioning units; some others (30–38%) use solar systems for domestic hot water. The light sources are mainly of domestic type, such as incandescent light bulbs or compact fluorescent lamps. All the buildings have a structure of reinforced concrete and exterior surfaces made up by hollow bricks and treated with plaster. Neither the flat roofs nor the vertical surfaces are insulated; in the majority of apartments the windows consist of aluminum frame systems and single glazed transparent components, prevailingly equipped with exterior shutters as windows' shading devices. The assumed operating program and the performance of the plant system do not affect the rightness of the conclusions because they have been kept constant in all simulations.

Table 2.3 The operating programs of the plant systems

Seasons	Heating	Cooling	DHW
Summer (Apr.–Oct.) (workdays and weekends)	Off (00.00–24.00)	On (00.00–24.00) (rooms) 07.00–23.00 (living, kitchen)	On/according to occupancy
Winter (Nov.–Apr.) (workdays and weekends)	On (00.00–24.00)	Off (00.00–24.00)	On/according to occupancy

Other common data inputs are: local weather data,[20] activity, occupation and ventilation,[21] which are characteristics of the plant systems. However, the existing differences among the scale, the consistency, and the building types of these urban contexts have implied a different approach in the degree of detail applied between them. Thus, different criteria have been conceived and adopted for determining the thermal zones in the different buildings in the energy analysis of the three case studies, following decremented level of details: from the more meticulous definition in the case

[20] A file containing all the hourly climatic data, as the dry bulb, temperature, wet bulb temperature, wind speed and direction, solar height, solar azimuth, atmospheric pressure, and direct and diffuse solar radiation is contained in the default weather data. These data have been compared with simulations performed in various climatic stations of the Athens metropolitan area (GRBES, Prof Santamouris) and, in particular with measures gathered in Peristeri urban compounds. Similarities among these data have lead to the conclusion that no necessary changes should be made at this stage of the analysis.

[21] The models reported the actual organization of the plan of the building, in the different level of detail specified after this note in the plain text. In each floor the different thermal zones assumed correspond to various rooms like kitchens, bathrooms, bedrooms, living rooms, and condominium areas like stairwells. The criterion chosen to make the partition into thermal zones was to differentiate rooms depending on the specific activities that are carried out in them. This made it possible to associate to each zone an activity program, built according to the average behavior of local users. The activity program defines the occupation times and ways, the use of appliances, the opening of the windows and the operation of manual or mechanical ventilation and it is used by Energy Plus to evaluate internal energy contributions. According to size and type, three up to six people usually occupies each apartment, thus a density of 0.05 people per m^2 has been assumed. The default occupancy program has been slightly changed to consider the specific composition of the dwellers: it has been estimated that about the 40% of the population is a non-working age residents, so most of them are present and carry out activities in their apartments during the whole day, while the 60% of the residents are working people and from Monday to Friday, between 9:30 am and 5:30 pm, are not at home. According to this, occupancy profiles were created for each thermal zone.

of the Aghia Varvara urban compound, through the intermediate level of detail energy investigations in Peristeri and finally, to a less definite scale in Evripidou central area.

Thus, for the case of Aghia Varvara,[22] the analysis has been conducted considering a *unit-oriented* subdivision of the main thermal zones, with the units relating to the representative apartments, which in turn have been evaluated in two separate environments consisting of day-living spaces and night areas.

Secondly, for the case of Peristeri, the analysis has been conducted considering both a *building type and apartments type* oriented focus, considering the main thermal zones corresponding to the single apartments, each one consisting in one single thermal zone.

Finally, in the Evripidou central Athens area, an *urban oriented analysis* has been developed to focus on the different energy performances of each building block. In this case, a more approximate subdivision has been operated, which consists of defining each floor in each building of the urban courtyard as a different thermal zone.

For both the two western urban settings of Aghia Varvara and Peristeri and the Evripidou urban courtyard, the following research tasks have been designed and developed:

- Evaluation of the overall energy performance and sensitivity analysis of its variability as a function of the different orientation of the buildings as well as of different internal distribution; simulations have been performed for each unit type.[23]
- Evaluation of the energy performance and sensitivity analysis of its variability related to the different arrangement of the surrounding buildings within the same urban compound.
- Results simulation and comparison in terms of comfort, energy contributions, and overall heat balance, in order to: (1) identify the main

[22] As described in the previous paragraph, Aghia Varvara RUE consists of a dense social housing urban complex of the 1960s, presenting a set of 15 similar residential buildings with different orientation and layouts; this aspect marks out this urban compound as the ideal candidate for a sensitivity analysis of energy simulations as a function of the different urban constraints and boundary conditions. In fact, all the buildings can be traced back to two similar layout models consisting of block buildings with two internal staircases serving four apartments in each floor per four floors.

[23] In Aghia Varvara units correspond to the residential apartments. Given the level of detail, in this case apartments have been subdivided in sleeping areas and living areas; in Peristeri units correspond to the apartments, while for the case study of Evripidou each floor in each buildings constitutes a different thermal zone.

critical problems affecting the building types, (2) classify them according to the their effects on the overall energy balance of the building, and (3) propose targeted solutions.

- Design of building refurbishment interventions and options in the selected reference building (presenting the worst energy and climatic response).
- Running a simulation of the effects that the alternative design scenarios would produce on the overall energy balance of the building, so as to understand the effectiveness of the different retrofitting options in terms of energy response.
- Final evaluation of the obtained results in terms of cost-effectiveness and quality.

2.3.1.1 Aghia Varvara

The main thermal parameters in Aghia Varvara are provided for the horizontal ground, intermediate, and roof floors as well as for the external building envelope. (Tables reporting the main thermal properties of the ground floor layers, the roofs, and the glazing components estimated for each building in each case study are reported in the Appendix)

As already specified, the real plan layout has been reported in the model, so on each floor, different zones have been created, which correspond to the various rooms: kitchens, bathrooms, bedrooms, living rooms, and condominium stairs.[24]

Detailed and articulated simulations have been performed for the "Type A" building,[25] considered in its four different configurations (named A1, A2, A3, and A4) and for all the types of residential units. These have been depicted and analyzed as a function of their different position in the building block (Fig. 2.33).

Thus, the energy performance has been evaluated both in relation to the buildings' orientation and the internal layout in each apartment type. In fact, to achieve a correct and efficient comparison, the different rooms have been considered merged in the two main categories of living space (consisting of a kitchen, a bathroom, and a living room) and sleeping area (bedrooms).

[24] The selected criterion to operate a proper partition into different thermal zones follows both geometrical and specific human activities carried out in each room.

[25] As described, the structural system of the similar buildings is very simple, made up by a regular grid of beams and pillars, with presumably prefabricated slabs as horizontal elements. The foundation mat leans on a concrete screed. The foundation has a waterproofing layer and no insulation is provided in the ground floor level.

Figure 2.33 ***Type A buildings in the Aghia Varvara urban compound.*** **Different subtypes (A1, A2, A3, A4) are identified as a function of different exposures in relation to the sleeping and living areas of the single residential units.**

This categorization, together with the existing different exposition of the building blocks A1, A2, A3, and A4 have formed a precise classification of main geometrical and thermal characteristics of each residential unit in relation to its position in the different building blocks. In particular, Table 2.4 shows the orientation of living and sleeping areas of the buildings as they are named, while, considering the type and the number of exposed surfaces (Table 2.5 and Fig. 2.34), the apartments can be classified according to common/recurrent characteristics. All the apartments present exposed surfaces as listed in Table 2.5, with an outdoor ambient area facing the external surroundings and one surface facing the common stairs, which can be considered as a semi-exterior unconditioned space (and they are represented as such in the model).

Simulations have been performed taking into account the main geometrical and construction building data (see the Appendix), the existing correspondence between different units' subtypes, and the different orientations (see Fig. 2.35).

Table 2.4 Classification of the building residential units as a function of different orientation of the internal thermal zones

Building	Living area facing	Sleeping area facing
A1	North	South
A2	South	North
A3	West	East
A4	East	West

Table 2.5 Classification of the building residential units as a function of different exposition and their different location within the building volume

Residential units	Exposed surfaces			
1	—	Side wall	Ground floor	2 facades
2	—	Side wall	—	2 facades
3	Roof	Side wall	—	2 facades
4	—	—	Ground floor	2 facades
5	—	—	—	2 facades
6	Roof	—	—	2 facades
7	—	Side wall	Ground floor	2 facades
8	—	Side wall	—	2 facades
9	Roof	Side wall	—	2 facades

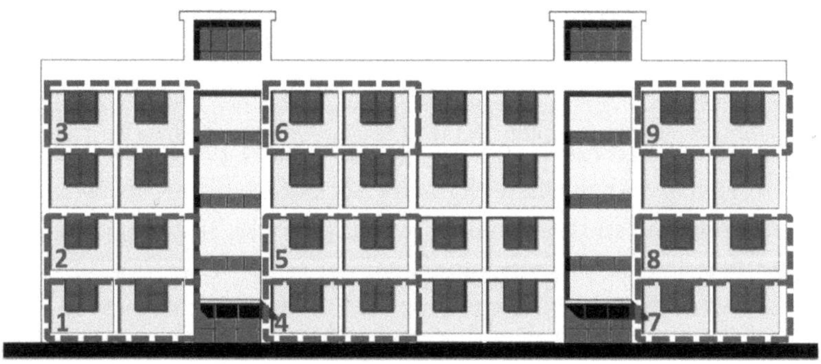

Figure 2.34 ***Different residential units considered in the energy analysis.***

By means of the previously described classification, it is possible to identify the differences in the energy balances of apartments having identical geometrical characteristics, but different orientations. These simulations also document the proper understanding of the variability of situations from unit to unit in the different building blocks. The first pair of charts, which shows the behavior of living and sleeping areas of apartment 1 in building A1, are reported in Fig. 2.36. The sleeping area of the building A1 is south-oriented, thus it receives large solar gains. On the contrary, the living area is north-oriented,

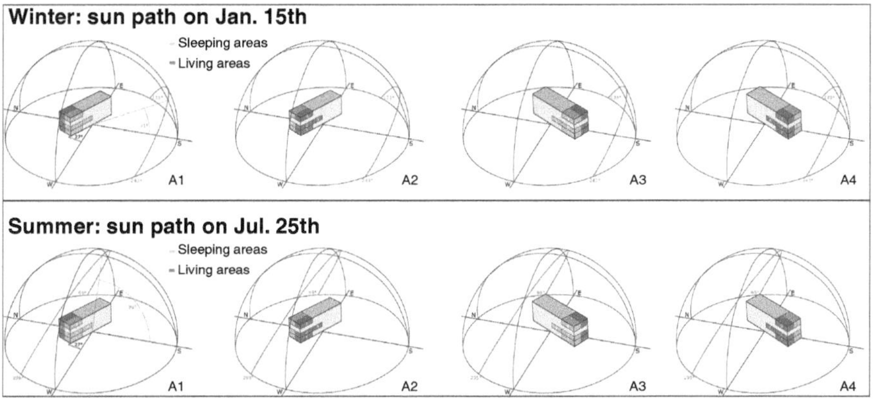

Figure 2.35 ***Solar path in cold winter season's and hot summer period's representative day in the four building types.***

thus it receives smaller solar gains through windows (about 40% of those received from the sleeping area). The high solar gains in the sleeping area help to lower the heating need of the area, which is still much higher than the other apartments and higher than the average of the building. This is due to the particular position in contact with the ground. However, in the living area, the solar gains are so small that they do not constitute a substantial reduction in the heating load. The second pair of graphs shows the energy balance of the same apartment type placed in the building with an opposite orientation, with the sleeping area facing north and the south-oriented living area. As observed, the behavior of the two zones is qualitatively similar to the previous case; in fact, from the graph, we see a similar trend, but results shown in Fig. 2.37 show important differences in absolute values.

All sleeping areas and living areas have the same geometrical and technical features, but the total exposed surface of the living area is approximately 40% greater than that of the sleeping area, and the total glazed surface of the living area is approximately 40% higher than the sleeping area in absolute value, as well. For this reason, in wintertime, the living area of building A2 receives 40% more solar gains through the windows compared to the sleeping area in building A1, although both of them are south-oriented; the opposite happens for the two complementary areas facing north. The case of building A2 presents a particularly favorable orientation because solar gains are concentrated in the living area, so they are properly exploited. In both cases, however, the higher solar gains imply a decrease of the heating; solar gains are obviously more effective in lowering the heating load of the whole apartment when they are concentrated in the living area. As shown in Fig. 2.37, in building A2, the 20% increase in daily solar gains led to a

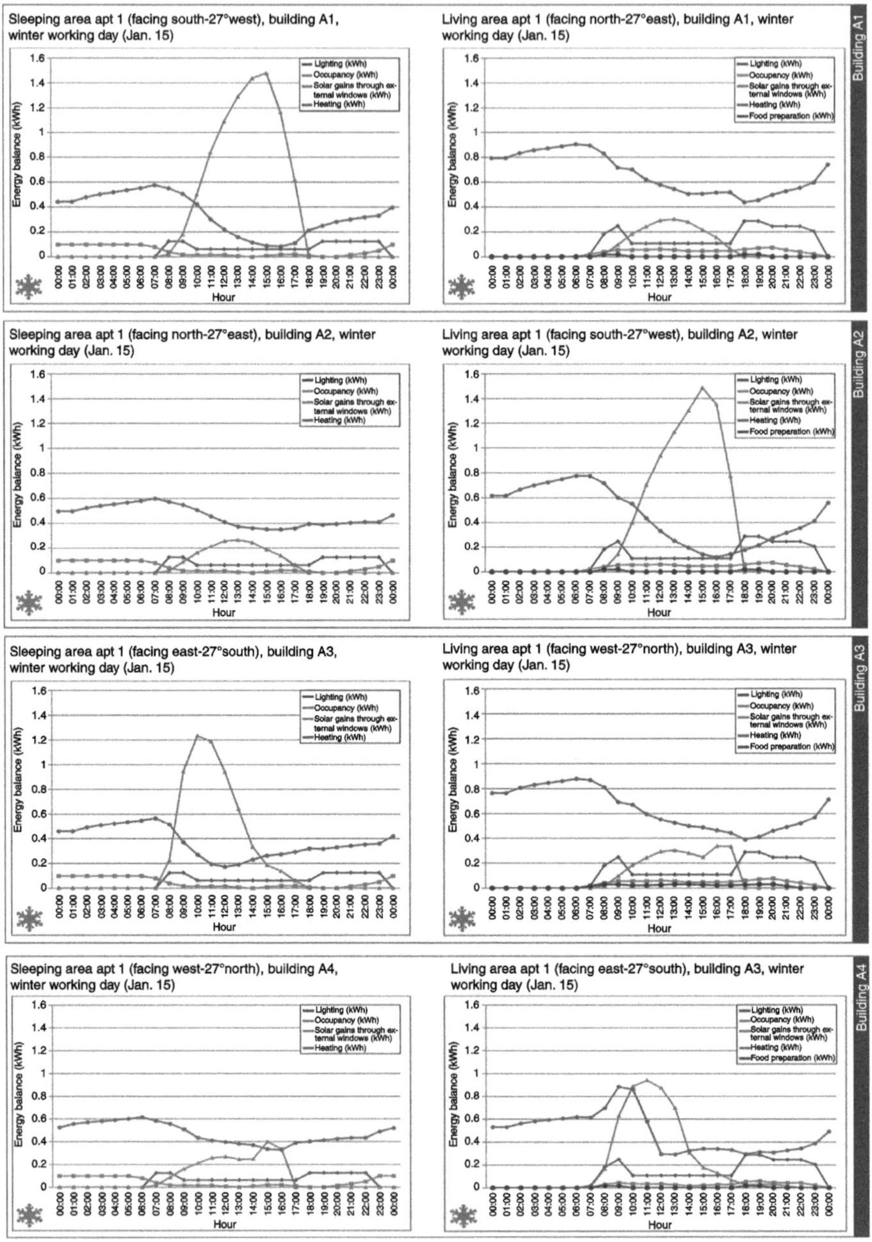

Figure 2.36 ***Charts of the simulation for the type unit 1 (ground floor) in the representative winter day.***

Apt1 (ground floor) winter		Facing	Exposed wall	Glazing surface		Glazing surface aligned with the outer wall		Lightning		Food preparation		Occupancy		Solar gains		Heating		
								max	daily tot	max	daily tot	max	daily tot	max	daily tot	max A.M.	max P.M.	daily tot
			m^2	m^2		m^2		kWh	kWh	kWh	kWh	kWh	kWh	kWh	kWh	kWh	kWh	kWh
A1	Sleeping	S-W	31.59	3.92	12.41%	3.92	100%	0.125	1.437	—	—	0.099	1.134	1.233	5.896	0.568	0.352	9.309
	Living	N-E	45.06	6.04	13.40%	1.96	32.45%	0.287	2.828	0.016	0.083	0.069	0.788	0.335	2.339	0.878	0.567	15.891
	Total		76.65	9.96	12.99%	5.88	59.04%		4.265		0.083		1.922		8.235			25.2
A2	Sleeping	N-E	31.59	3.92	12.41%	3.92	100%	0.125	1.437	—	—	0.099	1.134	0.266	1.630	0.598	0.410	11.505
	Living	S-W	45.06	6.04	13.40%	1.96	32.45%	0.287	2.828	0.016	0.083	0.069	0.788	1.486	8.239	0.774	0.408	11.413
	Total		76.65	9.96	12.99%	5.88	59.04%		4.265		0.083		1.922		9.869			22.918
	Δ(A2-A1)								+0.00%						+19.84%			−9.06%
A3	Sleeping	E-S	31.59	3.92	12.41%	3.92	100%	0.125	1.437	—	—	0.099	1.134	1.479	8.615	0.576	0.333	8.708
	Living	W-N	45.06	6.04	13.40%	1.96	32.45%	0.287	2.828	0.016	0.083	0.069	0.788	0.302	1.851	0.905	0.598	16.691
	Total		76.65	9.96	12.99%	5.88	59.04%		4.265		0.083		1.922		10.466			25.399
	Δ(A3-A1)								+0.00%						+27.09%			+0.79%
A4	Sleeping	W-N	31.59	3.92	12.41%	3.92	100%	0.125	1.437	—	—	0.099	1.134	0.398	2.227	0.614	0.488	11.826
	Living	E-S	45.06	6.04	13.40%	1.96	32.45%	0.287	2.828	0.016	0.083	0.069	0.788	0.941	4.991	0.886	0.326	12.05
	Total		76.65	9.96	12.99%	5.88	59.04%		4.265		0.083		1.922		7.218			23.876
	Δ(A4-A1)								+0.00%						−12.35%			−5.25%

Figure 2.37 ***Numerical values and main difference in the energy performance of apartment type 1 in the four building configurations during winter.***

correspondent decrease of the heating load by 9%, compared to the same unit in building block A1, which shows an opposite result. The third pair of graphs shows the energy balance of building A3, which is perpendicular to building A1, with an east-27°-south-oriented sleeping area and opposite-oriented living area. Once more, similar to building A1, the wide solar gains are not exploited in an efficient way to decrease the heating load, because they are concentrated in the sleeping area. Despite major solar contributions, the heating load is almost the same as that in case A1. Similar behavior has been recorded for case A4 (last pair of graphs in Fig. 2.36). Between unit 1 in building A3 and unit 1 in A4, there is a parallelism similar to the one existing between the cases in buildings A1 and A2.

The same simulations have been performed for the type 1 unit in the different building blocks, but instead for a representative day of the hot summer season (Fig. 2.38). While in winter solar gains are a positive resource that help to offset the required energy loads for the heating system, in the hot summer season they greatly exacerbate the energy balance, as they contribute to overheating the internal spaces. The south-oriented sleeping area of apartment 1, building A1 (first pair of charts, on the left in Fig. 2.36) receives solar gains similar to the amount it receives in the winter, but during the summer they are concentrated in the afternoon. The north-oriented living area instead receives solar gains during the morning. Both thermal zones overheated just when people started to use them, so the comfort conditions are hardly achieved.

In general, as solar gains are of a greater magnitude, the load for cooling is more relevant, but in all the considered cases concerning apartment 1, the cooling loads are very low; this is because the position of the residential unit in contact with the ground strongly influences the heat balance of apartment 1. While in winter conditions this position increases the need for heating, in the summer season, the same position reduces the cooling need. The effect of

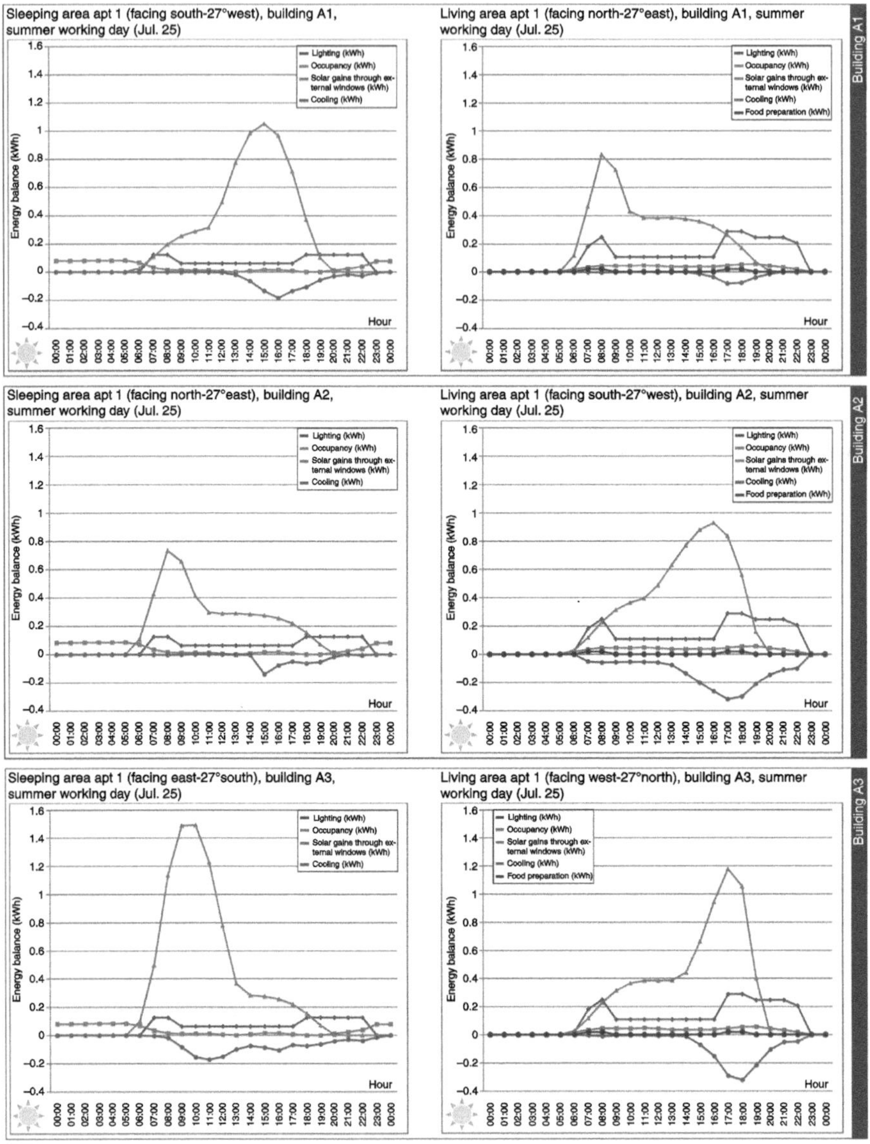

Figure 2.38 ***Charts of the simulation for the type unit 1 (ground floor) in the representative summer day.***

contact with the ground affects only the apartments situated on the ground floor and does not influence the other residential units (Fig. 2.39).

Conversely, larger differences are recorded due to the particular location of apartment 3, positioned just below the roof. Its position has direct consequences, especially in summer. In fact, from the graphs shown in Figs. 2.40

Apt1 (ground floor) summer		Facing	Exposed wall	Glazing surface		Glazing surface aligned with the outer wall		Lightning		Food preparation		Occupancy		Solar gains		Cooling	
								max	daily tot	max	daily tot	max	daily tot	max	daily tot	max	daily tot
			m^2	m^2		m^2		kWh	kWh	kWh	kWh	kWh	kWh	kWh	kWh	kWh	kWh
A1	Sleeping	S-W	31.59	3.92	12.41%	3.92	100%	0.125	1.437	—	—	0.099	1.134	1.052	6.659	0.185	0.799
	Living	N-E	45.06	6.04	13.40%	1.96	32.45%	0.287	2.828	0.016	0.083	0.069	0.788	0.831	5.294	0.082	0.284
	Total		76.65	9.96	12.99%	5.88	59.04%		4.265		0.083		1.922		11.953		1.083
A2	Sleeping	N-E	31.59	3.92	12.41%	3.92	100%	0.125	1.437	—	-	0.099	1.134	0.732	4.467	0.177	2.821
	Living	S-W	45.06	6.04	13.40%	1.96	32.45%	0.287	2.828	0.016	0.083	0.069	0.788	0.929	6.696	0.320	2.187
	Total		76.65	9.96	12.99%	5.88	59.04%		4.265		0.083		1.922		11.163		5.008
	Δ(A2-A1)								0.00%						−6.61%		+362%
A3	Sleeping	E-S	31.59	3.92	12.41%	3.92	100%	0.125	1.437	—	—	0.099	1.134	1.490	8.341	0.172	1.284
	Living	W-N	45.06	6.04	13.40%	1.96	32.45%	0.287	2.828	0.016	0.083	0.069	0.788	1.178	6.888	0.321	1.310
	Total		76.65	9.96	12.99%	5.88	59.04%		4.265		0.083		1.922		15.229		2.594
	Δ(A3-A1)								0.00%						+27.41%		+140%
A4	Sleeping	W-N	31.59	3.92	12.41%	3.92	100%	0.125	1.437	—	—	0.099	1.134	1.383	7.146	0.277	1.385
	Living	E-S	45.06	6.04	13.40%	1.96	32.45%	0.287	2.828	0.016	0.083	0.069	0.788	1.213	7.477	0.206	1.855
	Total		76.65	9.96	12.99%	5.88	59.04%		4.265		0.083		1.922		14.623		3.24
	Δ(A4-A1)								0.00%						+22.34%		+199%

Figure 2.39 ***Numerical values and main differences in the energy performance of apartment type 1 in the four building configurations during summer.***

and 2.41, it is immediately evident that there is high overheating generalized to all the environments, with a consequent high cooling demand. The first pair of graphs in Fig. 2.40 shows the energy balance of apartment 3 in building A1, with the sleeping area facing south, which receives significant solar gains through the windows during the afternoon and, vice versa, the living area faces north and receives them in the morning. Assuming that the trend of the direct radiation on the roof is similar to that of direct solar gains through the windows, it explains the curve of cooling load in relation to that of solar gains. The second pair of graphs shown in Fig. 2.38 shows the same apartment in building A2. The sleeping area exposed to the north and the south-facing living area receive a total solar input through the windows of about 8% less, compared to the previous case, even if the same change in the cooling load cannot be seen. This outcome, together with high dispersions and energy intakes across the surfaces, implies that the energy balance of the apartment located below the roof is much more influenced by the effects of direct radiation on the roof than from the direct radiation entering through the windows. The graphs in Figs 2.40 and 2.41 show the performance of apartment 3 in buildings A1, A2, A3 and A4, which have an orientation perpendicular to the two previous cases.

Apart from quantitative differences depending on the difference in the glazing surface, areas with the same exposure have a similar behavior. Thus, east-oriented areas receive large solar gains during the morning; similarly, west-oriented areas show the peaks of their curves of cooling load in the afternoon. Looking at the quantitative values, the behavior of the apartments that are east-west oriented is similar, since in both cases, the quantities involved have a gap between them of less than 5%.

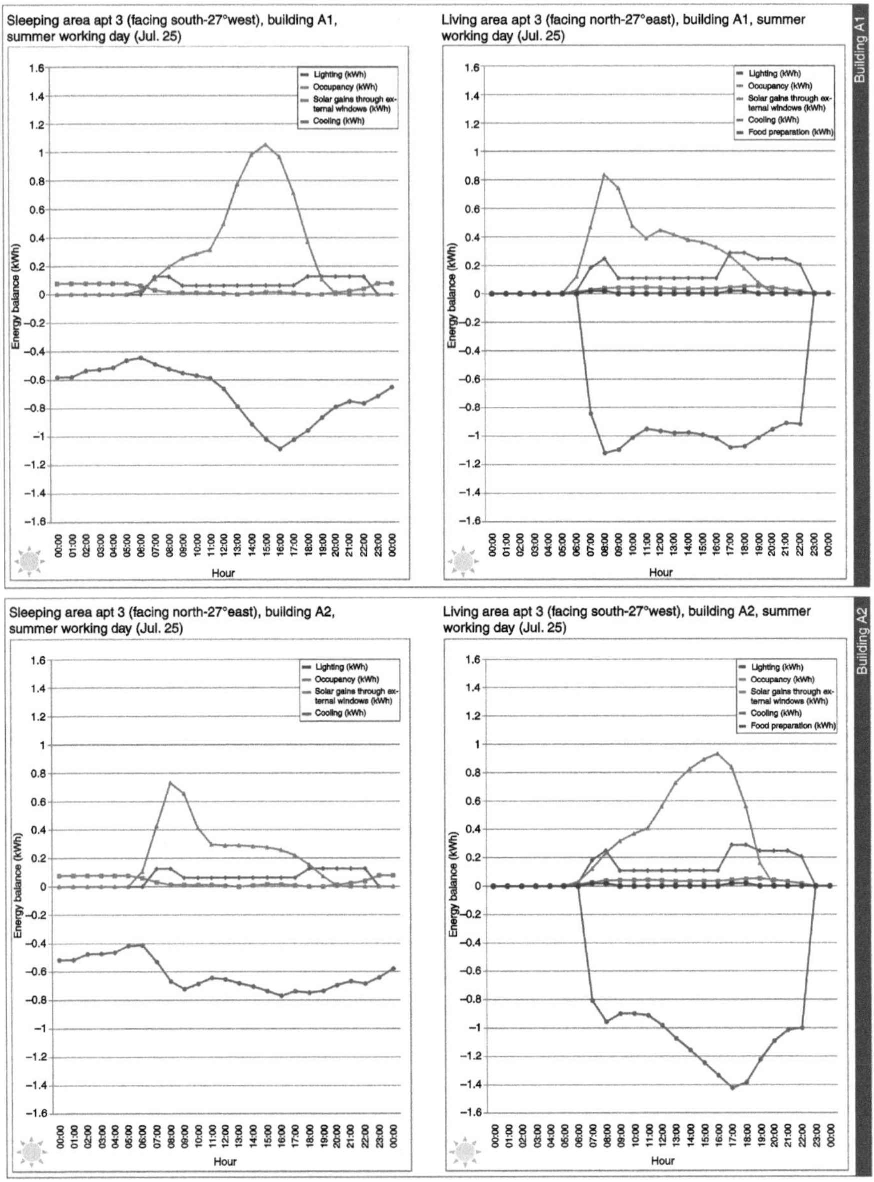

Figure 2.40 ***Charts of the simulation for the type unit 3 (under the roof) in building A1 and A2 in the summer representative day.***

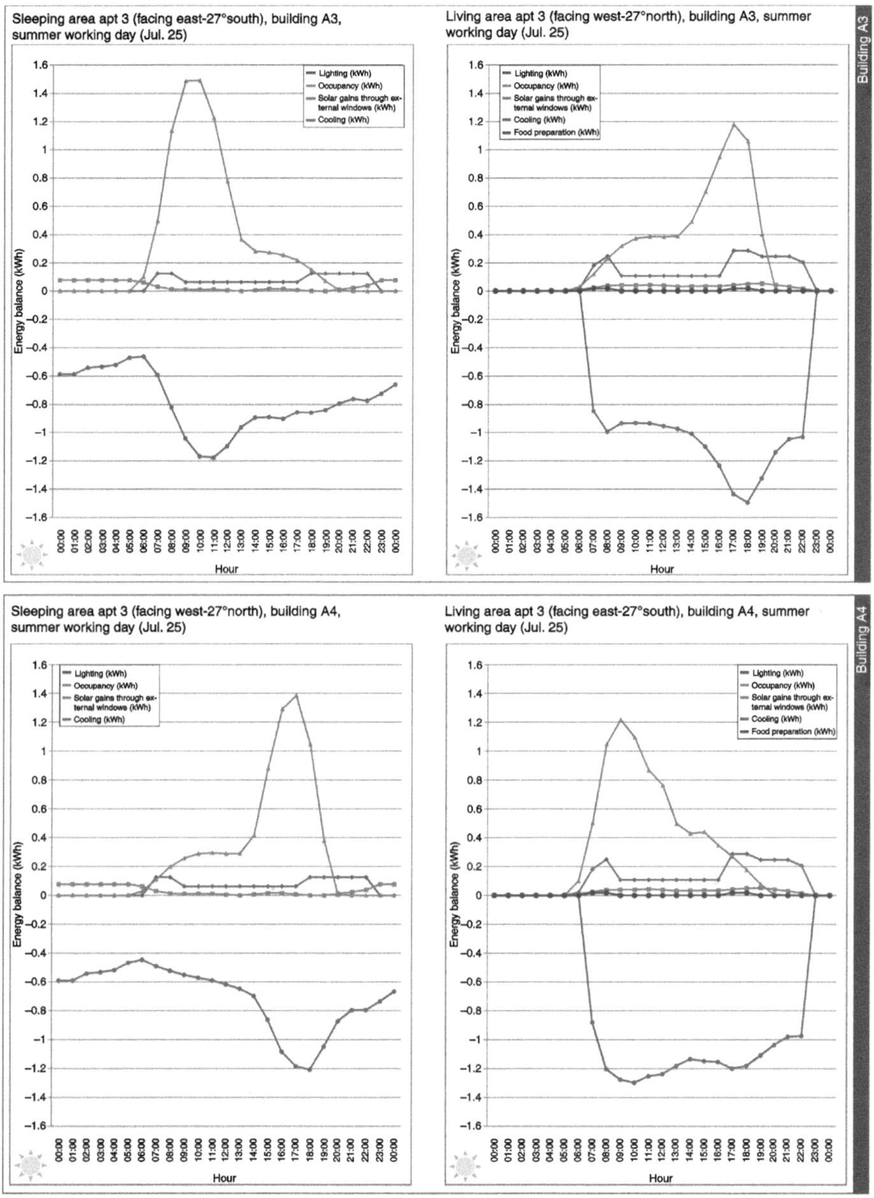

Figure 2.41 ***Charts of the simulation for the type unit 3 (under the roof) in building A3 and A4 in the summer representative day.***

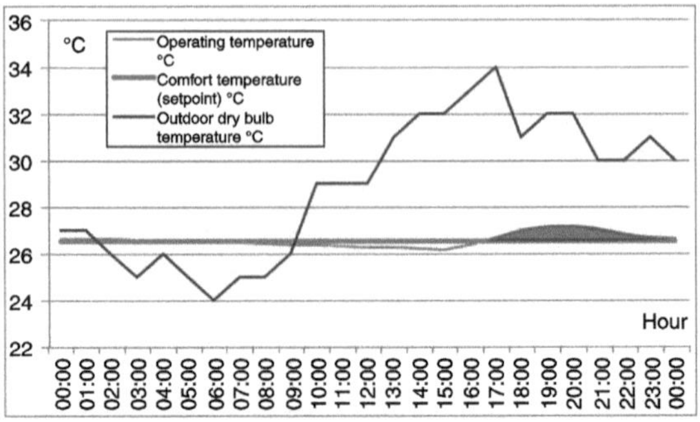

Figure 2.42 ***Overheating in the sleeping area of apartment 3, building A3.***

In the apartments located under the flat roof, despite the action of the cooling system, the overheating cannot be counterbalanced by the plant, which induces a state of high discomfort inside (Fig. 2.42). This occurs in all orientations, although some quantitative differences can be observed.[26]

Just as apartment 1 was strongly influenced by the contact with the ground, in the same way, apartment 3 suffers from its peculiar position. Apartments at the ground level are affected by aggravated winter conditions which are counterbalanced by improved comfort condition in summer; instead, the presence of the roof acts in both seasons, thus greatly exacerbating the balance both in summer and in winter conditions. Comparing the dispersion values and the contributions across the different surfaces, it is found that the roof generates higher dispersions in winter conditions and higher heat gains in summer conditions, compared to all other exposed surfaces. Comparing only the values of heating and cooling loads for apartments 2 and 3 in each orientation, there is a general increase of 20–25% for heating and 30–35% for cooling. This general consideration therefore applies to all the apartments placed on the top floor, arranged in any orientation. The same apartments in the building in the other two perpendicular orientations A3 and A4, while experiencing some slight advantage in winter conditions and in summer conditions respectively, are still less efficient; this occurs especially during summer, when the main facades are exposed to direct sunlight for most of the day, due to the building's axis inclination. The mutual comparison between the thermal balance and the energy

[26] The graph in Fig. 2.42 shows in detail the sleeping area of apartment 3 in the building A3: during the afternoon it can be noted that the internal temperature is still higher than the temperature of comfort, assumed to be 26°C in summer conditions (red area in the graph). In that time there has been a condition of discomfort and a feeling of warmth, but this difficulty cannot be overcome unless considering a general refurbishment program.

performance of different unit types in relation to the different building position clearly indicates that one of the most critical issues to be considered in the energy rehabilitation of the existing building is the roof level. Thus, we may assume that in the design phase of a refurbishment project, the evaluation of interventions aimed at improving the thermal performance in this specific area of the building is especially crucial.

The study carried out on building orientation allows researchers to assess the impact that a general choice like the orientation of the building has on the occupants' comfort and on the quality of the confined environments, as well as on the energy balance and on the future consumption, all as early as the design phase. The detailed information given in the simulation performed on a case-based and case-by-case variability basis is a necessary step toward producing tailored and more effective guidelines for energy-efficient retrofitting operations, in order to achieve nZEBs in existing buildings. Table 2.6 summarizes the obtained results in qualitative terms.

Considering *both* the winter and the summer conditions, the building oriented in a more efficient direction is the building A2, with the living area exposed to south-27° west and the sleeping area to north-27° east; this arrangement has been proved to achieve a fair energy performance by limiting the period of direct solar gains through the windows. As an alternative option, building A3, with a living area oriented at west-27° north and the sleeping area facing east-27° south, also showed fair performance. The east-west orientation is less energy-efficient than the previous orientation, but it still allows good utilization and control of solar radiation in both seasons. Throughout the analysis phase, the data were presented in terms of needs and contributions; however, according to the general aim, it is necessary to

Table 2.6 Qualitative evaluations of various orientations of the building residential units as a function of different orientations of the building blocks

	APT 1		APT 2		APT 3		APT 4		
Building	**WINT**	**SUM**	**WINT**	**SUM**	**WINT**	**SUM**	**WINT**	**SUM**	
A1	–	+	–	+	–	+	–	+	Generally negative
A2	+	+	+	+	+	+	+	+	Generally positive
A3	–	–	+	–	+	–	–	–	Generally negative
A4	+	–	+	–	+	–	+	–	Generally positive

deliver them in terms of indexes EPh (index of primary energy for heating in winter, which is the annual average energy requirements of the building for heating only per square meter) and EPc (index of primary energy for summer cooling), calculated according to their respective periods. They will then be used in the design phase to evaluate the effectiveness and performance of the interventions proposed.

The run simulations on the whole heating and cooling period shown are reported in Table 2.7; as can be observed, they are consistent with qualitative aspects reported previously in Table 2.6. These values have been calculated without considering the plants and thus represent the requirements of the building during the heated period (from Nov. 1 to Mar. 31) and the cooling period (that is the remaining period, from Apr. 1 to Oct. 31).

However, to properly assess the real energy consumption, systems cannot be ignored.

Considering the plants' contribution (assuming they have an efficiency $\eta = 0.85$, for the heating system plants and an energy efficiency ratio (EER) equal to 1,67 for the cooling system plant, consistent with the choices presumed so far), the amount of the corresponding indexes is reported in Table 2.8.

The simulations presented were carried out throughout the heating and cooling periods. In order to make the data more functional to the goal of

Table 2.7 Quantitative evaluations of the various orientations of the building residential units as a function of the different orientations of the building blocks

Building	EPh (kWh/m^2 year)	EPc (kWh/m^2 year)
A1	84	58
A2	78	59
A3	80	65
A4	77	63

Table 2.8 Qualitative evaluations of the various orientations of the building residential units as a function of the different orientations of the building blocks. Values include the plant systems' consumption

Building	$EPh_{(plant)}$ (kWh_t/m^2 year)	$EPc_{(plant)}$ (kWh_e/m^2 year)
A1	99	35
A2	91	35
A3	94	39
A4	91	38

this work, they have been presented in terms of daily data, using as a reference the days of Jan. 15th for the winter condition and Jul. 25th for the summer condition, as they have been identified as days having an average behavior for the relative seasons.

In addition, two weeks were selected as the weeks presenting the most extreme weather conditions (ie, they are the hottest and the coldest part of the year), because these weeks may point out particular aspects to be considered for the overall evaluation of building behavior under the worst climatic conditions. From the annual simulation, it is also noticeable that in the intermediate seasons near the period of switching on or off of the heating, the building as a whole requires a heating load and, at the same time, a cooling load. To reach the temperature and comfort set-points, there is the need to heat certain areas at certain times and to cool others (the data are obtained as sums of individual contributions and not as averages).

To obtain data comparable on an annual scale in order to numerically evaluate the performance of the various buildings and compare them with each other and with the results of similar studies, the indices EPh and EPc were chosen. They are calculated according to their respective reference periods and reported for the entire year. These benchmarks are useful to perform a comprehensive comparison between the buildings, but they incorporate a media operation that gives a good idea of the building's overall behavior, losing some detail in particular aspects. This means that, while it is correct to compare the performance of different buildings through their EPh and EPc, we can not expect to necessarily find the same values locally, that is, at the level of individual apartment.

Considering the presence of each apartment type within the same building block, the total EPh (kWh/m^2 year) and EPc (kWh/m^2 year) is obviously given by the mean value derived from the sum of all 16 apartments of the building. For the building block, these values are: EPh = 81.44 and EPc = 53.56.

Table 2.9 shows the value of EPh and EPc calculated for the different types of apartments, for example in building A1. The found values, and therefore the consequent consumptions, for the various apartments are very different from unit to unit and they may vary significantly from the value of the whole building. As shown, apartment 1, for example, is in contact with the ground and this greatly increases its heating requirement in winter conditions, but it also improve its summer energy performance; apartment 3, located on the third floor under the noninsulated roof, presents high heating and cooling load, especially in the summer because of the overheating of

Table 2.9 Qualitative evaluations of the mean energy performance indexes for the building residential units in building A1 for the winter and summer seasons

Apartment	EPh (kWh/m² year)	EPc (kWh/m² year)
APT 1	96	24
APT 2	71	66
APT 3	80	75
APT 4	96	24
APT 5	63	53
APT 6	80	75
APT 7	96	24
APT 8	71	66
APT 9	80	75

the roof due to direct radiation. The intermediate apartments such as 2 and 5 have some surfaces touching environments at their same temperature but also have a lower number of exposed surfaces, so their energy balances are lightened both in winter and summer conditions. Thus, to properly evaluate retrofitting interventions at building scale, by referring to the indices for the entire building, we should not forget the differences occurring at local level.

2.3.1.2 Peristeri

Taking into account the considerations developed for the previous case study, each building type has been split into different representative thermal zones corresponding to the real apartments of each building block.[27] The selection of the representative units (apartments) has been made according to both the position of the unit within the building block and with respect to the solar orientation, as shown in Figs. 2.43–2.45. As also described for the Aghia Varvara urban compound case study, the structural system of the buildings is very simple, made up by a regular grid of beams and pillars, with presumably prefabricated slabs as horizontal elements. The main thermal parameters provided for the horizontal ground, intermediate, and roof floors as well as for the external building envelope have been considered the same as the previous case study (see the Appendix). In the previous case study of the Aghia Varvara urban compound, the energy performance of residential units showed greater differences as a function of the position of the unit in relation to the building orientation.

[27] Tables reporting the main thermal properties of the ground floor layers, the roofs and the glazing components estimated for each building in each case study are reported in the Appendix.

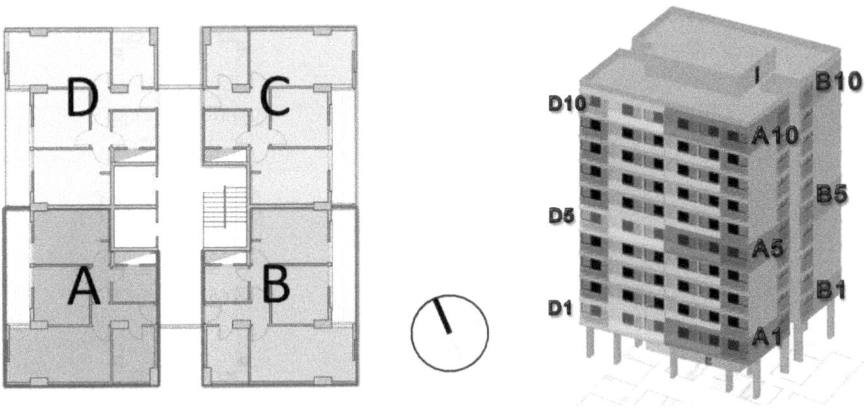

Figure 2.43 ***Apartments as thermal zones the tower building (T11) in the Peristeri urban setting.***

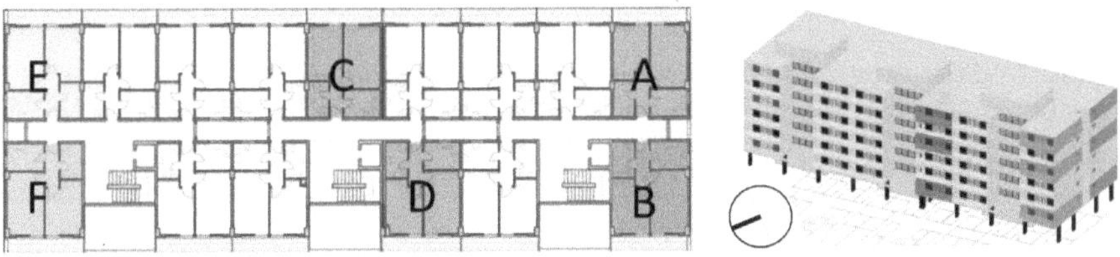

Figure 2.44 ***Apartments as thermal zones in the double block building (B6) in Peristeri.***

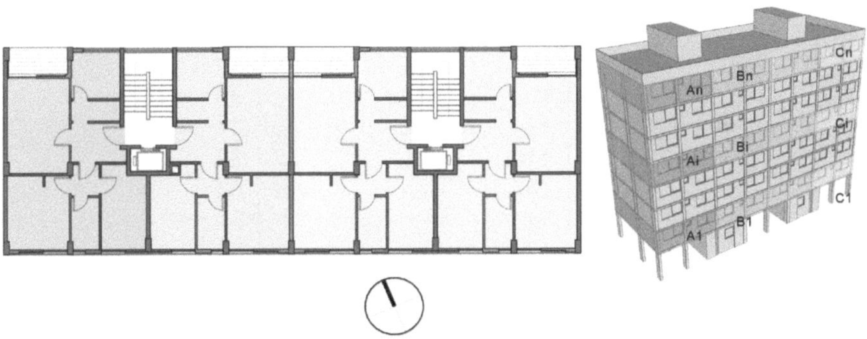

Figure 2.45 ***Apartments as thermal zones in the selected building block (T7) in Peristeri.***

Energy simulations have been performed for each building type, both in cold winter and hot summer seasons. The synthetic results for the tower building (T11), the double building block (B6), and the building block T7 are reported in Tables 2.10–2.12. The tables evaluate quantitatively the energy performance indexes of the single units as a function of their different locations within the specific building.

Table 2.10 Energy performance indexes of the different units in the tower building T11.

Apartment/thermal zone	EPh (kWh/m² year)	EPc (kWh/m² year)
A1	89	64
A5	75	54
A10	101	72
B1	100	67
B5	94	61
B10	113	73
C1	89	49
C5	75	43
C10	101	57
D1	99	52
D5	87	45
D10	107	62

Table 2.11 Energy performance indexes of the different units in the double building block B6

Apartment/thermal zone	EPh (kWh/m² year)	EPc (kWh/m² year)
A1	111	36
A4	89	42
A6	124	45
B1	115	40
B4	94	47
B6	129	49
C1	103	30
C4	82	34
C6	116	39
D1	107	33
D4	94	37
D6	129	43
E1	112	31
E4	90	31
E6	126	37
F1	107	29
F4	90	33
F6	125	38

To highlight the different energy demands of the single reference units as a function of both the position of the unit within the building block and the solar orientation, the graphs shown in Fig. 2.37 have been developed. As

Table 2.12 Energy performance indexes of the different units in the building block T7

Apartment/thermal zone	EPh (kWh/m² year)	EPc (kWh/m² year)
A1	113	89
A4	99	96
A7	167	106
B1	59	84
B4	71	89
B7	145	102
C1	84	95
C4	81	92
C7	170	102

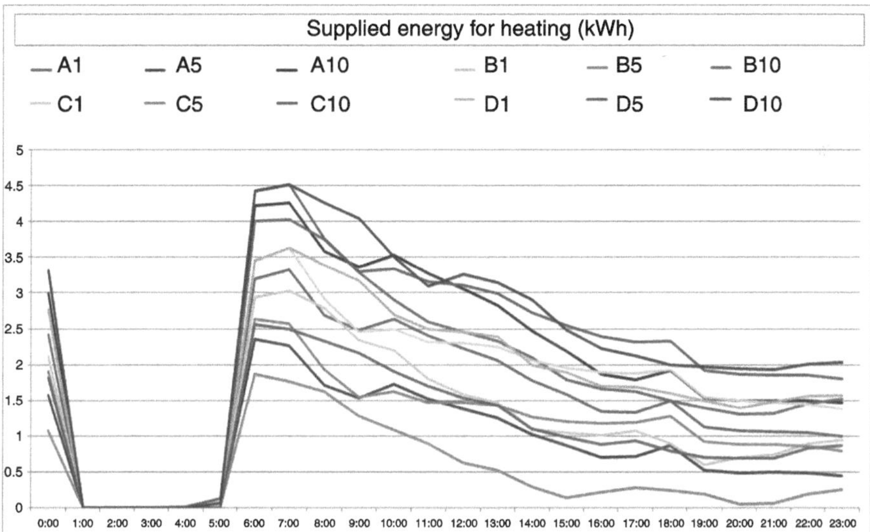

Figure 2.46 ***Different energy supplied in winter for the tower building.***

expected,[28] the most energy-consuming units (both in the summer and the winter season) are the apartments located on the upper floors, just underneath the flat roof of the building. Conversely, the less energy-consuming units (both in the summer and the winter season) are the apartments located on the intermediate floors. As an example, Fig. 2.46 reports the different levels of energy supplied for the representative day in winter time.

Again, substantial differences in the energy demand may also be observed by comparing the residential units in a same-floor location, but

[28] Similarly to the previous study case of Aghia Varvara.

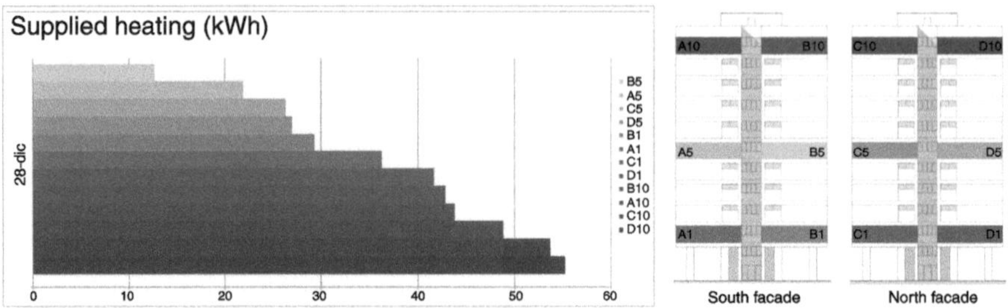

Figure 2.47 ***The graphs highlight the different levels of energy in the winter season of the reference units as a function of the position and orientation of the units within the building block.***

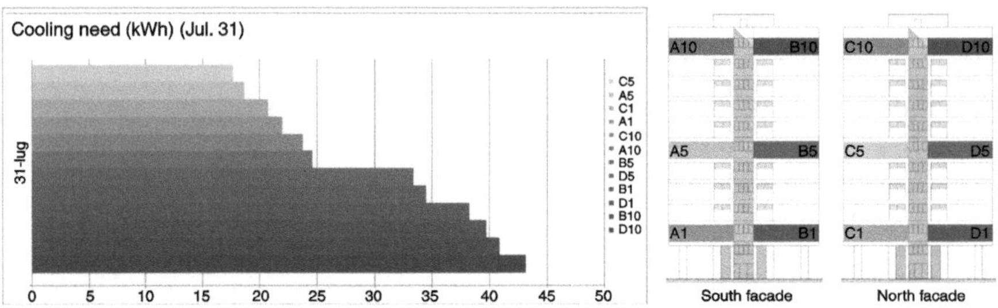

Figure 2.48 ***The graphs highlight the different levels of energy in the summer season of the reference units as a function of the position and orientation of the units within the building block.***

with different solar orientations. As shown on the right side of Figs. 2.47 and 2.48, south-oriented apartments are less energy demanding than the corresponding north-oriented ones, both in summer and winter conditions.

2.3.1.3 Evripidou

As indicated before, the case of Evripidou is different from the cases of urban stand-alone buildings like Aghia Varvara and Peristeri. In Evripidou, buildings were considered by their volumes, using some assumptions to simplify the implementation of the model as a whole courtyard at the level of the urban scale. In particular, it was decided to perform energy simulations for each floor of the building, with few considerations of the internal partitions of each of the residential units; however, all the partitions toward unconditioned spaces like stairwells and technical volumes were taken into account. Some additional partitions have been included inside the floor to simulate the existing thermal mass, thus achieving an approximate *facsimile* condition. Each floor in each building has been considered as a whole thermal zone with internal partitions assumed as internal mass. These hypotheses were chosen because they do not affect the objectives' results, since they are not considered a relevant error for the requested level of detail in the context of an urban energy investigation.

As mentioned, the subdivision of the planes was performed in a simplified way, identifying the unheated areas, such as stairwells and elevators, and realizing the approximate internal partitions of the different zones for each floor. Building construction elements, such as thickness and height of external walls and floors, have been imputed. Tables reporting the main thermal properties and the stratigraphic description of the building envelopes' components of the ground-floor layers, the roofs, and the glazing components estimated for each building in each case study are reported in the Appendix.

The input data described in the previous paragraphs have been included in the first dynamic simulation to determine the thermal energy demand and, subsequently, the energy required to maintain internal thermal and comfort conditions. Several physical factors have been considered in addition to the external air temperatures: solar gains, direct gains for occupants, lighting, computer and equipment contributions, and so forth. The final results are presented in graphical form: along the abscissa is the time interval and along the *y*-axis are the measurement units of the chart objects (temperature, radiation, etc.). The main data categories can be grouped by: location (site data), thermal comfort, internal contribution, annual heating, and ventilation systems.

To analyze these data, references have been made to the first building, simulated considering the whole annual period (see Fig. 2.48) summary data tables have been exported via Excel and calculations have been performed for all of the buildings, to be assessed on the total consumption. In Table 2.13, mean energy performance values are illustrated for winter and summer energy demand.

The results' analysis in the as built conditions is reported in Table 2.13.

2.3.1.4 Brief Notes on the Energy Demand in the Different Urban Settings

The performed analyses show the paramount importance of bioclimatic approach in the planning and design to pursue the goal of a low energy-consuming built environment. The site, the building orientation, the organization of the blocks, and the single units inside them play a very important role in determining the energy consumption and its variation. Unfortunately, when we deal with existing built environments and buildings, all

Table 2.13 Calculation of the winter and summer needs of all analyzed buildings (as built scenario) in Evripidou

Buildings	EPh (kWh/m² year)	EPc (kWh/m² year)
1	54	39
2	57	35
3	33	38
4	59	45
5	47	41
6	43	31
7	70	49
8	53	42
9	24	19
10	39	38
11	43	41
Mean values	47	38

these passive components are not variables to be defined, but they are already defined as critical features or potential strengths and they can only be used as input for the design stage of refurbishment intervention.

However, despite the very different contexts under study and the diverse levels of energy demand registered between the single units forming the building blocks and the urban settings, a general deduction arises from the comparison of the main physical and energetic variables we are facing with highly energy-consuming buildings. In the urban context, the energy performance indexes are variable between 44 and 88kWh/m^2 year for heating demand and between 36 and 100 kWh/m^2 year for cooling demand. Lowest values of energy demand in the heating and summer can be observed in the case of Evripidou, given by the dense urban form and by a very low the ratio (Ae/V)= 0.27, as seen in Section 2.1.1 at the beginning of this chapter

Overall, performed calculations denoted high-energy consumption as given by the sum of both cooling and heating demand, in turn derived from the typical climate of Athens (with very hot summers but also cooler winters) and the very low quality of buildings' construction components associated with the poor energy performance.

2.3.2 The Energy-Saving Potential of Retrofitting Options

Notwithstanding the limitations encountered in existing urban contexts, many are the possible retrofitting options to reach energy savings in the renovation process of existing buildings.

Generally, "energy retrofit" or "deep energy retrofit" to achieve nZEBs implies replacing existing systems in a building or in a set of buildings with similar, higher performing and higher quality systems to achieve a consistent reduction of energy consumption in the existing building.

In this framework, a profound and transversal debate is occurring among researchers, practitioners, urban planners, designers, energy managers, and main stakeholders on the building construction market on whether it is necessary that these buildings be renovated "deeply" or not. Agreed definitions of deep renovation (retrofit/refurbishment) and the major challenges surrounding this issue are discussed by Shnapp et al. (2013) and Bettgenhäuser et al. (2014). According to these sources, it is a prerequisite that existing buildings are renovated "deeply" for the building sector to meet global energy reduction objectives.[29] Based on the findings collected in these research studies, definitions relating to deep renovations were established by Shnapp et al. (2013).[30] *Deep renovation* or *deep energy renovation* is a term for

[29] The global building performance network's (GBPN) mission is to dramatically reduce the energy use of existing buildings and consequently reduce the GHG emissions associated. GBPN facilitates this by following a "deep energy efficiency scenario."

[30] Deep refurbishment or deep energy refurbishment means to bring something back to its original good state and if such a process should be deep, it will include a very substantial improvement of the energy use and will bring the building beyond the original energy efficiency. Many possible definitions are reported in Shnapp et al. (2013): (1) deep reduction or deep energy reduction is a term used in the United States for a deep renovation or a deep refurbishment, which aims at more than 75% reduction in energy use in comparison with that prior to the improvement; (2) zero-carbon-renovation: A deep renovation with large-energy consumption reductions, where the energy needed to supply the resisting need is carbon neutral; (3) zero-energy-renovation: A deep renovation with large-energy consumption reductions, where the energy needed to supply the resisting need is supplied as renewable energy on site. Furthermore, other definitions based on relative targets are reported to support the clarification of deep renovation projects and can help to separate the level of ambition in DR projects. They are: Factor Two or Factor 2 Renovation: A renovation with energy consumption reductions of 50% compared to prerenovation performance; Factor Four or Factor 4 Renovation: A deep renovation with energy consumption reductions of 75% compared to prerenovation performance; Factor Six or Factor 6 Renovation: A deep renovation with energy consumption reductions of 84% compared to pre-renovation performance. The most commonly used expressions are: deep renovation, deep retrofit, deep refurbishment, and to a lesser extent, deep reduction. The problem is that these expressions mean different things to different analysts. In the report a "standard renovation" or refurbishment is defined as the one harvesting the minimum possible energy savings, ranging between 20% and 30%. Conversely, by renovating deeply, using state-of-the-art technologies, it is possible to reduce the energy consumption of a building by more than 75%. The primary energy consumption includes energy used for heating, cooling, ventilation, hot water, lighting, installed equipment and appliances. After the deep retrofit the buildings energy reduction is 50% or more compared to the status of the existing building/s the retrofit. (GBPN/definition mainly used in the United States).

a renovation that captures the full economic energy efficiency potential of improvement works, with a main focus on the building shell, of existing buildings that leads to a very high-energy performance. The renovated buildings energy reductions are 75% or more compared to the status of the existing building/s before the renovation. The primary energy consumption after renovation, which includes, *inter alia*, energy used for heating, cooling, ventilation, hot water, and lighting after the deep renovation of an existing building is less than 60 kWh/m^2/year (GBPN/definition often used in Europe).

Recalling the EU EPBD introduced in Chapter 1, a new "nearly zero energy" building[31] is assumed as a building that has a "very high energy performance," with a nearly zero or very low amount of energy required to be covered by energy from renewable sources. Following this assumption, the basic principle inspiring retrofitting solutions toward nearly zero energy appears to consist of a threefold direction, which in turn follows three specific, but concurring, objectives:

1. The reduction of energy requirement for heating and cooling through passive interventions on the building envelope
2. The use of renewable energy sources to cover consumption and to feed plant systems
3. The reduction of consumption by improving the efficiency of the plant systems

The order of these listed actions (1, 2, 3) is not a secondary element if we consider the common building practice. As a matter of fact, because of a shorter life cycle time compared with other architectural and constructive components or solutions[32] and lower payback time,[33] plant systems are commonly seen as the first option to be considered in the retrofitting of existing

[31] Article 2(1a) gives a qualitative definition of what a "nearly zero energy building" is: "A 'nearly zero energy building' is a building that has a very high-energy performance. The nearly zero or very low amount of energy required should be covered to a very significant extent by energy from renewable sources (RES), including energy from RES produced on-site or nearby."

[32] While architectural and technological solutions may have a life-cycle time close to 30–50 years, plant systems' life is commonly assumed to be close to 10 years. Thus, after this time they are usually considered obsoleted. Furthermore, generally mechanical plant systems are part of an industrial production sector affected by faster and shorter renovation processes compared to typical industrial components of the building sector market, which usually suffer from more inertia toward changing.

[33] Economic aspects and consequent economic feasibility of solutions is discussed in the Chapter 3.

buildings. Thus, taking into account these practical considerations, in a possible step-by step renovation, a more probable order could be the following: 3, 2, 1 (namely: 3. reduction of consumption by improving the efficiency of the plant systems; 2. the use of renewable energy sources to cover consumption and to feed plant systems; 1. reduction of energy requirement for heating and cooling through passive interventions on the building envelope). Nonetheless, considering our final mission to achieve nearly ZEB in urban settings, this last hypothesis would lead to the unavoidable oversized dimensioning of the plant systems. In the present work, preference has been given to the technical consistency and efficiency of a one-step renovation (or a step-by step renovation based on logical, technically-based, follow-on steps) regardless of other practical-based aspects. We do believe that this approach is the most consistent with respect to the final mission of progressing toward nearly zero energy built environment.

Thus, to evaluate the technical feasibility of the possible energy retrofitting options in our urban settings, a series of transformation – gradually increasing, from the sectorial single intervention up to the far-reaching and combined scenarios – have been considered on the different building types. These transformations produce the following possible scenarios:

Scenario 1: Wall insulation (external walls, ground floor, and lower floors on "pilotis" when applicable)

Scenario 2: Window replacement with high-performing glazing components

Scenario 3: (3 = 1 + 2): Combined scenario

Scenario 4: Roof insulation[34]

Scenario 5: Roof insulation and green roof[35]

Scenario 6: (6 = 3 + 5): Combined retrofitted scenario

Scenario 7: New plants and energy generation by renewable energy sources

Scenario 8: (8 = 6 + 7): Complete retrofitted scenario

We will limit our investigation to the scenario number 6, thus considering the technological modifications applied at the building shell level. New

[34] This is generally achieved using layers of polystyrene to avoid additional loading on the existing structures.

[35] The retrofitting operation consist in the positioning of: (1) a layer of anti-root membrane, (2) a drainage layer and a filter layer, and (3) the moist soil that can be planted with grass or small shrubs, with underground watering system. Table 2.14 shows the change in building performance through indexes EPh and EPc and the resulting changes in terms of consumption and consumer spending, in reference to the apartment type for the whole year.

plant systems associated with the renewable sources to set to zero the energy balance of existing buildings are described next.

Notwithstanding slight and negligible differences depending on the specificity of building types and geometry, in all the case studies, simulations performed have suggested that architectural and technological components are the building sections where the possible energy retrofitting operations should be concentrated, namely, the roof level, the windows, the facades, and the ground floor.

In a Mediterranean climate like Athens, the passive solutions selected have to be able to address both insulation requirements and the building's need for thermal inertia. Hypothesized solutions for opaque envelope components (wall, ground floors, and roofs) should therefore consist of retrofitting both dense and highly massive coating materials to achieve the improved thermal performance in both hot summer seasons and cold winter periods.

To evaluate the effectiveness of the proposed solution, a simulation of the building before and after the intervention was carried out annually, and indexes of primary energy for heating in winter (EPh) and for summer air conditioning (EPc) were calculated and used as a benchmark for assessing the effectiveness of the intervention. These interventions also have been evaluated from an economic point of view, considering the amortization time of the investment (see Chapter 3).

For all these evaluations, no changes in the systems or in the user profiles have been introduced to properly and equally compare the consequences of the variable design choices as assumed in the design retrofitting operations. Before proceeding with the choice of the insulation and window frames, some tests were performed to determine the appropriate thickness. Finally, the option of a heat-proofing, 10-cm thick insulation system for the external vertical walls and on the roof has been selected. Higher performing windows with 4-mm thick double glass and an air layer of 12 mm were considered.

Two different options also have been considered for the insulation of opaque elements (walls and floors): polyurethane foam panels on the one side and, on the other side, more dense autoclaved aerated concrete (AAC) and mineral-based, nonfibrous panels, ideal for the thermal insulation of external walls in the winter and summer seasons. These panels also present more stability and fire resistance; they are more transpiring, thus avoiding the risk of mold growth by internal humidity. The choice of these kinds of panels is mainly due to the improved thermal behavior in summer conditions with respect to the polyurethane foam expanded panels, which, on their side, present some advantages in cold winter conditions. This is illustrated by the fact that the chosen insulation panels have a high density, which increases the mass of the opaque envelopes. Thus, the resulting construction will increase its performance in summer by restraining the effects of oscillation of external surface temperature caused by solar radiation and external air temperatures.

In existing buildings, while for vertical walls denser and highly thermal-resistant materials (like AAC panels) are preferable, for flat-top roofs, polystyrene panels that avoid additional loading on the existing structures are the more desirable option. A detailed stratigraphic description of the building envelopes' components applied for the different retrofit options is reported in the Appendix.

2.3.2.1 Aghia Varvara: Energy Retrofitting Scenarios

As a result of the simulations performed in the analysis phase, building B has been identified as the worst in terms of orientation and layout of the surrounding buildings, and therefore it has been used for further investigation of the energy retrofitting options. Following the unit-oriented energy analysis adopted for the energy analysis of the as-built state in the case study of Aghia Varvara, with the units consisting of single apartment, to investigate the energy saving of each retrofitted option, we should have referred to the different type of apartments.

Nonetheless, although some retrofitting operations may be done in the single apartments, the majority of them (namely the new building coatings for the opaque envelopes, the insulation of roofs, or green roofs, etc.) have to be considered as interventions to be applied on whole buildings.

For all the proposed intervention, a table showing the indices EPh and EPc (summer and winter consumptions before and after the intervention and the cost for consumption) has been reported. As shown in the energy analysis, for the case study of Aghia Varvara, the major critical surfaces are the external flat roofs; they receive the larger amount of solar radiation during summer period and are responsible for the majority of heat loss during winter. Thus, in this case, the proper insulation of the roof has been proposed as a step assessed individually, because of: (1) the simple technical implementation of the material operation; (2) the small economic entity involved; and (3) the great potential to lead to great advantages in terms of energy savings.

As an alternative to the choice previously selected, a further simulation has been performed to assess whether a green roof could yield better performance, by being able to decrease the thermal transmittance both by decreasing the thermal capacity and increasing the mass. It has been found that intervention leads to an improved building performance in both winter and summer conditions and leads to the consequent decrease of energy costs. However, the energy performance improvement appears to be similar to that obtained by the roof insulation, although some benefits of this solution, both to the heat island reduction as well as the environmental benefit, have been disregarded by the performed calculations expressly related to the building's energy demand. Different costs between the two options are discussed in Chapter 3.

An optimized solution could be to contemplate using a combination of insulating polystyrene panels and a green roof. Of course no single retrofitting operation is able to achieve a substantial energy improvement. Thus, the combination of multiple interventions on the building envelopes has to be investigated.

Since thermal dispersions/gains through the transparent surfaces are very relevant and comparable to those through the roof, a further scenario has been hypothesized that considers the replacement of the existing aluminum frames with single glass, along with the new window components described in the previous paragraph. The replacement of windows, together with the roof insulation, has been also investigated.

The last retrofitting scenario consists of the combination of thermal insulation on vertical surfaces, roof insulation, and window replacement; the retrofitting operation is based on the construction of a thermal insulation made by external panels of stone wool, with settlement of the ground floor, and including the subsequent restoration of the exterior plaster and window thresholds.

EPh and EPc indexes (kWh/m^2 year) for summer and winter consumptions before and after each retrofitting intervention are reported in Table 2.14.

2.3.2.2 Peristeri Energy Retrofitting Scenarios

Similar investigations have been performed for the possible energy retrofitting scenarios of existing buildings in the Peristeri urban compound. In this case, consistent with the building-type oriented energy analysis adopted for the as-built state in the specific case study, to investigate the energy savings of each retrofitted option, we referred mainly to the different building types. Results of performed simulations are synthetized in Figs. 2.49–2.51.

In Peristeri, the three different building types have been analyzed with different levels of simulations quantifying all the encountered variants in

Table 2.14 Total energy demand in the different urban contexts

Urban setting	Aghia Varvara	Peristeri	Evripidou
Number of buildings	15	12	11
Total square meters (m^2)	18,396	27,368	12,233
EPh (heating) (kWh/m^2 year)	78	88	44
EPc (cooling) (kWh/m^2 year)	55	99	36
Thermal energy consumption (winter) (kWh year)	1,677,282	2,817,456	533,043
Electric energy consumption (winter)	0	0	0
Electric energy consumption (summer)	600,348	1,651,705	438,063
Total thermal energy consumption	1,677,282	2,817,456	533,043
Total electric energy consumption	600,348	1,651,705	438,063

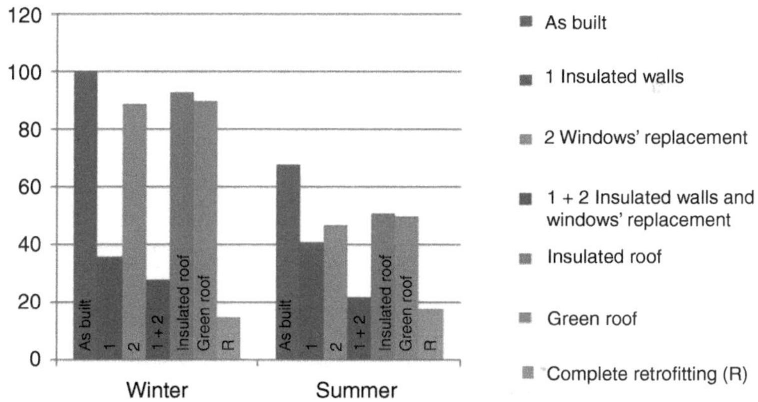

Figure 2.49 ***Possible energy retrofitting options in the tower building and correspondent savings in terms of energy performance indexes (kWh/m^2 y).***

buildings and apartment types, in order to identify the most appropriate retrofitting actions to achieve reduction of energy consumption up to nZEB in existing building blocks; the detailed diagnosis of the individual residential units, similar to the previous case study in Aghia Varvara, have shown that, in some cases, it is possible to reach an average energy performance (EP) of up to 20–15 kWh/m^2 y by insulating opaque surfaces – roof and walls – and replacing existing windows.

2.3.2.3 Evripidou: Energy Retrofitting Scenarios

As for the previous case studies, for Evripidou the single retrofit options also have been selected and combined in an attempt to find the best compromise

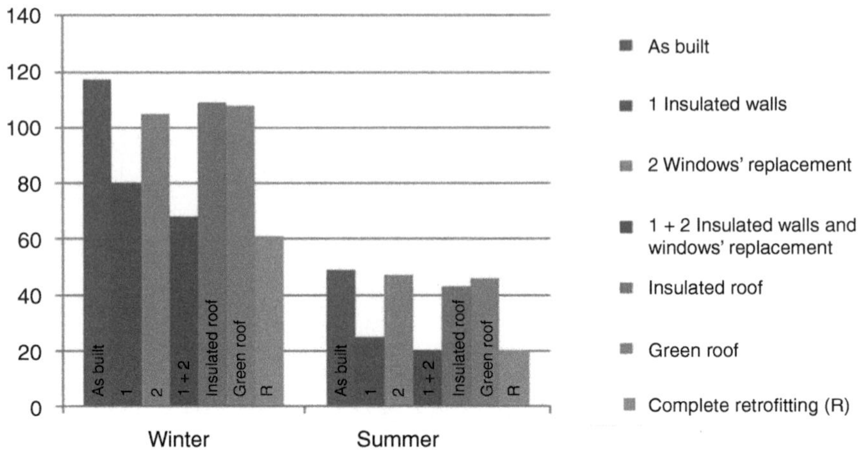

Figure 2.50 ***Possible energy retrofitting options in the double block and correspondent savings in terms of energy performance indexes (kWh/m² y).***

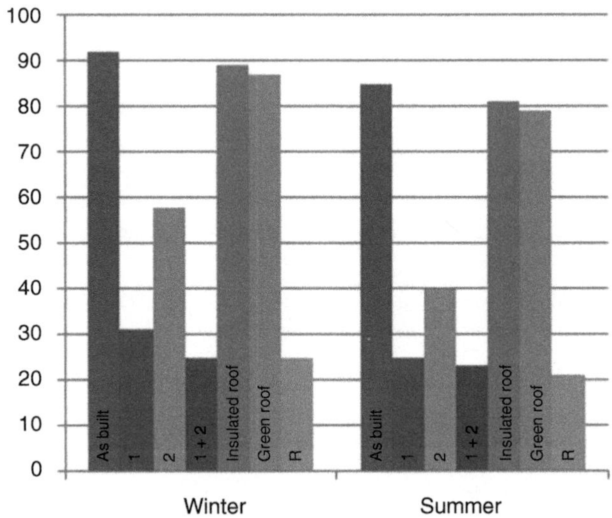

Figure 2.51 ***Possible energy retrofitting options in building block (T7) and correspondent savings in terms of energy performance indexes (kWh/m² y).***

between the different (and often conflicting) winter and summer demands, with existing constraints limiting the possible retrofitting actions in the existing buildings.[36]

High performing windows with 4 mm double glass and an air layer of 12 mm have been hypothesized in the Evripidou case study.

[36] In example, the polyurethane insulation on the roof while having characteristics of low thermal inertia has been considered preferable for the reduced weight with respect autoclaved aerated concrete (AAC).

Before proceeding with a simulation including all the designed retrofitting options, the same options have been analyzed separately, in order to understand and select the highest performing solutions. A detailed analysis for each scenario has been performed for all 11 buildings in the selected area, thus obtaining, for each building, the energy requirements in both the summer and winter periods. Furthermore, an economical analysis taking into account the cost of operations and retrofitting options has been developed with reference to each building (see Chapter 3). The tables mentioned later report the performed calculations for building 1.

The same calculations have been performed for all 11 buildings in the selected urban courtyard in Evripidou. From the mutual comparison between the tables, the following conclusions can be drawn:

- The insulation of the vertical walls through external coatings is more or less convenient depending on the orientation of the individual buildings. It does not present the greatest advantage during the hot summer season.
- The roof insulation always represents an excellent intervention.
- The green-insulated roof produces a slight disadvantage in winter.

Among the interventions, combination number 3 produces the greatest benefits: compared to the insulated roof, or the isolated green roof, consumption is much lower and savings are increased by up to three times as much.

2.3.2.4 Comparing Energy Retrofitting Scenarios

In all the case studies considered, the calculations have shown that it is possible to reach an average EP of up to 15 kWh/m^2 y by insulating opaque surfaces – roofs and walls –- and replacing existing windows (Tables 2.15–2.19).

A complete and comprehensive calculation has been performed for obtaining the global energy demands in the urban complexes considered as case studies in the Athens area.

Table 2.20 illustrates the different components of energy consumption in the scenario as built and after the complete renovation of all the buildings given by the retrofitting scenario number 6 (6 = 3 + 5), the combined scenario consisting of wall insulation, window replacement with high performing glazing components, and the insulated and green roof. A reduction in energy consumption that varies between 33 and 31% is the result. It shows how a huge potential for energy savings can be realized by renovating the existing building stock.

Table 2.15 The indices EPh and EPc (summer and winter consumptions before and after each retrofitting intervention)

Building type 1 in Aghia Varvara: gross surface 928 m²

	WINTER		SUMMER	
Retrofitting option	**Q (kWh/y)**	**EPh (kWh/m² year)**	**Q (kWh/y)**	**EPc (kWh/y)**
As built	78,880	85	62,176	67
1. Insulated walls	64,029	69	55,632	60
2. Windows' replacement	68,680	74	52,352	56
3. 1 + 2	53,829	58	45,808	49
4. Insulated roof	72,379	78	57,855	62
5. Insulated green roof	73,318	79	57,857	62
6. (6 = 3 + 5)	48,266	52	41,488	45

Table 2.16 Retrofitting options in the tower building (T11) and corresponding energy savings

Building type: tower (T11): gross surface 3560 m²

	WINTER		SUMMER	
Retrofitting option	**Q (kWh/y)**	**Epw (kWh/m² year)**	**Q (kWh/y)**	**Ep (kWh/m² year)**
As built	302,600	100	205,768	68
1. Insulated walls	108,936	36	124,066	41
2. Window's replacement	269,314	89	142,222	47
3. 1 + 2	84,728	28	66,572	22
4. Insulated roof	281,418	93	154,326	51
5. Insulated green roof	272,340	90	151,300	50
6. (6 = 3 + 5)	45,390	15	54,468	18

The calculations illustrated in Table 2.20 result in the global annual electric energy demand, which is the total amount of energy considered for the energy generation to be provided to achieve the zero energy balance of all the buildings in the considered urban settings.

2.3.3 Energy Generation by Renewable Energy Sources

Given the existing constraints of the built environment in the majority of the Mediterranean cities, we have to consider the unavoidable need to

Table 2.17 Retrofitting options in the double building block (B6) and corresponding energy savings

Building type: double building block (B6): gross surface 6200 m²				
	WINTER		SUMMER	
Retrofitting option	Q (kWh/y)	Ep (kWh/ m² year)	Q (kWh/y)	Ep (kWh/ m² year)
As built	354,042	117	148,274	49
1. Insulated walls	242,080	80	75,650	25
2. Window's replacement	317,730	105	142,222	47
3. 1 + 2	205,768	68	60,520	20
4. Insulated roof	329,834	109	130,118	43
5. Insulated green roof	326,808	108	139,196	46
6. (6 = 3 + 5)	184,585	61	60,520	20

Table 2.18 Retrofitting options in the building block (T7) and correspondent energy savings

Building type: building block (T7): gross surface 3560 m²				
	WINTER		SUMMER	
Retrofitting option	Q (kWh/y)	Ep (kWh/ m² year)	Q (kWh/y)	Ep (kWh/ m² year)
As built	278,392	92	257,210	85
1. Insulated walls	93,806	31	75,650	25
2. Windows' replacement	175,508	58	121,040	40
3. 1 + 2	75,650	25	69,598	23
4. Insulated roof	269,314	89	245,106	81
5. Insulated green roof	263,262	87	239,054	79
6. (6 = 3 + 5)	75,650	25	63,546	21

Table 2.19 Different retrofit options calculated for building 1 in the Evripidou urban courtyard

Building type 1 in the Evripidou courtyard: gross surface 930 m²				
	WINTER		SUMMER	
Retrofitting option	Q (kWh/y)	EPh (kWh/ m² year)	Q (kWh/y)	EPc (kWh/y)
As built	46,442	47	41,353	41
1. Insulated walls	42,727	43	38,045	38
2. Windows' replacement	43,191	43	38,458	39
3. 1 + 2	39,940	40	35,564	36
4. Insulated roof	36,689	37	32,669	33
5. Insulated green roof	37,154	37	33,083	33
6. (6 = 3 + 5)	27,865	28	24,812	25

Table 2.20 Comparison of total energy consumption in the as-built scenario and after the complete energy renovation (scenario 6) in the three urban settings

		Urban settings	Aghia Varvara	Peristeri	Evripidou	
		Number of buildings	15	12	11	AS BUILT
		Total Square metres	18,396	27,368	12,233	
		Qh (winter)	1,563,660	3,219,845	442,425	
		Qc (summer)	1,232,532	1,833,656	731,565	
A		Thermal energy consumption (cold season)	1,839,600	3,788,053	533,043	
B		Electric energy consumption (cold season)	0	0	0	
C		Thermal energy consumption (hot season)	0	0	0	
D		Electric energy consumption (hot season)	738,043	1,097,998	438,063	
E	A + C	Total thermal energy consumption	1,839,600	3,788,053	533,043	
F	B + D	Total electric energy consumption	738,043	1,097,998	438.,063	
G		Thermal energy reduction (cold season)	714,198	3,219.,765	306,644	RETROFIT SCENARIO 6
H		Electric energy reduction (cold season)	0	0	0	
I		Thermal energy reduction (hot season)	0	0	0	
L		Electric energy reduction (hot season)	242,343	360,537	166,464	
M	E − G − I	Final thermal energy demand	1,125,402	568,288	226,399	
N	F − H − L	Final electric energy demand	495,701	737,461	271,599	
O		Percentage of reduction of thermal energy	−39	−85	−70	
P		Percentage of reduction of electric energy	−33	−33	−38	
Q		Electric energy supply for heat pump system (cold season)	673,894	340,292	183,619	
R	N + Q	Annual energy demand for heating and cooling	1,169,594	1,077,754	455,218	
S		Electric consumption for domestic appliances	643,860	957,880	393,715	
T	R + S	Global annual electric energy demand	1,813,454	2,035,634	848,933	

actively supply energy in order to meet heating and cooling requirements and reach the ambitious goal of nZEBs. Obviously, simply reducing the requirements obtained with passive systems cannot be sufficient, thus the energy source feeding the heating and cooling residual requirements that still exists after renovation must be addressed. In Mediterranean climates, the most immediately available renewable source is certainly the sun. Thus, in the evaluation of retrofitting operations, it is necessary to consider possible absorbing surfaces or space that might be needed to cover the building's energy consumption.

The relationship between the entities involved in response to external environmental conditions requires a certain amount of energy in order to establish and maintain the comfort conditions expected in indoor environments. The equipment needed to produce this amount of energy requires a power source. Therefore, the interventions on the final consumption or on systems may not change the overall energy requirement balance of the building, but should aim to change the way in which the requirements are met. As has been already stated, to minimize energy consumption, careful bioclimatic analysis and consistent design are of strategic importance; if wisely used, they can lead to the construction of totally passive buildings.

The technical feasibility of nZEBs in new building construction can be partially achieved by using passive techniques in the Mediterranean climate, as examples of traditional architecture have shown; but existing buildings, and especially the buildings of the 1960s and 1970s, do not respect the rules of the bioclimatic design, for example, with regard to the choice of the site and the orientation and arrangement of the blocks, and also with regard to the construction characteristics and materials. Thus, we need to understand and evaluate how to reduce consumption and transform existing buildings into nZEBs. More specifically, we are attempting to answer whether adopting the previously proposed passive solutions will produce small enough energy consumptions to cover the energy requirement via the sole use of renewable energy.

Therefore, we considered combining passive retrofitting interventions with the goal of realizing thermal insulation on vertical surfaces, together with the insulation of the roof and replacement of windows. The combination of these passive solutions allows a substantial reduction of the energy requirements (reduction varies between 22 and 85% in terms $kWh/m^2/$year in winter and summer, respectively).

To evaluate the feasibility of nZEBs in the selected buildings, the installation of a photovoltaic system has been assumed, considering it as the renewable source for the production of electrical energy to supply the HVAC plant systems, which consist of (1) the cooling system made up by air conditioning units, (2) the heating system with heat pump, and (3) the electricity for other domestic uses. Given the objective of achieving nearly zero energy in the selected RUEs, the need to compare different situations and building components and to avoid excessive fragmentation of data in the calculations, we assumed the use of PV plants to supply energy in the heat pump system for all domestic uses, including the production of domestic hot water (DHW). The thermal requirements currently produced through a central heating system using gas/oil to meet the electricity demand have been turned into electrical requirements through the conversion multiplicative factor of 0.46 to transform kWh_t into kWh_e.

Photovoltaic and solar thermal systems on the flat roofs of buildings have been hypothesized, considering the possible construction of a grid-connected plant, that is, an installation connected to the public network. A single panel of 1.6 m^2, in polycrystalline silicon, with a peak power of 315 W, has been chosen in order to assess the feasibility of net zero energy buildings. We have assumed a photovoltaic and solar system as part of the complete retrofitted option, namely the installation of thermal insulation on the vertical surfaces, roof insulation combined with a green roof, and the replacement of regular windows with high-performing double-glazed windows.

The installation of a photovoltaic system for the production of electricity has been assumed to power the following: the cooling system with AC units, the heating system with heat pump, and electricity for other domestic uses. Furthermore, the evaluation of costs and payback times has been conducted; once again, it showed that although the nZEB goal is technically achievable, it remains ambitious due to the high cost of technology for the photovoltaic system. Thus, it needs to be tackled in conjunction with the "passive" retrofit interventions aimed at reducing as much energy consumption as possible.

The following steps illustrate the calculation of the photovoltaic (PV) and solar systems to balance the energy requirements in the different urban compounds. The calculation of the PV system that would be necessary to cover the total electrical consumption in the different urban settings has been developed considering the following data:

- Athens average total global annual radiation on a horizontal surface: 1800 kWh/m^2 year
- Estimated performance of the PV system = 19.53%

Given the previous data, the total power required in the case of Aghia Varvara is as follows:

			Aghia Varvara
T	R + S	Global annual electric energy demand	1,813,454

Needed power required (peak power) = 1,813,454/1800 = 1007.47 kW_p

Given the estimated performance of the PV system = 19.53%,

the consequent total area of PV system surface needed to produce 1007.47 kW_p is as follows:

$$1007.47 / 0.1953 = 5158.34\,m^2$$

A photovoltaic pergola (providing an additional shading over the green roof surface) on the building blocks in Aghia Varvara, covering the total roof's surface of each building has been hypothesized. In this case study, the roof surface area available for PV panel installation on the flat-top roof is about 330 m^2. Thus, by covering the existing roof, we almost completely cover the total building energy demand. Additional overhangs could be hypothesized to cover the remaining 20 m^2. Fig. 2.52 is a schematic representation of the energy balance of the buildings.

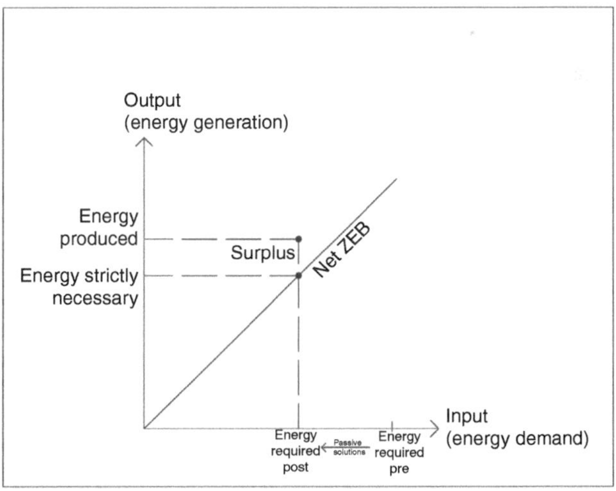

Figure 2.52 ***The PV system adopted by the retrofitting scenario could cover the energy demand in Aghia Varvara.***

As for the previous case of Aghia Varvara, a simple calculation has been made for the urban setting of Peristeri. Given the previous data, the total power required in this case is as follows:

			Peristeri
T	R + S	Global annual electric energy demand	2,035,634

Needed power required (peak power) = 2,035,634/1800 = 1130.91 kW_p

Given the estimated performance of the PV system = 19.53%,

the consequent total area of PV system surface needed to produce 1130.91 kW_p is as follows:

$$1130.91 / 0.1953 = 5790.32\,m^2$$

The photovoltaic pergola hypothesized on the green roofs of building blocks in Peristeri will not necessarily cover the entire available surfaces (about 6800 m^2). Thus, by covering just part of the existing roofs, we cover the total building energy demand.

Finally, the same calculation has been made for the urban courtyard in the Evripidou Athens central area.

Given the previous data the total power required in this case is as follows:

			Evripidou
T	R + S	Global annual electric energy demand	848,933

Needed power required (peak power) = 848,933/1800 = 471.62 kW_p

Given the estimated performance of the PV system = 19.53%, the consequent total area of PV system surface needed to produce 1130.91 kW_p is as follows:

$$471.62 / 0.1953 = 2414.90\,m^2$$

In the case of Evripidou, the photovoltaic pergola (providing an additional shading over the green roof surface) will cover all roof surfaces. In fact, in this case study, the roof surface area available for PV panel' installation on the flat-top roofs of the urban courtyard is about 2.100 m^2; thus, by covering the existing roofs and adding shading devices equipped with PV on the last floors of the buildings, we might cover the total buildings' energy demand.

We may conclude that nearly zero energy in the selected existing buildings of Athens has been proven to be a technically feasible goal even for high energy-consuming and energy-inefficient housing stock. In Chapter 3 we will analyze the costs associated with these goals.

2.3.4 The Energy-Saving Potential of Greenery and Passive Techniques in the Selected Urban Environments

As shown in Chapter 1, the radical transformation of vegetated landscapes as they are replaced by constructed cityscapes, accelerated by increased industrialization and urbanization in recent years, has dramatically affected the urban environment and its atmosphere, causing the thermal degradation of urban climate and hindering the environmental efficiency of buildings.[37] The resulting evolution of urban areas created cities that are getting progressively hotter than the surrounding areas. From the beginning of this century, the average annual temperatures in many cities have increased by as much as 2.8°C.

Denuded landscapes, impermeable surfaces, massive buildings, heat generating cars, and air pollutants have all helped to increase the well known urban heat island effect with higher summer temperatures in urban areas than in the rural surroundings. Morphological and spatial geometry of buildings, together with the thermal properties of surface coatings and, most of all, green surfaces have a strong potential to improve the energy performance and bring about a reduction in cooling demand in urban areas. Thus, planning strategies to reduce the cooling demand in Mediterranean cities should include greenery and natural components as the main tools for improving climatic conditions.

At the territorial and urban scale, it has been largely demonstrated that plants have a strong effect on climate: trees and green spaces can help cool our cities (Santamouris, 2001) and save energy (Environmental Protection Agency, 1992). Trees also help mitigate the greenhouse effect, filter pollutants, mask noise, and prevent erosion (Ferrante and Mihalakakou, 2001). Results of computer simulations aimed at studying the combined effect of shading and evaporative transpiration of vegetation on the energy use of several typical one-story buildings in US cities have showed that by adding

[37] As Zinzi (2010) has pointed out, there are two main challenges concerning the building sector: (1) reducing the energy consumption in European countries and (2) preventing less developed countries from following the same development patterns of the most developed ones, as is happening in recent years. These tasks are not easy to achieve in the Mediterranean region, especially for inland locations, with highly variable conditions throughout the year. Here the buildings have to respond to such variations and be flexible both in hard winters and oppressive heat in the summer.

one tree per house, the cooling energy savings varied from 12 to 24%, while adding three trees per house can reduce the cooling load from 17 to 57%. According to this study, the direct effects of shading account for only 10 to 35% of the total cooling energy savings (Santamouris, 2001; Gaitani et al., 2007). Other prominent studies (Akbari et al., 2001) have verified that the density of the tree crown would reduce the air temperature under the tree between 2 and 4°C.[38] Therefore, considering vegetation in urban areas can help alleviate the greenhouse effect. In the Mediterranean hot climate, vegetation planted around buildings can alter the energy balance as well as provide cooling energy benefits to buildings by sheltering windows, walls, and rooftops from strong solar radiation and radiation reflected from the surroundings. Beyond energy considerations, the redesign of outdoor spaces (streets, squares, and courtyards) is very important for cities (Tapias and Schmitt, 2014), as these provide daily pedestrian traffic and different outdoor activities contributing to urban livability and vitality (Chen and Ng, 2011).[39] Also, it has been found that increased urban temperatures[40] affect the concentration and distribution of urban pollution because heat accelerates the chemical reactions in the atmosphere leading to high ozone

[38] Besides energy consideration, high temperatures could increase health risks and atmospheric pollution, as it is known that the greenhouse effect contributes towards global warming, higher energy demand and emission. Hence, the urban green areas and infrastructure should be maintained in tropical hot and humid cities to guarantee the well-being of the urban communities (Mansor et al., 2010). Trees could help reduce the CO_2 level, increase O_2 and the quality of life (Borhan et al., 2013).

[39] Promoting the use of streets and outdoor spaces by pedestrians will benefit cities from physical, environmental, economical, and social aspects. In this way, ensuring that people are comfortable in outdoor spaces is essential to a better quality of urban life. Over the past few decades, making outdoor spaces attractive to people, and ultimately used by them, has been increasingly recognized as a goal in urban planning and design. The complexity of the physical phenomena in the urban layouts derives from interactions among urban geometry and climatic physical phenomena: the layouts of existing buildings and volumes act as ground obstacles and therefore they usually generate turbulence, thermal airflow instability, shade for solar access, vertical air movements. For example, narrow passageways, comers of buildings close together, expose pedestrians to gusting air movements. Furthermore, interactions among the above mentioned physical effects and current negative urban conditions such noise and pollution should to be considered.

[40] M. Santamouris, *POLIStudies – a Multimedia Educational Structure on The Energy Efficiency of Buildings in Urban Areas*, Department of Applied Physics-laboratory of Meteorology of Athens, 1997. In this frame it should also be considered that higher urban temperatures increase the electricity demand and the consequent production of carbon dioxide and other pollutants, in turn increasing pollutants' concentration (CO_2, NO_2).

concentrations. Moreover, if pollutants land in sheltered areas such as street canyons, they may reside longer than they normally would in a windy environment. The roughness of buildings and urban structures may therefore increase pollutant concentration (CO_2 NO_2).

However, the urban microclimate represents the key indicator among the many factors that determine the quality of outdoor spaces. Pedestrians are directly exposed to their immediate environment in terms of variations of air temperature, relative humidity, wind speed, and solar radiation. Therefore, people's sensation of thermal comfort is greatly affected by the local microclimate (Chen and Ng, 2011).[41] The microclimatic conditions have significant impact on urban dwellers, as the higher temperatures in cities will result in increased energy use for indoor and cooling and higher water demand for landscape mitigation. The high humidity in cities, combined with higher air temperature, create discomfort and undermine the function of outdoor places. As a result, the demand for a better microclimate condition is more pronounced today than decades ago.

The ability to understand and especially to be able to predict and manipulate urban microclimates may help improve aspects on pedestrian activity in urban spaces, and also may improve the performance of buildings, especially with respect to energy conservation (Erell et al., 2011; Kolokotroni et al., 2006; Santamouris, 2006).

Realization and awareness of the climate conditions, seasonal variations, and climate change bring additional demands on the planning and design of urban developments. In this context, an urban "climate-sensitive" design may be defined as a process that considers the fundamental elements of microclimates (eg, sun, wind, temperature) for design purposes (Grimmond et al., 2010). This concept is applied to benefit from the positivity of natural sources in urban microclimate conditions and to mitigate, decrease, and counterbalance the negative effects of urban disfunctionality through appropriate design and planning options.

[41] The outdoor thermal condition is impacted by various elements in the built environment such as: anthropogenic heat (Ichinose et al., 1999) ground surface covering (Lin et al., 2007) evaporation and evapotranspiration of plants (Robitu et al., 2006) and shading mainly level, which is affected by the canopy layer heat island (CLHI). As shade can block incident solar radiation, some studies have discussed mainly the shading effect on thermal environments. For example, street orientation and the height/width (H/W) ratio have been measured to assess the shading levels in some studies (Emmanuel et al., 2007; Johansson 2006). In the context of urban planning, how outdoor thermal conditions influence thermal sensations of people and their behavior (use of outdoor spaces) is of great interest for designing urban spaces.

Nowadays urban planners who intend to create comfortable microclimates can take advantage of easy methods of assessment of the thermal component of climate (Höppe, 1999). The degree of impact of the outdoor thermal environment on thermal comfort varies with the thermal requirements of people in different climatic regions.[42]

Improvement of the ambient microclimate in the urban environment requires the use of more appropriate materials, increased use of green areas, use of cool sinks for heat dissipation, appropriate layout of urban canopies, and the like, all of which help to counterbalance the effects of increased energy consumption in the urban areas. Construction of new generating plants may help generate the needed energy to achieve nZEBs, but it cannot be the sole option to solve outdoor climatic issues. Thus, the adoption of measures to decrease the energy demand in the urban areas, like the aforementioned use of more appropriate materials, increased plantation, use of sinks, and so on, seems to be a much more reasonable option.[43] In this framework, the Athens central region has to be considered a very special pilot study. Measurements developed by the group Building Environmental Studies have shown dramatically higher temperatures in the central Athens area than in the suburban areas (by around 10–15°C). In the same research context, work developed by the research project POLIS in Athens (Ferrante et al., 1998) has shown some effective uses of natural components – such as green roofs and pedestrian permeable surfaces in the streets – within urban canyons. It is evident that microclimatic enhancements in urban scale involving water systems, planting of trees, and lightening of colors in urban surfaces may decrease pollution loads and save huge amounts of energy, consequently improving both indoor and outdoor comfort conditions. Unfortunately though, open spaces among buildings or free courtyards are not always available; therefore the use of passive natural elements can be reduced in dense urban environments. Therefore, an urban energy concept has to address these problems; it must investigate, on a

[42] A number of biometeorological indices have been developed to describe human thermal comfort level by linking local microclimatic condition and human thermal sensation (Chen and Ng, 2011). A major group of such indices are the so-called "steady-state" models. These models are based on the assumption that people's exposure to an ambient climatic environment has, over time, enabled them to reach thermal equilibrium, and they provide numerical solutions to the energy balance equations governing thermoregulation.

[43] Such a strategy, adopted by the Sacramento Municipal Utility District, (SMUD), has proved to be very effective and economically profitable, (Flavin and Lenssen, 1995). It has been calculated that a megawatt of capacity is actually eight times more expensive to produce than to save it. This is because energy saving measures have low capital and no running cost, while construction of new power plants involves high capital and running costs.

smaller scale, the building envelope and the energy-saving potential of selected passive techniques such as building materials and components, shading devices (including green ones such as pergola), and low albedo surfaces, in order to minimize thermal discomfort both in the open spaces and inside the buildings.

The design of outdoor spaces – even if reduced to the envelope of the buildings because of existing urban constraints within thickly built urban areas – as well as the use of natural components have been regarded as key means to improve urban conditions in relation to both microclimate and reduction of pollutants. By "making up" the building's surfaces and elevation facades with green components or shading devices, different scenarios have been proposed in different case studies in downtown Athens. Experimental software research models have been used to quantify the positive effects of these selected passive techniques. Obtained results clearly indicated that alternatively designed outer surfaces act as a prior microclimate modifier and deeply improve outdoor air climate and quality (up to 2/3°C reduction in ambient temperature) (Santamouris, 2001).[44]

2.3.4.1 Previous Studies on Heat Island's Mitigation

Extended analyses on urban layout and passive cooling in the Athens area have been carried out throughout more than 20 years' worth of research studies by Santamouris et al. (2014).[45] The optical characteristics of materials used in urban environments and especially the albedo-to-solar radiation and emissivity-to-long-wave radiation have a very important impact to the urban

[44] Other significant physical factors in the thermal performance of urban environments are wind flows and air circulation, (Ricciardelli et al., 2006) as well as air stratification within urban canyons. In particular, the heat island effect and the microclimatic conditions typical of urban canyons (Bitan, 1992) appear to be strongly influenced by thermal properties of the materials and components used in the buildings and on the streets. Comparative research studies demonstrated that the use of cool colored materials (Synnefa et al., 2007) and thermo-chromic building coatings can contribute to energy savings in buildings, providing a thermally comfortable indoor environment, and, at the same time, highly improve the urban microclimatic conditions (Karlessi et al., 2009).

[45] M. Santamouris, *Passive Cooling and Urban Layout*, Physics Department, University of Athens, 1997. Physical and mathematical definition of the air space above the city, as well as equations defining wind distribution, the urban temperature field and the impact of the main design parameters such as the role of the surface albedo, of green spaces, of anthropogenic heating systems have been described in his work. His work describes the mean features by which the urban climate differs from the climatic conditions of the surrounding rural areas, defining specific parameters and conditions with a physical and climatic approach. Nonetheless, studies on urban temperature have been conducted by many authors in literature, starting form the last three decades, like Oke (1976) and Johnson et al. (1991).

energy balance. Yap (1975) has reported that systematic urban–rural differences of surface emissivity have the potential to cause a portion of the heat island. Using high albedo materials reduces the amount of solar radiation absorbed through building envelopes and urban structures and keeps their surfaces cooler. Materials with high emissivity are good emitters of long-wave energy and readily release the energy that has been absorbed as short-wave radiation. Lower surface temperatures contribute to decreasing the temperature of the ambient air, as heat convection intensity from a cooler surface is lower. Such temperature reductions can have significant impacts on cooling energy consumption in urban areas, a fact of particular importance in hot climate cities (Lin, 2009). Trees and green spaces contribute significantly to cool our cities and save energy (Shahidan et al., 2012). Trees can provide solar protection to individual houses during the summer period while evapotranspiration from trees can reduce urban temperatures.

The causes and effects of urban climates and heat islands are diverse and their interactive are complex, as discussed previously. Urban design, and thus energy consumption in urban spaces, can be influenced if the impact of the main urban design factors is well understood. The optical characteristics of materials used in urban environments and especially the albedo-to-solar radiation and emissivity-to-long-wave radiation have a very important impact on the urban energy balance.

The albedo of a surface is defined as its hemispherical and wavelength integrated reflectivity. Use of high albedo materials reduces the amount of solar radiation absorbed through building envelopes and urban structures and keeps their surfaces cooler. Asaeda and Ca (2000), has experimentally tested the impact of various pavement materials used commonly in urban environments during the summer period. They found that the surface temperature, heat storage, and its subsequent emission to the atmosphere were significantly greater for asphalt than for concrete and bare soil.[46]

[46] At the maximum, asphalt pavement emitted an additional 150 W/m^2 in infrared radiation and 200 W/m^2 in sensible transport compared to a bare soil surface. They also found that the rate of infrared absorption by the lower atmosphere over asphalt pavement was greater by 60 W/m^2 than that over the soil surface or concrete pavement.
Albedo and emissivity for selected surfaces have been classified by Bretz and colleagues (1992). Concrete: albedo 0.3; Emissivity 0.94; Red Brick: 0.3, 0.90; Building Brick: emissivity 0.45; Concrete Tiles: emissivity 0.63; Wood (freshly planted): 0.4, 0.90; White paper: 0.75, 0.95; Tar paper: 0.05, 0.93; White Plaster: 0.93, 0.91; Bright Galvanized Iron: 0.35, 0.13; Bright Aluminum foil: 0.85, 0.04; White pigment: 0.85, 0.96; Grey Pigment: 0.03, 0.87; Green Pigment: 0.73, 0.95; White Paint on Aluminum: 0.80, 0.91; Black paint on Aluminum: 0.04, 0.88; Aluminum paint: 0.80, 0.27–0.67; Gravel: 0.72, 0.28; Sand: 0.24, 0.76.

Increasing the surface albedo has a direct impact on the energy balance of a building. Large-scale changes on urban albedo may have important direct and indirect effects on the city scale.[47] Various studies have been performed that examine the relationship between the canyon layout and especially the sky view factor, with the heat island intensity as well as with surface temperatures.[48]

Other studies have been developed in relation to anthropogenic heat in urban areas, mainly related to transportation systems, power generation, and other heat sources (Johnson et al., 1991).[49]

In this framework, we should point out that plants and vegetation also play a key role in CO_2 absorption. For example, Steemers (1993) has shown high levels of CO_2 absorption by plants in relation the rate of woody growth.[50]

2.3.4.1.1 Mitigations on Urban Canyons in Athens

Among previous studies and experiments on urban mitigation, it is worthwhile to report experiments on urban canyons developed in Athens, which investigated the effect of urban canyons' characteristics on the temperature

[47] Using computer simulations and actual measurements, Bretz et al. (1992) report that roof albedo increase strongly reduces the cooling demand. Other studies have demonstrated how trees and green spaces can significantly contribute to cool cities and save energy. Trees provide solar protection during the summer period, mitigate the greenhouse effect, filter pollutants, mask noise, prevent erosion and calm their human observers; evapotranspiration from trees further reduces urban temperatures. As pointed out by Akbari (1992), "the effectiveness of vegetation depends on its intensity, shape, dimensions and placement. But in general, any tree (...) can have a noticeable impact on energy use." Numerical studies trying to simulate the effect of vegetation have been performed by various researchers. Huang et al. (1986) report that computer simulations predict that increasing the tree cover by 25% in Sacramento and Phoenix, USA, would decrease air temperatures at 2:00 pm in Jul. by 6.0°F to 10.0°F.

[48] Yamashita et al. report a clear correlation of urban air temperature and sky view factor for some Japanese cities. Bärring et al. (1985) has studied the relationship between the street surface temperature and the sky view factor in Malmoe, Sweden. They report a correlation of the surface temperature pattern on the street geometry and the highest the sky view factor the lowest the surface temperature. Higher surface temperatures are recorded in low sky view factor canyons also outside the city.

[49] In different kind of street canyon geometry it has been found that heat island intensity is much higher at low ambient temperatures and can reach values between 2 and 8°C. (Source: Johnson et al., 1991.)

[50] This study refers to the rate of absorption per square meter of planted area. Results can be used to determine plants required to absorb certain amounts of CO_2, in an aim to ensure sustainability of urban developments. The CO_2 absorption (kg/m^2/year) for different plant's types are: Trees (average): 1.0; Hawthorn (*Crataegus macrocarpa*): 1.9; Blackthorn (*Prunus spinosa*): 1.4; Field Maple (*Acer campestre*): 1.2; Beech (*Fagus sylvatica*): 0.4.

distribution in the urban environment[51] and proposed mitigations to improve microclimate conditions.[52] Experiments in the Athens central area have been performed in ten different canyons presenting dissimilar layout, orientation, anthropogenic heat, and vegetation. These measurements have been performed taking into account the surface temperature,[53] air temperature,[54] and wind speed.[55]

The distribution of the surface ground temperature of pavements and roads have been measured in seven different canyons on an hourly basis time-lapse and for the period of a whole summer in 1997.[56] Asphalt temperatures during the day period reach peak temperatures of up to 57°C, while the corresponding maximum temperature for white and dark slab pavements are close to 45 and 52°C, respectively. The mean temperature of all the materials during the night period is about 23–25°C. The orientation

[51] M. Santamouris, POLIStudies – a Multimedia Educational Structure on the Energy Efficiency of Buildings in Urban Areas –, 1997. EU Project.

[52] Ferrante, A., 1997–1999. Water, Green, and Selected Passive Techniques to improve microclimate and reduce pollution in the urban contexts. Research project developed during the 24 months of PostDoctoral Research Activity in the University of Athens, Department of Applied Physics. Programme work selected and funded in the framework of EU TMR Research Programme. Directorate G. XII, Science, Research and Development.

[53] Surfaces receive short wave radiation as a function of absorptivity and exposure to solar radiation, receive and emit long wave radiation as a function of their temperature, emissivity and view factor, transfer heat to or from the surrounding air and exchange heat via conduction procedures with the lower material layers. Surface temperature measurements performed from the bottom to the top of both facades of the canyon using a step of 3–3.5 m. Additional measurements have been performed in some cross sections of the canyon where different materials are used. All measurements have been performed from the street level. Temperatures of pavement, road and 5 additional points along the width of the canyon were measured.

[54] It is worthwhile remembering that temperature of the external materials in a canyon is governed by its thermal balance. The thermal balance of a surface in a canyon can be expressed as follows: $Q^\star = Q_H + Q_G$; Where $Q^\star$ is the net radiation, Q_H represents the convective heat exchanges and Q_G are the conductive heat exchanges with the substrate.

[55] Wind speed measurements by means of a three-axis anemometer to measure the three components of the wind speed inside the urban canyons.

[56] Existing knowledge on the temperature distribution of materials used for pavements and roads has been implemented by Asaeda and Ca (2000), texting experimentally the impact of various pavement materials during the summer period used commonly in urban environments. They found that the surface temperature, heat storage and its subsequent emission to the atmosphere were significantly greater for asphalt than for concrete and bare soil. At the maximum, asphalt pavement emitted an additional 150 W per square meter in infrared radiation and 200 W/m^2 in sensible transport compared to a bare soil surface. They also found that the rate of infrared absorption by the lower atmosphere over asphalt pavement was greater by 60 W/m^2 than that over the soil surface or concrete pavement.

of the streets, the H/W ratio,[57] as well as the type of used materials define the surface temperature of the materials.

The impact of the absorbed solar radiation on the temperature increase of the materials used in pavements and roads has been found to be very important. It has been shown that for white and dark slab pavements, the instantaneous temperature difference across the street varies from 10 to 22°C as a function of street layout and orientation; for asphalt, the instantaneous temperature difference goes up to 27°C. The effect of the street orientation on the temperature of the street and pavement materials can be understood if the daily peak temperatures of same materials across the canyon are compared. Usually the southern pavement slabs present a higher daily peak temperature, to about 8–15°C, compared to the peak daily temperature of the northern pavements. Similar values are observed for asphalt streets between the southern and northern part of the road across the same section of the street. To understand the impact of thermal and visual characteristics of the various materials used in a road on their surface temperature, the temperature differences of these materials across a street have been compared.

Surface temperatures of the buildings' envelope have been measured for seven canyons. Temperatures vary as a function of all parameters defining their thermal balance, but mainly as a function of the received solar and emitted infrared radiation.[58] As expected, in a canyon, the south-oriented or almost south-oriented surfaces present much higher temperatures than the opposite surfaces. The maximum daily simultaneous temperature difference between two opposite panels was close to 19°C. The highest recorded difference of the daily maximum temperature of two opposite surfaces was 14°C. In almost all canyons, there was a period during the day at which each of the two facades presented a higher temperature than the opposite

[57] Urban canyons are characterized by three main parameters: H, the mean height of the buildings in the canyon, W, the canyon width, and L the canyon length. Given these parameters, the geometrical descriptors are limited to three simple measures. These are the ratio L/H and the building density J=Ar/A1 where Ar is the plan of roof area of the average building and A1 is the "lot" area or unit ground area occupied by each building.

[58] Minimum observed surface temperatures during the night period varies between 18 to 27°C, while the maximum temperature during the day period varies between 25 to 50°C. To evaluate the impact of the building's orientation and relative position in the canyon, on its surface temperature, the temperature of building panels facing each other, from the ground up to the upper height, have been compared for the day and the night period. Taking into account that the received solar radiation by a surface varies mainly as a function of its orientation and its relative height in the canyon, that defines the degree of shading from the adjacent buildings. The study has been concentrated on two specific parameters.

surface. This time period is a certainly a function of the canyon orientation, layout, and characteristics of the material.[59]

Since distribution of the ambient air in a canyon greatly influences the energy consumption of the buildings, important deductions can be derived from these experimental activities. It is important to specify that the orientation of streets determine the amount of solar radiation received by the canyon surfaces; in other words, measurements taken at the southwest facade present a higher peak temperature during the day period, up to 2°C, compared with the corresponding northeast-oriented facades.

During the night period, all temperatures were almost similar. The orientations of the streets do not influence the bulk air temperature in the canyon; however, orientation has a rather strong influence on the air film developed in the proximity of the facades; close to the facade of the buildings, an air film is governed by the temperature of the building surface and the vertical air transport is developed.[60] Measured air temperature differences between the two facades vary as a function of the canyon layout and surface characteristics: as expected, south, southwest, or southeast facades presented higher air temperatures. In all cases, during the night, higher temperatures were measured at the ground level, and temperatures were found to decrease as a function of height. During the night, the temperature of the air in the middle of the canyon was found to be higher than that of the air film close to the facades of the canyon.

The daytime intensity of the heat island for all canyons and measuring points at the ground level varies between 4 and 15°C. The highest heat island intensities have been found for the very central canyons characterized by high circulation and high H/W values. In particular the Solonos, Ippokratous, and Mavromichalis, located at the very central Athens area, presented a heat island intensity between 10 to 15°C. The highest intensity, close to 15°C, has been found on Solonos Street, a south-southwest-oriented canyon having a very high concentration of cars. Also in the Valaoritou canyon, a pedestrian street located at the central area of the city and in the neighborhood of high-concentration main roads, the absolute heat island

[59] In northeast-southwest and southeast-northwest-oriented canyons, where the solar surfaces are of southeast or southwest orientation, the simultaneous temperature difference between opposite surfaces during the day is lower at the ground level and increases as a function of canyon height: as a matter of fact, in this type of canyon, the lower height surfaces of the southeast-southwest facades receive much less radiation than the upper level ones and thus the upper southwest-southeast floors present higher surface temperatures.

[60] Obviously, it has been found that the middle canyon temperature is very different from the mid temperature of the air film temperature close to the facade of the buildings (in most of the cases, temperature at the middle of the canyon is lower than the corresponding film air temperature).

intensity was high (11°C), similar to the intensity found in other high-concentration canyons. In the south-southwest-southeast-oriented facades and during the day period, the temperature gradient varied between 0 and 14°C, while the ground floor presents the lower surface temperature, because of the lower incident solar radiation. Intermediate floors presented the highest temperatures, because they received almost the same solar radiation as the upper floors, as well as the infrared radiation from the opposite buildings, while convective fluxes were lower because of the lower wind speed at this level. In the north-northeast-northwest-oriented facades and during the day period, the temperature gradient varied between 2 and 8°C.[61]

To evaluate the energy-saving potential of greenery and passive techniques in these urban environments considering the specific restrictions encountered in the AMA, two different rehabilitation procedures have been considered:

1. Passive technological measures related to the envelope of the buildings in existing dense thickly built urban areas close to central high circulation streets.
2. Passive technological measures related to the envelope of the buildings added to (alternative) use of outdoor green spaces, courtyards, and pedestrian streets like places able to improve microclimate in existing urban areas, where a relative amount of open space was available.

A climatic conscious design of outdoor spaces as well as an appropriate use of natural components are key elements to reducing the effect of unsound evolution of urban areas where impermeable surfaces, denuded landscapes, and air pollutants determine higher summer temperatures and unhealthy life conditions.

Courtyards, streets, and outdoor spaces (even if reduced to the shell and envelope of the buildings within existing thickly built urban areas) can be redesigned by using natural elements (water, green) to improve urban conditions in relation to both microclimate and reduction of pollutants (Nikolopoulou et al., 2001).

[61] However, comparison of simultaneous measurements with the high circulation Solonos street, taken at the northern facade of the streets, clearly shows that this pedestrian street has to about 2°C lower temperature. Taken into account that both streets are parallel, have the same layout and almost same H/W ratio, it can be concluded that higher recorded temperatures are mainly due to the anthropogenic heat and the use of more absorbing street materials like asphalt. Lower heat island intensities, from 4°C to 8°C, have been found in canyons located in the surroundings of Athens characterised by much lower circulation and lower H/W ratios. Heat island intensity during the day period is mainly a function of the thermal balance of the urban area. In particular, in Solonos canyon it is found that mean daily heat intensity is close to 5.7°C.

The architectural rehabilitation design, by making use of passive cooling techniques like water, appropriate plantings, and vegetation, proposes alternative scenarios in the main canyons to improve thermal comfort conditions and decrease pollution loads in the high temperatures of the central Athens areas. The proposals are based on the following strategic objectives:

1. Promotion of all forms of passive natural devices according to the potential effects on minimizing the localized pollution problems and to enhance the urban microclimate. The first step to avoiding the problem of poor air quality is to reduce pollution at the source by improving efficiency and extracting pollutants. The next step is to ensure that unavoidable pollutants are dispersed safely. A planning strategy to minimize pollution impact is to provide spatial separation of environmentally incompatible activities. Where open spaces and buildings are naturally ventilated, ensure that ventilation air is drawn from a source that is clear and green.
2. Maximum increase of permeable surfaces in the outdoor spaces.
3. Improvement of natural cooling, reducing solar heat gains both in the open spaces and on building facades (allowing protection for southeast and southwest facades by means of shading devices and "natural filters"). The evaporative and transpiration cooling effects of plants are used to precool ventilation air, both for the buildings and the open spaces. Furthermore, dense vegetation helps to filter particles from the air.
4. Ensure CO_2 absorption by plants and urban pollutant dispersion by natural ventilation and night cooling.

To reach these strategic objectives, the study has theoretically envisaged the possible redesign of these urban contexts, trying to use and consider as much as possible the existing constraints of the as-built scenarios. Thus, different scenarios have been designed by using new materials and green components onto building surfaces and elevation facades, and in the street layouts.

In the redesign of the urban canyons, shaded, under-planted roof gardens have been regarded as a common feature.[62] In fact, the technical tool of

[62] Planted green roof on the bare roofs of the buildings may greatly reduce the high thermal load, both from the single buildings and from the city as a whole. The factors behind temperature reduction are evaporative cooling and shading of the ground, whereas temperature increase during night is the result of the reduced sky factor within the canopy. Results of the simulations show that a vegetative cover of 30% could produce a noontime oasis of up to 6°C, in favorable conditions and a nighttime heat island of 2°C. Other simulations indicated that the net effect of the increased urban vegetation is a decrease in ozone concentrations. Hader (1970) summarizes the conclusions of different studies on the distribution of dust in and out of urban green areas. He found that inside the green area, a diminution of dust is

(*Continued*)

an additional pergola over the roof garden represents a simple but effective architectural measure. Hence, in all the flat surfaces of the roofs, an additional pergola layer has been combined with the planted roof layer. By using this device, the hypothesized results suggest that the air temperature in the roof garden should be even lower than the shade temperature simply provided by trees. This solution could also represent a key element to address the static problems derived from the additional soil weight. In fact, it is estimated that a 0.8–1.00 m depth layer of soil is able to support the growth of medium-high vegetative plants. Thus, a garden with a wide variety of plants, including bushes of up to 1.5 m in height, would imply a considerable soil weight, with consequent additional load bearing on existing buildings.[63]

Shaded roof gardens under-planted with shade-tolerant plants are an architectural measure that has been widely employed throughout all the studied urban canyons. This is a sustainable solution that contributes to a thermal load reduction of the building's shell and to the improvement of microclimatic conditions of densely built urban centers with no natural environment.[64]

During the rehabilitation process for four different canyons, specific proposals for canyons with different geometrical features, orientation, and types of buildings have been hypothesized. In brief, the following strategies were proposed:

1. Provide spatial separation between polluted areas and open public spaces by means of green pedestrian streets (proposed for Mavromihali, existing in Valaoritou). These main streets, rehabilitated by trees, permeable pavements, and watercourses, will provide evaporative cooling and ensure clearer air free of noise and cars.

noticeable, while he describes the processes of filtration of dust by vegetation. The processes of heat transfer into the planted roof has quantitatively and qualitative differences in comparison with the normal processes in bare roofs. The solar radiation, the external temperature and the relative humidity are reduced as they pass through the foliage coverings the roof. The plants, because of the biological functions of photosynthesis, respiration, transpiration, and evaporation, absorb a significant proportion of the solar radiation. Therefore the creation of gardens on the roofs of buildings could represent one of the key steps to be considered for the environmental and climatic improvements in the city of Athens. Offering large surfaces with vegetation, a garden blocks the infiltration of solar radiation by creating a protective layer, reducing range of temperature in winter and summer.

[63] Additional technical tools to contain the soil thickness, involving the external facades of the buildings and extra economic costs are also implied by using soil and greenery.

[64] In this contribute, the role of planted roofs in the improvement of the thermal performance of the buildings has been deeply investigated (Eumorfopoulos and Aravantinos, 1995).

2. Position openings from internal courtyards (where a higher portion of fresher and cleaner air is available) toward open spaces and building facades.
3. As much as possible, increase permeable surfaces on the pavements.
4. Provide shadow and filter for building facades (especially the southwest and southeast-oriented ones) by means of selected passive techniques such as pergolas and winter gardens.
5. Create planted roof gardens on the bare roofs of the buildings and shaded roof gardens planted with shade-tolerant plants.

In canyon Hyppokratous, the increase of permeable surfaces has been obtained by removing one car gangway in front of the southeast-oriented blocks of buildings to allow filtered and shadowed areas for pedestrians, as shown in the figures mentioned later. In Mavromihali, a multi-story green building has been proposed in the existing open area to provide a green "oasis" between the existing buildings along the canyons. Finally, since canyon Valaoritou is the only existing pedestrian street, the alternative scenario is to consider adding shading devices between the buildings (to shade the southwest-oriented facades of the canyons and the open spaces). Fig. 2.53 shows a map of the central Athens area, giving the position of the three

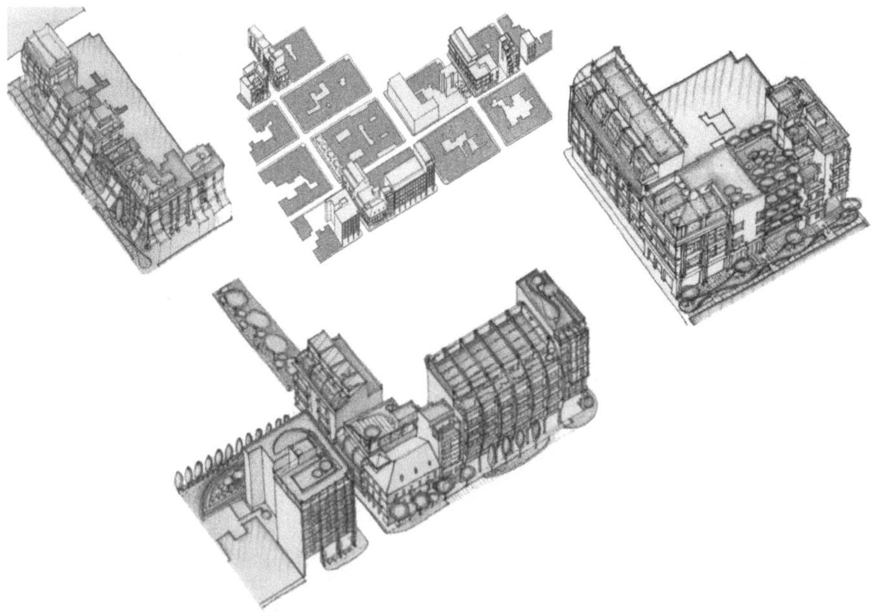

Figure 2.53 ***A schematic 3D model of the central Athens area showing the position of the main canyons (in the center) with the envisaged scenarios (left, right, bottom in the figure).***

(a) (b)

Figure 2.54 ***Canyon Valaoritou (a) and envisaged scenario with greenery and shading devices (b).***

canyons. In the downtown city quarter, the three canyons Hyppocratous, Solonos, Mavromihali are concentrated, while Valaoritou is a pedestrian street.

To assess the microclimatic improvements of the proposed architectural and technological measures, selected software simulation tools have been used,[65] selecting the pedestrian street in canyon Valaoritou Fig. 2.54. In this case, air temperature distribution inside the canyon calculated for 15th July, selected as the reference day. Simulation results are shown in Fig. 2.55.

[65] The cooling potential of the proposed techniques has been investigated for the Canyon Valaoritou, by means of a theoretical software model developed in the MATLAB program. The overall energy exchanges and internal heat flows to be considered in an urban canyon thermal model are: reception, absorption and reflection of short wave radiation, as well as absorption and emission of long wave radiation by involved surfaces and facades. In addition, the heat transfer toward or from the surrounding air via convection and conduction with the lower materials layers has to be fixed. The thermal energy balance for transient conditions at any surface of a canyon can be expressed by the following differential equation: $\rho C\delta Tsi/\delta T = = hc(Tsi - Toi) + \lambda/\Delta x(Tsi - Tmi) + \Sigma Fij\sigma(Tsj4 - Tsi4) + asqs + Fi/sky\,(\varepsilon c\sigma Tc4 - \varepsilon\, i\sigma Tsi4)$; (1) where: $hc(Tsi - Toi)$ is the heat rate transferred by convection between the surface at temperature Tsi and the adjacent air layer. In this case the convective exchange coefficient hc is considered to be a function of the air velocity near the building facade and it is expressed by: $hc = 4.1\,V + 5.8\ W/m^2K$. $\lambda/\Delta x(Tsi - Tmi)$ is the heat rate transferred by conduction into the wall or the ground. The temperature Tmi is defined as the temperature inside the wall at a distance Δx from the surface; λ is the conductivity constant and Δx has been set up by the operator; $\Sigma Fij\sigma(Tsj4 - Tsi4)$ represents the rate heat exchange by long wave radiation between a surface i and the n surrounding surfaces ($j=1{:}n$). The view factors between every element of surface i and the surrounding elements are calculated by the equation: $Fij = \cos(\theta)\sin(\theta)/r^2$ where θ is the angle between the line connecting the centers of the two segments at distance r and the normal line to the plane containing the segment; asqs represents the short wave solar irradiance absorbed by the surface element, as is the absorption coefficient, while qs is made up of the direct solar radiation, and the diffused irradiance from the sky and the surrounding surfaces. It has been calculated for each segment as a function of their partial shaded area (PSA).

(Continued)

For the basic equation describing thermal energy balance for transient conditions of a canyon, to investigate the proposed alternative scenarios, a series of transformations have been considered.

First, regarding the scenario that includes shading devices, an additional energy balance between the proposed curtains and the ground has been assumed, which will simulate the new thermal performance of a theoretical first scenario modified by the curtains in the canyon. Second, to assess microclimate modifications brought about by the use of greenery, the mean energy and vapor balance has been inserted into the MATLAB equation. Finally, in order to assess the microclimate modifications brought about by

The total solar radiation (direct, diffuse, and reflected) incident on each segment by the following equation:
$I(t)=I\text{dir}\ (t)(1-\text{PSA}(t))+I\text{dif}(t)\text{SVF}+\Sigma\text{I dir+dif}(t)i\ (1-\text{mi})\text{FiA}$, Where: PSA= partial shaded area of the segment; $I(t)$ is the mean hourly total solar radiation incident on the segment (W/m^2); Idir(t) is the hourly mean unobstructed direct solar radiation (W/m^2); Idif(t) is the hourly mean unobstructed diffuse solar radiation on a horizontal plane, (W/m^2); Idir+dif(t) is the hourly mean direct and diffuse solar radiation; F=view factor for two segments; m= surface absorptivity; SVF= sky view factor for the segment. Subscript i is the ith surface type; the direct radiation is corrected for specific solar zenith angle; the incoming diffuse radiation is assumed to be isotropic and the first reflection is only included as it is very difficult to model multiple reflections. Direct and diffuse radiations are derived from observed values at the meteorological site and assumed specifically in the urban site. Fi/sky $\varepsilon c\sigma Tc4$ represents the diffused atmospheric long wave heat contribution; it is the long wave irradiance emitted by the atmosphere and received by the surface element i. $\varepsilon c\sigma Tc4$ corresponds to the diffuse flow emitted by the atmosphere where Tc is the sky temperature given as: $Tc = To - 6$; Fi/sky $\varepsilon\ i\sigma Tsi4$ represents the heat rate as long wave radiation exchanged between the surface and the sky.The general form of equation (1) works the same for any segment on the ground as well as on the walls except for qs, which is considered as the solar total irradiance on a horizontal plane. The surface temperature of each urban unit cell is used as the boundary condition in the determination of the ambient air temperature. This equation has been inserted in the MATLAB, and different simulation of initial scenario have been performed, as showed in the following graphs. Respectively the ambient air distribution, calculated at the height of 2 meters, the air temperature distribution and the surface distribution on SW Facade have been simulated.

The main output matrixes describing the variation of temperature distribution inside the canyon are: Tout = air temperature inside the canyon; Tout1 = air film temperature on the right wall (south-west oriented); Tout2 = air film temperature on the left wall (north-west oriented); TM = surface temperature in the middle of every segment of the left wall; TMR = surface temperature in the middle of every segment of the right wall; TMW = surface temperature in the middle of every segment of the ground; As a reference temperature for the air temperature inside the canyon the whole fifth column of the matrixes has been considered for the main parameters above mentioned. As far as it concerns the air temperature distribution inside the canyon the fifth row of the fifth column has been considered. The fifth element of the fifth column of Tout has been selected as the representative value of the temperature in the middle of the canyon for the pedestrians (at the height of 2 m).

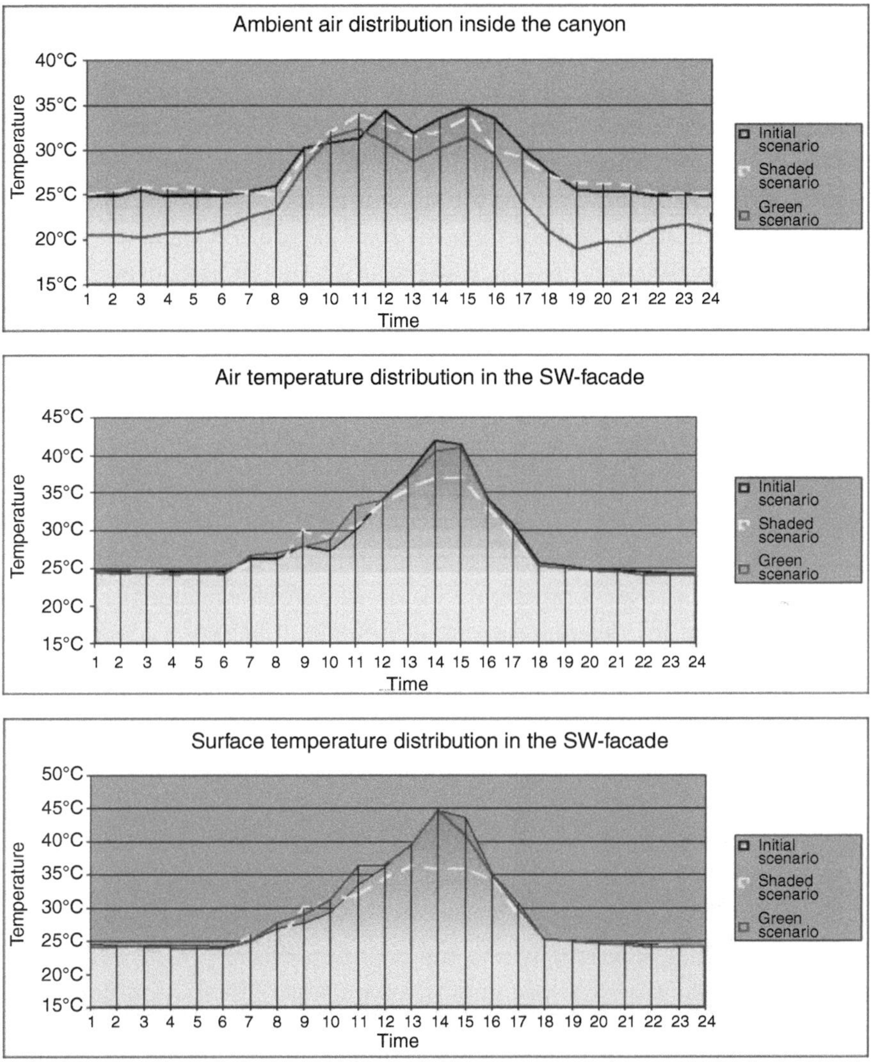

Figure 2.55 ***Ambient air distribution in the urban canyon; air temperature distribution and surface temperature distribution for the southwest facade.*** The dark gray refers to the green areas that have been inserted in place of street in the urban context.

the use of permeable surfaces and greenery (in place of existing stone slabs) in the pedestrian street, equations representing the macroscopic or "mean" energy and vapor balance in a canopy have been considered.[66]

[66] (ρc)ρ d LAI dTp/dt = φ rad.sol + φ rad, TIR. + φ conv, ρ−a + φ trans, ρ−a; $(\rho c)_a$ H $dT_a/dt = \varphi_{conv,a\text{-}\rho} + \varphi_{conv,a\text{-}g'} + \varphi_{conv,a\text{-}\infty}$ ρa H $d\theta_a/dt = \varphi_{vap,a\text{-}\rho} + \varphi_{vap,a\text{-}g'} + \varphi_{vap,a\text{-}\infty;}$ where: T_p= Leaves temperature (average in control volume), (°C); T_a= Air temperature (average

(*Continued*)

Fig. 2.55 describes the ambient air distribution in the urban canyon, the air temperature distribution, and the surface temperature distribution for the southwest facade. The air temperature distribution refers to the variation of the temperature calculated in the middle of the urban canyon, while the air temperature distribution and surface temperature distribution refer to the southwest facade (namely, the air film of the southwest-oriented facade and the surface temperature on the same facade). As expected, air film and surface temperatures are much higher than the air temperature in the middle of the urban canyon, with the surface temperatures in turn higher with respect to the air film temperature. Results also show that the proposed shading devices do not affect the air temperature inside the canyon, while strong temperature reductions can be observed for film air and the surface temperature on southwest-facing facades.

Conversely, the proposed greenery on the ground level of the urban canyon does not seem to affect the film air and the surface temperature on the southwest facades, while a strong air temperature reduction can be observed for the air temperature inside the urban canyon (the air canyon temperature for the initial scenario fluctuated between 25–35°C; for the greenery scenario, the air temperature inside the canyon varied between 20–31.5°C). Therefore, the green scenario contributes significantly to the temperature reduction, especially during nighttime, because of the increased long-wave heat losses from the planted surfaces. This can be associated by the different times when maximum is observed at the two canyon's walls.

Regarding the air temperature distribution and the surface temperature distribution for the southwest facade, the most improved situation is the theoretical shaded scenario where the air temperature profile varied between 24 and 42°C, while the surface temperature distribution fluctuated between 24 and 45.5°C.

in control volume), (°C); θ_a= Air specific humidity (average in control volume), (kg.kg^{-1}); d= Average leaves thickness, (m); $(\rho c)_\rho$, $(\rho c)_a$ = Leaves and Air, respectively, specific thermal capacity, (J/m^3/K); ρ_a = Air density (kg. m^{-3}); H = Canopy layer thickness (m); $\varphi_{rad,sol}$, $\varphi_{rad,TIR}$ = Solar radiation absorbed by leaves and Net thermal radiation flux on leaves, respectively, (W.m^{-2}); In addition the equations describing: – the convective heat transfer between leaves and air ($\varphi_{conv,\rho-a}$), - the heat and vapor transport between the air within the canopy and the outdoor air ($\varphi_{conv,a-\infty}$ and $\varphi_{vap,a-\infty}$), – the transpiration flux, the energy flux consumed to let water evaporate in leaves (φtrans,ρ-a), – the heat and vapor flux between the ground surface and the air ($\varphi_{conv,a-g}$) have been introduced in the software model developed in POLISTUDIES Multimedia Tool.

To sum up, the obtained results and the relative mutual comparison clearly indicate that:

- The air temperature inside the canyon is much lower than the film air temperature and the surface temperature on the southwest-oriented facades.
- The air temperature in the middle of the canyon is strictly influenced, governed, and improved by the presence of greenery (which acts as the microclimate modifier), while the proposed shading devices do not seem to importantly affect the air temperature inside the canyon.
- The air film temperature on the southwest facade as well as the surface temperature distribution are much lower in the theoretical shaded scenario (see broken line in Fig. 2.55) than in the initial one, while, as expected, the presence of green does not influence the surface temperature distribution on the facades of the building.

As a whole, the research study has demonstrated that the redesign of urban surfaces (both in relation to the building envelope and the ground level – streets, courtyards, squares, etc.) acts as the prior microclimate transformer of urban conditions and can deeply improve outdoor air climate and quality. Nonetheless, the positive effects of each different scenario, as simulated so far, seem to address specifically and differently the various climatic parameters inside the canyon. But, as a matter of fact, the energy potential of proposed architectural measures cannot be analyzed separately from an overall comprehensive investigation that addresses the combination of the building with the surrounding urban boundary conditions.

This deduction, together with the encouraging results, led us to other exploratory exercises aimed at:

- Investigating the potential of these alternative scenarios from the outdoor urban spaces to the indoor thermal performance of the urban buildings
- Assessing the energy potential of combined scenarios by integrating the green component with the shading devices in the different urban environments.

2.3.4.2 Modeling the Urban Microclimate (Annarita Ferrante and Anastasia Fotopoulou)

It is well known that modeling a built environment project may help to visualize its future impact and is an important part of responsible design decision-making (Erell et al., 2011; Kanton and Unger, 2010). There are three types of modeling methods used in urban climatology; physical scale models, integrated open-air models, and mathematical models. Computer programs simulate the physical environment and produce knowledge that may say something about real-world behavior. However, all types of models need validation, and because of the deficiency of well-documented, high-quality data from field studies, this remains a major problem for most urban climate models (Erell et al., 2011). In terms of outdoor thermal comfort, there is the need for a predicting tool by means of urban climate models and tools. Although people's subjective perceptions and responses to the urban environment are various and not yet well understood, simulation and scenario-testing tools are always of particular importance in an assessment framework because they provide a platform for the integration of knowledge from various perspectives and comparisons of various design scenarios (Chen and Ng, 2011). These "predicting tools" can support the research on how changes in design details influence outdoor thermal comfort. The statement applies with equal force to the more general context of research in this area, which is how urban design can influence the microclimate of an urban environment and people's outdoor thermal comfort and, in turn, how people's thermal comfort can influence their use of outdoor urban spaces. Design regulations and guidelines in this respect require comprehensive assessment before they are adopted. City planners and decision makers, when faced with the task of designing urban spaces that are desirable and thus used rather than abandoned, will be better informed with a predicting tool that allows various design alternatives to be compared and tested in terms of attractiveness and effectiveness. In particular, a testing tool is needed that can provide both quantitative and qualitative understanding of the relationships among microclimatic environment, subjective thermal assessment, and social behavior. Such a tool should have the ability to process detailed environmental information according to time and location variations and to generate analytical results to reveal the relationship.

To simulate the cooling potential of greenery in the urban setting of Peristeri, ENVI-met,[67] an environment and micro-climatic simulation

[67] The ENVI-met software uses input values for buildings, vegetation, ground surfaces, climatic conditions, soils, and then simulates the modifications from the proposed building form, additional shading, alternative orientations, etc. ENVI-met is a three-dimensional computer model, which analyzes micro-scale thermal interactions within urban environments. The software uses both the calculation of fluid dynamics characteristics, such as air flow and turbulence, as well as the thermodynamic processes taking place at the ground surface, at walls, at roofs and at plants. It calculates the dynamics of microclimate during a diurnal cycle [24–48 h]. Main prognostic variables of the program are wind speed and

(*Continued*)

software, has been used. The main goal here is the investigation of microclimatic variations linked to modified parameters and their impact on buildings. The focus of the analysis is, in particular, on the possible replacement of parking areas with green parks and pedestrian streets, of standard roofs with green ones, of vertical walls to green ones, and of the creation of roof shading systems with the use of a photovoltaic system.

A packet of simulations has been modeled for this elaboration, based on reliable standards from the urban compound of the Peristeri area. Variations in the models concern the replacement of standard roofs with green ones (from 0 to 100%) and the reorganization of the car circulation in order to leave open asphalt areas that become, in a second step, permeable and planted green surfaces. In addition, one of the most important steps for the simulation of the new scenario is the creation of

direction, air temperature and humidity, turbulence, radiative fluxes, bioclimatology and gas and particle dispersion. The basic data structure of ENVI-met is described by Ozkeresteci et al. (2003). ENVI-met takes into account all types of solar radiation (direct, reflected, and diffused) and calculates the mean radiant temperature. The calculation of radiative fluxes includes the plant shading, absorption and shielding of radiation as well as the reradiation from other plant layers. ENVI-met has two basic steps before the simulation is run. The first is editing the input of the urban area to be tested. For this task, one needs the horizontal and vertical dimensions of the architectural environment along with any specific design features such as open breezeways, overhangs, horizontal surface materials, ground cover, vegetation size and coverage, and so forth. The input is designed in a 3D setting where the buildings, trees/vegetation, and the various surfaces are placed. These elements are represented by various size grid cells. The smaller the cell is, the finer the resolution (as small as 0.5 m). The cell area can be designated at any dimension from 0.5 to 10 m. For example, a 100×100 m^2 area can be represented in a 100×100 grid cells of 1×1 m^2 each, or it can be represented by a 20×20 grid cells of 5×5 m^2 each, depending on the size of the test area and the desired resolution. This stage can be quite complex depending on the environmental domain that is chosen to work. The high-resolution aspects of the program enable the user to go into finer details in smaller scale. The second step is editing the configuration file, where the information about the site location, temperature, wind speed, humidity, PMV parameters, and databases for soil types and vegetation are entered. The simulation is then processed using both the input and configuration files. The output data then has to be imported and visualized in LEONARDO 3.0. ENVI-met outputs binary files (.EDI/.EDT) which have to be imported into a visualization program – LEONARDO 3.0. Each output file has a multitude of information, which has to be translated into different layers in Leonardo. The following main layers are typically used to visualize the output data: Data layer: Displays continuous data (eg, temperatures); Special: Displays singular data (eg, buildings, plants); Vector: Displays vectors such as wind. Visualization is then configured to display the urban environment with the desired section of data cut horizontally (plan view), vertically (section view), or in a 3-D axonometric view. ENVI-met however has certain limitations. The building blocks have no thermal mass.

roof shading (from 0 to 100%). Investigations are focused on the thermal variation resulting at the scale of the urban contest. Introducing parks instead of asphalt areas leads to fresher air (temperature decreases generally in a range set from 0.25 to 1.5°C, locally the difference can be 3°C). The replacement of standard roofs with green ones does not lead to significant variations in temperatures; the maximum decrease is of 0.7°C). These conclusions focus attention on the interactions between green area and air temperature, but also look at the impacts of these variations on project choices at and the urban building level.

2.3.4.2.1 Peristeri Urban Setting: Measured Climatic Data and Simulation on the Scenario as Built (Annarita Ferrante and Anastasia Fotopoulou)

Peristeri is located in the western part of Athens, as this area along with the center presents the highest heat island (HI) intensity of the whole city. The blocks are typical social housing from the 1960s constructed by pillars and beams, and the basic structure material, concrete, was used. Its parts consist of: the three tower buildings, the three double block buildings south-north oriented, and four east-west oriented blocks. Focusing on urban areas, the proposed research considers the buildings and the related space as a whole. The urban compound extension is 37,820 m^2, of which 25,713 (68%) is occupied by building construction and impermeable surfaces (parking areas and streets); the remaining (32%) is a green space, or open area. The built area represents the 29% (7,504 m^2) of the total impermeable surfaces.

In order to achieve the goal of zero energy, a strategy of five basic steps has been followed:

- Selection, analysis, and simulation with ENVI-met of a characteristic urban compound
- Selection of different urban sustainable design scenarios, with new simulations
- A comparison of the results and selection of the best possible scenario for the improvement of the HI effect and the microclimate in the urban compound
- Simulation with "Energy Building" to check how much influence the improvement of the microclimate has on the existing buildings
- Analysis and collection of the buildings' energy-performance results after the redesign of the urban compound and the improvement of the microclimate

For the simulations, the area of interest has been transformed into a model grid with the dimensions 130 × 120 × 30, with a resolution of

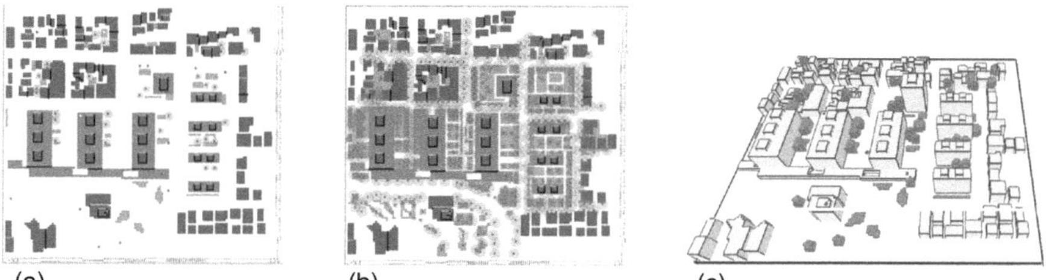

Figure 2.56 ***The Peristeri urban compound in the ENVI-met 2D and 3D interpretation in the as-built and alternative scenario.*** (a) Existing situation (b) Proposal (c) 3D map.

2 m × 2 m × 3 m, resulting in a total area of 260 × 240 m in the horizontal extension (see Fig. 2.56).[68]

These databases (soils, plants, buildings, surface materials) are used as generic types in the first stage, so a comparative match to the closest resemblances are chosen for soil, plant, and building characteristics. These data can also be localized and additional plants, soils, and source information can be added to the model. The outputs are basic 3D information on atmosphere, surface, and fluxes and soils. The meteorological input for a one-year period (2014) recorded by the urban climate station in Peristeri plus measurements performed by the University of Athens Department of Physics (Santamouris, 2014) was applied. For the simulation, the hottest day of the summer of 2014 (9th of July) was selected, with a mean temperature of 30°C, while the maximum was 35.7°C. The wind was in a westerly direction with an average speed of 3.1 m/s. From the previous analysis, the simulation results showed drastic measurements and the need for action. The lack of green spaces, lack of parking, lack of pedestrian streets, the high temperatures measured at the ground surface as well the high air temperature, plus the indication of uncomfortable thermal sensation are some of the problematic issues that the urban planners must consider.

The next proposed step is the analysis and simulation of alternative sustainable scenarios that could help improve the existing situation and come a step closer to the goal of this research, which is zero-energy urban settings. In brief, these actions are:

- Addition of trees and green surfaces on the ground level
- Green roof and green walls
- Shading system with photovoltaic elements on the rooftops

[68] The digital map data for the area was obtained by Google earth but additional surveying took place to correct changes. ENVI-met uses its own graphic interface to plot the layout of the area, and its own configuration editor for the climatic data plus the plants, soils, and sources databases. Therefore all cartographic information had to be gathered on a digital map and redrawn in ENVI-met graphic editor for the modeling language.

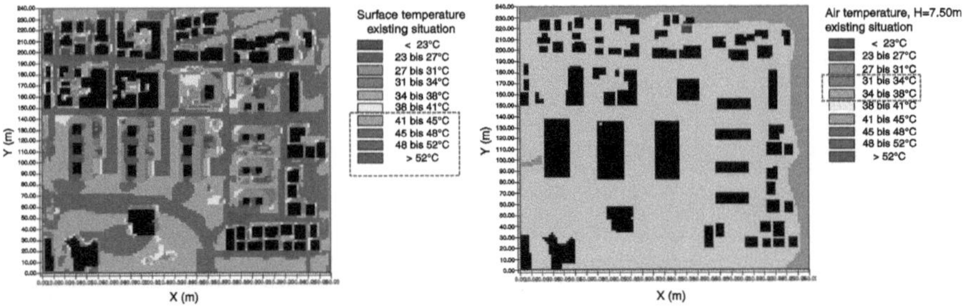

Figure 2.57 ***Comparison between ENVI-met output of surface ground temperature (on the left) and the air temperature at 7.5 m height in the existing situation at 14.00 pm.***

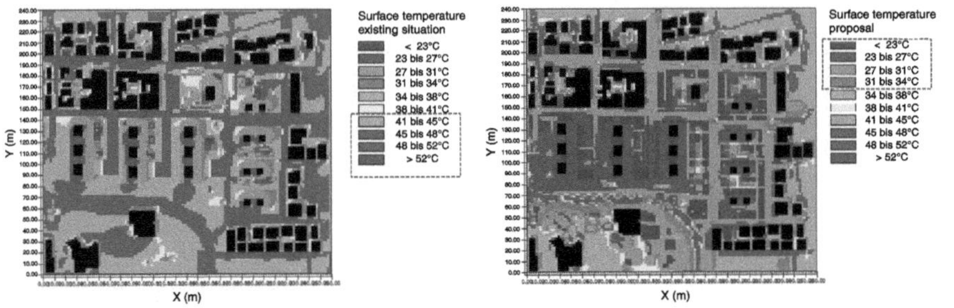

Figure 2.58 ***ENVI-met output of surface ground temperature in the existing situation (on the left) and for the designed scenario at 14.00 pm.***

- Reorganization of the parking and creation of pedestrian permeable streets
- Linear park (green roof) on the existing roof of the market

The research conducted consisted of two brief studies of the area. Figs. 2.57–2.61 visualize the model of the area in different conditions.[69] This stage is used to explore calculation and visualization potentials of the model plus it also helped the researcher to hypothesize on certain factors of the model itself and on certain characteristics of the urban compound.[70]

[69] Data of external temperature in the model of the as it is condition have been verified with measured data by the team GRBES (Prof Santamouris).

[70] The processes of working with the models in the research are described by Ozkeresteci et al. (2003). These are summarized as follows.•Model shows a fine detail in the analysis of the atmospheric data form the user's point of view.Model shows a fine detail in the analysis of the atmospheric data form the user's point of view.•It provides area-based information about critical areas in the study domain like very hot sports, problematic areas.•It plots in 3D and 4D the path and the direction of the wind movement and particle diffusion around the landscape elements therefore critical areas of polluted spots in the study domain.•The previous inferences are critical design and planning data either from an urban designer's point of view preparing a development project in the area or from the point of a planner's working in the city reviewing and communicating the development in different stages.

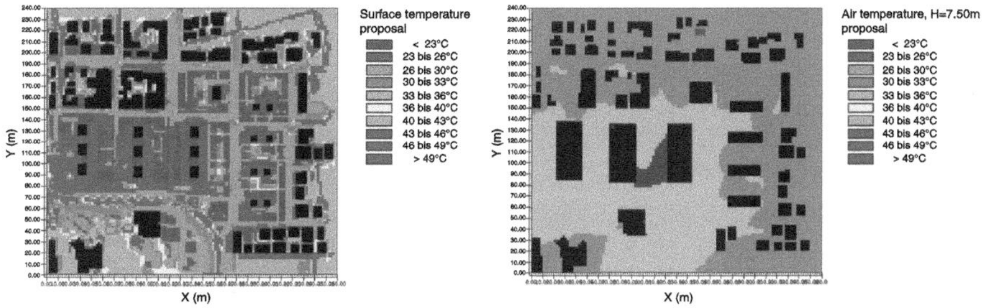

Figure 2.59 ***ENVI-met output of surface ground temperature (on the left) and the air temperature in the proposed scenario at 14.00 pm.***

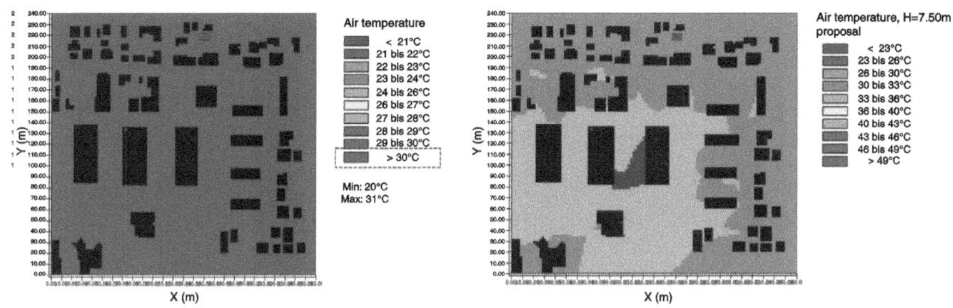

Figure 2.60 ***ENVI-met output of air temperature in the as-it-is scenario (on the left) and the air temperature in the proposed scenario at 14.00 pm.***

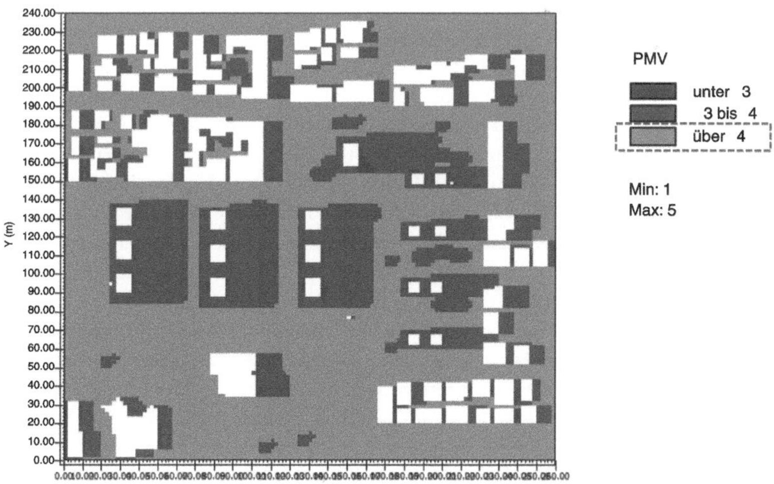

Figure 2.61 ***Thermal comfort interpretation in the existing area at 16.00 pm.***

For the environmental simulation with ENVI-met, the same climatic data and maps were taken into account. To compare and discuss in further detail here, the results of the simulations are provided for four of the five scenarios: green elements and trees, green walls and roofs, pedestrian streets, and design of a linear park on the roof level of the market. Variations imply a different microclimatic behavior in the models. Introducing parks and pedestrian streets in place of asphalted areas led to fresher air (temperature decreases generally in a range from 0.25 to 1.5°C; locally, the difference can be as much as 4°C). Results show consistencies with the previous studies discussed at the beginning of this section (Figs. 2.62).

Selection of trees and vegetation should be made according to the warm, dry climate of Athens. Plantings with big crowns provide adequate shade while requiring less water. The replacement of standard roofs with green roofs does not seem to lead to significant variations in temperatures, as maximum decrease is only equal to 0.7°C; conversely, measurements performed locally, close to one of the buildings, with a shading system on the rooftop in place, show a significant decrease (about 2.5°C).

Furthermore, by using the simulation tool it is also possible to define the best area of thermal comfort.[71] A method of describing thermal comfort

[71] According to the ANSI/ASHRAE Standard 55-2010, thermal comfort is defined as "that condition of mind which expresses satisfaction with the thermal environment and is assessed by subjective evaluation." Also known as human comfort, thermal comfort is the occupants' satisfaction with the surrounding thermal conditions and is essential to consider when designing a structure that will be occupied by people. A cold sensation will be pleasing when the body is overheated, but unpleasant when the core is already cold. At the same time, the temperature of the skin is not uniform on all areas of the body. There are variations in different parts of the body, which reflect the variations in blood flow and subcutaneous fat. The isolation potential and quality of clothing also has a marked effect on the level and distribution of skin temperature. Thus, sensation from any particular part of the skin will depend on time, location and clothing, as well as the temperature of the surroundings.

There are six factors to take into consideration when designing for thermal comfort. Its determining factors include the following: Metabolic rate (met): The energy generated from the human body; Clothing insulation (clo): The amount of thermal insulation the person is wearing; Air temperature: Temperature of the air surrounding the occupant; Radiant temperature: The weighted average of all the temperatures from surfaces surrounding an occupant; Air velocity: Rate of air movement given distance over time; Relative humidity: Percentage of water vapor in the air. The environmental factors include temperature, radiant temperature, relative humidity, and air velocity. The personal factors are activity level (metabolic rate) and clothing. Thermal comfort is calculated as a heat transfer energy balance. Heat transfer through radiation, convection, and conduction are balanced against the occupant's metabolic rate. The heat transfer occurs between the environment and the human body, which has an area of 19 ft.^{2}. If the heat leaving the occupant is greater than the heat entering the occupant, the thermal perception is "cold." If the heat entering the occupant is greater than the heat leaving the occupant, the thermal perception is "warm" or "hot."

Figure 2.62 ***The urban compound in the existing situation and in the proposed alternative scenario.***

was developed by Ole Fanger and is referred to as predicted mean vote (PMV) and predicted percentage of dissatisfied (PPD). The PMV refers to a thermal scale that runs from cold (−3) to hot (+3), originally developed by Fanger and later adopted as an ISO standard (Table 2.21).[72] Figs. 2.63 and 2.64, show the comparison between the temperatures and PMV of the existing situation versus the transformed scenario.

Actually, in the current as-it-is scenario, shade is indeed provided by existing trees in the center of the courtyards between the buildings. The presence of the trees is assumed to be increased and enhanced by the redesign of

[72] The original data was collected by subjecting a large number of people (reputedly many thousands of Israeli soldiers) to different conditions within a climate chamber and having them select a position on the scale the best described their comfort sensation. A mathematical model of the relationship between all the environmental and physiological factors considered was then derived from the data. The result relates the size thermal comfort factors to each other through heat balance principles and produces the following sensation scale.

Table 2.21 Predicted mean vote sensation scale

Value	Sensation
−3	Cold
−2	Cool
−1	Slightly cool
0	Neutral
+1	Slightly warm
+2	Warm
+3	Hot

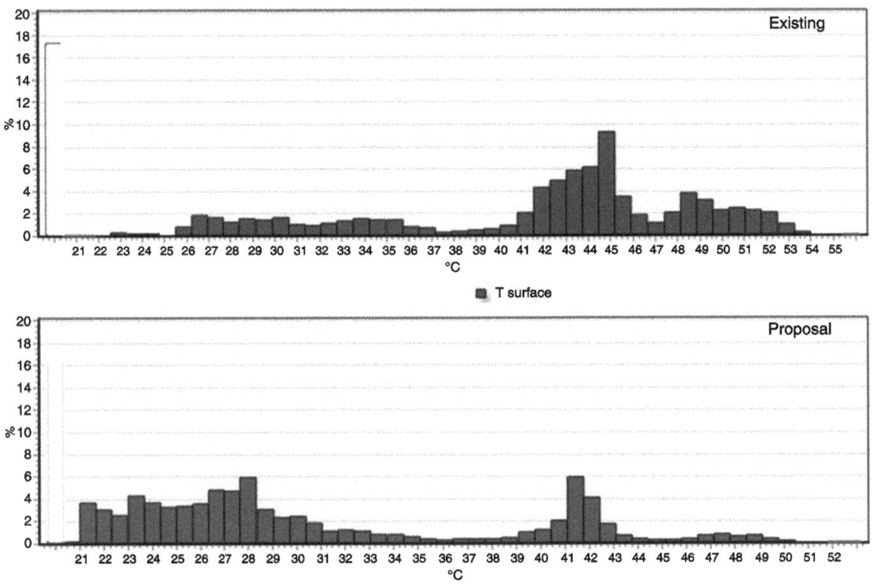

Figure 2.63 ***Temperature comparison of the existing situation and the alternative scenario.***

external surfaces, increasing the permeable surfaces and reducing the asphalt area as much as possible. A great impact in local decrease in temperature is brought about by the linear park and the shading of the new plantings in that area. There is a significant temperature reduction on the surface of the ground level due to the use of new materials and the shading provided by the vegetation. According to the measurements around the central building, the resulting air temperature reduction is between 0.5 and 2.9°C, while a lower reduction is observed to the west and south.

A further step is needed in the effort to understand the potential effects of modifying external conditions and what improvements that will make

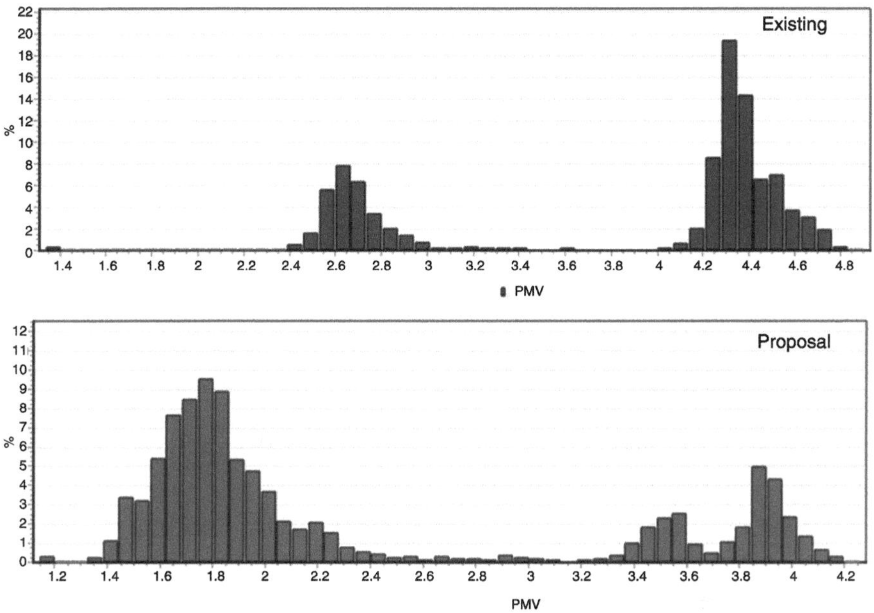

Figure 2.64 ***PMV comparison of the existing situation and the alternative scenario.***

on the energy performance of existing buildings. To provide an idea of the potential energy performance improvements in this urban compound, taking into account the mutual interactions between buildings and open areas, further simulations have been run.

The new climatic data of the area simulated by ENVI-met have been used in conjunction with the simulation by Design Builder (IWEC climatic data file), allowing new input for the re-calculation of the energy performance of the building. Fig. 2.65 shows an area of 4 × 4 × 4 m around the main central double block building. A vertical division in five critical points was also performed (ground level, pilotis level, first floor, middle floor, and

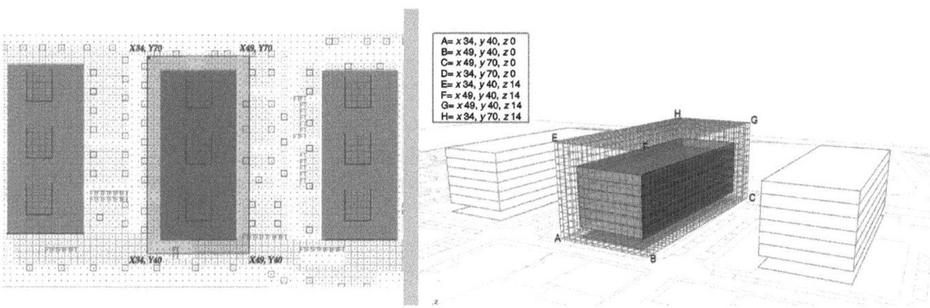

Figure 2.65 ***The defined area around the double block building for the new climatic data.***

last floor including 4 m above the rooftop) so as to subtract results from all the points referred to in these elevations.

The air temperature of these points has been used for the simulation of the corresponding apartments. South-oriented residential apartments at the first, third, and last floors of the double building block B6 of Peristeri have been used as reference units. Simulations have been performed for the summer period, from Jun. 1 until Aug. 31.

The standard climatic data used by Design Builder have been modified in two different and opposite ways. On the one hand, they have been decreased according to the corresponding value of the external temperatures from the ENVI-met calculations; on the other hand, the same data have been increased as a function of measured data in the corresponding meteorological station of Peristeri (Giannopoulou et al., 2011). Results of the energy demand for cooling (kWh) these apartments for the whole summer period are reported in Fig. 2.66. As shown, these results are highly variable as a function of external temperatures (Table 2.22).

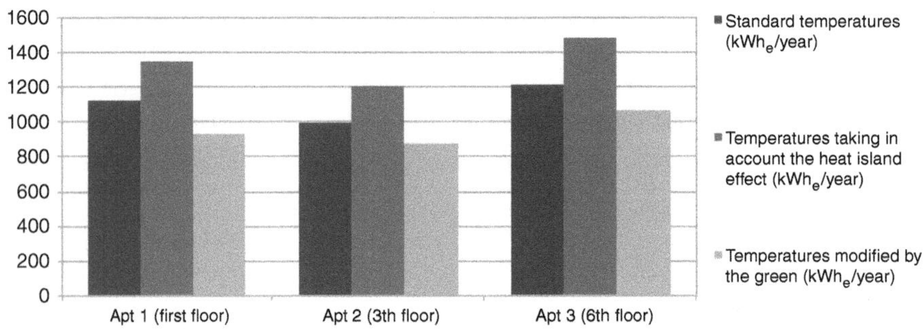

Figure 2.66 ***Cooling demand in three reference apartments of the building block B6 in Peristeri.***

Table 2.22 Cooling demand values in three reference apartments of the building block B6 in Peristeri

	Energy demand for cooling (kWh) (standard air temperatures) (kWh_e/y)	**Energy demand for cooling (kWh) (air temperatures HI effect) (kWh_e/y)**	**Energy demand for cooling (kWh) (temperatures modified by green) (kWh_e/y)**
Apt 1 (first floor)	1122.55	1348.37	926.54
Apt 2 (3th floor)	990.98	1202.45	874.06
Apt 3 (6th floor)	1211.22	1483.49	1067.35

As it can be observed, values of energy demand for cooling the selected apartment on the first floor vary between 1348.37 kWh, which is the worst condition (due to the HI effect), to 1122.55 kWh, which is the best scenario. Note that air temperatures are presumed to be "under standard condition"; the reduction to 926.54 kWh in the improved scenario is due to the cooling effect of the greenery.

Corresponding values of energy demand for cooling the apartment on the intermediate floor vary between 1202.45 kWh (worst condition, a result of the HI effect) to 990.98 kWh (best condition). Again, air temperatures are presumed to be "under standard condition"; the reduction to 874.06 kWh in the improved scenario is due to the cooling effect of the greenery.

Corresponding values of energy demand for cooling the apartment on the top floor vary between 1483.49 kWh (worst condition, a result of the HI effect) to 1211.22 kWh (best condition). As per the previous two examples, the same assumption about air temperatures holds; the reduction to 1067.35 kWh in the improved scenario is due to the cooling effect of the greenery.

A 20% decrease among each of the three scenarios can be seen. Globally, approximately a 40% reduction in energy consumption can be observed, demonstrating the huge potential of environmental modification at the urban scale on the energy performance of existing buildings. A very significant, twofold impact on HI reduction at the urban scale and energy consumption at the building scale can be seen. We have deduced that this is the direct result of shading and greenery in the model.

Results of these simulations are highly consistent with data and corresponding values reported by Santamouris (2014). In fact, as illustrated by Santamouris and colleagues and already reported in Chapter 1, the energy penalty of ambient overheating is quite high and depends mainly on the characteristics of the building stock, the climate zone, the urban form (and its boundary conditions), as well as the type of the energy services provided. A significant increase in the energy consumption of buildings caused by the ambient temperature intensification in metropolitan areas was calculated for the period 1970–2010. In particular, for the Athens case study, a percentage increase of the base electricity load per degree of temperature increase equal to 4.1% was found. In particular, in Athens, Santamouris (2014) found that the cooling load increased from 99.5 kWh/m^2/y in 1970 to 124.8 kWh/m^2/y in 2010, while the corresponding heating load decreased from 39.4 kWh/m^2/y in 1970 to 31.7 kWh/m^2/y in 2010.

The urban HI effect and global warming increase the temperature of metropolitan cities like Athens and exacerbate the energy demand of buildings. In particular, Santamouris (2014) found that energy increase is very significant in cooling dominated zones, where summer energy needs are much higher than the possible decrease of heating needs in winter, as can be observed by the figures reported. Although the available studies comparing the energy consumption of similar buildings located in urban and rural zones are quite limited, existing data reveal that the average increase of the cooling load, because of the HI effect, is statistically significant and averages about 13%. So far, limited studies are available regarding the global energy penalty induced by urban warming on the total building stock of a city.[73]

It can be concluded that in order to reduce the specific impact of urban HI and global overheating on electricity consumption, buildings and urban structures have to be adapted to the specific climatic conditions, taking into consideration the built environment in its different scales, and conceiving the retrofitting options in the building block and in its urban boundary conditions. In fact, energy demand and energy retrofitting options should be observed, quantified, and designed not only with reference to the building blocks and urban textures that are considered to be the "solid" part of the city, but also with regard to the open street environments, thus considering the buildings and the related open spaces as a whole; in fact, energy investigation should be conducted taking into account the cumulative effects between the buildings and the open areas.

The development of low energy or close-to-zero energy buildings may significantly reduce the energy needs and thus the resulting stress to the

[73] As shown by Santamouris (2014), calculations performed for a set of different cities around the world have demonstrated that the global energy impact of higher urban temperatures is very important. The average global energy penalty per unit of city surface is significant (2.4 kWh/m^2), while the average global energy penalty per unit of surface and degree of UHI intensity is close to 0.74 $kWh/m^2/K$. In parallel, the average global energy penalty per person is close to 237 kWh/p and the global energy penalty per person and per degree of the UHI intensity is around to 70 kWh/p/K. These figures consist of preliminary information may change when more data are available. Based on a significant number of studies, it is calculated that the average increase of the cooling demand of representative buildings during that period 1970–2010 was close to 23%. In parallel, the corresponding average reduction of the heating load is around to 19%, while during the same period, the average total energy load of representative buildings spent for heating and cooling purposes increased by 11%. The specific studies and analysis make clear that urban warming has a very significant impact on the global energy consumption of buildings.

utilities and the consumers would also be lessened. In parallel, the development and use of advanced urban adaptation and mitigation techniques and technologies with the potential to decrease the ambient temperatures in the built environments may also considerably reduce urban temperatures.

2.4 ON THE TECHNICAL FEASIBILITY OF NEARLY ZEUS

Achieving a nearly zero energy building in these three urban compounds has been proven to be a technically feasible goal; in particular, a large set of possible solutions from the best available products and components is technically applicable to achieve nZEBs, even in buildings that presently consume a lot of energy. In all the considered cases, to reach zero energy, passive retrofitting interventions and the production of energy from renewable sources are both necessary.

The effective reduction of energy consumption remains a necessary objective that should be combined with alternative uses of open spaces, where water, greenery, and selected passive techniques may foster and accelerate the significant reduction of energy consumption. This is especially important in the Mediterranean urban areas, where the dramatic combination of HI phenomena and global overheating is severe. This chapter has discussed and demonstrated the technical feasibility of these goals, using examples from the urban setting of Athens.

Nonetheless, so far, as reported by several prominent studies, only about 1.2% of European existing buildings are renovated every year.[74] This means that a series of barriers must be overcome to facilitate the adoption of measures to reach the target number of nZEBs. Chapter 3 will discuss these barriers as well as the estimated costs that must be taken into account when proposing energy retrofitting options. It will also address the two following important questions:

- Is technical feasibility associated with economic feasibility when retrofitting existing buildings toward becoming nZEBs?
- To what extent is deep renovation and high transformation of buildings toward becoming nZEBs competitive with conventional retrofitting?

[74] The demolition rate is about 0.1% per year and highly energy efficient new-build reach 1% additions/year. JWG: Toward assisting EU member states on developing long-term strategies for mobilizing investments in building energy renovation (EU EED Article 4), see the Composite Document of the Joint WG of CA EED, CA EPBD and CA RES, Nov. 2013 (http://www.ca-eed.eu/reports/art-4-guidance-document/eed-article-4-assistance-document).

REFERENCES

Asaeda, T., Ca, V.T. 2000. Characteristics of permeable pavement during hot summer weather and impact on the thermal environment. Building Environ. 35, 363–375.

Akbari, H., Pomerantz, M., Taha, H., 2001. Cool surfaces and shade trees to reduce energy use and improve air quality in urban areas. Sol. Energy 70 (3), 295–310.

Bärring, L., Mattson, J.O., Lundqvist, S., 1985. Canyon geometry, street temperatures and urban heat island in Malmö, Sweden. J. Climatol. 5, 433–444.

Berger, M., 2014. The unsustainable city. Sustainability 6 (1), 365–374.

Bettgenhäuser, K., de Vos, R., Grözinger, J., Boermans, T., 2014. Deep renovation of buildings. An effective way to decrease Europe's energy import dependency, Project number: BUIDE14901 © Ecofys 2014.

Bitan, A., 1992. The high climatic quality of the city of future. Atmos. Environ. 26B, 313–329.

Borhan, H., Ahmed, E., Hitam, M., 2013. CO_2, quality of life and economic growth in East Asian 8. J. Asian Behaviour. Studies 3 (8), 13–24.

Bretz, S., Akbari, H., Rosenfeld, A., Taha, H. 1992. Implementation of solar reflective surfaces: materials and utility programs. LBL Report 32467, University of California.

Bruegmann, R., 2005. Sprawl: A Compact History. University of Chicago Press, Chicago, IL.

Caniggia, G. Maffei, G.L., 1984. Il progetto nell'edilizia di base, Marsilio Editore, Venezia.

Chen, L., Ng, E., 2011. Outdoor thermal comfort and outdoor activities: a review of research in the past decade. Cities 29, 118–125.

Chow, W.T.L., Brazel, A.J., 2012. Assessing xeriscaping as a sustainable heat island mitigation approach for a desert city. Build. Environ. 47 (1), 170–181.

Conzen, M.R.G., 1988. Morphogenesis, morphological regions and secular human agency in the historic townscape, as exemplified by Ludlow. In: Denecke, D., Shaw, G. (Eds.), Urban Historical Geography: Recent Progress in Britain and Germany. Cambridge University Press, Cambridge, pp. 253–272.

Conzen, M. R. G., 2004. Thinking about Urban Form: Papers on Urban Morphology, 1932–1988, Peter Lang, Oxford.

Conzen, M.P., 2009. How cities internalize their former urban fringes: a cross-cultural comparison. Urban Morphol. 13 (1), 29–54.

Curdes, G., 2010. Urban morphology and climate change. Which morphology can survive? Proceedings of the 17th Conference and International Seminar on Urban Form, ISUF – "Formation and Persistence of Townscape," Hamburg and Lübeck.

Donadieu, P., 1998. Campagnes Urbaines, Actes Sud, E.N.S.P.

Emmanuel, R., Rosenlund, H., Johansson, E., 2007. Urban shading – a design option for the tropics? A study in Colombo, Sri Lanka. Int. J. Climatol. 27, 1995–2004.

Environmental Protection Agency, 1992. Cooling Our Communities: A Guidebook on Tree Planting and Light Colored Surfacing, Environmental Protection Agency, Washington.

Erell, E., Pearlmutter, D., Williamson, T., 2011. Urban Microclimate: Designing the Spaces Between Buildings. Earthscan, UK.

Eumorfopoulos, E., Aravantinos D., Numerical approach to the contribution of the planted roof to the cooling of buildings, Proceedings of the International Symposium Passive Cooling of Buildings, Jun. 19–20, 1995, Athens, Greece.

Ferrante, A., Santamouris, M., Koronaki, I., Mihalakakou, G., Papanikolau, N. 1998. The design parameters' contribution to improve urban microclimate: an extensive analysis within the frame of POLIS Research Project in Athens, The Joint Meeting of the 2nd EU Conference on Energy Performance and Indoor Climate in Buildings and 3rd International Conference IAQ, Ventilation and Energy Conservation. Lyon, 19-21/11/1998.

Ferrante, A., Mihalakakou, G., 2001. The influence of water, green and selected passive techniques on the rehabilitation of historical industrial buildings in urban areas. J. Sol. Energy 70 (3), 245–253.

Ferrante, A., 2014. Energy retrofit to nearly zero and socio-oriented urban environments in the Mediterranean climate. J. Sustain. Cities Socie. 13 (2014), 237–253.
Fishman, R., 1987. Bourgeois Utopias: The Rise and Fall of Suburbia. Basic Books, New York.
Flavin, Lenssen, 1995. Power Surge: A Guide to the Coming Energy Revolution. Earthscan, London.
Gago, E.J., Roldan, J., Pacheco-Torres, R., Ordóñez, J., 2013. The city and urban heat islands: a review of strategies to mitigate adverse effects. Renew. Sust. Energy Rev. 25 (0), 749–758.
Gaitani, N., Mihalakakou, G., Santamouris, M., 2007. On the use of bioclimatic architecture principles in order to improve thermal comfort conditions in outdoor spaces. Building Environ. 42 (1), 317–324.
Garreau, J., 1991. Edge City: Life on the New Frontier. Doubleday, New York.
Giannopoulou, K., Livada, I., Santamouris, M., Saliari, M., Assimakopoulos, M., Caouris, Y.G., 2011. On the characteristics of the summer urban heat island in Athens, Greece. Sustain. Cities Soc. 1, 16–28.
Grimmond, C.S.B., Roth, M., Oke, T.R., Au, Y.C., Best, M., Betts, R., Carmichael, G., Cleugh, H., Dabberdt, W., Emmanuel, R., Freitas, E., Fortuniak, K., Hanna, S., Klein, P., Kalkste in, L.S., Liu, C.H., Nickson, A., Pearlmutter, D., Sailor, D., Voog, J., 2010. Climate & more sustainable cities: climate information for improved planning and management of cities (producers/capabilities perspective) world climate conference – 3 better climate information for a better future. Procedia Environ. Sci. 1, 247–274.
Hader, F., 1970. The climatic influence of Green Areas, their property as air filters and noise abatement Agents. Climatol. Buildings. Commitee Internationale de Batiment, Wien.
Hayden, D., 2003. Building Suburbia: Greenfields and Urban Growth, 1820–2000. Pantheon, New York.
Hillier, B., 1996. Space is the Machine. Cambridge University Press, Cambridge.
Höppe, P., 1999. The physiological equivalent temperature – a universal index for the biometeorological assessment of the thermal environment. Int. J. Biometeorol. 43, 71–75.
Huang, Y.J., Akbari, H., Taha, H.G., Rosenfeld, A.H. 1986. The potential of vegetation in reducing summer cooling loads in residential buildings. LBL Report 21291, Berkeley, CA, 1986.
Ichinose, T., Shimodozono, K., Hanaki, K., 1999. Impact of anthropogenic heat on urban climate in Tokyo. Atmos. Environ. 33, 3897–3909.
Johansson, E., 2006. Influence of urban geometry on outdoor thermal comfort in a hot dry climate: a study in Fez, Morocco. Build. Environ. 41, 1326–1338.
Johnson, G.T, Oke, T.R., Lyons, T.J., Steyn, D.G., Watson, I.D., 1991. Simulation of surface urban heat islands under ideal conditions at night. Boundary layer meteorology 56, 275–294.
Karlessi, T., Santamouris, M., Apostolakis, K., Synnefa, A., Livada, I., 2009. Development and testing of thermo chromic coatings for buildings and urban structures. Solar Energy 83 (4), 538–551.
Kolokotroni, M., Giannitsaris, I., Watkins, R., 2006. The effect of the London urban heat island on building summer cooling demand and night ventilation strategies. Sol. Energy 80 (4), 383–392.
Lang, E., 2003. Edgeless Cities: Exploring the Elusive Metropolis. Brookings Institution Press, Washington, DC.
Lin, T.P., 2009. Thermal perception, adaptation and attendance in a public square in hot and humid regions. Build. Environ. 44, 2017–2026.
Lin, T.P., Ho, Y.F., Huang, Y.S., 2007. Seasonal effect of pavement on outdoor thermal environments in subtropical Taiwan. Build. Environ. 42, 4124–4131.
Mansor, M., Said, I., Mohamad, I., 2010. Experiential contacts with green infrastructure's diversity and wellbeing of urban community. Asia. J. Environ.-Behav. Studies 1 (2), 33–48.

Mozas, J., Fernandez, P., 2004. Densidad/Density, Nueva Vivienda Colectiva, a + t ediciones, Vitoria Ga-Gasteiz.
Nikolopoulou, M., Baker, N., Steemers, K., 2001. Thermal comfort in outdoor urban spaces: understanding the human parameter. Sol. Energy 70 (3), 227–235.
Oke, T.R., 1976. The distinction between canopy and boundary layer urban heat islands. Atmosphere 14, 268–277.
Ozkeresteci, I., Crewe, K., Brazel, A.J., Bruse, M., 2003. Use and evaluation of the envi-met model for environmental design and planning: an experiment on linear parks. Proceedings of the Twenty First International Cartographic Conference (ICC). Aug. 10–16, 2003, Durban, South Africa, Cartographic Renaissance. Hosted by The International Cartographic Association (ICA) ISBN: 0-958-46093-0.
Phelps, N.A., Wood, A.M., Valler, D.C., 2010. A post-suburban world? An outline of a research agenda. Environ. Plan. A 42, 366–383.
Reale, L., 2008. Densità, città, residenza. Tecniche di densificazione e strategie anti-sprawl. Gangemi Editore, Rome.
Ricciardelli, F., Polimeno, S., 2006. Some characteristics of the wind flow in the lower Urban Boundary Layer. J. Wind Eng. Ind. Aerodyn. 94, 815–832.
Robitu, M., Musy, M., Inard, C., Groleau, D., 2006. Modeling the influence of vegetation and water pond on urban microclimate. Sol. Energy 80, 435–447.
Santamouris, M. (Ed.), 2001. Energy and Climate in the Urban Built Environment. James and James Science Publishers, London.
Santamouris, M., 2006. Environmental Design of Urban Buildings: An Integrated Approach. Earthscan, UK.
Santamouris, M., 2014. On the energy impact of urban heat island and global warming on buildings. Energy Build., 100–113.
Segal, R., Verbakel, E. (Eds.), 2008. Urbanism without Density: Cities of Dispersal. Wiley, West Sussex.
Shahidan, M.F., Jones, P.J., Gwilliam, J., Salleh, E., 2012. An evaluation of outdoor and building environment cooling achieved through combination modification of trees with ground materials. Build. Environ. 58 (0), 245–257.
Shnapp. S., Sitjà, R., Laustsen, J., 2013. What Is a Deep Renovation Definition? Global Buildings Performance Network, Technical Report, Feb. 2013, info@gbpn.org, www.gbpn.org, @GBPNetwork.
Steemers, K.A., 1993. Environmental Urban Planning, Proceedings of the International Conference on Solar Energy in Architecture and Urban Planning, Florence, Italy, pp. 505–510.
Stirling, R., Carmody, J., Elnicky, G., 1981. Earth Sheltered Community Design: Energy Efficient Residential Development. Van Nostrand Reinhold Company, New York.
Synnefa, A., Santamouris, M., Apostolakis, K., 2007. On the development, optical properties and thermal performance of cool coloured coatings for the urban environment. Solar Energy 81, 488–497.
Tapias, E., Schmitt, G., 2014. Climate-Sensitive Urban Growth: Outdoor Thermal Comfort as an Indicator for the Design of Urban Spaces. WIT Transactions on Ecology and the Environment, Vol. 191, www.witpress.com, ISSN 1743-3541 (on-line), WIT Press.
Whitehand, J.W.R., 2009. The structure of urban landscapes: strengthening research and practice. Urban Morphol. 13 (1), 5–27.
Yap, D., 1975. Seasonal excess urban energy and the nocturnal heat island – Toronto. Arch. Meteorol. Geoph. Bioklima. B23, 68–80.
Zinzi, M., 2010. Cool materials and cool roofs: Potentialities in Mediterranean buildings. Adv. Building Energy Res. B, 201–266.

CHAPTER 3

Understanding Economic Constraints to Achieve Nearly ZEUS

3.1 OVERCOMING EXISTING CONSTRAINTS TOWARD ZEUS

To reach the ambitious target of nZEBs (EPBD recast 2010/31/EU), the technical feasibility in general is not sufficient to widely spread nZEB into building current practice, especially considering the quality and the amount of existing building stock, which represents a huge potential in terms of carbon reduction. As a matter of fact, the representative case studies selected for this work appear to be consistent with the quality of the existing building stock at the European level, made up in the larger majority, from about 60% of buildings built after World War II and between the 1960s and the 1980s[1] (see Fig. 3.1). In fact, the current age of existing building stock derives from the political and social events that occurred just after the WW II.[2] This percentage further increases (about 70–75%) if we confine the analysis within

[1] In fact, a high number of quite drastic measures have been undertaken to vastly and quickly combat the great post-war housing shortage in the EU countries. These measures were mainly conceived through standardization of plans and building elements, as well as setting up of a central organization for the distribution of building material and labor and the creation of a building industry connected to the industrialization. This happened identically even if with different technologies and industries involved, between the two wars and after 1945. In particular, the social housing stock has been conceived everywhere in Europe according to the modernist idea of the minimal residential units. This solution has been then applied – with few considerations on the context – all over Europe, often resulting in housing complexes that are now facing very similar problems: small dwellings, mono-functional districts, concentration of weak minorities and immigrants, low quality of public spaces. This monotony of endless rows of identical houses is threatening the identity of urban dwellers and inhabitants. On the other side, the industrialized system applied for rapid building edification was without any doubt effective and has proven its potential, rapidity, and capacity of serving the purpose of providing a shelter for everyone throughout the decades until the present time.

[2] This was not only a problem related to the post-war reconstruction but also to the emergency caused by the phenomena of internal migration. In many Mediterranean countries

(*Continued*)

http://dx.doi.org/10.1016/B978-0-08-100735-8.00003-9

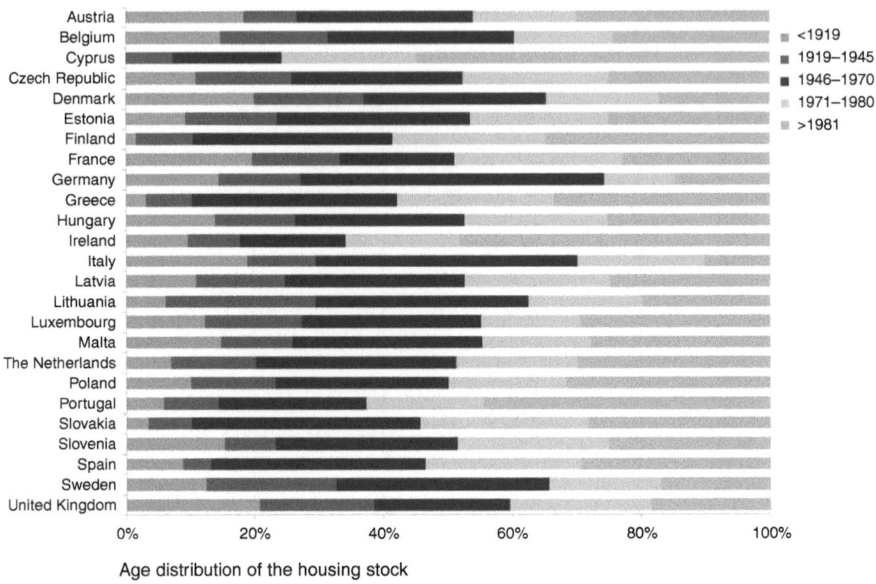

Figure 3.1 ***The age distribution of EU existing building stock shows that the larger majority of buildings (about the 60%) have been constructed after World War II (1960s and 1980s).*** Housing Statistics in the European Union, 2010. *(Source: Boverket and MMR, 2005).*

the boundaries of the Mediterranean European countries (Greece, Cyprus, Spain, Portugal, etc.) and slightly increases for Italy, as well (about 65%).

This means that the majority of existing buildings in the European Union (EU) have been built well before the design and application of

of the current European Union like Italy, France, and Spain the extent of the migration from the south to the north, the return of a large number of citizens from the colonies and the migration from the countryside to the cities, resulted in a consistent request for housing which became national issues in these countries, causing legislation answers, dedicated urban plans and several special measures (ie, Fanfani Plan, Gescal, Casa Plan, etc.). The objective of those actions was to combine the interests of developers and construction companies to generate cheap and fast-to-build dwellings that could be rented or sold to the inhabitants or allocated upon social conditions. In the Eastern Europe countries the situation was similar: the organization of socialist states after the defeat of the Axis powers, centralization of planning and powers according to the soviet model lead to the idea that housing, as social good, is a good of collective consumption and the Institutions had to provide for their construction, favoring multifamily homes organized in large settlements. Architects and urban planners tried to find the solution to the housing emergency, reinterpreting the principles of the modern movement and its results from the experiences carried between the two wars. The increase in the Housing production created a vast and fairly uniform patrimony of buildings that are now facing similar conditions.

regulatory measures on energy consumption reduction, because the energy requirements applicable to housing before 1995 were absent or very limited; as a result, the existing building stock across EU presents very low standard energy performance (Housing Statistics in the European Union, 2010). In fact, an awareness regarding energy consumption related to buildings' performances arose only after the energy and oil crises in the 1970s: these buildings that were once built to answer to the housing emergency represent now the largest energetic emergency. In fact, from illustrated figures, almost two-thirds of the buildings that make up the European real estate assets have been designed in total absence of specific regulations and with very few considerations on energy efficiency. As a result, the existing buildings' stock is responsible for over 40% of the total European energy consumption, of which residential use represents two-thirds of the total building sector's consumption; this data is not surprising looking at the percentage of elements, within the building envelope, lacking insulation, and or the necessary requirements.

Furthermore, housing in the EU represents a huge part of the building stock. EU dwelling stock accounts for about 200 million units, representing around 27% of energy consumption in the EU: the potential reduction in CO_2 emissions that energy-efficient housing would provide cannot be underestimated. But, of Europe's existing buildings, only about 1.2% per year are renovated.[3]Nonetheless, research studies and analysis performed in different EU contexts have shown wide margins of energy efficiency (EE) in retrofitting through standard technical interventions such as renewal of plants or U-value reduction of building components (Ferrante et al., 2011).

Within the frame of social housing sector, many experiences have highlighted the coincidence between low-quality and high-energy management costs, often associated with the deterioration of both buildings and urban

[3] JWG: Towards assisting EU Member States on developing long term strategies for mobilising investment in building energy renovation (per EU Energy Efficiency Directive Article 4), Composite Document of the Joint Working Group of CA EED, CA EPBD, and CA RES, Nov. 2013 (http://www.ca-eed.eu/reports/art-4-guidance-document/eed-article-4-assistance-document). A survey of social housing providers across EU (Cecodhas, 2013) has identified five key categories of barriers (economic, technical, credibility, social, legislative) in delivering new construction and retrofit to nZEB standards; in particular, among others, the lack of access to available and affordable finance to retrofit existing stock toward nZEBs is one of the major barriers.

contexts (Santamouris et al., 2007). Thus, promoting energy improvement measures for the housing stock is an urgent and challenging objective.

3.1.1 Is Technical Feasibility Associated with the Economic Feasibility in ZEUS? Important Economic Barriers for ZEUS

To respond to this important question, cost analyses and simple payback times in the different cases have been calculated. Generally, the basic costs to be considered in a preliminary analysis of the economical feasibility of retrofitting actions are:

- Current prices of energy for heating and cooling demand
- Current prices for electric appliances
- Costs of designed retrofitting interventions

In this work, to avoid excessive fragmentation of data and achieve comparable results, consumptions are assessed under the same plant systems' conditions and the cost for consumption was estimated applying the average energy tariffs in force in Greece for domestic residential users, currently 0,102 €/kWh for gas and 0,145 €/kWh for electricity. The electrical equipment in the buildings is assumed close to 35 kWh/m^2 per year.[4] A summary evaluation of the costs of the various interventions has been calculated for each case study in order to be able to compare the cost of the refurbishment with the savings and payback time of the initial investment.[5]

General assumptions have been made for costs for the retrofitting options in the possible scenarios:

Scenario 1: Wall insulation

Scenario 2: Windows' replacement with high performing glazing components

Scenario 3: (3= 1 + 2): Combined scenario

Scenario 4: Roof insulation

Scenario 5: Roof insulation and green roof

[4] While the electrical need to meet thermal demand can be reduced, there are still other demands to account for and the electrical equipment in the building could easily double the electrical demand. An equipment demand of 35 kWh/m^2 of floor area per annum is not unusual (Dunster et al., 2009). This estimation is consistent with other studies' evaluations reporting an annual consumption of about 4000 kWh for a family consisting of four persons.

[5] In all performed calculation simple payback period in capital budgeting refers to the period of time required to recoup the funds expended in an investment, or to reach the break-even point. The time value of money is not taken into account.

Table 3.1 Mean costs assumed for each retrofitting option

Retrofitting options	Cost per unit EUR
Wall insulation (per square meters)	50.00 – 110.00
Windows' replacement with high performing glazing components, including costs for removal (each window)	250.00
Roof insulation and waterproofing (per square meters)	60.00
Insulated green roof (per square meters)	125.00
New plants (per square meters of net residential floor surface)	30.00
Energy generation by PV panels per kW_p	1,800.00

Scenario 6: (6= 3 + 5): Combined retrofitted scenario

Scenario 7: New plant system associated with scenario 6

Scenario 8: Energy generation by RES based on scenario 7

Scenario 9: Complete retrofitting scenario to nZEB

To identify the economical feasibility of the energy retrofit scenarios, the cost-benefit analysis was conducted by means of a market survey, determining, for each different design option, the evaluation in terms of energy performance improvement and the related cost estimations.

The general assumptions are summed up in Table 3.1

3.1.1.1 Cost-Benefit Analysis in Aghia Varvara

Considering these assumptions, for each proposed intervention for the Aghia Varvara case study, the relative cost and the consequent savings have been reported.

Table 3.2 shows the different retrofitting options in the Aghia Varvara urban setting. Results show that a significant decrease in terms of cost of energy may be achieved.[6] The increasing progression of hypothetical retrofit options show that some interventions, if taken alone, do not significantly affect the energy savings and the consequent energy costs. This is the case, for example, of the insulated green roof. Here, the benefits to the contribution of this solution to the heat island (HI) reduction and its overall environmental benefits have been disregarded by the performed calculations, expressly related to the building's energy demand; in this calculation, the energy performance improvement appears to be similar

[6] The calculated consumption does not include the consumption for domestic appliances, but only the one for electric energy for cooling and for gasoil for heating.

Table 3.2 Costs and simple payback time for the different retrofitting options in the Aghia Varvara urban setting

		Retrofitting option cost (EUR)	Annual cost of thermal energy (EUR)	Annual cost of electric energy (EUR)	Total annual cost for total energy consumption (EUR)	Payback time (years)
	As built	0	9,465	5,391	14,864	-
1	Wall insulation	47,122	7,683	4,830	12,513	20.0
2	New windows	24,280	8,241	4,545	12,787	11.7
3	1 + 2	71,402	6,459	3,977	10,436	16.1
4	Roof insulation	9,150	8,685	5,023	13,708	7.9
5	Insulated green roof	18,471	8,798	5,023	13,821	17.7
6	1 + 2 + 5	89,873	5,791	3,602	9,394	16.4
7	Plants systems	27,840	0	4,190	4,190	2.6
8	Renewable plants (PV)	122,400	0	0	0	8.2
9	nZEB-nZEUS = 6+7+8	240113	0	0	0	16.2

to that obtained by the roof insulation. The cost of the green roof is higher when compared to the cost of the simple insulation; therefore, the payback time in the green roof option is longer. Nonetheless, since the extension of roof surfaces in building blocks and in densely built urban compounds is usually limited with respect to the amount of surfaces of other building components like walls or windows, the up-front costs remain acceptable in absolute value.

Performed evaluations on cost analysis in Aghia Varvara essentially show a long payback time of the different retrofitting scenarios: the cost-benefit assessments of these interventions vary from 8 to up to 20 years. New plant systems associated with renewable energy systems (RES) given by the photovoltaic installations have been evaluated as the consequent options of the complete retrofit scenario modification. In particular, results show the high costs for photovoltaic (PV) panels to set to zero the energy balance of the building. Nonetheless, by coupling the RES in a complete retrofitting scenario, the payback time becomes shorter, while achieving almost null costs for the total energy consumption.

In this, as in all the urban settings, results are oriented at achieving net ZEB, net zero energy building (Net ZEB), conceived as a grid-connected

Table 3.3 Costs and simple payback time for the different retrofitting options in Peristeri urban setting

		Retrofitting option cost (EUR)	Annual cost of thermal energy (EUR)	Annual cost of electric energy (EUR)	Total annual cost for total energy consumption (EUR)	Payback time (years)
	As built	0	371,849	192,284	564,133	
1	Wall insulation	2,893,434	221,512	111,039	332,551	12.5
2	New windows	998,295	337,066	180,935	518,001	21.6
3	1 + 2	4,091,910	196,929	95,000	291,930	15.0
4	Roof insulation	393,129	342,857	179,951	522,808	9.5
5	Green roof	625,263	341,002	179,281	520,283	14.3
6	1 + 2 + 5	4,474,764	170,719	80,657	251,376	14.3
7	Plants systems	1,089,360	0	167,550	167,550	2.7
8	Renewable plants (PV)	2,635,034	0	0	0	4.7
9	nZEB (6 + 7 + 8)	8.199.158	0	0	0	14.5

building or groups of buildings in the whole urban settings that generates as much energy as it uses over a year. The "net zero" balance is attained by applying energy conservation and efficiency measures and by incorporating RES.[7]

3.1.1.2 Cost-Benefit Analysis in Peristeri

Similar cost-effective analyses have been developed for the whole urban setting in Peristeri (Table 3.3). Furthermore, they have also been developed for each building type and for each different scenario in the urban context of Peristeri (Tables 3.4–3.6). Simple payback time of the investments has been put in relation using the synthetic results of these calculations in Fig. 3.2.

Generally, the double building block and the tower building type present the higher payback times in all the retrofitting options; in the tower building, the sole exception to this rule is embodied in the intervention

[7] While based on annual balances, a complete description of a net ZEB should require the detailed analysis of the system at smaller time-scales, evaluating, in example, the relationship between power generation and building loads to evaluate the resulting interaction with the power grid. These detailed aspects are disregarded in the present work, whose aim is a simplified, yet effective, comparison among different contexts and different building types in the aim of ZEUS.

Table 3.4 Costs and simple payback time for the different retrofitting options in the tower building in Peristeri

	Tower Building T11	Retrofitting option cost (EUR)	Annual cost of thermal energy (EUR)	Annual cost of electric energy (EUR)	Total annual cost for total energy consumption (EUR)	Payback time (years)
	As built	0	30,865	15,186	46,051	
1	Wall insulation	236,736	20,062	12,060	32,122	17.0
2	New windows	111,636	25,309	13,400	38,709	15.2
3	1 + 2	348,373	17,902	8,710	26,612	17.9
4	Roof insulation	23,455	27,161	14,070	41,231	4.9
5	Green roof	37,299	26,853	13,846	40,699	7.0
6	1 + 2 + 5	371,595	14,815	6,923	21,738	15.3

Table 3.5 Costs and simple payback time for the different retrofitting options in the double building block B6 in Peristeri

	Double block B6	Retrofitting option cost (EUR)	Annual cost of thermal energy (EUR)	Annual cost of electric energy (EUR)	Total annual cost for total energy consumption (EUR)	Payback time (years)
	As built	0	36,112	10,943	47,055	
1	Wall insulation	318,394	24,692	5,583	30,275	19.0
2	New windows	135,769	32,408	10,496	42,905	32.7
3	1 + 2	454,163	20,988	5,136	26,125	21.7
4	Roof insulation	65,700	33,643	10,273	43,916	20.9
5	Green roof	104,396	33,334	10,273	43,607	30.3
6	1 + 2 + 5	558,559	18,828	4,467	23,294	23.5

Table 3.6 Costs and simple payback time for the different retrofitting options in the building block T7 in Peristeri

	Block T7	Retrofitting option cost (EUR)	Annual cost of thermal energy (EUR)	Annual cost of electric energy (EUR)	Total annual cost for total energy consumption (EUR)	Payback time (years)
	As built	0	28,396	18,983	47,379	
1	Wall insulation	204,674	14,488	9,685	24,173	8.8
2	New windows	42,680	27,237	18,208	45,445	22.1
3	1 + 2	280,717	13,329	8,910	22,239	11.2
4	Roof insulation	20,944	26,657	17,821	44,478	7.2
5	Green roof	33,363	26,657	17,821	44,478	11.5
6	1 + 2 + 5	280,717	11,590	7,748	19,338	10.0

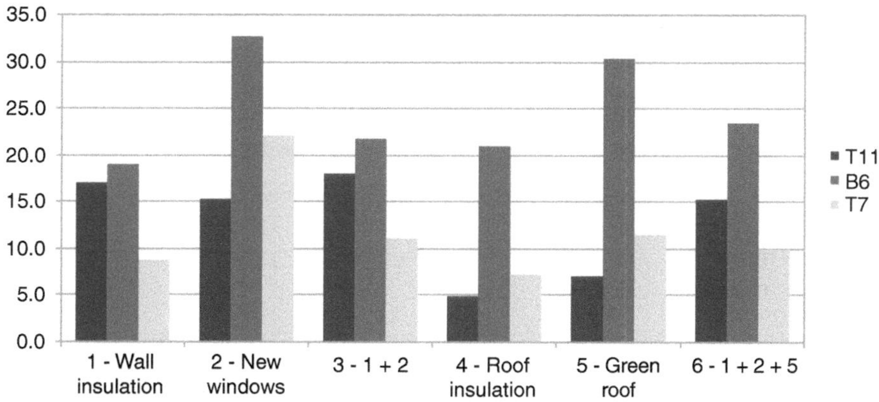

Figure 3.2 ***Payback times of the investments range widely: a variation between 30 and 4 years can be observed as a function of the different hypothesized scenarios in the different building types (B6 = double building block T6= single building block).***

of the roof (scenarios 4 and 5 of the roof insulation and green roof), given the limited extension of the roof surface in this building type. All performed evaluations on cost-benefit assessments of these interventions generally show excessive payback times (from 5 up to 32 years). As shown, variations essentially depend on building type and on the number of glazed surfaces, since window replacement is by far the most expensive retrofitting option.

The costs in Tables 3.4–3.6 do not yet include the installation of RES plant and the plant interventions that are in any case necessary to set to zero the energy balance of the selected buildings. They have been incorporated in Table 3.3, hypothesizing only one PV system network serving the whole urban setting. However, costs and simple payback time calculated in the different retrofitting options of the whole urban compounds in Peristeri and Aghia Varvara (Table 3.3) show how the hypothesis of the complete retrofit scenario equipped by new plant systems and PV sources to set to zero the energy balance can imply a significant reduction in payback times with respect to single retrofitting options and are similar to the complete retrofit scenario 6 while providing null costs for total energy consumption. Nonetheless, the option nZEB (6 + 7 + 8 = 9) requires very high up-front costs. Thus, as for the Aghia Varvara case study, the energy retrofitting in the existing building of Peristeri urban compound is proved to be a technically feasible goal, but it is necessary to face the problem of very high up-front costs, due to the need to apply both active and passive interventions to achieve nZEBs; these high costs are amortized

Table 3.7 Different retrofit options and relative payback time calculated for building 1

	Building 1	Retrofitting option cost (EUR)	Annual cost of thermal energy (EUR)	Annual cost of electric energy (EUR)	Total annual cost for total energy consumption (EUR)	Payback time (years)
	As built	0	6,300	11,206	17,506	
1	Wall insulation	87,218	5,796	10,309	16,106	62.3
2	New windows	61,531	5,859	10,421	16,281	50.2
3	1 + 2	148,749	5,418	9,637	15,055	60.7
4	Roof insulation	34,727	4,977	8,852	13,830	9.4
5	Green roof	40,370	5,040	8,964	14,005	11.5
6	1 + 2 + 5	189,119	3,780	6,723	10,504	27.0

Table 3.8 Different retrofit options and relative payback time calculated for building 2

	Building 2	Retrofitting option cost (EUR)	Annual cost of thermal energy (EUR)	Annual cost of electric energy (EUR)	Total annual cost for total energy consumption (EUR)	Payback time (years)
	As built	0	6,489	11,411	17,899	
1	Wall insulation	101,074	5,970	10,498	16,467	70.6
2	New windows	87,631	6,035	10,612	16,646	69.9
3	1 + 2	188,705	5,580	9,813	15,393	75.3
4	Roof insulation	12,337	5,126	9,014	14,141	3.3
5	Green roof	14,342	5,191	9,129	14,320	4.0
6	1 + 2 + 5	203,046	3,893	6,846	10,740	28.4

over a relatively long time, generally comparable with the life of the systems of energy production from RES.

3.1.1.3 Cost-Benefit Analysis in Evripidou

A detailed analysis for each scenario has been performed for all 11 buildings in the selected area of the urban courtyard in Evripidou, thus obtaining, for each building, the energy requirements in both the summer and the winter periods. Furthermore, an economical analysis taking into account the cost of operations and retrofitting options has been developed with reference to each building. Thanks to this analysis, the annual savings and simple payback times can be assessed. Tables 3.7–3.18 show the performed calculations for all 11 buildings.

Table 3.9 Different retrofit options and relative payback time calculated for building 3

	Building 3	Retrofitting option cost (EUR)	Annual cost of thermal energy (EUR)	Annual cost of electric energy (EUR)	Total annual cost for total energy consumption (EUR)	Payback time (years)
	As built	0	5,907	11,821	17,728	
1	Wall insulation	82,100	5,434	10,875	16,310	57.9
2	New windows	111,230	5,493	10,993	16,487	89.6
3	1 + 2	193,330	5,080	10,166	15,246	77.9
4	Roof insulation	18,277	4,667	9,339	14,005	4.9
5	Green roof	21,247	4,726	9,457	14,182	6.0
6	1 + 2 + 5	214,577	3,544	7,093	10,637	30.3

Table 3.10 Different retrofit options and relative payback time calculated for building 4

	Building 4	Retrofitting option cost (EUR)	Annual cost of thermal energy (EUR)	Annual cost of electric energy (EUR)	Total annual cost for total energy consumption (EUR)	Payback time (years)
	As built	0	3,028	5,282	8,310	
1	Wall insulation	39,960	2,786	4,859	7,645	60.1
2	New windows	15,533	2,816	4,912	7,728	26.7
3	1 + 2	55,493	2,604	4,542	7,146	47.7
4	Roof insulation	9,139	2,392	4,172	6,565	5.2
5	Green roof	10,624	2,422	4,225	6,648	6.4
6	1 + 2 + 5	66,116	1,817	3,169	4,986	19.9

Table 3.11 Different retrofit options and relative payback time calculated for building 5

	Building 5	Retrofitting option cost (EUR)	Annual cost of thermal energy (EUR)	Annual cost of electric energy (EUR)	Total annual cost for total energy consumption (EUR)	Payback time (years)
	As built	0	4,737	8,678	13,415	
1	Wall insulation	69,966	4,358	7,984	12,342	65.2
2	New windows	106,997	4,405	8,071	12,476	113.9
3	1 + 2	176,963	4,074	7,463	11,537	94.2
4	Roof insulation	10,509	3,742	6,856	10,598	3.7
5	Green roof	12,217	3,790	6,943	10,732	4.6
6	1 + 2 + 5	189,180	2,842	5,207	8,049	35.3

Table 3.12 Different retrofit options and relative payback time calculated for building 6

	Building 6	Retrofitting option cost (EUR)	Annual cost of thermal energy (EUR)	Annual cost of electric energy (EUR)	Total annual cost for total energy consumption (EUR)	Payback time (years)
	As built	0	10,769	20,106	30,875	
1	Wall insulation	152,372	9,907	18,498	28,405	61.7
2	New windows	168,229	10,015	18,699	28,714	77.8
3	1 + 2	320,600	9,261	17,292	26,553	74.2
4	Roof insulation	41,581	8,507	15,884	24,391	6.4
5	Green roof	48,337	8,615	16,085	24,700	7.8
6	1 + 2 + 5	368,938	6,461	12,064	18,525	29.9

Table 3.13 Different retrofit options and relative payback time calculated for building 7

	Building 7	Retrofitting option cost (EUR)	Annual cost of thermal energy (EUR)	Annual cost of electric energy (EUR)	Total annual cost for total energy consumption (EUR)	Payback time (years)
	As built	0	1,811	3,069	4,881	
1	Wall insulation	16,166	1,667	2,824	4,490	41.4
2	New windows	17,033	1,685	2,855	4,539	49.9
3	1 + 2	33,199	1,558	2,640	4,198	48.6
4	Roof insulation	13,251	1,431	2,425	3,856	12.9
5	Green roof	15,404	1,449	2,455	3,905	15.8
6	1 + 2 + 5	48,603	1,087	1,842	2,928	24.9

Table 3.14 Different retrofit options and relative payback time calculated for building 8

	Building 8	Retrofitting option cost (EUR)	Annual cost of thermal energy (EUR)	Annual cost of electric energy (EUR)	Total annual cost for total energy consumption (EUR)	Payback time (years)
	As built	0	2,890	5,157	8,047	
1	Wall insulation	41,972	2,659	4,745	7,404	65.2
2	New windows	23,132	2,688	4,796	7,484	41.1
3	1 + 2	65,104	2,485	4,435	6,921	57.8
4	Roof insulation	10,966	2,283	4,074	6,357	6.5
5	Green roof	12,748	2,312	4,126	6,438	7.9
6	1 + 2 + 5	77,852	1,734	3,094	4,828	24.2

Table 3.15 Different retrofit options and relative payback time calculated for building 9

	Building 9	Retrofitting option cost (EUR)	Annual cost of thermal energy (EUR)	Annual cost of electric energy (EUR)	Total annual cost for total energy consumption (EUR)	Payback time (years)
	As built	0	2,866	5,043	7,909	
1	Wall insulation	33,579	2,637	4,639	7,276	53.1
2	New windows	24,332	2,666	4,690	7,355	44.0
3	1 + 2	57,911	2,465	4,337	6,802	52.3
4	Roof insulation	11,423	2,264	3,984	6,248	6.9
5	Green roof	13,280	2,293	4,034	6,327	8.4
6	1 + 2 + 5	71,191	1,720	3,026	4,745	22.5

Table 3.16 Different retrofit options and relative payback time calculated for building 10

	Building 10	Retrofitting option cost (EUR)	Annual cost of thermal energy (EUR)	Annual cost of electric energy (EUR)	Total annual cost for total energy consumption (EUR)	Payback time (years)
	As built	0	4,555	8,320	12,874	
1	Wall insulation	54,886	4,190	7,654	11,844	53.3
2	New windows	106,163	4,236	7,737	11,973	117.8
3	1 + 2	161,050	3,917	7,155	11,072	89.4
4	Roof insulation	11,880	3,598	6,572	10,171	4.4
5	Green roof	13,811	3,644	6,656	10,299	5.4
6	1 + 2 + 5	174,860	2,733	4,992	7,724	34.0

Table 3.17 Different retrofit options and relative payback time calculated for building 11

	Building 11	Retrofitting option cost (EUR)	Annual cost of thermal energy (EUR)	Annual cost of electric energy (EUR)	Total annual cost for total energy consumption (EUR)	Payback time (years)
	As built	0	5,018	9,364	14,382	
1	Wall insulation	81,159	4,617	8,615	13,232	70.5
2	New windows	84,264	4,667	8,709	13,376	83.7
3	1 + 2	165,423	4,316	8,053	12,369	82.2
4	Roof insulation	15,079	3,965	7,398	11,362	5.0
5	Green roof	17,529	4,015	7,491	11,506	6.1
6	1 + 2 + 5	182,952	3,011	5,618	8,629	31.8

Table 3.18 Costs, savings, and payback time for the urban courtyard in Evripidou

		Retrofitting option cost (EUR)	Annual cost of thermal energy (EUR)	Annual cost of electric energy (EUR)	Total annual cost for total energy consumption (EUR)	Payback time (years)
	As built	0	54,370	99,456	153,827	
1	Wall insulation	760,452	50,021	91,500	141,521	61.8
2	New windows	806,075	50,564	92,494	143,059	74.9
3	1 + 2	1,566,527	46,758	85,532	132,291	72.7
4	Roof insulation	189,169	42,953	78,571	121,523	5.9
5	Green roof	219,909	43,496	79,565	123,061	7.1
6	1 + 2 + 5	1,786,434	32,622	59,674	92,296	29.0
7	Plants systems	387,016	0	82,722	82,722	5.4
8	Renewable plants (PV)	633,888	0	0	0	4.1
9	nZEB (6 + 7 + 8)	2,807,338	0	0	0	14.5

The same calculations have been performed for all 11 buildings in the selected courtyard. From the mutual comparison between the tables, the following conclusions can be drawn:

- The insulation of the vertical walls through external coatings is more or less convenient depending on the orientation of the individual buildings. It does not present the greatest advantage during the hot summer season.
- Given the very high payback time of the high performing triple-glazed windows, the double-glazed windows are always convenient in winter; conversely, during the summer period, in particular for the additional costs of the shading devices, they present very high payback time (especially in those buildings with large windows).
- The roof insulation always represents an excellent intervention.
- The green insulated roof produces a slight disadvantage in winter, but the costs still appear balanced.

Among the combined interventions, combination number 3 produces the greatest benefits: although the payback times are higher compared to the insulated roof or the insulated green roof, consumption is much lower and the resulting savings increase up to three times as much. The overall cost assessment of the whole set of buildings is reported in the Table 3.18.

Fig. 3.3 shows how payback times vary in a wide range. In particular, the very high payback time due to the wall insulation and the window

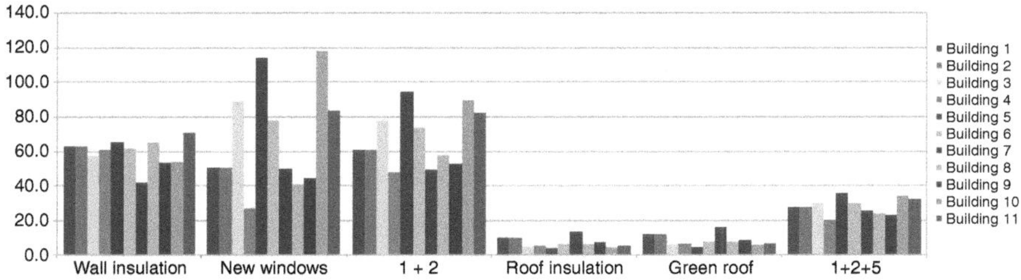

Figure 3.3 ***Payback times of the investments in Evripidou show a variation between 75 and 4 years can be observed as a function of the different hypothesized scenarios in the different building types of the urban courtyard.***

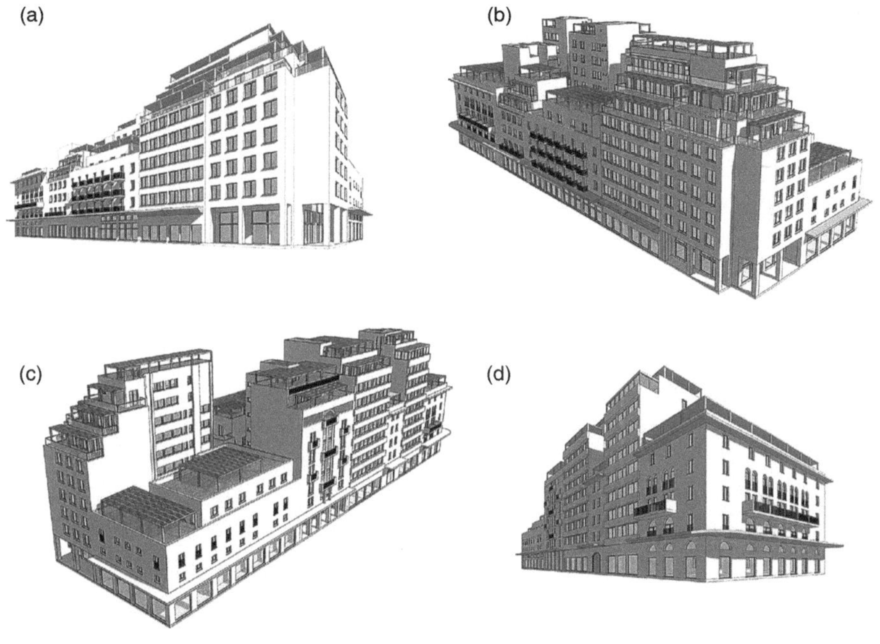

Figure 3.4 Existing buildings retrofit in Evripidou: energy retrofitting up to nearly zero buildings can be hardly visible at the urban scale, especially from the street level perception (a–d).

replacement is shown. This is particularly true for those buildings with large glazing surfaces. As can be observed in Fig. 3.4, also in this phase the volumes do not undergo visible changes in the architectural and urban form.

3.1.1.4 Conclusions on Energy Costs

To assess the technical and economical feasibility of retrofitting actions, different urban settings in Athens have been investigated, calculating and comparing costs and energy savings connected with different retrofitting actions. To identify the economical feasibility, a cost-benefit analysis was conducted by means of a market survey, determining, for each different design option, the evaluation in terms of energy performance improvement and the related cost estimations. The same options have been analyzed both individually and in combination, according to an increasing degree of transformation up to nZEBs in existing building blocks.

So far, our work has been focused on quantifying different variants in retrofitting actions of existing housing; simulations showed that it is possible to reach an average energy performance (EP) of up to 35–50 kWh/m^2 year, by the insulation of opaque surfaces and the replacement of existing windows. This energy index may decrease to 15 kWh/m^2 year in the complete retrofit scenario. However, the cost-benefit assessments always show excessive payback times (from 35 up to 70 years) for these interventions.

The set of all considerations emerged by the retrofitting options as developed so far lead us to the conclusion that, although it is possible to transform highly consuming existing buildings to achieve nZEB, the transformation requires necessarily high up-front costs, resulting in very long paypack times.[8] These high costs are due to the high cost of the technology for the renewable energy sources to be realized jointly with other costs occurring by the passive retrofitting operations to reduce energy consumption. Furthermore, both passive and active solutions require maintenance costs and present a certain decrease in performance over time; in particular, the PV system presents the higher management fees and is subject to a decrease in performance over time.

Thus, nearly zero energy in the existing buildings of these urban settings have been proven to be a technically feasible goal, but it is necessary to face the problem of very high up-front costs due to the need to operate both active and passive interventions, which are amortized over a relatively long time, generally comparable with the life of the systems of energy production from RES.

As a preliminary conclusion, we may state that the nZEB goal remains highly ambitious and exceptionally challenging, especially in the Mediterranean urban areas, considering the context of economic crisis, fuel poverty, unemployment, poverty, and social exclusion that these areas of the world are currently experiencing. Nonetheless, the effective reduction of energy consumption toward zero energy buildings and districts is and remains an

[8] Both passive retrofitting interventions and the production of energy from renewable sources are necessary to achieve nZEBs in the retrofitting of highly energy consuming building stock.

unavoidable objective, just right in the Mediterranean areas, where the dramatic combination of the HI phenomena and global overheating is severely threatening humans and the environment.

The reflection on the estimated costs of proposed energy-retrofitting options clearly indicate the need for additional tools and measures to be developed to counterbalance the large payback times of energy retrofitting measures aimed at achieving nZEBs in existing urban buildings. In particular, it is necessary to activate new forms of incentives in order to attract potential investors in the uptake of deep renovation toward nZEBs and ZEUS. This process should also be set by considering the users' and urban dwellers' contributions in the energy retrofitting of their urban environments.[9]

3.1.2 Why Do We Believe in the Economic Feasibility of nZEBs in Existing Urban Environments? Energy and No-Energy Related Issues in the Energy Retrofitting as Possible Incentives toward ZEUS

Housing represents a huge part[10] of the building stock in the European Union and the potential reduction in CO_2 emissions that energy-efficient housing would provide cannot be underestimated.

[9] According to these partial results and to the general objective of comparing various urban contexts analysis and design scenarios, in the project *Urban Recreation*, the larger and more densely built area of the workers' houses in Peristeri has been selected for further investigations with regards to social aspects; in this context we studied the energy demand/potential of real urban environments in a socio-oriented perspective by designing – in different steps – retrofitted scenarios in order to achieve both zero energy, and socially inclusive solutions, by means of an interdisciplinary research approach. We aimed at: (1) developing, performing and evaluating different possible solutions to achieve a sustainable and spatially localized distribution of energy, by means of technologies to be diffusely integrated in the urban buildings and open areas by means of a similar process with respect to the Aghia Varvara complex; and (2) promoting further actions to contribute to fill the knowledge gap existing among low carbon passive techniques and the adoption of these technological solutions at social and community city level. In particular, *Urban Recreation* has moved to envision the activation of new forms of incentives in order to attract potential investors as well as dynamically involve the social participation to motivate urban dwellers in the energy retrofitting of their urban environments. In this context, energy retrofitting should focus its attention on the roles of energy and environmental re-qualification of existing buildings, by a more appropriate use of technological options in architecture and urban/spatial planning. The functionally oriented approach and the mono perspective use of architectural buildings have to be considered a past attitude. In this effort, we also considered the necessity to find possible solutions by suggesting a multi-fold approach to stimulate the process of renewal also according to a nonenergy-related use of technology in urban environments.

[10] Dwelling stock accounts for about 200 millions units, representing around 27% of energy consumption in the EU.

Three quarters of the buildings standing today, including the residential stock, are expected to remain in use in 2050 (COM, 2012). But, so far, only about 1.2% of Europe's existing buildings are renovated every year.[11] In fact, most improvements in building energy performance are modest – usually in the range of 10–20% for specific energy efficiency interventions. Until recently, those improvements have generally been considered as enough. However, if the energy and climate goals of achieving almost nearly zero energy have to be achieved by 2050, we have no choice but to dramatically upscale the energy performance target of the existing building stock. The implementation of single measures, such as application of insulation layer, replacement of windows, improved operation and maintenance, lighting upgrades, or roof space insulation, constitutes only the first level of intervention. The combination of single measures (which can be termed standard renovation) involves the simultaneous and integrated implementation of a few individual energy-saving measures. Deep renovation or deep energy renovation refers to renovations that capture the full economic EE potential of improvements. A renovation or refurbishment will often harvest the minimum possible energy savings, ranging between 20 and 30%, sometimes even less. However, by renovating deeply as is shown by using state-of-the-art technologies, it is possible to reduce the energy consumption of a building by 75%. This typically includes actions on the building envelope; on the contrary, the definition of a retrofit focuses mainly on the building's mechanical systems.[12]

However, these considerations regarding the urgency of moving toward a "deeper" approach are related merely to the energetic aspect of the problem. Given the various characters of the existing barriers illustrated previously, it has emerged that the multiobjective and integrated character of the intervention has to face the problematic issues through a more all-inclusive approach. The methodology presented in this section seeks to include in

[11] The demolition rates is about 0.1% per year and highly energy efficient new-build reach 1% additions/yr. JWG: Towards assisting EU Member States on developing long term strategies for mobilising investment in building energy renovation (EU EED Article 4), Composite Document of the Joint WG of CA EED, CA EPBD and CA RES, November 2013 (http://www.ca-eed.eu/reports/art-4-guidance-document/eed-article-4-assistance-document).

[12] As observed, the primary energy consumption after renovation, which includes, inter alia, energy used for heating, cooling, ventilation, hot water, and lighting after the deep renovation of an existing building is less than 60 kWh/m^2/year. However, as the level of achievable savings will vary depending on climate conditions and on the level energy performance of the building prior to renovation.

the deep renovation process additional tools that could help to overcome the existing economical, legislative, technological, and even social barriers. Thus, the energy efficiency challenge in buildings mainly concerns energy efficient refurbishment and investments in its existing building stock (EEFIG, 2015).

The simple payback time of retrofitting interventions examined so far has shown how the economic feasibility of nZEBs in existing urban environments is a very challenging objective.

Given the challenging character of nZEB in existing built environments, how can we still believe in the feasibility of achieving this objective using current building practices? In the next paragraph, we will discuss a possible series of energy and non-energy-related issues in energy retrofitting as possible incentives toward the challenging goal of nearly zero urban settings.

3.1.2.1 Barriers and Opportunities at the Economic and Technical Levels toward Achieving nZEBs

Modeling retrofitting hypotheses through technical interventions such as plants renewal and u-value reduction of building-envelope components (opaque envelope insulation and window replacement) have shown great potential for energy efficiency (EE) achievements in very high energy consuming building blocks.[13] However, the cost-benefit assessments always show excessive payback times (of up to 35–45 years, without considering possible incentives).

Besides our case studies, studies and surveys by housing providers, financial institutions (Cecodhas, 2013) and concerted actions documents (BPIE, 2013) identified key categories of hurdles (economic, technical, social, legislative) in delivering deep energy renovation for buildings.[14] Among them, the high up-front costs and long payback times of retrofitting interventions are indicated as the major barriers. Yet, there is a clear investment gap in the retrofitting of existing buildings and, in particular, in the

[13] Providing the energy redevelopment of an average existing building up to 25 kWh/m^{2}year it could be obtained a reduction of primary energy demand by over 90%; in Bologna, an estimation of 0.23 m^2 of PV panel for each square meter of residential net surface will set to nZEB the energy balance of this reference building.

[14] When referring to deep renovation, most of the EU discussions and the consequent definitions on the subject focus on HVAC and the general understanding is that these deep renovation interventions should lead to an improvement of at least 75% after the building has been renovated, with an absolute target between 15 and 60 kWh/m^2/year (Shnapp et al., 2013).

housing market. This has proven to be due to the fact that high investments are required up front and are characterized by a large degree of risk with potential limited return on investments; furthermore, there is a generalized lack of confidence by both public and private investors (including owners and financial institutions). Therefore, it is necessary to develop harmonized actions that can help unlock the needed public and private finance sources, fill the energy efficiency investment gap, and ultimately contribute to re-launching the construction market.[15] Moreover, these operations do not fully compensate the different EP of the units within the building; this can produce additional barriers toward the adoption of monitoring systems to control energy waste[16] and social conflicts in relation to the sharing of costs and benefits among owners or between them and tenants in multi-owner properties.

In the search for possible encouragements toward our aim, the opportunity given by factors able to reduce energy costs can be found in other non-energy-related factors influencing the renovation and valorization of buildings, which are corresponding key benefits. In fact, operating a mere energy retrofitting on individual residential units or building blocks in urban contexts in severe degradation conditions may imply a devaluation of the potential efficiency of the same retrofitting actions; in fact, building renovation actions are always connected with and intrinsically involve a larger spectrum of quality related aspects. In other words, we can state that energy retrofitting of existing buildings, being connected with technical and architectural modifications of building's components, may imply a strong influence on the overall quality of the built environment and that this quality can not be ignored, since it may be a key added value in terms of technical and economic feasibility.

Multilayered components and stratification of elements within the building envelopes have been presented so far as an effective solution in the energy retrofitting of existing building stocks of urban environments.

[15] In example, EU Plans expect to deliver 2.000.000 jobs, increased industrial competitiveness, annual financial savings/emissions reductions of 740 million tons of CO_2e (BPIE, 2013; EEFIG, 2015; EU Report, 2014, Assistant documents by JWG).

[16] As an example, the adoption of thermostatic valves on radiators in centralized heating plants causes the unbalanced distribution in the energy bill and possible problems of negotiations among the dwellers. In case of a proportional redistribution of the total cost of the energy retrofitting measures, the disadvantaged units would incur significantly higher expenses.

But sustainability in architecture and building construction also implies the achievement of other technical requirements like comfort, safety, aesthetic quality, and user's needs, while using minimal amounts of energy and environmental resources. The possibility of achieving this goal in a building is mainly given by acting on its envelope. In this context, producing light prefabricated elements for the facade composition could become an important goal to implement and foster all the highly advanced existing technologies by simply increasing the efficiency in the production and assembly phases. We should observe that the idea of a multi-task and highly energy performing facade is nothing new: well before the energy efficiency and renovation emergency, the goal of producing an integrated facade system was envisaged by Davis (1981), who formulated the idea of a polyvalent wall. In this wall, several functional layers within a glass element were to provide sun and heat protection, and to regulate the functions automatically according to current conditions. The wall itself was to generate the necessary energy and perform at its best for the different housing needs and climates.[17] Again, the main problem is that for the production of prefabricated panels to be economically justifiable, it is necessary to produce them in a large scale and this is highly related to the different morphology of the facade and the opening types. This is a minor problem in case of a new building but becomes important when trying to apply this technology to the renovation of an existing building: the enormous variability of the existing housing stock makes it impossible to develop a catalog of fixed options since every panel should be customized and produced according to the information of the single surveys. The process would be highly un-economical, especially considering the social housing market and the low economical possibilities of these interventions. A possible way to reduce costs and implement efficiency in the production cycle today is given by the movement to establish a strong relationship between the design and the production phases. In fact, the design and production of the panel can be developed today through computer-aided tools that are based on the definition of precisely those geometrical parameters that can vary and represented the aforementioned problem, allowing the producer to simply set these parameters

[17] This idea, not yet realized today, still acts as a driving force for new facade technologies, and many researchers have been engaged in this topic over the last two decades. The idea was born as highly innovative technology for the façade of the future, but what are the benefits that such technology and system could bring when applied in the energy renovation field?

differently every time and send the 3D model to the computer numerical control, or CNC, printer.[18]

To do so, the starting point consists of analyzing the envelope as an architectural and technological complex system, suggesting a shift in coding the facade and its elements, changing the focus from the cell-scale or the building-scale to the module-scale. The process starts from a discretization and division of the facade in a grid, and finding out the original unit and its variants, which is very easy in buildings like the ones considered in this book, that are generally characterized by repetition and rhythm in the facade division and drawing. The same facade element is often repeated and replicated throughout the facade. The division elements are characterized by the staircases and/or the toilet rooms where the opening of the facade as well as the balconies or loggia disappear to leave space for a different geometry of the openings or a change in material, or again, a volumetric transition. But how shall the facade be divided? The approaches to renew and, as far as possible, to prefabricate energy-efficient new envelopes for existing buildings can be very different. This is not only because buildings themselves are divers. Even for identical buildings, there are various options to achieve practical solutions for retrofitting and to find a way to produce prefabricated modules for facades and roofs. For example, the integration of a ventilation system can be very differently addressed and will strongly influence the whole

[18] To better clarify the high potential of this refurbishment method, it is important to underline the following three aspects: (1) The facade modules are standardized in construction, layers, and joints; (2) The facade modules are flexible in architecture, form, and cladding; (3) The facade modules can be combined with each other with non prefabricated (conventional) retrofit options. The application of this process (typical of the mass industrialization trend), the idea of increasing the customization and the specificity of the final product, could drive the production toward unexplored paths. The production chain remains the same, while the parameter of the product can change and follow the specific geometrical characteristic of each single building and the specific wishes and need of the users, thus shifting toward a mass prefabrication system based on modularity and variability in the preliminary design phase; the tools available today (especially considering the potential of laser scanner, CNC machines and 3D printing systems) allow to set up a unique process to produce multiple designs, hundreds of different types of the same piece of facade. Thanks to the transfer of the customization and prefabrication logics in the field of housing refurbishment, it could be possible to re-interpret the residential artefact, no longer in its dimension of building object with a close and rigid configuration, but in the adaptive nature typical of the housing function, subjected to time transformations and users changes.

Figure 3.5 ***Van Shagen in Amsterdam and Kolpa Architecten in Rotterdam (on the right): examples of volumetric compensation (roof-top extension) and creation of new units in the energy retrofitting of existing buildings.***

planning process. Also, the country specific and well-established process chains should not be bypassed.[19]

3.1.2.1.1 Economic, Financial, and Social Potential of Add-ons

The high investment costs in energy retrofitting have driven some experiences at EU level,[20] as well as pilot cases in Austria and The Netherlands (Fig. 3.5), to focus on the strategy of rooftop extensions as additional volumes whose economic value in the housing real estate market may counterbalance the energy retrofitting costs. Examples of volumetric compensation (rooftop extension) in the energy retrofitting of social housing have been adopted from Van Shagen in Amsterdam and Kolpa Architecten in Rotterdam (Fig. 3.5).

In the following work illustrated for the Athens metropolitan area case studies, we will base our effort on the prior assumption that

[19] They set basic conditions from planning to working on site. Four different approaches have been analyzed according to the IEA ECBCS Annex 50, by national teams from Austria, France, Portugal and Switzerland: The Swiss solution is based on small modules with a high degree of standardization. The prefabricated façade modules concentrate on the window area as the area with the highest concentration of technical details, whereas the plain façade parts are insulated traditionally. Also the façade finish – ventilated cladding or rendering – is done on site. This concept has also been used for the development of roof elements, for flat or sloped roofs. The Austrian solution is based on large glazed façade modules, which are fully prefabricated. They are as high as the building story and up to 12 m long. They have been already tested at various demonstration buildings in Graz. The French development was focused on large vertical façade elements with metal under construction. Special attention was given to the avoidance of thermal bridges. Prototypes were produced and tested. The Portuguese study concentrated also on fully prefabricated but smaller sized modules. They are based on easy mountable metallic and of course insulated façade panels. Considering the entire lifecycle of the building and the sustainability of the intervention as a matter that involves the entire life cycle of the building the suggested proposal is based on the assumption that light prefabricated elements that use wood as main material are the favorable option.

[20] For example, European activities and projects like SuRE-Fit, Reshape, Solar Decathlon 2014, and so forth.

non-energy-related benefits play a key role in the deep renovation of existing buildings. In particular, our work will focus on the following main benefit: the generation of a substantial increase of the real estate value of the buildings through significant energy and architectural transformation (mainly integration of RES systems with new volume additions or new building construction) to go beyond the minimum energy performance and aim at achieving nZEBs. Thus, we will try to explore the potential of energy retrofitting actions contemplating a higher level of transformation in the existing buildings and urban settings. Our efforts will concentrate on the following question: To what extent can a deep renovation through higher transformation be competitive with respect to a more standard retrofit?

3.1.3 Deep Renovation and Volumetric Additions in Energy Retrofitting Actions toward ZEUS

It is well known that current energy policies in the European Union, such as EPBD and EED, contain provisions to increase the energy performance of existing buildings and to encourage member states to convert building stock through the development of a marketplace for cost-effective deep renovation. But deep energy renovation of buildings is a composite process that needs to be explained and made attractive to key financial overseers and decision makers (financial institutions, developers, managers, householders, policy makers, building owners, and associations). In order to do so, there is a need to share knowledge, build capacity, and capitalize on all possible win-win solutions that may arise from the mutual collaboration of architects, urban planners, financial actors, decision makers, and social stakeholders. This is the main path toward grasping the existing obstacles to deep renovation and defining effective solutions to overcoming them.

In the search for possible encouragement toward our ZEUS objectives, we should seek out other technical opportunities that reduce energy costs in other non-energy-related factors influencing the renovation and valorization of buildings. In fact, energy retrofitting in real urban environments (RUEs), being connected with technical and architectural modifications of building's components, may involve other qualities that cannot be ignored, since they may be a key added value in terms of technical and economic feasibility.

We may therefore search for a higher degree of transformation; to move toward this, the levels of adaptability of existing buildings and their connected pluses and minuses need to be explored as well. In this context, the positive benefits of prefabricated facade modules as described in the previous section for energy renovation of existing buildings are given by different aspects involving ecological, architectural, engineering, and plant systems.[21] Here the major aids are given by the economic gain and the longer lifecycle of the entire building. In particular, regarding the economic gain, in the real estate market, the integration of spatial and volumetric additions within the same construction system and intervention time frame may represent an important payback accelerator for the investment in the renovation process.

The possibility of integrating both benefits in one technology (the facade renovation and the extension of surface as a volumetric bonus to the energy renovation) embodies great potential, considering the payback time reduction and the increase of value in the real estate market.

In order to support the construction of a new facade (when this can not rely on the existing structure) or the new facade extension that could be part of the intervention, additional structures have to be envisaged. These structures should be entirely prefabricated and mounted on site. The load transfer scheme is mostly thought to be independent from the original structure in order to avoid the worsening of existing static conditions that would not be possible to evaluate with the necessary accuracy in seismic areas.[22] The proposed methodology will focus on the validation of a multitask and highly energy performing facade and its possible integration of space (volumes and surfaces), in addition to the energy renovation of the existing building, to evaluate the possibility of this process overcoming the existing barriers to the deep renovation in order to create nearly zero energy buildings.

As mentioned previously, the basis for the use of prefabricated retrofit building elements could be given by a frictionless digital workflow from survey, planning, offsite production, and mounting onsite, based on a precise initial 3D scanning measurement. This would allow a high degree of

[21] The underling basic principle of using and recycling a building or a building structure is undoubtedly an ecological-related aspect. Architectural quality can be achieved by the possibility of redesign of the façade and entire building envelope. Under the engineering and structural point of view the system could provide the possibility of improving the structural elements and integrate new prefabricated structure that could cooperate with the existing in case of seismic event, finally the substitution of façade by new envelopes can also provide the integration of the heat ventilation air conditioning (HVAC) directly in the façade and the integration of systems for solar components. As far as it concerns energy efficiency, the increase of the performance of the existing envelope is accomplished by: adding extra insulation; removing existing envelope components that are no longer meeting the standards (window frames, old insulation, cladding); integrating innovative systems such as dynamic façade, ventilated and mechanical systems, and so forth. These are the main benefits. The possibility of volumetric addition and spatial transformation within the same construction system is also a great plus.

[22] Further research on the topic could also demonstrate the positive impact and effect that these systems could have as aid to the existing structure when involved in a seismic event.

precision and correspondence between the new facade modules and the old facade, reducing time waste, errors, and risks in the construction phase.

This type of expansion could involve the increase of one room size, or the entire rooftop level, the underground level, the side, or the in-between space at the urban level as well. Moreover, a combination of building system and space components would help to standardize and minimize the connections between space modules. "Nearly zero energy or energy-plus solutions" introducing space modules that are self-sufficient in energy could be one solution to simplifying the connection of extensions to existing building systems. Town planning measures have been developed to increase the density of existing urban structures, which can be achieved by adding roof floors, neighborhood-specific infill developments, or extension of building spaces in a framework that envisages the refurbishment need as an opportunity for reinterpreting and redesigning the periphery of our cities.[23] Therefore, it is not surprising that the last European edition of the Solar Decathlon (Jun. 2014, Versailles) has identified the five central topics of the competition in relation to the main issues of sustainability in European cities: density, mobility, sobriety, innovation, and affordability. This important discussion on sustainability and architecture reflects a clear shift from the search of a formal application of green active energy systems in figurative architectural connotation, to the urban application of housing prototypes that are capable of answering the energy demand. In one sentence, the project has to be designed in its environment and it develops around, above, and within the built environment. Many of the highly efficient home prototypes presented have been designed to be directly integrated in the urban tissue as volumetric additions, rooftop extensions, and/or infill between existing buildings. The prototypes built in Versailles as well as the preliminary studies and similar research experiences show the possibility of building this addition through the utilization of light and prefabricated systems (wooden panels). This type of construction system will ensure the rapidity of realization, the reversibility of the intervention, and will avoid structural interference with the static condition of the existing building that "hosts" the addition.

[23] In many countries of the European Union, the political parties are already moving toward a more flexible framework for energy renewal interventions. The open discussion on the potential of increasing the volume of existing building is an important occasion to review the current legislative framework in favor of an indispensable liberalization of the normative. The urgency for densification policies that Europe is facing requires a direct engagement from both the research and development sectors and the industrial sector.

Therefore, energy renovation through volumetric additions can be divided into three consequential and incremental steps, each independent of one another, but in order of impact and consistency of the transformation: facade transformation, horizontal extension, and vertical extension.[24]

3.1.3.1 Expandable and Adaptable Architecture: Add-ons and Tailored Solutions to Increase the Techno-Economic Feasibility of ZEUS (Annarita Ferrante and Elena Cattani)

The concept of adaptability refers to the capacity of buildings to accommodate substantial changes (Kobayakawa et al., 2006). Adaptability is broken down into three aspects: flexibility, convertibility, and expandability (IEA, 2001). The possibility and the degree of transformation in a building and, generally, in a built environment, is essentially linked to its degree of adaptability. The main consequent and related quality is the durability of the same built environment. A construction can be divided into a structural part and the envelope part, whose life cycle is about one-third of the forecasted life cycle of the structure. It follows that, if the building presents a certain degree of adaptability, the most convenient renovation strategy lies in the

[24] Level 1: *Façade transformation*: According to the architectural and structural type of the existing building it is possible to apply the new façade in addition to the existing one (this is especially true in case of external load bearing walls or façade whose structure and state is still in good static conditions and cooperate with existing building frame) or façade in substitution of the existing one (in cases where the existing façade has reached the end of its life cycle or can no longer meet the minimum requirements). Moreover, the redesign of the façade offers a possibility to engage a dialog with the inhabitants and involve them directly.

Level 2: *Horizontal extension*: Incorporating volumetric addition, new rooms or transformation of the existing balconies, loggias, and ballatoires. The transformation in surface increases the real estate value of the apartments involved in the transformation and also the degree of participation of the inhabitants called to decide upon the transformation according to their needs.

Level 3, *Vertical extension*: The volumetric addition can involve also the rooftop and generate entire new housing units. The economical payback of the intervention can be reduced and the roof surface can be used for production of solar energy.
The following characteristics have to be met in order to be able to proceed from level 1 to level 2 and 3: Reserves in distance spaces and heights according to the existing building regulations; Reserve in permitted density of buildings according to development plan and need for urban densification of the compound; Existing circulation system adapted for an extension (change of building class, escape routes, and accessibility); Simple building geometry (few setbacks or cantilevers, openings of simple geometry); Load reserves in supporting structure (up to the foundation) would represent a plus by reducing the cost of additional structure; Original building with an anyway need for refurbishment; Ownership favorable for investment and/or inhabitants willing to participate in the initiative.

potential exchange of the envelope and those components that have already reached the end of their life cycle. Being that the built environment is already used and lived in by its inhabitants, any process of adaptation cannot ignore them; on the contrary, transformation should be done according to the current user's needs, but also with a high enough degree of flexibility in order to allow for future transformation in case of a change in use or users of the building.

While traditional construction methods were based on physical requirements, minimal space units, and healthy standards values, today the processes of transformation and adaptability increasingly looks at "differentiation, not uniformity in designing the built environment" (Preiser et al., 1991); for the emphasis is on the customization of the products through mass production rather than standardization.[25] Adaptation is also defined as the means by which we can begin to address the urgent challenges of the contemporary age (not only in the building and construction field) and enter a new era of innovation. In this context, preliminary questions to be answered are:

How can we evaluate and calculate the adaptability degree?

What are the factors that influence it and how can they be improved?

According to several previews of research and literature (Dimitrijevic et al., 2002), there are six assessed spatial, structural, and services design features: (1) site (possibility of expansion, access for pedestrians, access for services); (2) interior layout and design (completeness of brief, flexibility of layout, grouping of functions, average main room size, provisions for disabled); (3) facade (window size, balconies, loggias, "ballatoires"); (4) structure (strength of columns/walls, column density/span, floor-to- ceiling height, floor loading, floor structure, removability of partitions); (5) HVAC system (plant location, plant size space wise, access for people, access for equipment, ducting access); (6) services and elevators (capacity, extra space).[26] By intervening on the aspects connected to the adaptability, the consequent result is an increase in the durability of the listed design features: durability in fact is defined as service life, that is, the actual period during which no

[25] As industry and academic leaders discussed already in the early 1999, during the CERF/CIB Symposium, "sustainable" means being able to meet the multiple requirements of society through the life cycle of a building or structure (Bernstein, 1999).

[26] In order to be able to evaluate levels of adaptability, a scale 1–5 has been developed by (Dimitrijevic et al., 2002): scores from 1 to 2 (Low adaptability: design features are appropriate for minor changes); score 3 (Medium adaptability: design features are appropriate for more complex changes). Scores 4–5 (High adaptability: design features are appropriate for complete change and transformation).

unacceptable expenditure on maintenance or repair is required (Dimitrijevic et al., 2002).

In energy retrofits of existing buildings, the approach to durability practiced usually leads to regulations and discussions about the degree of insulation and the materials that are to be used. Although there is no doubt about the value of these aspects, durable building includes facets that have not received much attention so far (Vreedenburgh and Melet, 2012). Vreedenburgh and Melet (2012) defined durability as the ability of a building to last in time, despite the change in number or needs of the people that live in the building itself. When we interpret durability as making use of the full potential of what has already been built, then the possibility of increasing density within the built environments in contemporary cities by adding space may provide a useful opportunity to increase durability while adapting the cities to current energy and non-energy-related needs.[27]

Thus, the possibility of intervening through the adaptation of the total volume of the building (eg, by increasing the number of storeys, transforming the balconies into closed extra rooms, or adding new space on the side of the building) hides some great potential in terms of durability and overall value of the existing building. The densification strategy through punctual volumetric addition could become a key node to overcome existing economic and social barriers.

Indeed, the practice of adding, transforming, changing, or extending existing buildings in different periods of time is everything but new in our building heritage.[28] Formal and informal versions of this may often coexist. While informal is essentially linked to spontaneous bottom-up processes of

[27] The relationship between the adaptability quality and the durability quality of an existing building is thus a cause-effect relation: by increasing the adaptability of the construction, its life cycle increases and the real estate value can rise of significant figures. The application of construction systems and technologies that incorporate directly the concept of variability among the intrinsic characteristics of the project (for new construction but also and more specifically for renovation) can be able to transform according to the needs the building throughout its lifecycle. This could also guarantee final products and urban compounds with increased efficiency and durability at the building and urban scale, including sociological, environmental, and economic systems; in fact, the contribution of an integrated action would produces an overall benefit in all the spheres involved.

[28] Examples of housing addition on historical tissue can be found in Ponte Vecchio, Florence, in Teatro Marcello, Rome; in the northern countries many traditional houses from the 16th century with 17th have been intertwined with wooden frame additions, and so forth. Furthermore, different forms of spontaneous settlements as well as all types of illegal occupations, invasions, and squatter settlements are generally identified as informal addition.

appropriations by users, "formal" addition is considered to be that which is planned through rational processes and under legal institutional frameworks (Hutchison, 2010). So far, the architecture of the addition has been limited to occasional manifesto experiments or largely and commonly applied to transform and evolve the dwelling environment in informal settlements. This occurred commonly in the history of European architecture, but today, in light of the so-called shanty towns (Klanten and Feireiss, 2007) it has received renewed interest. The informal is synonymous with continuous transformation but also with the potential for variety and adaptability, characteristics that current residential areas often lack.[29] In this perspective, it is possible to integrate much of the hidden knowledge, techniques, and systems typical of the informal processes in the formal process of reconstructing our suburbs. The volumetric addition is a possible bridge between experimental design practice and the self-made practice of the informal architecture (Marini 2008).

Some of the projects presented by the Convertible City exhibition in the German Pavilion (Convertible City, 2006) have demonstrated the possibilities of architectural addition onto existing buildings: superimposition, shifting, and penetration make it possible to extend existing building structures and combine and define them in new ways.[30] According to the type and strength of the interconnection that can be introduced between the new and the pre-existing – in conceptual, formal, and technological terms – different relationships can be found. The architectonic and constructive result of addition is, in fact, given by the system of relationships among the different technological components and layers; thus, this system is closely linked with and can reinterpreted to serve the needs of energy renovation of existing buildings.

A categorization of typology of the addition may be also attempted by basing the increasing subversive level of the possible synergies between the pre-existing and the new-addition element(s): starting from a subdued role

[29] Informal architecture can be assumed as a model for the experiments in the field of addition elements for the existing built environment, since the strategy of high transformation through addition joins and connects different expertise and technologies, at the different scales, of the housing system in the common effort of finding regeneration processes starting from densification as a mean for change.

[30] For the first time, the addition in architecture gained a space and identity of its own, rather then a manifestation of informal practice. As formal and theoretical ground for the architecture of the addition, a selection of the 20th century's most significant precedents has been carried. This led to a categorization of the different relationships that could be created between the addition and the existent building. In fact, by the analysis of the previews significant architectural experience and established connections, it has been possible to also draft theoretical assumptions and considerations upon which the innovation aspects could further be developed.

of the addition, moving to a reflexive role, up until a subversive act. The resulting possibilities for addition are: (1) *Completing* (filling in the existing empty spaces that make the volume "incomplete." In this case the addition does not overcome the boundaries of the original volume's projection); (2) *Adding* (side or front or facade integration of extra new elements. Here the addition stands outside the original perimeter and exceeds the boundaries of the original volume's projection); (3) *Topping*[31] (Addition on the rooftop by adding a volume or new element on top of the existing one); (4) *Translating* (reconfiguration and transformation – through addition, subtraction, and replacement of the original components – of the existing building, with the goal of entirely redefining the original, preexisting space).[32]

Furthermore, adaptability and flexibility are a must for any project that hopes to attain a longer life cycle and subsequently, to attain sustainability. In the same way that doctors do not care only for patients' long life but also for their quality of life, a building should guarantee high-quality comfort for its users during its life cycle. However, requirements constantly change due to new society patterns. Predicting future needs is an exercise in speculation.

[31] The topping should be considered a special typology of addition considering the special attention given both in the practice and in literature to the rooftop extension. The authors believe it could represents a stand-alone category with a specific character of peculiarity and criticalness, also linked to the limited possibility of application in seismic areas or where poor static conditions occur.

[32] The main difference in the architectural outcome of these possibilities stands in the type of relationship that the addition created with the preexistence, whether it creates a fracture and declares its own identity or rather hides behind the preexistent formal composition and integrates with it. In the first case, the addition represents an act of subversive transformation that underlines the moment and space, the upcoming of a new time for the building. In the second case the continuity between past and present is emphasized and the continuity between before and after is preserved. The first strategy is the most largely used, mainly because, until now, the role of the architect has never been consistently discussed and readjusted toward a social-oriented and user-tailored dimension. Still, the architecture of the addition did not come as the answer to a social need but rather as a new way for architects to express their style and vision upon the existing, another act of over-imposed formalism that our society can no longer afford. It is important to define the framework of the addition and the typology of the proposed intervention in order to insert the design proposal – which will focus exclusively on the existing housing stock – in a broader architectural scenario. This analysis underlines also the existing contradiction between the eccentric character of these punctual injections, completions, or additions and the rigid status of the current building blocks. The potential of addition is here analyzed as a subject with its own identity in the architecture scenario, claiming the right of overcoming the marginal boundaries and be legitimate as a densification and renovation element whose specific function is that of answering to an urgent energy revision of the building stock and a necessary return to the final user-oriented dimension of architecture.

The more reliable solution is to enhance flexibility in the building system in order to guarantee adaptability. When the world changes, buildings should be able to change with it (Plagaro and Schwehr, 2008). Thus, applying the typology of the addition should imply the understanding of the existing flexibility or adaptability to the transformation of a given system. Those areas, paths, or options where flexibility and adaptation can occur represent spaces and potential for the extension. As clearly exposed by Habraken (1999) in his definition of *open building*, the built environment can be assumed as the product of an ongoing, never-ending design process in which environment transforms part by part. The question that stands behind this process is: how can architecture anticipate diversity and change and make this part of the renovation? According to Price (2001), this process can be described as "thinking the unthinkable," the unpredictability of human behavior, and finally, the substantial illusion of planning.

The programmatic revision of energy renovation practice should by closely connected to the opportunity for addition and transformation aiming at increasing the flexibility of the structure and of the typology of the existing buildings, with particular reference to those housing structures whose rigidness and fixity represent a critical aspect. Flexibility and adaptability may be conceived and interpreted in the construction process by making use of prefabrication of modular retrofit systems that allow optimized construction and easier replacement of parts. Consideration of such systems and the optimization of the operation process and its relevance for the building cycle and user's cycle allow a harmonized building system that can react flexibly during the whole life cycle of the building. In fact, some building parts age more quickly than others; these differences hinder coordinated maintenance and a damaged building part can lead to faults in the whole building system. Based on concepts described by the Fraunhofer Institute (Fraunhofer, 2002) and supported by typology-based building evaluation (Fischer and Schwehr, 2008; Plagaro and Schwehr, 2008), different building flexibility types have been identified. First of all, we should differentiate the external and inner capacity of a building to assume different configurations.[33]

[33] The external flexibility refers to and involves the capacity of positioning and achieving structural improvement from extensions and retrofit systems. The internal flexibility defines the internal adaptability of the building: in which degree are modifications within an existing structure possible. How does the extension influence the building and the internal layout together is a very challenging and interesting aspect connected with the possible and new relationships between the inside and the envelope.

Extension and internal are building characteristics that enable high-value stability over the entire (renovation) life cycle (Schwehr, 2010) and are intrinsically related to the possibility of having an *autonomous* extension. It is evident that the typology of the addition is a result of the potential in terms of the external and internal flexibility of the existing building. Buildings with *a priori planned flexibility* (Ferrante et al., 2011) have a structure, a mechanical engineering system with joints and space, which has been planned to allow future changes. This allows the development of *integrated additions* (which rely on the existing structure and distribution scheme).

However, even considering the possibility of integrating the extension in the existing building, the new regulations (especially considering the seismic and safety regulations in Mediterranean countries like Greece) make it generally complicated and expensive to proceed down this path. When this flexibility is not found – as happens in the majority of the apartment blocks that were built during the 1950s, 1960s, and 1970s – *ad hoc flexibility* can be integrated into the renovation phase, generating the option of a *semiautonomous addition* (that uses the original distribution and layout scheme but is independent structurally and mechanically). Another way to do this is to create an independent and *autonomous addition* (not connected to the existing building), which is associated to the original building solely by energy, economic, and procedural relationship.[34] In this case, we have assistant building(s). However, energy and economically related aspects are the most important for the purpose of this work and have the full potential of enriching the theoretical strategy of addition with further connotative issues.[35] In fact, the rereading and rewriting processes applied

[34] The main differences between the previous experimental experiences of parasite architecture and the methodology of the addition here suggested for the renovation of building stands in the programmatic character of the proposed strategy and in the direct participation of the inhabitants. As illustrated, the addition is used as an instrument and a tool to incrementally foster the willingness from an economical point of view and to provide the inhabitants with additional spaces or rooms. The intervention of addition and extensions stand here as a need to balance the economical and payback circle of the renovation process together with the need of answering to the technical question of the renovation of the facade. But, the addition can also be used to express and translate the need of the inhabitants directly in the facade, on the external layer of the building and consequently on the urban level.

[35] In example, from the energy point of view we should consider that the higher energy consuming apartment units are the ones located on the last levels, on the side of the building or on the first level due to the higher amount of surfaces directly exposed to the external air temperature (see also Chapter 2 on the energy demand of selected buildings in the Athens areas).

on the renovation practice represents an important opportunity to develop new or renewed technological, constructive, and functional models, shifting toward the simplified application of prefabricated and highly adaptable products that may efficiently produce the open redefinition of the building as a whole. In many cases, the building structure can not be modified in an economically sustainable manner but it is possible to integrate secondary systems, extend the existing structure, and redefine the layouts and functions; thus, retrofitting becomes a smart alternative to total demolition in favor of a metamorphosis that responds to the current and future needs of the users. Theoretical architectural speculations on the reinterpretation and the right of the parasite intervention become secondary and merely formal issues. Nonetheless, they can be used as a basis to understand and even anticipate the formal outcome paradigms of the architecture of addition in the design process.[36]

3.1.3.1.1 Expandable and Adaptable Architecture in Athens: Peristeri Case Study

Starting from these considerations, higher transformation scenarios have been conceived and developed for the energy retrofit toward zero energy of the urban compounds of Peristeri and Evripidou, in order to check the technical and economic viability of these solutions in the retrofitting of existing densely built environments in the Athens area. The necessary integration between the need for retrofitting and the opportunity to assume a wider rehabilitation process scenario have found further pilot experiments in the energy retrofit design of the public residential district in the Peristeri urban compound; within this urban context, different reference buildings have been used to test and compare four possible energy retrofit options according to a socio-oriented approach. The different structures in this case study show numerous possibilities and combinations. The design solutions have not been constrained to a simple prefabricated structure; indeed the need for developing modular and variable solutions has lead to a wide set

[36] It is important to underline here that the freedom of the design should not be thought to be entirely left to the inhabitants; they are expected to select different options while control and supervision of critical formal boundaries should be set by the architect, as the style should be in any case anticipated in a preliminary vision of metadesign. The balance between order and disorder: a controlled balance of the diversity and variation of the addition will be guarantee by the planning team and the computer-aided tools will be used as interface to control the variation.

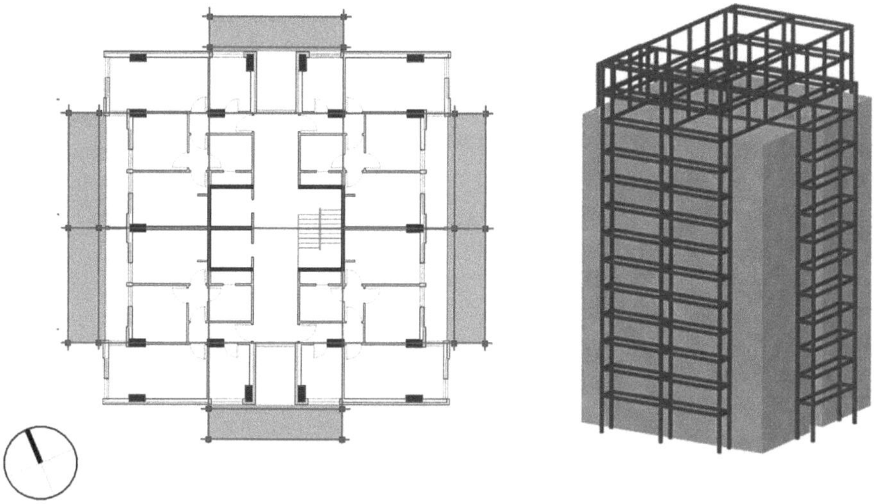

Figure 3.6 ***Structure frame in the Peristeri case study for the tower typology (building T11).*** Technical study of the structural feasibility to combine additions on the top of the buildings realizing new residential units.

of possible urban configurations. The improvement of the interventions in terms of energy performance and structural safety have been investigated and have proven the great potential cost-effectiveness of expandable architecture.

The technical and structural reliability of hypothetical volumetric addition on existing buildings is based on the possibility of considering external structures to be put alongside the existing buildings.

In the majority of illustrated cases, additions have been made on the rooftops of existing buildings: in this case, the risk of exceeding the load bearing of these buildings, which were not conceived for these additional volumes, necessitates a limit of add-ons because adding too much is very dangerous, especially in seismic areas like Athens. This is the main reason why the hypothesized add-ons have been conceived with external support structures[37] in our work. Here, the technical and structural feasibility of potential add-ons relies on the hypothesis of an external supporting frame (Figs. 3.6–3.8). These structures have been

[37] The external structures can be also applied for the purpose of improving the mechanical, static and seismic performance of existing structures, providing an additional element of construction quality and durability.

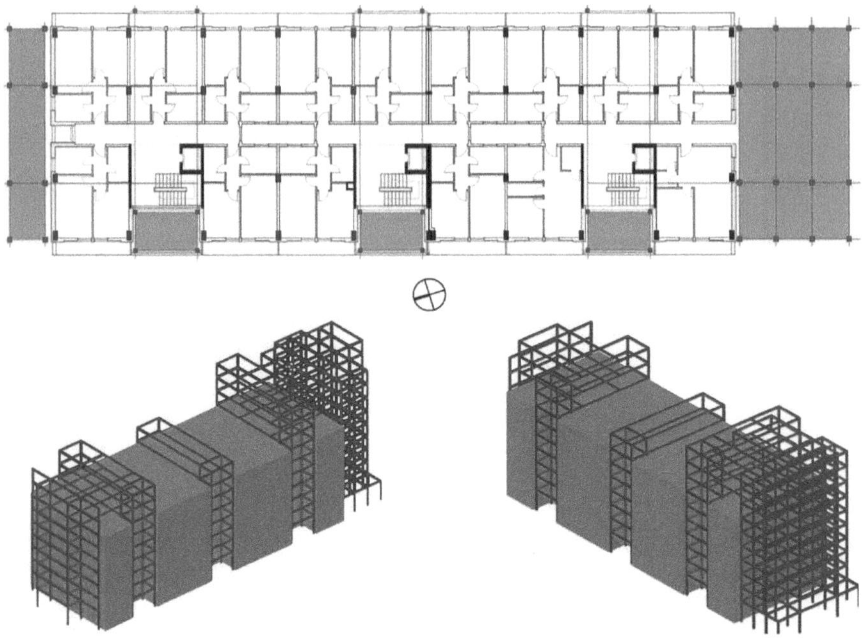

Figure 3.7 ***Structure frame for the B6 block type.*** Technical study of the structural feasibility to combine additions on the top and aside the buildings realizing new residential units.

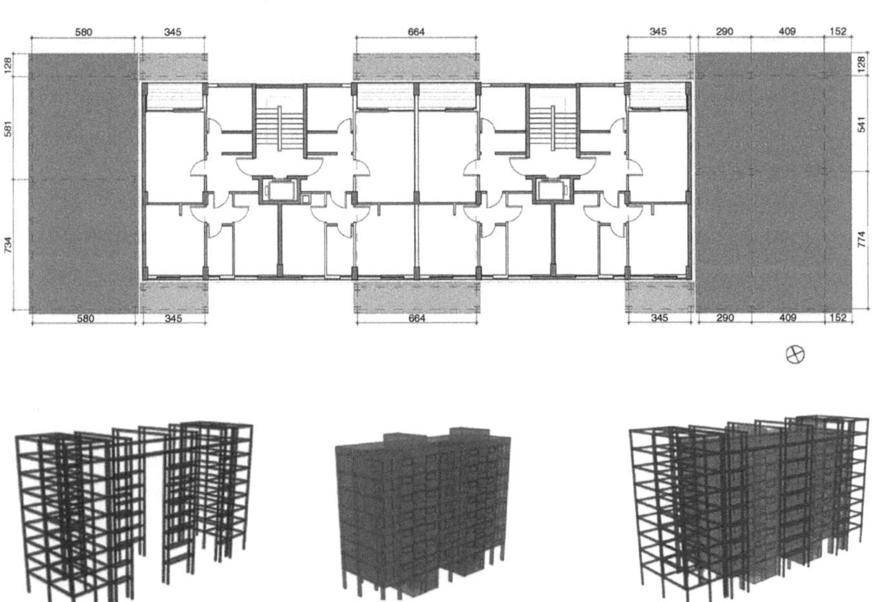

Figure 3.8 ***Structure frame for the T7 block type.*** Technical study of the structural feasibility to combine additions on the top and aside the buildings realizing new residential units.

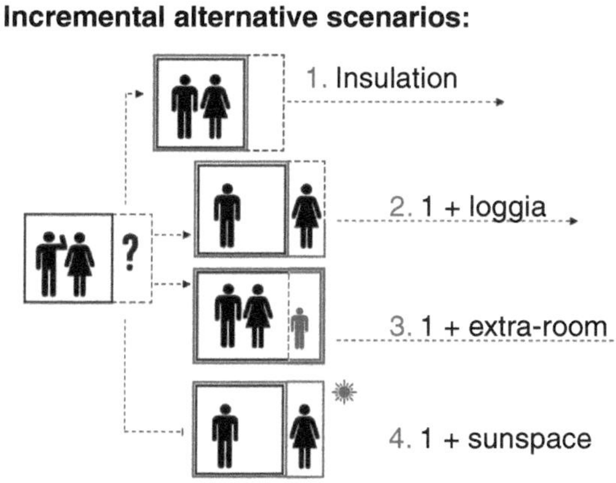

Figure 3.9 ***New volumetric facade solutions for the existing residential units.***

tailored and customized as a function of different constructive elements and architectural types.

In this scenario, the hypothetical structural frame as designed also may be used to combine new residential units with facade additions aimed at improving the quality and the space of existing residential units. So, on existing facades, the possible expansions of the existing envelopes may be thought of as extensions of the existing internal spaces. Whether occupants desire a bigger balcony rather than a loggia or an extension of the existing room, this system could allow people the ability to choose the use of their outside and inside spaces, according to various alternative and incremental hypotheses, as shown in the scheme reported in Fig. 3.9.

Also, the new envelope of these facade solutions may be equipped with a variety of coatings, obscurants, and shadings according to different technical, aesthetic, and socially oriented requirements. Thus, we may create a kind of "abacus" of the possible technological and formal solutions. This abacus can be conceived as the main instrument to give the inhabitants the ability to express their needs within a common order respecting the structure, the functionality, and the guidelines of a renovation strategy involving architectural modifications. Since the "letters" of this alphabet have been studied as a modular system and can be joined together in multiple combinations, the "discourse" that it will produce from inhabitants could have a positive result in terms of aesthetics and functionality.

The suggested technology can be based upon the utilization of prefabricated facade components and extensions that could easily be produced through a CNC machine.[38] The number and size of panels and openings can vary and is a function of horizontal and vertical coordinates (for example, the distance of the openings from the borders of the panel and among each other). By mapping and coding the generic facade panel as per the listed variables, it is possible to define an open geometry through the utilization of software tools.[39]

[38] As previously explained, the possible interventions have been categorized into three different levels, at each level corresponds a different scale of the renovation element. Level 1 is based upon the facade element, level 2 upon the horizontal extensions, and level 3 upon the vertical extensions. In order to group all the possible typological and geometrical variables that can be produced with the same technology the abacus chart can illustrate the range of options. When considering the design and production relationship, it is necessary to exactly and preliminary understand the morphology of the buildings, the most representative types of openings in the facade and define the possible transformation that each facade module could encounter in the design proposal. It is then fundamental to define the different shapes of the panels that could fit within a large number of scenarios and at the same time could be produced in opened forms and configurations so to answer to a vast number of possibilities, as required by the vast housing stock and various user needs. In this terms, the abacus becomes one of the main design tools for planners and professionals involved in the process: it the catalogue of options among which the inhabitant can choose and will enrich the possibilities of synergies between the technical and sociological proposed strategies. In order to do so the starting point is the data collection through the utilization of a 3D scanner enabling the planner to collect all the geometrical information regarding the existing façade. The mere collection of points is anyway not sufficient, it is in fact important to study and map the original façade, the main axes, the possible symmetries and the division of the envelope in parts that can then be separately reproduce, built and mounted. The planner has to define where are the boundaries of the base element and the possible subdivision of the façade in the most efficient and effective base component. The idea is that to figure out which is base element whose dimension can vary in infinite numbers but whose typology stays fixed. Thus, a façade type can be divided into regular modules. Every dimension of the base facade module has then been given a parameter: the height of the floors, the window dimension, the balconies or loggias outline, the ratio between opaque and transparent surfaces, and so forth. The planners can directly introduce the data that come from the survey on the existing building, each and every time different, each panel and each building, but the definition of the module and the production machine will be the same reasoning on the same parameters. This module is generally the one that is then repeated in the original pattern of the façade. It is however true that the façade chosen for the above example is a simplification and can not represent the large number of cases that could be encountered in the housing building stock. Broader can be studied but also open catalogue continuously updating and enlarging considering a broader range of possibilities and façade element typologies might be foreseen. The focus is here once again on the methodology proposed rather than on the completeness of the survey upon existing buildings' elements and their typology.

[39] Rhinoceros (MacNeel) as three-dimensional modeler and the Grasshopper plug-in to govern the parametrical variation of the geometrical dimensions can be the assisting tools of this process. The Grasshopper plug-in is a graphical algorithm editor that allows designers with no formal scripting experience to quickly generate parametric forms. Therefore, each of the panels can be modeled and its parameters (geometrical dimensions) can be left unknown and changed time by time according to each different panel and each different building.

The additions can be thus grouped in an abacus of possibilities that will become one of the main design tools for planners and professionals involved in the process. In fact, the abacus may represent the communication instrument between technicians and inhabitants, which would enrich the possible synergies between the proposed strategies.[40] In this perspective, the potential add-ons can be also considered as a tool to combine the observed ethnographic aspects and end user needs or expectations, thus fostering their potential inclusion.

Up to six modifications of building envelope for each span have been considered in the reference case studies for the Peristeri urban compound. The hypothetical building transformation has been based on the assumption that additional spaces for users and inhabitants would be gained in the current residential units, and the additional units would generate added real estate value.

The technical and structural feasibility of the abacus relies on the hypothesis of an external supporting frame (Figs. 3.6–3.8), and has been tailored and customized as a function of different constructive elements and architectural types; the volumetric additions and options (including the zero-option one = no intervention) have been further combined with different options of materials, according to the construction and types of the different reference buildings.

The current condition of modernist buildings and urban settings of the 1960s and 1980s present clear evidence that the building techniques and the market have not been able to take into account any possible further development and changes, neither in the environment nor in the buildings themselves. The strategy of architectural modification of external envelopes proposes a reverse methodology: by defining the main design principles and

[40] Façade add-ons may be a strategy aimed at delivering forms of customized and variable components of self-expression in operations of energy retrofitting of existing building stock, with respect to urban dwellers and users' expectations. To achieve this, it is important to consider participatory and socio-integrated policies during the abacus definition phase. The research goal of the case studies is to study the possible evolution of the buildings in these specific post-war districts, by envisioning and simulating how inhabitants could take over the architectural role in the future and possibly transform their own houses. In this context, ethnography is not considered as a mere techniques toolbox directly borrowed from social sciences; on the contrary, design rhetoric refers to the ability and the purpose of design thinking to engage and foster transformations. In example, in Peristeri, there is a high percentage of elderly people and a continuously increasing number of young couple becoming part of the community. The traditional approach for multicultural society based on an unsubstantiated and narrow perspective would be ineffective in these urban contexts, therefore inhabitants should be given structural and technological solutions on which they may base their own personal changes and design processes.

parameters according to their variability, it is possible to develop a modular system that incorporates multiple design solutions where the single components can be interwoven, changed, or transformed, while guaranteeing the coherence of the urban configuration in its evolving process.[41] Thus, it becomes important to go back to the initial user analysis, on the collective spaces of the context, and to consider the interrelations between users and architecture without looking for a universal solution, but instead to search for new and dynamic methods and processes that can actively respond to the changeable nature of the surrounding urban, social, and environmental contexts.

Abacuses of modular options as a function of possible energy retrofitting scenarios and users' requirements in the different building types of the Peristeri urban setting have been developed.

Hypothetical incremental scenarios for the existing residential units in the tower building block are summarized in Fig. 3.8 illustrates the first hypothetical retrofit scenario on a residential unit type and the consequent effect produced on the map and the west facade of the buildings. Considering that symmetrical conditions do exist for the east facade of the same building type, hypothesized scenarios may apply to the east facades as well. Fig. 3.10 shows the comparison between the scenario as built (on the left part of the figure) and the first retrofit option (on the right).

As another example, Fig. 3.11 shows the possibility of combining facade insulation and window replacement with additional space for balconies (on

[41] Forecasting the possibility and studying the pattern as a modular system, keeping in mind the development of different combinations, it is possible to let everybody do what they want with their home while still leaving open the possibility for self expression, without clashing with construction codes and regulations, according to a metabolic approach (Macguirk, 2011). Autonomy in the design or production of space means that people involved in designing and building need to have access to knowledge of design and building processes and components in order to discern and enact. But at the same time it means that those processes have to be open enough to increase autonomy and adaptability in time, instead of limiting it or even turning it impossible. Design here is more related to a means for people to experiment different spatial possibilities, so they can evaluate, for example, where to place the openings, the size of the rooms, and so on (Ferrante et al., 2011). Concepts like durability, adaptability, and energy efficiency have been increasingly taken into account in the fields of building and urban research. Especially when considering the high standards and requirements for the retrofitting actions, one of main goals consists of the development of long-lasting building products that can adapt to different needs and environmental conditions throughout their life-cycle and that could guarantee a high level of building envelope energy performance.

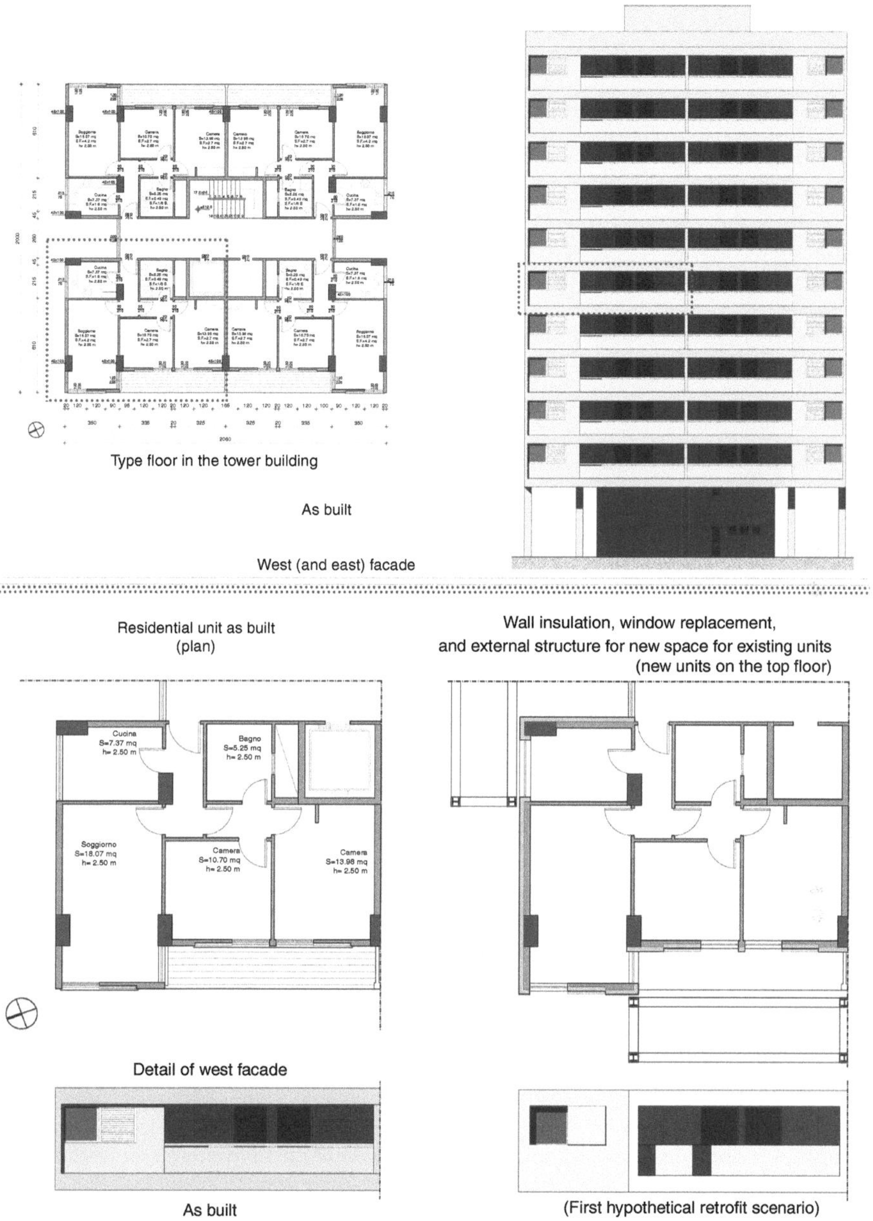

Figure 3.10 ***Tower building type and hypothetical retrofit scenario considering wall insulation, window replacement, and structural addition of the external frame.***

Figure 3.11 ***Abacus of modular options as a function of possible energy retrofitting scenarios and users' requirements in the west and east orientation of tower building type in Peristeri.*** (Facade insulation, window replacement combined with additional space for balconies, on the left, or additional space both for the room and for the balcony, on the right).

the left) or additional space both for a room and for the balcony (on the right part of the figure). Each solution is provided with a set of options/details for finishings and coatings. Fig. 3.12 shows the combination of facade insulation and window replacement with additional space for the other room (on the

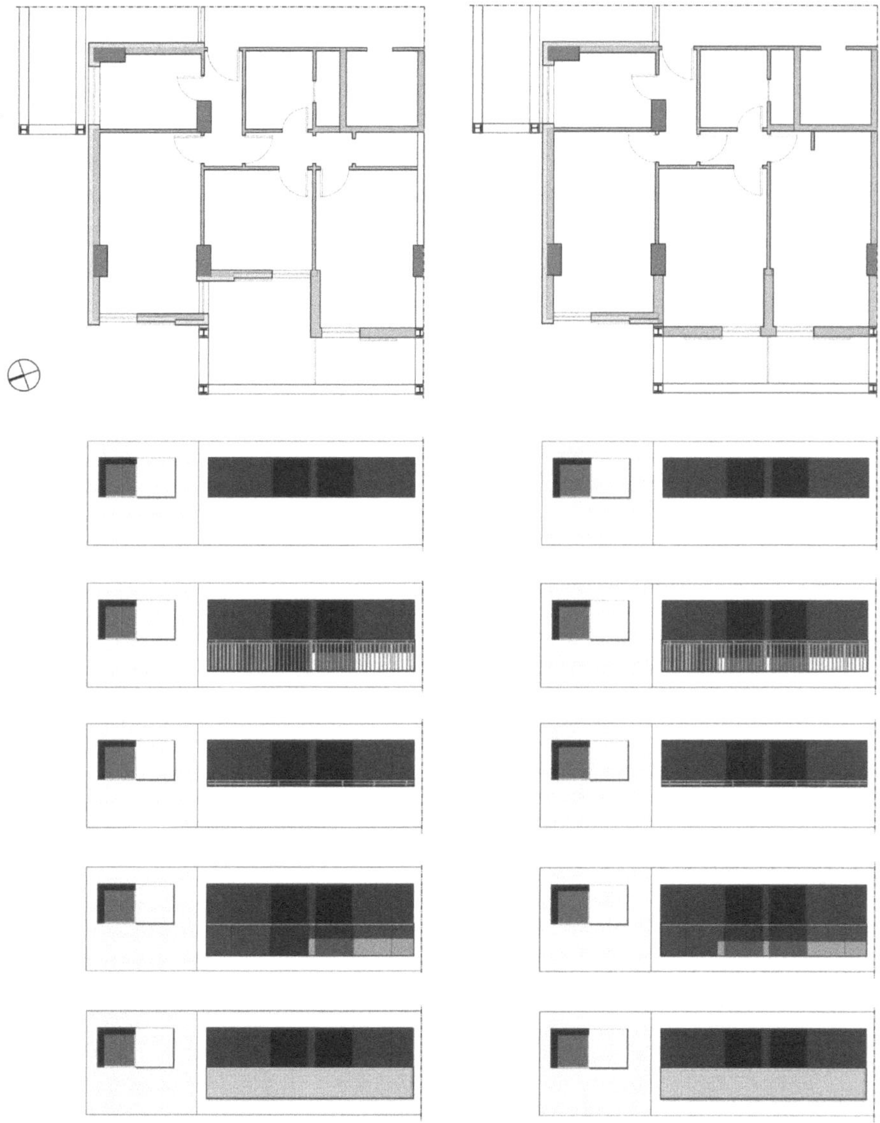

Figure 3.12 ***Abacus of modular options as a function of possible energy retrofitting scenarios and users' requirements in the west and east orientation of tower building type in Peristeri.***

left side of the figure) or additional space for both rooms, maintaining the same size of the as-built balcony (on the right part of the figure). Again, each solution is provided with a variety of options for finishings and coatings.

Similar abacuses have been hypothesized, assessed, and verified for all the facades in all the different building types of the urban compound in Peristeri.

The general results demonstrate that the abacus can be used as leverage to incentivize inhabitants in their decision to renovate their apartments, since it represents a possible user-driven design scheme to let inhabitants adopt tailored solutions according to their needs and requirements. This overall strategy may signify an innovative and different design approach toward the energy-saving retrofit of existing buildings. By using and combining different options, the outcome is far from being a single solution project; rather, it may represent the projection of a large set of possible projects, all different but similar at the same time, and therefore closer to the inhabitants' desired uses and habits.

The outcome of this strategy is not the design of a model, but the design of a process, as shown in the following different options and envisaged for the different building types in the urban compound (Fig. 3.13). Furthermore, the possibility of combining add-ons on top of and next to the buildings (and thus new residential units) with facade solutions was explored (Figs. 3.14 and 3.15) for the double block building type. Fig. 3.14 illustrates the possibility of combining new residential units on the top floor and in south side (right part of the facade) with new facade solutions in the existing residential units of the block building B6. Fig. 3.15 illustrates the possibility of combining new residential units on the top floor and in the south and north sides (right and left sides of the west facade) with new facade solutions in the existing residential units of the block building B6.

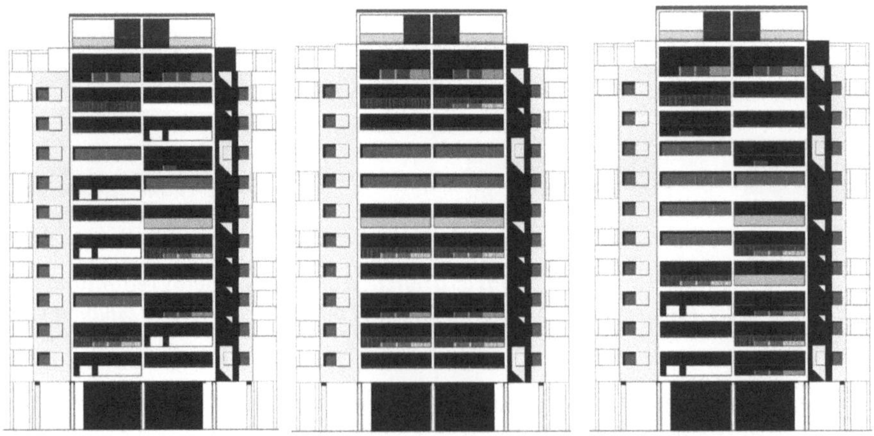

Figure 3.13 ***Combination of additions on the top (new residential units) and on the elevation of the buildings with new façade solutions selected from the abacus for the west façades in the tower building type.***

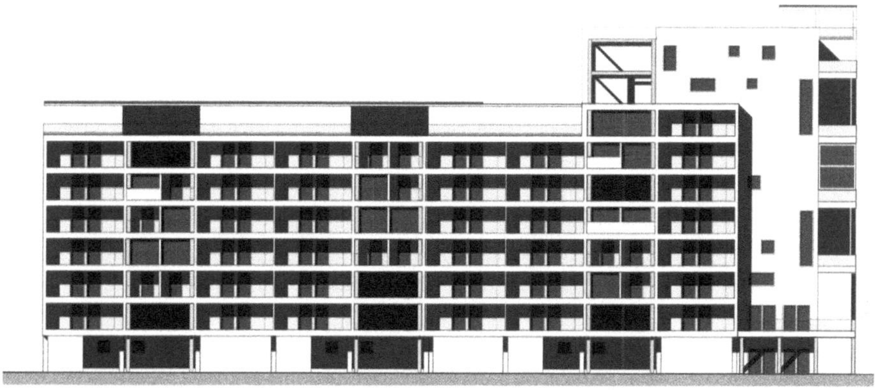

Figure 3.14 ***Designed scenarios on the west façade of the double building block B6.*** New residential units are combined with add-ons and new façade solutions in the existing residential units.

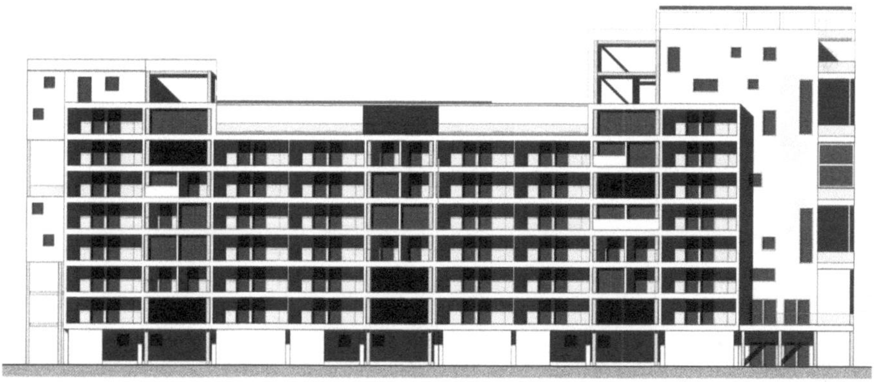

Figure 3.15 ***Designed scenarios on the west façade of the double building block B6.*** New residential units both on the south and north side of the block building are combined with add-ons and new façade solutions in the existing residential units.

The abacus-tool could be conceived as a "real-time, interactive tool" to engage inhabitants and owners in the selection process of the required solutions, giving life to a creative, variable yet controlled, self-expressed possibilities within the retrofitting process of urban buildings.

Finally, the same strategy has been applied to the building block T7 case study and some of the possible configurations are reported in Figs. 3.16–3.18. As shown, for each building, the first step consists of the definition of the facade element type and geometry. The second step consists of defining the possible variation that each of the panels can have in the energy renovation phase. This is strictly related to the type and level

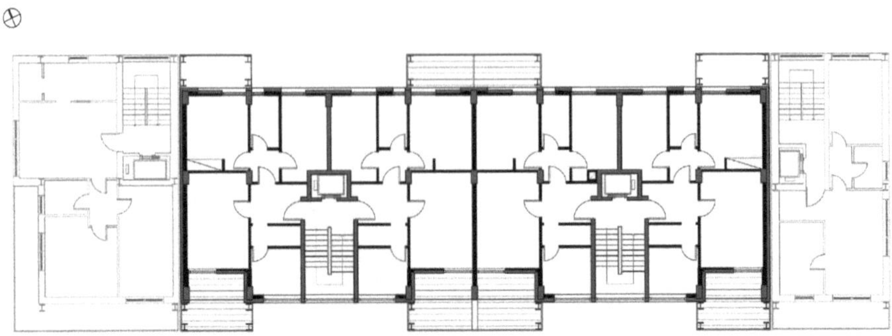

Figure 3.16 ***Plan type of the building block T7.*** Renovation scenario considering wall insulation, window replacement, and addition of the external frame at the west and east side of the building creating new residential units.

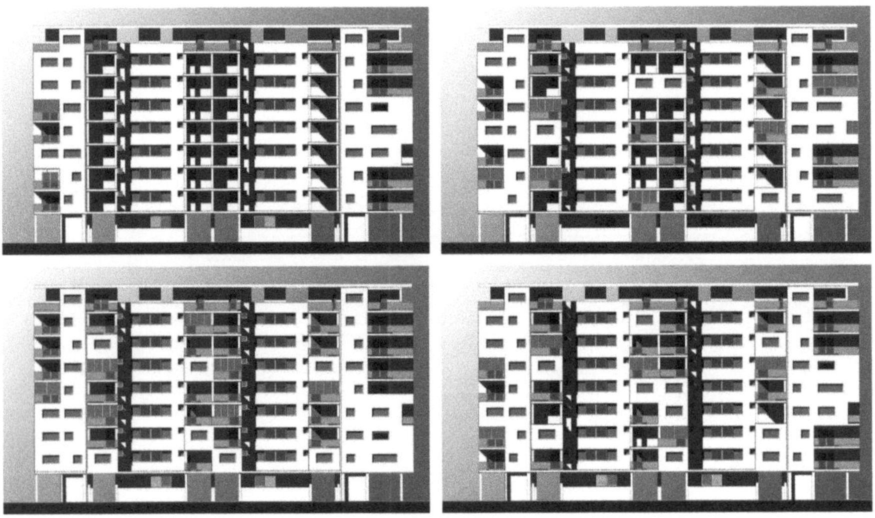

Figure 3.17 ***South façade of the block building T7.*** New residential units combined with façade solutions for existing residential units.

of intervention planned. In the hypothesized process, different levels of transformation can be envisaged:

- Level 0 is considered to be the stage where the inhabitant does not want to undertake the facade upgrade and there is no alteration.
- The Level 1 intervention envisioned is the facade transformation; the existing facade component could be switched to a component with identical shape and high-energy performance, with proper insulation. The grade has increased in terms of performance and improvement given to the envelope, calculated on the basis of the envelope performance index.

Figure 3.18 ***North façade of the block building T7.*** New residential units combined with façade solutions for the existing residential units.

- The same can be observed for Level 2, where the horizontal/vertical extension could simply be generated by the extrusion of the existing façade, or rather, the geometry of it could be changed.
- As for level 3, in case of a vertical extension, the prefabricated module that is added on the roof can reproduce the facade of the existing building, continuing the alignment of the facade and the openings, or it can signal an interruption or a break with the existing building. Since Level 3 is the definition of a new prefabricated housing unit module, the typology of the extension has not been grouped in the chart illustrating the levels of increasing upgrades that could be applied to the facade.

Up to six possible modifications for each of base modules have been envisaged, but the number of variations can be increased case by case, according to the planners' desires.

The design process is therefore based upon the idea of developing a collection of possible technological and formal solutions for the new envelope that should give the inhabitants the ability to express their needs within a common order, respecting the guidelines of the project. The three levels of action are characterized by the use of the same technology that could therefore apply to all the incremental stages of the renovation design. The catalog that will be delivered to the inhabitants will show all the possible transformations. For each option, together with the information related to the costs of the intervention, a detailed evaluation of the energy savings and increasing envelope performance will be provided. This information will be illustrated and it will be easy to comprehend the positive impact as it is meant mostly for a nonexpert audience. The details provided are fundamental in order to enable the inhabitant to make the right choice regarding his or her own needs and economic feasibility; the information disseminated should be based upon the conviction that an informed choice is the best one. Together with the performance aspects, it is also important to set the variables related to the aspect of the facade and the personalization degree. Whether the abacus promotes the breaking of the original geometry or maintains the original rigor, or if the main characterizing object of the project is based upon the usage of a special material, (ie, its pattern or the tiling of the façade) must be determined by the planners according to their analysis and their own sensibilities. It is therefore impossible to define a standardized catalog of choices and possibilities that could fit every circumstance, but the suggested method guides the planner in the definition of a case-specific set of elements (facade components, horizontal extensions, and vertical extensions) among which the inhabitants could choose and express their preference.[42]

[42] From a geometrical, typological, and formal point of view, it is important to set the rules for the definition of the new facade outcome. Should the style be defined by the inhabitants, directly giving them a high degree of freedom in the selection of material, colors, shape, and function or shall the architect impose his own interpretation? As anticipated, the sociological component of sustainability is central and plays an important role in the feasibility of the intervention. The topic is once again not new, the problematic has been faced often in the history of architecture and has often landed in simplified answers. Friedman (1975) states that "an architectural style can arise from inhabitants' instinctive and automatic decisions, it is a randomness in order (...) design is a game with no explicit rules". "Because the architect of the past served a single client, he was able to become thoroughly acquainted with the client's individual tastes and way of life – he could make the client's decisions for him. The majority of architects designing housing today do not work for millionaires, but for millions of individuals who will work or live in the architects' projects. The architect cannot study the behavior of each user: instead he constructs an ideal user and plans for this ideal. [...] He should, instead, devise methods of promoting choice among the users themselves". The same idea is supported from Price as well that writes: "One uses a tight, carefully designed technology to achieve a loose, free-will social patterning. I don't think this sort of variation necessarily requires any compromise on the validity of the first design intention." It is, in other words, a search of randomness in order: the idea is that it can be possible to control the chaos, with the aid of a computer and interface, without the need of defining a formal style and any control on the outcome. Upon this aspect the author wants to stretch a clear distinction between the proposed methodology and the related design outcome of the theories above mentioned. According to the

(*Continued*)

The division in several base modules allows the planner to control the facade composition and assemblage at all the scales, but how is it possible to combine the different facade elements, leaving the possibility for the inhabitants to choose, and at the same time, prevent chaos from governing the entire facade system? The question is strictly related to the idea of defining a set of rules and combinations that are possible and letting the final outcome be open. The designed approach aims at defining the energy renovation practice by seeking the definition of a design method that could lead to multiple design outcomes and not yield one prefixed and overimposed result. The variability of the different elements of the new facade are numerous. The previously mentioned variables are only the primary ones; a broader range can be implemented and can be increased or changed according to the specific case study. In fact, all the possible design solutions can be studied and tested on the primary module According to the suggestions of the planners and the inhabitants, a first set of options is produced and the abacus can work as a catalog of possible solutions among which the inhabitant can choose. Theoretically, the approach is based on the assumption that it is possible to develop multiple variant designs and multiple criteria analyses of a building refurbishment in order to enable from one to an infinite number of alternative versions; in this manner, the architecture is an unfinished work of art and the structure is the support, which leaves blank spaces for the user to participate in the decision-design process. Since a formal design is necessary to guarantee a structured and organized process that corresponds coherently with current energy and safety regulations/requirements, fixed components (structural and functional invariants) must be applied in the (re)design processes of the buildings. On the other hand, a degree of adaptability/flexibility and the adoption of processes that directly engage the inhabitants could offer a real solution both to the anonymity and standardization of these housing complexes. Defining the main design principles and parameters according to their variability leads to the possibility of understanding and controlling a "genotype" that incorporates multiple design solutions in itself. The single components (the phenotypes) can be interwoven, changed, or transformed, but could still guarantee the

here presented process the architect is and should always be the person that stands behind and controls the entire system. The scheme provided is only a reference illustrating the method that stands behind the creation of the abacus. It is clear that, once the base modules have been found and defined, the formal, geometric, material, and chromatic variation possibilities will be determined by the architect. In fact, all the aspects that overcome the merely typological analysis have to be evaluated and based upon geographical and climatic consideration, coherently with the planners' sensibility, the context, and a large number of variables that are not part of the core of this research.

coherence of the final composition as an open and in fieri composition. By iteratively changing the parameters used to generate models, the planner can create multiple possibilities for consideration.[43]

Current experiences on social or participative design have shown great potential and interesting aspects on the theoretical side, but often have displayed a high degree of chaos in the final formal outcome. The proposed approach here is that the rules and the style must be set and defined in order to avoid a deconstructionism ruled only by randomness, within which there is very little beauty. In this frame, we can state that innovation is capable of finding the set of rules that can control the process and its outcome, guaranteeing a formal coherence of the system and of the architectural result as well. The task of the architect is therefore to draw the pieces of the mosaic and the rules for the possible combination, until such time when the final output, which will be unforeseeable until the completion of the choices from the inhabitants, will be determined.

3.1.3.1.2 Energy and Cost Benefits of Volumetric Addition in Energy Retrofitting Actions: Low-versus-High Transformation Retrofitting Options toward ZEUS

Further economic analysis has been developed to test the feasibility of the previously envisaged scenarios. Results show that in each building type, the option of incremental units and add-ons represents an effective strategy to decrease the long payback time of the energy retrofitting.[44] As shown in Tables 3.19–3.21, the potential gains related to the construction of new residential units greatly decrease the payback time of all the initial costs of investment, including the

[43] The possible solutions and the formal suggestions for the facade transformation and extensions should come from the architect that sets the pattern and the series of proposal collected in the abacus, together with the rules that stand behind the computer aided program that is then used as an interactive interface to handle the dialogue with many final users. This ensures the coherency of the final outcome, proposing only fitting solutions and reiterating the trial-and-error loop, narrowing down the possible choices, still defined by the architect, that therefore plays the role of moderator and definitions controller. It is once again the architect that sets the rules of the game, iterates the formal guidelines, and guarantees the correspondence of the project with the general needs, requirements, and pre-existing context. Within this open form and supports scheme, the inhabitants can chose the most fitting option among the possibilities defined by the planner. Interaction and clashing of choices from one inhabitant to the other are also controlled and handled by the program: the options proposed to one inhabitant are influenced by and different than the wishes and decisions made by another inhabitant that, for example, lives next door or above; the neighboring options in the facade will not collide, and the risk of turning the job into a chaotic mosaic of uncontrolled design can be avoided.

[44] For the economic analysis of these hypotheses, we have assumed a price € 1,400 per m^2 per net floor area of new construction (add-ons) and a correspondent sale price of 2,200 €.

Table 3.19 Cost-benefit assessment of the proposed "deep retrofitting," including the creation of new units in the tower building type T11

T11 (40 units)			
Retrofitting options	**Costs/gains related to the option (€)**	**Savings (€/year)**	**Payback time (year)**
A – Complete retrofitting	−371,595	18,120	20.5
B – Metal structural frame	−173,910		
C – PV and plants	−196,000	16,500	11.9
D – Construction of new residential units	−264,000		
Total costs/investments	−1,005,485		
G – Gain from new units	600,000		
Total	−405,485	34,620	11.7

Table 3.20 Cost-benefit assessment of the proposed deep retrofitting including the creation of new units in the double block building type B6

B6 (90 units)			
Retrofitting options	**Costs/gains related to the option (€)**	**Savings (€/year)**	**Payback time (year)**
A – Complete retrofitting	−558,559	39,015	14.3
B – Metal structural frame	−323,897		
C – PV and plants	−440,000	30,800	14.3
D – Construction of new residential units	−1,331,000		
Total costs/investments	−2,653,456		
G – Gain from new units	2,420,000		
Total	−233,456	69,815	3.3

Table 3.21 Cost-benefit assessment of the "deep retrofitting," including the creation of new units in the block building type T7

T7 (28 units)			
Retrofitting options	**Costs/gains related to the option (€)**	**Savings (€/year)**	**Payback time (year)**
A – Complete retrofitting	−280,717	18,704	15.0
B – Metal structural frame	−174,319		
C –PV and plants	−236,800	30,800	7.7
D – Construction of new residential units	−1,582,000		
Total costs/investments	−2,273,836		
G – Gain from new units	1,924,000		
Total	−349,836	49,504	7.1

costs for RES (row C in the tables) that are required to set to zero the energy balance of the buildings. Furthermore, if we consider the hypothetical investments of developers in add-ons, the gains obtained by sales of the new apartments would come close to counterbalancing both the standard energy retrofit cost and the cost of RES (PV and solar panels) to set to zero the energy demand of the whole building. Noticeably, these kinds of transformations go well beyond the concept of deep renovation and imply non-energy-related factors. Next we will move from deep renovation to a radical, major transformation of the existing built environment.

3.1.3.1.3 From Deep to Major transformation in the Retrofitting Options toward ZEUS

These hypothetical interventions designed for the urban compound of Peristeri do not negatively affect the availability of open green areas. As was forecast and indicated in Chapter 2, to properly address the objective of low-carbon, zero energy in existing urban environment, we also need to consider the buildings and the related open areas; this is especially true in the Mediterranean warm regions of the EU, where the snowballing effects of all the buildings and open areas is very important, especially in the summer season. Furthermore, a more holistic vision of the urban energy demand, which considers outdoor spaces and buildings as a whole, can better drive urban planning decisions and bottom-up, grass-roots initiatives toward the adoption of alternative residential configurations/redevelopment of existing urban areas, thus motivating main stakeholders like public bodies, policy makers, urban dwellers, and administrative bodies in the adoption of participative processes or green behavior. This is particularly challenging in urban contexts of "modernism," like the district of Peristeri,[45] where the

[45] Most of the buildings of European suburbs have been designed according to modernist principles and concepts and conceived to be universal. Solutions have been applied with few considerations on cultural and social characteristics all over Europe, generating, in different sites, very similar problems In brief, the functionally oriented vision of architecture has produced dull blocks in a dull environment. Nowadays, these buildings are in urgent need of environmental and energy retrofit. In a complexly structured context, where the number of decision-makers and cultural scenarios overlap; and where the temporal dimensions and social background of the citizens are dissimilar; and where local and global, physical, and virtual dimensions coexist and where, finally, it is no longer possible to ignore the direct and indirect relations with the context, it is necessary to identify design procedures that can quickly adapt to environmental variations and new requirements. Succeeding in this endeavor requires more than getting the engineering right (Webler et al., 2010) and focus should be placed on the social aspects at community level. In fact, developing more sustainable environment depend upon consumers' willingness to engage in greener and more collective behaviors (Peattie, 2010).

majority of the blocks are currently characterized by the presence of pillars and a narrow and dark basement level. Cars are everywhere and block the possible fruition of the street/garden/park level from the pedestrian perspective. A proper renewal action has to be developed to guarantee a higher degree of freedom so that the inhabitants will be able to choose the best solution according to their needs.

(Re)design processes of existing urban buildings are called upon to respond coherently to social and end-users' requirements, to current energy and safety regulations/requirements, and more permanent components (structural and functional invariants) as well. In this process, a certain degree of adaptability/flexibility and the adoption of processes that directly engage the inhabitants could offer a real solution both to the anonymity and standardization of these housing complexes. This approach also should be applied to the ground-floor level, incorporating different levels of privacy in the collective areas, emanating from the direct and active participation of the inhabitants.[46]

A new spatial organization for the outdoor spaces in the Peristeri urban compound has been outlined. To reach the savings potential indicated in Chapter 2, a feasibility study was developed to maintain and increase the green and permeable surfaces as much as possible. Access to the open spaces was reorganized to let people enter the area from the renewed retail and commercial areas into "semi-private" green courtyards, to give a warm and welcoming feeling that would enrich the quality of the communal spaces. Thus, a five-step progressive design scenario was developed for the transformation of the existing open area in Peristeri, taking into account the incremental volumes as designed in the major transformation envisaged so far. The final configuration combines the need for a more inclusive environment for urban dwellers and the cooling potential of increased green and permeable surfaces (see Table 3.22 and Fig. 3.19). The results show how the potential increase of green surfaces in the area spreads to 80% of the existing permeable areas. Thus, notwithstanding a significant increase of existing building volumes, there is still a wide potential for recovering green space and natural open areas in this urban context. In Fig. 3.20 the final master plan, including green surfaces and areas shaded with greenery and PV pergolas, is shown.

[46] As Gehl points out, the majority of social residential areas have been built according to principles and ideologies that give very low priority to outdoor activities, to the connection between distance, intensity, to the closeness in various contact settings has an interesting parallel in decoding and experiencing cities and city spaces (Gehl, 2010).

Table 3.22 Calculation of progressive increase of green and permeable surfaces in Peristeri urban compound

Intervention	Addition of green surfaces (m^2)	Green surfaces (m^2)	Increase compared to the as-built case (%)
1 – As built – (Existing green surfaces)	—	12,100	0
2 – 2 Multistory parking building and reduction of parking areas at ground level	800	12,900	7
3 – Green roof on the existing commercial porch	1,500	14,400	12
4 – Green roof on the top floors of existing buildings	3,680	18,080	30.5
5 – Green vertical surfaces on buildings	3,700	21,780	30.5
Total amount of potential green surfaces	9,680	21,780	80

3.1.3.1.4 Existing Limits for Add-ons in Dense Built Areas: The Case Study Evripidou

As for the previous case studies, further hypotheses of transformation have been envisaged to understand whether it is possible and cost-effective to realize new construction elements connected with the existing urban setting. Given the location (the historical center) and the urban density, the procedure to perform this analysis in the central Athens area needs to be articulated according the following successive steps:

- Assessment of constraints (historical, structural, morphological, etc.).
- Realization and evaluation of the possible design scenarios.
- Energy analysis and the economic impact of designed solutions.

To determine whether it is feasible to implement the transformation and expansion in the RUE, it is necessary to establish which of the 11 buildings may be subjected to such a transformation, individually and in their mutual relation.

In Table 3.23, two types of possible intervention (addition to the existing buildings and new construction) have been set in relation to the main constraints. The new building hypothesis (demolition of the old building and construction of a new one) can be considered in the following cases:

- The existing buildings do not present historical or architectural features to be maintained/preserved.

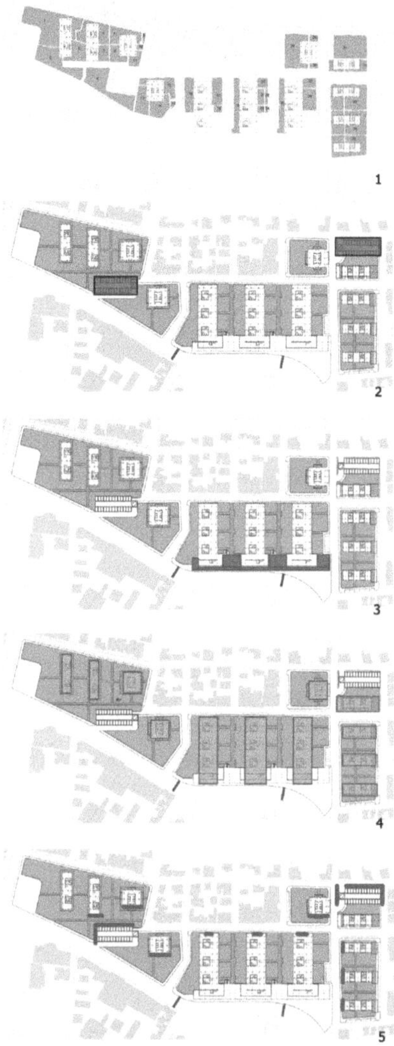

Figure 3.19 ***Maps of the five-step progressive design scenario were developed for the transformation of the existing open area in Peristeri.***

- The boundary conditions, consisting of adjacent buildings with blank walls, do not generate property or rights conflicts among dwellers.

In the specific context of the Evripidou central RUE, for the plot area of building 7, which consists of only two floors with no historical value, the demolition and subsequent construction of a new building has been hypothesized. Fig. 3.21 illustrates the analysis of resulting constraints and the possible densification in the Evripidou urban courtyard. Furthermore,

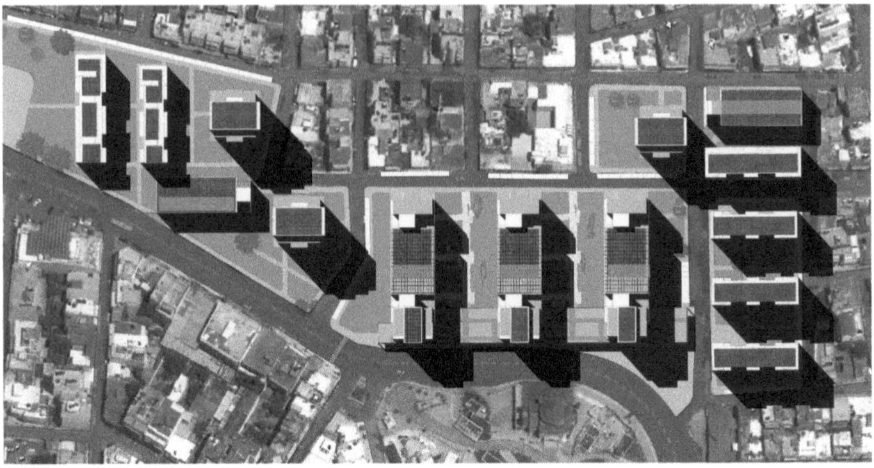

Figure 3.20 ***Plan view, final scenario in the urban compound in Peristeri.***

Table 3.23 Main existing limits/constraints in the analyzed context

Limits/constraints

Historical and architectural constraints	The buildings with historical and architectural features can only be preserved and in no case can change of volumes be admitted
Constraints border	Inconvenient situations related to the reciprocal views of the various buildings (limited or avoided in the presence of blind walls)
Morphological constraints	Most of the buildings present the smaller floors located at the roof level. This makes spaces too small to be suitable for add-ons
Structural constraints	Do not exceed the load bearing of the existing buildings
Existing legislative constraints	• Avoid "pilotis" • Maximum height of 22 m • Not exceeding six floors • Floor area covering ratio within 60% of the surface area • Maximum size achievable for each building is four times its surface area in plan

Table 3.24 illustrates the possible incremental additions in terms of maximum possible surfaces. Calculations to evaluate how potential gains from new additions can counterbalance the retrofitting operation costs have been performed and the results, including and calculated simple payback time, are shown in Table 3.25. Additions and energy retrofit to increase the market

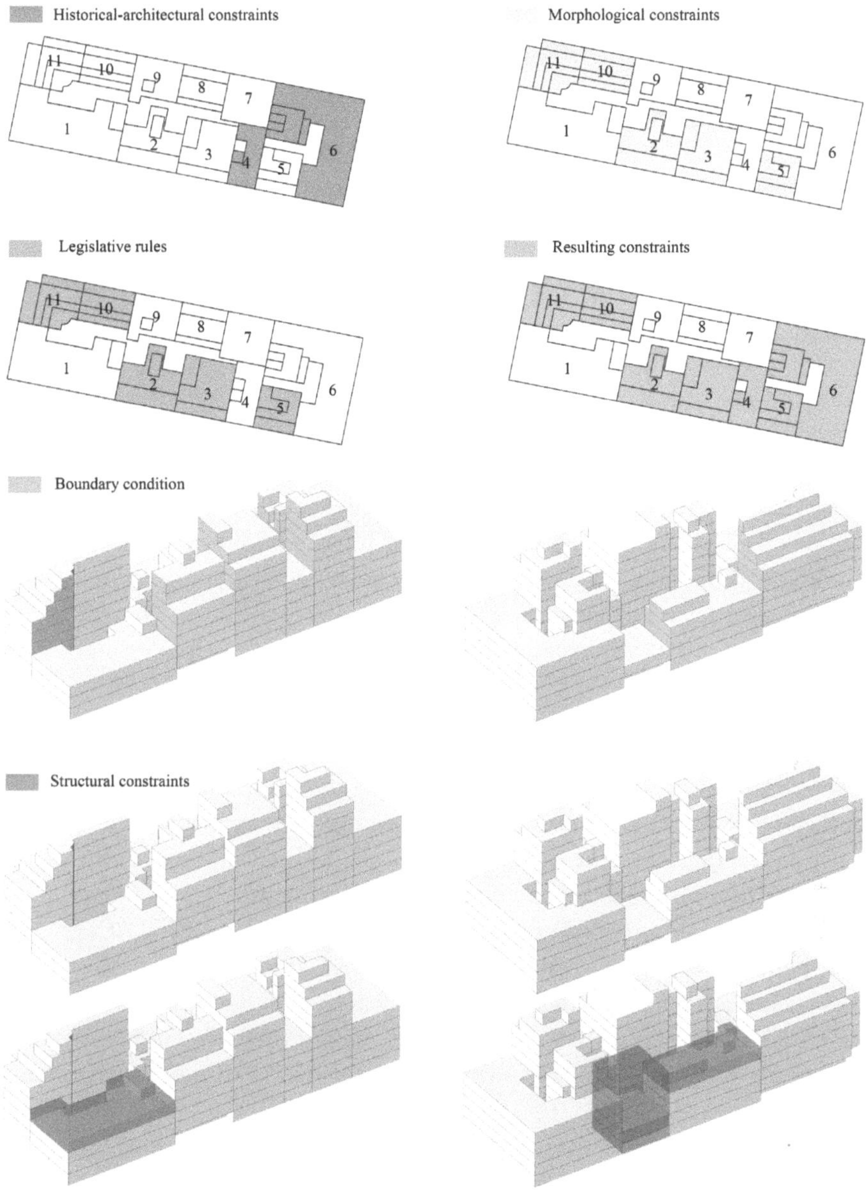

Figure 3.21 ***The resulting constraints and the possible densification in the Evripidou urban courtyard.***

Table 3.24 Admitted incremental volumes as a function of the emerged limits/constraints in the Evripidou urban courtyard

Buildings where possible addition/new construction is feasible	Surface in plan (m²)	60% surface (m²)	Max admitted surface (m²)
1	410	246	1640
7	160	96	640
8	130	78	520
9	140	84	560
8+9	270	162	1080

value and the standards up to nZEBs in this urban courtyard can provide an economical advantage, bringing to 15 years the simple payback time of retrofitting costs, which was initially estimated to equal 20 years (see Table 3.25). Negative values in building 7 (*) means that there is an immediate gain received by the sale of a new building in place of the existing one-story building (7) (Table 3.25). Different views of this final scenario, where RES are again readjusted to new energy demands, are reported in Fig. 3.22.

3.1.3.1.5 Energy and Cost Benefits of Add-ons in Densely Built Areas: Evripidou Case Study

In this last phase, we proceed with the creation of a new model to obtain new results through energy simulations with the program DesignBuilder.[47] In the following tables the new dimensioning of these facilities, the total costs, the savings, and the payback times of the buildings are illustrated; a final assessment is made on the entire urban courtyard.

Studies and simulations performed for the Evripidou central Athens area have demonstrated that zero energy buildings are also achievable in the dense central area of the AMA. However, from the economic point of view, the initial up-front costs are still very high, even if they do sensibly decrease after using deep renovation to bring the standards up to nZEB with volumetric addition.

For a concrete implementation, the energy efficiency of existing buildings at the urban scale should also be considered for the relations (potential

[47] For the modeling of new volumes the assumptions chosen for the previous phases have been used. Regarding the cost analysis of the hypotheses, a price of € 1400 per m² has been assumed for the construction of new expansions and a sale price of 2200 €/m² has been estimated and applied for calculations. For Building 7, the cost of a building replacement also included the cost of a controlled demolition (estimated equal to € 295.8/m²), to avoid damage to the adjacent buildings.

Table 3.25 Hypothesized costs for retrofitting operations including new additions and calculated simple payback time

Buildings	Insulation costs (€)	Window replacement (€)	Insulated green roof (€)	Plants and PV costs (€)	Construction costs (€)	Sell (€)	Total costs (€)	Total savings (€)	Payback time
1	87,218	61,531,00	40,370	90,641	344,400	541,200	82,960	9,039	9
2	101,074	87,631,00	14,342	102,161	0	0	305,208	12,094	20
3	82,100	111,230,00	21,247	157,209	0	0	371,786	21,914	13
4	39,960	15,533,00	10,624	40,986	0	0	107,103	6,275	14
5	69,966	106,997,00	12,217	87,370	0	0	276,550	13,188	17
6	152,372	168,229,00	48,337	216,655	0	0	585,593	29,310	16
7*	16,166	17,033,00	15,404	62,916	772,024	1,331,000	−447,457	4,027	0
8	41,972	23,132,00	12,748	55,816	98,000	154,000	77,668	6,999	11
9	33,579	24,332,00	13,280	55,284			126,475	6,914	10
10	54,886	106,163,00	13,811	93,305	0	0	268,165	10,329	21
11	81,159	84,264,00	17,529	118,262	0	0	301,214	12,497	18
Total	760,452	806,075	219,909	1,080,605	1,214,424	2,026,200	2,055,265	132,586	15

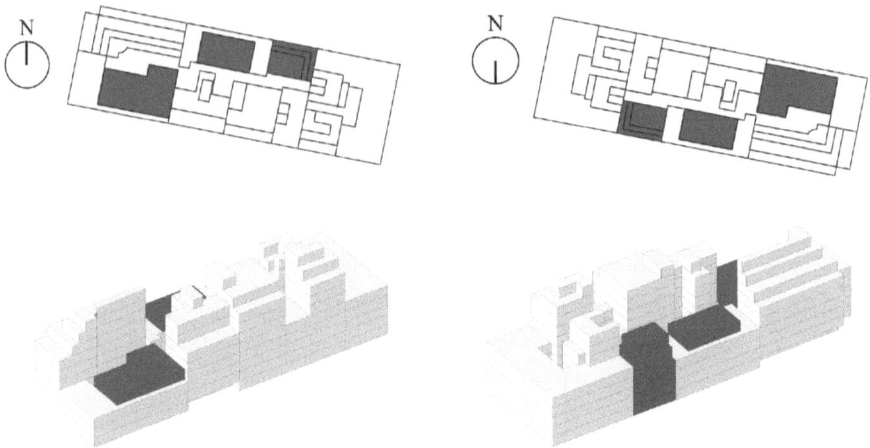

Figure 3.22 ***The considered final configuration for add-ons and densification in Evripidou.***

synergies/constraints/barriers) that may arise in the transition from the single buildings to the urban courtyard. These interrelations open up different scenarios of technical and economic feasibility: social and user-oriented, together with economic factors, become determinant and imperative in this context. For this reason, further possible synergies in the central urban compound can be envisaged.

In other words, we attempt to consider the single buildings as a unique urban service and house settlement; among the many possibilities, considering the building as a single element, it can be hypothesized that all owners can share the total cost of operations. Obviously, this could lead to very lengthy negotiations and nontimely agreements between urban dwellers, considering the disadvantage of those buildings having limited or no additions.

Another possible scenario could be "joint liability," or support, to be established, thanks to or in force of possible recommendation by public administrations and municipalities. For example, again, the owners (or the potential investors) of the buildings where a major transformation is admitted and where major gains can subsequently be achieved, can be charged a quota for the energy retrofitting of the adjacent buildings. This can be done in exchange for possible tariff incentives, incremental volumes, feed-in tariffs, and so on, in order to achieve a fair, balanced and win-win sharing of the burden among different dwellers. In this scenario, shorter and more effective negotiation/agreement times can be obtained, as only the property owners or the investors in the incremented buildings (and not all of them) are the main stakeholders, so they become the negotiators of the retrofitting operations.

In the case in question, for example, in the building undergoing major expansion (building 7), major gains could cover a portion of the expenditure for the roof insulation (the higher cost-effective retrofitting option) and the photovoltaic system.

More generally, considering the payback time of 15 years as in the final scenario, it could be extremely interesting to evaluate the possibility of further incentives in terms of volumetric compensations, to be used in exchange for energy, environmental, and social performance.

3.1.3.2 Discussion

A significant part of the major Athens metropolitan area has been mapped and three significant RUEs have been selected to investigate the energy demand, and to propose hypothetical design scenarios to set to nearly zero the local energy demand, according to incremental assumptions of transformation to achieve nearly ZEUS. The assumed transformations moved from standard retrofit actions to be applied on the envelopes of existing urban fabrics, up to their full mutual combination to reach the level of deep renovation and easily achieve nZEB standards using RES. In particular, the energy analysis on various building blocks in the different RUEs of Athens has demonstrated the energy-saving potential of different standard retrofitting operations (insulation and coatings of opaque elements and envelope surfaces, window replacement, green surfaces). Given the representativeness of Athens' climatic conditions, the availability of data, measurements, and building type descriptions, the urban compound of Peristeri has been selected for further investigation including physical/climatic, architectural, and economic aspects. To identify the economic feasibility of the standard energy retrofit, a cost-benefit analysis was conducted by means of a market survey, determining the evaluation in terms of energy performance improvement and the related cost estimations for each different design option. Simulations have shown that it is technically possible to reach an average EP index of around 25–60 to kWh/m^2 year with the insulation of opaque surfaces and the replacement of existing windows. However, cost analysis and cost-benefit assessments of these interventions have always shown high up-front costs for investments and excessive payback times.

These transformation hypotheses were compared with the hypotheses of a deep renovation, to drastically reduce the energy performance indexes of the buildings up to the target of a passive house, which can easily achieve the nZEB target by employing a reduced use of RES. Furthermore, the different EP indexes (kWh/m^2 y) of the deep renovation options have been compared to the relative options' costs. As expected, the higher the building's transformation, the higher the costs. We found that the economic barrier is the most threatening obstacle toward the adoption of ZEUS in the real estate building practice.

In the search for possible encouragements toward the objective of achieving ZEUS, we found the opportunity given by factors able to reduce energy costs in other nonenergy-related factors influencing the renovation and valorization of buildings, which are corresponding key benefits. We stated that energy retrofitting in RUEs, being connected with technical and architectural modifications of building's components, may involve other qualities that can not be ignored, since they may be a key added value in terms of technical and economic feasibility. Thus, we moved from the concept of *deep renovation* to a *major renovation,* including other energy retrofitting architectural and technical modifications such as the use of new building coatings, sunspaces, buffer zones, and extra rooms (add-ons) integrated with RES. Feasibility studies of these volumetric additions to concentrate them in the most available (and often more dispersive portions) of the existing buildings, like rooftops and blind facades, have been developed. The various design assumptions and the consequent variations in the economic viability were compared. Results showed that in the case of a hypothetical investment by developers or by the same inhabitants in add-ons, the gains obtained by sales of the new apartments would come close to counterbalancing both the standard energy retrofit and the cost of RES (PV) to set to zero the energy demand of the whole building.

It can be concluded that, in the hypothesis of a higher transformation of selected building blocks, the volumetric additions may be conceived as powerful, energy generating, and insulating buffer zones able to set to zero the energy demand of the original building. Furthermore, this effective strategy may help to balance the different EP of the units within the building blocks. As for economic feasibility, the incremental transformation may produce an interesting opportunity to counterbalance the high investment costs of energy transformation. Thus, in the search for additional forms of compensation and incentives for nZEB in existing buildings, volumetric addition, densification, and/or "infill" may represent crucial tools to enhance the technical and economic feasibility of energy retrofitting operations.

The costs of these interventions are not lower than the standard solutions, but their payback (considering roof extensions and construction of new units), together with the direct benefit for potential investors who profit and/or inhabitants who gain space or money the increased asset value that results from the add-ons, have a very positive effect on the technical and economic feasibility of buildings retrofit toward nZEBs.

Other advantages, which are not always quantifiable and are not necessarily related to technical-economic estimations, confirm the strategic role of compensation of volume additions. These are: (1) the possibility of using the add-ons to optimize the thermal operating conditions in summer conditions (due to the reduction of incident solar radiation, for the vertical air flow, etc.). These new structures can also accommodate or act as support for new/renewed ducts for plants and lighting and so forth; (2) the possibility of using the support structures of the volumetric additions for the supplementary purpose of improving the conditions of the static and mechanical performance of existing structures. These "structural invariants," if properly sized and fixed to the frames of existing structures (often made up of reinforced concrete), provide an additional element of construction quality and durability; and last, but not least, (3) the use of these invariants generates a higher degree of adaptability and variability that facilitates the transition from standardized solutions toward custom-made and tailor-designed responses.

The highly flexible and adaptive character of the add-ons gives an extra chance to the nZEB process: different possible add-ons can be studied to meet multiple needs and respond to the heterogeneous demand of users/tenant/dwellers. In this context, it is evident that involving the users represents a necessary step toward the achievement of effective nZEB energy retrofitting.[48]

As stated, we moved from the concept of deep renovation toward a major transformation of the built environment. Major renovations of this kind

[48] This is why a closer collaboration has been developed with the municipality of Peristeri and a questionnaire has been delivered to the inhabitants of the case study urban compound. Results have been used to readjust the feasibility and design scenarios. Outreach initiatives with the inhabitants and the municipality have been undertaken in order to maximize the social participation in the process of retrofitting the whole urban compound in the Peristeri area. The final results of the study were presented on Jan. 16, 2015 to the inhabitants in the municipal hall. Furthermore, contacts with potential investors and with the public district body in charge of regulation and policy recommendation have been established.

have higher costs than "lighter" standard solutions (new plants, insulation, new windows) and deep renovation consisting of the complete retrofit scenarios as presented in Chapter 2, as well. Nonetheless, the reduced payback may have a very positive effect on the technical and economic feasibility of major renovation toward nZEBs. More specifically, major renovations may own the potential to activate the self-financing cycle through add-ons for deep renovation.

Thus, the reflection on economically quantifiable values, whether they are energy-related, economic, or technical, must find the intersection with other aspects, not quantified here. More holistic perspectives are needed to properly assess the competitiveness of deep energy renovation with respect to shallow or more conventional retrofit. In particular, a set of nonenergy-related factors should be considered, consisting of the overall retrofit of the building and addressing other physical, social, and environmental characteristics, as listed in the following points:

1. The possibility of achieving additional and new spaces with volumetric additions without exceeding the load-bearing of the buildings and the possible contribution of external structures to the seismic retrofitting of the building; as a matter of fact, a large majority of existing building stock in the European Union consists of buildings built well before the standardization of regulatory requirements for structural reliability in seismic areas. In many urban areas, especially in the Mediterranean, it is becoming imperative to restructure and retrofit existing buildings to improve their seismic reliability and safety. As a result, when retrofitting the built environment, the adoption of seismic provisions like energy dissipation through external structures and other safety protection devices could be more easily provided by using external structures and high transformation scenarios.
2. The influence of the new external structure as a buffer zone that protects the building from atmospheric agents, thus contributing to the improved maintenance of the existing building envelope and to a longer lifetime of the building as a whole.
3. The possibility of using an external addition for stack ventilation through vertical chimneys and a new pipe network for heating and DHW plant, centralized mechanical ventilation and, generally, for the plant system retrofitting to have less impact on existing residential units and dwellers during retrofitting construction. This may imply a consequent enhanced social inclusiveness for the dwellers (often low-income tenants/owners) who will not be forced to abandon their dwellings

during refurbishment, and will allow flexibility in the internal restructuring of each apartment.

Furthermore, in environmental terms, it is important to highlight the lower environmental impact of using the retrofitting option instead of demolition and reconstruction. The extraction of raw materials for construction, production, processing of materials and their transport, as well as the demolition and removal of the buildings would require high energy and would consequently produce high carbon emissions.

To sum up, major energy and architectural renovation, by incorporating nonenergy-related benefits, might use them to increase the motivation and willingness of main stakeholders (inhabitants, homeowners, public bodies, real estate agencies, ESCOs) in the energy retrofit uptake. In other terms, it can be concluded that retrofit is a multiobjective optimization problem, and the energy and economic benefits are indeed the main objectives but cannot be the sole criteria for the selection of retrofit options; nonenergy-related aspects could help to increase the feasibility of deep renovation toward nZEBs in current building practices.

From a market perspective, the real estate incremented value that potentially could be produced by the volumetric additions, along with the new building envelope and the achievement of nZEBs, are important parameters in estimating the commercial value of the buildings, as will be discussed in more detail in the following paragraphs.

This work is aimed at presenting, investigating, and demonstrating alternative ways of achieving nearly zero energy in urban environments, by exploring the possibility of using energy retrofitting options as form of creating additional value and quality in existing buildings; the scenarios and actions proposed envisaged a series of possible technical actions involving business investors, public bodies, and local communities in the common effort to achieve the decrease of energy consumption up to zero carbon emission in existing urban settings.

The connecting theme of the proposed path is that the crises of energy supply and global warming need to be tackled with an interdisciplinary approach where technical and economical aspects can be intertwined with social and aesthetical qualities of the building fabrics and urban settings.

In particular, the design study and the performed technical-economic evaluation demonstrate that energy efficiency in residential urban complexes can be considered as an extraordinary opportunity to restore environmental, social, and urban quality as a whole.

The technoeconomic feasibility assessment, the proper identification of the types of interventions and their combination in possible scenarios, must be investigated and estimated on a case-by-case basis, with an effective and interdisciplinary design approach. As stated before, environmental sustainability and energy efficiency in urban settings are more than technical problems. In this context, a renewed role of architects and planners is what is needed for a real shift in the building practice. Instead of trying to structure the informal through the architectural production based on authorship, architects should consider the users' perspective and their need as self-organized processes of negotiation. In other words, the architect should not act not as a "Prima Donna," but as a human being willing to donate her/his energy and her/his ideas to society. While during the Modernism period, architects were the absolute form-givers of an over-imposed process of construction, now they must become the interpreter of the process. Thus, instead of trying to structure the informal through the architectural production based on authorship, architects should consider both the technical requirements and the users' perspective, and engage in a participative processes of negotiation.

Even though the approach is based on specific case studies of the city of Athens, the underlying principles will be applicable to other districts and other cities as well. This kind of approach brings about a real shift from the current practice toward a socially sustainable process where inhabitants and designers work together to find effective and real solutions to social and technical questions.

The urban and technological strategies presented here suggest a multifold approach that could stimulate the process of energy renewal according to a multi-oriented combination of energy retrofitting practice up to nearly zero energy with a wider set of elements, components, and tools to solutions belonging to urban environments.

Furthermore, the envisaged solutions are based on the prior assumption that non-energy-related benefits play a key role in the deep renovation up to zero energy of existing buildings.

In particular, our actions have been focused on the following main benefit: the generation of a substantial increase of the real estate value of the buildings through significant energy and architectural transformation (mainly integration of RES systems with new volume additions or new building construction) as an essential process to go beyond the minimum energy performance and aim at achieving nZEBs.

This results in an important reduction of the payback time of the interventions, a strengthening of the key investors' confidence, a higher quality and attractiveness of existing buildings and, finally, concrete market acceleration toward energy-efficient buildings and nZEBs. Thus, the comprehensive renovation process shown is essentially linked to the economic, financial, environmental, and social cobenefits of deep renovations, taking into account the payback time and net present value of investments, by bringing the existing building to a very high-energy performance level.

3.1.4 Potential Impact of Designed Solutions

Is the overall proposal of energy/architectural change with add-ons feasible throughout the existing building stock in diverse areas of cities? Is the concept designed here for the Athens case study feasible in practice? To respond to these questions, we should consider different issues found in the existing building stock and the urban environments.

First, we need to recall the aspects of energy retrofitting related to the current age of our building stock; we have demonstrated the feasibility of the ambitious target in energy performance level (nZEBs), by applying hypothetical energy retrofit to nearly zero in very inefficient buildings like the building stock from the 1950s to1960s onward; in Athens as elsewhere, as already illustrated in Chapter 2, these buildings represent 60–70% of the current building stock.[49]

Second, peripheral areas also represent a large majority of urban areas in European cities; as reported in notable studies (Nilsson et al., 2013), the peripheral urban areas represent the larger majority with respect to central urban areas. Indeed, for the Peristeri case have revealed how dense peripheral contexts do preserve large areas for possible densification. These additions and densifications, if combined with a re-organization of open spaces, may even result in the increase of permeable surfaces. Furthermore, by addressing this type of urban setting, we are concentrating on built environments where change and transformation is feasible (because of free or available areas) and beneficial for energy, architectural, economic, and social reasons.

Finally, by grounding the main actions on the generation of a substantial increase of the real estate value of the buildings through significant energy and architectural transformation, we can largely aim at addressing almost the

[49] These buildings represent the large majority of the EU building stock. As a matter of fact, about 70% of buildings in the EU have been built after World War II (1950s to 1990s) and well before the application of regulatory measures on energy consumption reduction, as reported in prominent studies (EU Report, 2008).

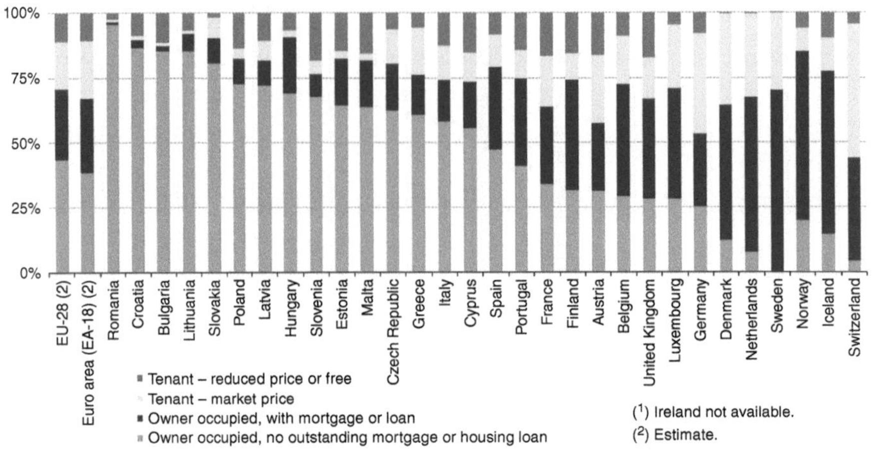

Figure 3.23 ***Ownership is the larger occupancy regime in the EU 28 (more than 75%).*** **Tenure split in the member states as a percentage of the total housing stock. Housing statistics in the EU 2012.** ***(Source: EUROSTAT [online data code: ilc_lvho02]).***

entire residential stock: those owned by owner-occupiers, private landlords, and social housing companies. In other words, we could potentially address the accommodations of almost all EU citizens. Indeed, just over seven out of every ten (70.6%) persons in the EU lived in owner-occupied dwellings, while 18.5% were tenants with a market-price rent, and 10.9% were tenants living in reduced-rent or free accommodations (see Fig. 3.23).

While generic conclusions can not be drawn from case studies, the cases presented for the urban settings in Athens present a remarkable series of repeatable factors at the larger scale of the European urban context. Considering the case study's significance, it should be considered that Modernism in architecture has conceived buildings to be universal; thus, similar buildings in the different contexts of Western civilization and the European Union as well were produced, starting from the 1950s. As a consequence, the relatively recent European building stock presents similar energy and nonenergy problems and many common solutions can be identified, which result in a potentially large impact.

Amongst others, one common solution can be identified in the proposed external structure, representing a powerful tool incorporating other aspects of building performance improvement such as safety, seismic reliability, and market attractiveness.[50] In fact, the volumetric addition presented here is a design answer with high level of technical feasibility, whose novelty relies in the original integration of the main requirements of a building (energy, safety, comfort, attractiveness, etc.) in a single technological solution.[51]

Using a wider perspective, we should also consider that add-ons as additional spaces can then be distributed or combined in the three following ways:

1. As extra living space for the existing units (a bonus for the inhabitants that contribute to the investment)[52]

[50] The new facade layer can be placed in addition to the existing facade when the quality has been tested and proved to be still within high standard levels. In case the existing facade is no longer suitable the living and energy standards, the second option, is to remove (totally or partially) the existing envelope and replace it with the prefabricated facade modules. For load bearing structures, generally, it is favorable to prefer the first option although the wall thickness increases considerably and the light gains are reduced. On the other hand, for columns and beams structure or load bearing panels structure, the possibility of removing the existing facade and replace it with a high energy performing one is the preferable option. The base façade module can essentially accommodate all window sizes and various numbers of ventilation duct runs. The size of the individual components follow a regular pattern established by the planner at the beginning of the project phase. The module size is normally not longer then 10 m and not higher than 3.5 m. Within the module it is possible to open and adapt all sort of windows and doors according to the original facade. In case of existing balcony or loggias it is possible to enclose the area in the new facade and turn it in livable space or follow the original contour of the building and leave it as it is. Although the exploded drawing of the module and the above explained methodology might appear complex at first sight, this merely reflects the attempt to collect and tackle many "problems" simultaneously within a single base module. All components can be factory-assembled. The assembly procedures, together with the joint typologies can be tested in a full-scale working model.

[51] To successfully implement this kind of energy solution in the deep renovation of buildings, further research are under development. They are mainly focused on: (1) the evaluation and assessment of the seismic performance given by the external structures as designed; (2) the lifecycle analysis of the environmental benefit due construction activities in the different design options; and (3) strategies and applications to increase the economical competitiveness of the volumetric additions, using them as a firmer leverage to reach marketable zero-energy buildings within 2020.

[52] This is the case of horizontal extension, especially feasible in the case of existing buildings characterized by a low depth; here is possible and desirable to consider also to extend the façade by adding extra space, increase the size of the existing balconies/loggias and/or add new rooms to the existing unit according to the needs of the inhabitants, as shown for the case study of Peristeri. The ratio of opening and the light gain should be enough to allow the intervention in this direction. The height of the rooms as well as the depth of the building plays a central role in this evaluation. By acting through horizontal extension it is possible to also reconfigure the internal layout and circulation of the building by adding a new elevator for example and proving ballatoires to access the unit or a walkway aside the building. The horizontal extensions can result in punctual addition elements or in a uniform offset of the existing façade. The structure to support the new buffer surface can be entirely independent from the existing one or collaborate. The following schemes illustrate the possible substructures of the new façade and the horizontal extension.

2. As additional units and surface adjoining the blind facades of the buildings, and/or the rooftop addition[53]
3. As an "additional building," a new construction that we may even call the "assistant building", which can be built next to the existing buildings to economically support the renovation investment and help counterbalance the investment costs of retrofitting by reducing payback time; the assistant building may consist of a bonus, a complementary economic instrument for the investors (real estate investors, construction companies in conjunction with ESCO, etc.).[54]

The proposed design scenario can be replicated within different neighborhoods. In particular, the design solutions adopted for potential add-ons, although constrained into a simple prefabricated structure, have lead to a wide set of possible architectural, technical, and structural configurations that develop modular and variable solutions. The different facade abacuses that have been produced for these case studies represent the basis for the development of a new construction system for the facade production. The design becomes a game with no specific rules between technicians and inhabitants.[55] The flexibility and interactive character of the proposed system indicate significant potential impact on the social housing renovation market: through the application of one process, the revision of millions of buildings would be possible, not through standardized intervention but

[53] The possibility of creating additional living space on the rooftop of the existing building is strictly related to the structural safety. In general terms it is possible to elevate the existing building of at least one story in nonseismic contexts. In seismic areas the situation is more complex and the extra building on top has to be structurally independent from the pre-existent or rather be tested for behavior under seismic event. Moreover, the layout and circulation of the building has to allow the extension. As far as it regards the plant systems, the existing cavities and canalizations have to be able to reach the top level. The benefit of combining this type of renovation with the previously mentioned ones has effect on the economic and energetic performance; in fact, the most disadvantage and dispersive units are generally those located on the top level, on the ground floor or on the side. By creating an extra floor, the top units' energy losses are automatically reduced and the roof surface can be used to integrate solar panels. Independently from the geometrical and design purpose of the vertical extension, the structure is strictly related to the original structure of the existing building and the evaluation regarding the reserve in static terms of the existing roof.

[54] It may be used as a possibility of creating a risk fund with the real estate surplus generated by the new building that could cover the risk of arrearage from the inhabitants in paying the bills.

[55] Ethnographic and cultural differences can be considered in the development of customized solutions to revision the envelope toward a more culturally responsive skin, keeping into account not only the performances aspects but also the cultural background of the inhabitants.

with user-driven and tailored solutions, different for each building and each user.[56]

In this context, the availability and accessibility of design software, the growing culture of "making" (or "hacking"), as well as a demand for product personalization and user-oriented design reinforce of the idea of collaboration through hardware and software online communities. The increasing number of open-source platforms for sharing and confronting the ideas, along with growing concerns about the increasing alienation of architecture from its social connotations, bring the question of computer-aided participatory design back to the forefront, and initiate a demand for redefinitions and reformulations.[57]

[56] Illustrated case studies show how the abacus can generate infinite number of facade configuration, even when limited to a few set of possibilities. The rule behind the process has to be set by the planner that is the director of the 'game' within which the inhabitants can personalize their own apartment. Still, in the illustrated projects, there is a character, a driving idea that remains and guides the elevation design, whether the new facade will be based upon a contraposition relationship with the previous one or rather mime the preexistence trying to reproduce its order, or it is playing on the chromatic transition between one architecture span and the other or rather is based on one monochromatic color. The abacus is the instrument to achieve the purpose and the method ensures the possibility by the inhabitants to act and choose, decide, and personalize their own façade ambit, within a certain set of rules. Already several researches have shown the potential of interactive systems to incorporate a high degree of adaptability within the renewal for office or retail spaces, investigating the possibility from the users to actively intervene on the functional distribution and layout of the building. Still very few has been said about how to incorporate the multiplicity of users needs and habits to the complex sector energy refurbishment of existing residential buildings. The machines and the computers can help us in controlling the wishes of inhabitants and integrate them in the decision phase, directly allowing the user to control each one his one part of envelope.

[57] Friedman (1975) inserts the idea of the open form into a programmatic approach of interactions between the users and the architect. His idea is based on a trial and error reiterative process that would lead to a continuous enlightenment of the users as nonexperts actors of the design process but as the best knowledge carrier about what is suitable for them. The dweller becomes the designer through "domesticated" architecture machines that permit to each resident to overlay his architectural need upon the changing framework of the city and fit his own module, within the infrastructure, in the urban context. The suggestion and inspiration for using a machine for this dialog came directly from the research work developed in the 1950s by Yona Friedman who also invented the "Flatwriter," a computer program conceived to enable the process of self-planning. He explained this principle in a lecture that was broadcasted by the French television in 1969. "I call this a choice machine." In practice, we say that there are an incredible number of possibilities for apartments. You can make millions and millions. And in this machine, there is a keyboard which shows the different configurations, the different forms, the different positions for the kitchen, the bathroom, the toilet, and everyone can use different keys on the keyboard,

(*Continued*)

The importance of the continuous transformation and evolution of architecture – through consecutive development phases – is not new and it has been broadly investigated and attributed with important theoretical meaning as a gesture that, in itself, is significant (Eisenman, 1971). The current normative and political revision of the European measures/policies to reduce the increasing consumption of agricultural land and to contain the urban sprawl brought the idea of architecture as means to rewrite the city back to the center of the discussion. It is an alternative solution to the premade recipes of *leaving it as it is* or the *tabula rasa*. The design phase should be no longer conceived as an act of art, a stand-alone gesture of the architect called to interpret the context and society according to his or her own singular sensitivity. The urgent need to act on the existing buildings by rewriting and reinterpreting the original architectural connotation opens fascinating terrains for technological and procedural innovation in the field of architecture. Architects can now take advantage of this opportunity to retrieve architecture's social connotation and explore a new discourse on densification and/or transformative actions.

Coming back to the problem related to the economic barrier toward nearly zero energy retrofitting in existing buildings, these schemes for add-ons and deep renovation could also be implemented on a step-by-step scale, if pre-financing opportunities are limited or not available to cover the up-front costs. Hence, new assistant buildings could be considered as the trigger for a renovation roadmap that would suggest the logical and subsequent (for technical and/or economical reasons) renovation steps that should be undertaken.

To continue modeling in this framework, four hypotheses have been envisaged for the comprehensive renovation of a building of about 3.000 m^2 with 30 residential units of 100 m^2 each (600 net m^2 per floor) and a time lapse of 40 years. The different scenarios consist of:

1. The "as-built" case of a building block consuming 180 kWh/m^2 year.
2. A step-by-step renovation where the plant system renewal is undertaken during the first year with a consequent 30% energy consumption reduction (ECR), and after the first 10 years, the replacement of

and print for themselves the apartment that they prefer. The machine at the same time, checks that this person's choice doesn't block the access to anyone else's apartment. And does not block the light and the ventilation. The utopic envision of Friedman can be an inspiration for continuing the work and translate the questionnaire into a programmatic system that will enable inhabitants to express their desires and receive a direct feedback within possibilities and solutions that will be incorporated and will directly compose the over-whole renovation project.

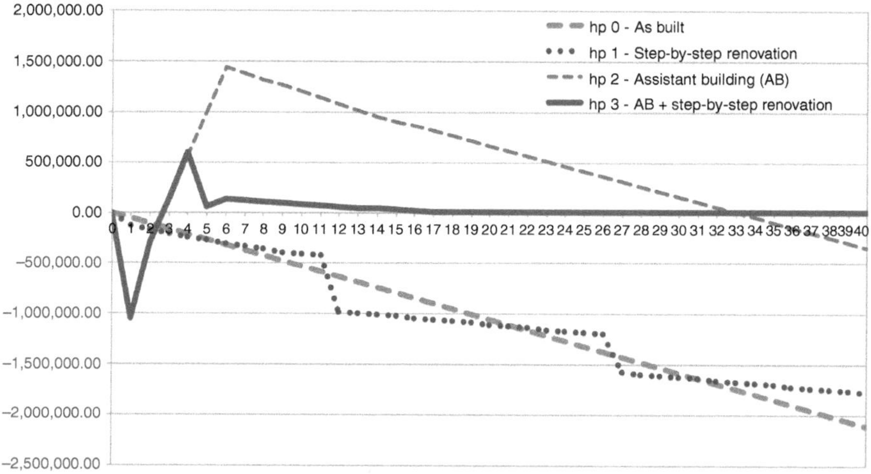

Figure 3.24 ***Calculations performed for the cost benefit analysis in the building model.*** Graphs show how the construction of an "assistant building" can be integrated in the process of a step-by-step renovation.

windows (+15% ECR) with high energy-performing windows is realized; after 20 years, all the surfaces of the building envelope undergo renovation (+25% ECR).

3. The construction of a new building (assistant building) aimed at reducing only 20% of the actual building.
4. Finally, the combination of a step-by-step renovation with the construction of an assistant building of about 1000 m^2 built with passive standards (15 kWh/m^2 year) and producing energy from RES for its own requirements as well as for the requirements of the "adopted building" (Fig. 3.9). In this simulation, the assistant and assisted building are equipped with a PV system to set to nearly zero the energy consumption of both buildings.[58]

Cost-benefit analyses performed for the building models are illustrated in Fig. 3.24. Results show how the construction of an assistant building can be integrated in the process of a step-by-step renovation. Fig. 3.24 derives from the calculated cash flows calculated for the different hypotheses; this figure also shows how the assistant building acts as a trigger in a stepped renovation process. It can be designed to act as the catalyst or attractor for private-sector financing, playing an extremely important role in the context

[58] A total 21,231 kW_p PV plant for the assisted and assisting building has been considered to cover 187.600 kW_e for the total net residential surface of 4000 m^2.

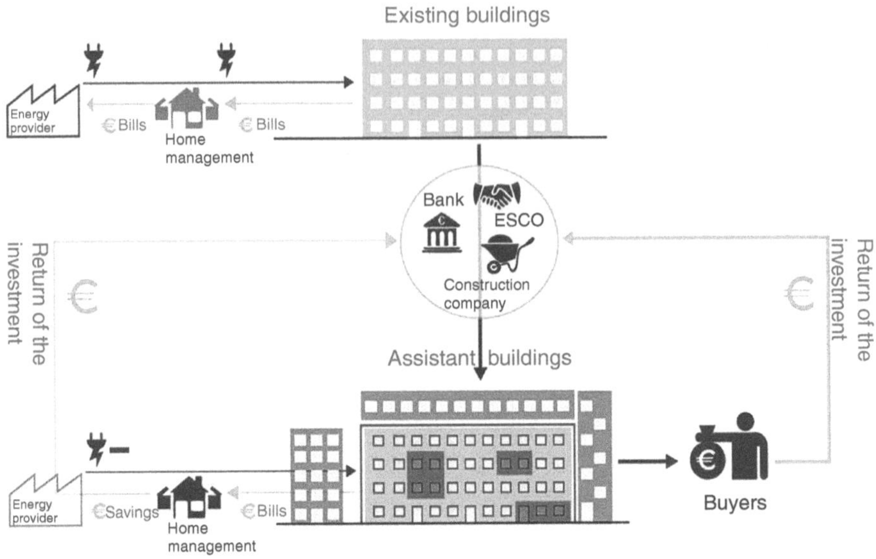

Figure 3.25 ***Scheme of potential actions to be undertaken by an ESCO or a construction company CoCo (or a joint venture between them) in the construction of add-ons and/or assistant buildings to retrofit the "assisted" building.***

of scarce private finance where the search for affordable up-front investments is crucial. Of course, it is important to specify that in the hypothesized scenario, we need to consider a new (lighter) investment after a time lapse of 35–40 years.

The simulated scenario in Fig. 3.24 considers the hypothesis of the inhabitants/property owners as direct investors: it is the basis to calculating the revenues of a potential ESCO[59] or a construction company (or a joint venture between them), as simplified in Figs. 3.25 and 3.26.

Fig. 3.25 illustrates a possible scheme for the creation of assistant buildings (add-ons and new buildings are both represented). In this case, the potential

[59] ESCO is an acronym indicating an energy service company providing a broad range of energy solutions including designs and implementation of energy savings projects, retrofitting, energy conservation, energy infrastructure outsourcing, power generation and energy supply, and risk management. Currently ESCO's business focuses on innovative financing methods including a range of applicable equipment configured in such a way as to reduce the energy cost of a building. The building occupants, or landlord, then benefit from the energy savings and pay a fee to the ESCO in return. At all times, the saving is guaranteed to exceed the fee. The savings in energy costs are often used to payback the capital investment of the project over a 5–20 year period, or reinvested into the building to allow for capital upgrades that may otherwise be unfeasible. Generally, the ESCO is often responsible to pay the difference in the case of a running project that does not provide returns on the initial investment.

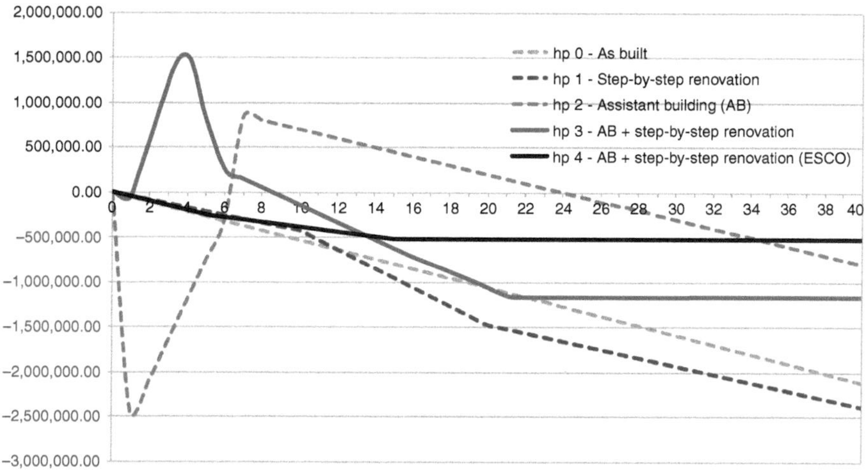

Figure 3.26 ***ESCO/CoCo have return on investments from the combination of both add-on revenues and energy bills from the existing buildings' inhabitants or house manager.***

investors' ESCO, in combination with a construction company, would have a return on investment from the combination of both add-on revenues and energy bills from the existing building inhabitants or the house manager. Combinations of different forms of incentives may result in win-win solutions: on the one hand, the urban dwellers (or property owners, landlords, etc.) may benefit from a reduced fee or a reduced number of years to reimburse the potential investors (ESCO, or ESCO in combination with a construction company [ESCO/CoCo]) while, on the other hand, ESCO/CoCo will receive revenues that can be immediately available. In this case, the financial interests for banks and ESCO have been considered in the diagram on the left.

Results show evidence that the provision of a new building with few or no links with the assisted building does not produce long-term investments (the scenario has considered a reduction of 20% of the energy consumption of the actual building; in other words, it is a case of investing only a part of the revenues from the selling of the new units in the plant retrofit of the assisted building).

On the contrary, in the hypothesis of establishing a binding contract between the construction of an assisted building and the renovation up to nZEB of the assisted or adopted building, a more stable investment climate for market and social actors can be achieved. In fact, on the one hand, the occupants or owners may have a reduced fee or a reduced number of years to reimburse the investor (ESCO or ESCO/CoCo), but on the other hand, the same investor(s) will receive revenues, which can be available within 5 years. This can also result in the co-related benefits of reducing risks and of releasing funds for further investments, thus stimulating other deep renovations in a faster, but well-planned, long-term market strategy (JWG, 2013). In addition to the evidence of the resulting increase of scale and impact of the major renovation processes, it is important to highlight how the scheme minimizes the disturbance for inhabitants. They are generally able to remain in their apartments during the renovation work and, where possible, they will keep paying the bills through the usual system, thus being the provider part of the process.[60]

Completing the financial scheme related to deep renovation, the creation of new buildings would increase the interests from financial actors and banks.[61] In fact, it could help to guarantee the solvency of the investments by consistently reducing the payback times. Of course, this is strictly related to the scale of investment.

Overall, parameterizing with respect to the renovation costs and the estimated value of the assistant building, we could aim at working on the basis of the following simplified formula:

$$\frac{C_r \times y + C_c \times x - P \times x}{R \times y} = PBT$$

where C_r = renovation costs including RES to set to nearly zero the existing buildings; y = surface (m^2) of existing building; C_c = Construction costs of the assistant building (add-ons) per m^2 (€); x = additional surfaces of add-ons (m^2); P = assistant building's price point per m^2 (€) in the reference market of the specific target region; R = cost saving per m^2 (€), PBT= payback time. To this basic formula, it is necessary to add all the financial, commercial, and legislative expenses that will occur in the construction and real estate phases. Using the previous formula, it is possible to calculate the simple payback time for the above-mentioned parameters. Indeed, to make the intervention attractive for owners (who make the ultimate decision to renovate) and for the banks and the financial agencies, we can, for example, decrease the return of the investment up to 10 or even 8 years and still derive the necessary amount of square meters that are necessary to cut the payback time, given a specific real estate market (C_r, C_c, P values). In real market conditions, financial costs and revenues for the ESCO/CoCo have to be considered to produce reliable financial and business plans.

[60] As discussed in the public workshop on innovative financing for energy efficiency and renewables, PWIF, 28 April 2015, Brussels. Executive Agency for SMEs.

[61] Strelnikova, L. European Energy Efficiency Fund, PWIF, Session subject: How to improve the bankability of investments in energy efficiency and renewables?

In fact, the confidence in a win-win perspective can drive stakeholders to work together to find common solutions, thus producing a structured, long-term financing scheme for deep renovation that will be beneficial to market actors (in terms of profit of investments), policy makers, and occupants (in terms of environmental/social benefits). In general, many scenarios can be modeled and combined in the energy renovation of existing buildings and the volumetric additions may assume different configurations, depending on building type, structural conditions, and urban capacity.

The cost-benefit analysis developed in this project demonstrates that the hypothetical cost of additions can be counterbalanced by the increased real estate value of the building on top of reduced energy losses and increased energy savings. Thus, this strategy may generate a self-sufficient business model whose high potential to deliver long-term financing for energy efficiency is undeniable.

3.1.5 Potential Constraints of Add-ons and Urban Densification toward nZEBs

Nonetheless, potential constraints may limit the application of these hypotheses to the wider contexts of the existing built environments. This is due to many shortages.

First (1), we are perfectly aware that major legal obstacles exist in relation to add-ons or assistant buildings. They are notably linked to property rights, tenancy, and condominium law as well as local and urban planning rules. For example, the creation of extra units on the roof of a multi-apartment building needs to overcome legal planning and urban rules at the local level, such as condominium decision-making rules and tenancy contractual arrangements. On top of that, and maybe even more importantly, the question of ownership of the added volume needs to be solved,[62]

Second, (2) there is a lack of confidence about the technical feasibility of add-ons,[63] notwithstanding the fact that additions may assume

[62] In a wider context with respect to the examples reported in this book, we should consider that Add-ons might be also grouped in an additional building, the Assistant Building, which can be built next to the existing to financially support the renovation investment and reduce the payback time.

[63] In the meetings with the inhabitants of Peristeri this aspect has dramatically emerged. But, this is also due to the fact that so far, in the majority of cases, additions have been realized on the top roof of existing buildings; here, the risk of exceeding the load bearing of these buildings, which were not conceived for these additional volumes, is especially thorny, especially in seismic areas. Owners and tenants of highly seismic urban areas like Athens, appeared to be very concerned and anxious about safety requirements.

different configurations, depending on the building type, structural conditions, and urban capacity (a side or facade addition [add-ons], or entire new buildings in the form of assistant buildings). Conversely, as shown in the Peristeri case study, the same additions can be conceived with support structures, which allow additions to be used for the purpose of improving the mechanical, static, and seismic performance of existing structures, providing an additional element of construction quality and durability.[64] Furthermore, (3) current regulatory environments often do not permit these kinds of interventions. However, in some countries both public-private ad hoc agreements and legislative rules have been introduced,[65] thus showing a public sector effort/demand in the search for ways to incentivize the private market uptake of major renovation.[66] Finally, (4) there is a lack of understanding on the balance of the costs and financial benefits of these measures by potential investors and financing institutes; so far, no or little attention has been paid to: the great benefits and effects of the additions in the payback time for a potential ESCO; the possibility of creating a risk fund with the real estate surplus generated by the addition of new buildings; the potential for increasing the bankability by standardizing financial schemes based on volumetric addition and urban densification.

To sum up, there is a general lack of dialog among stakeholders about the possible implementations of this strategy. Further policy actions are thus necessary in order to drive the process of innovation in the construction and real estate sectors by improving the strategy of major renovation through densification and creating a new and attractive market in the deep renovation of existing buildings toward nZEBs.

[64] Structural and safety evaluation will be carried to demonstrate the potential positive impact of the add-ons to the existing structures, since if properly anchored, the additional structure could even provide existing buildings with extra resistance to horizontal stresses, typical of the seismic events.

[65] Public-private ad-hoc agreements have been set for various retrofitting interventions in social housing, like volumetric compensation adopted from Van Shagen in Amsterdam and Kolpa Architecten in Rotterdam. From the legislative point of view, that is, Italy has approved the Law Decree n. 47, 28/03/14 converted in Law n. 80, 23/05/14, the "Piano Casa" which, *inter alia*, sets out measures to give extra volume benefit in deep energy renovation interventions of existent buildings.

[66] At the urban and city district level, an urban policy based on the punctual densification has to provide a correspondent implementation of existing infrastructure like system plants, streets, parking number, and services to compensate the increase in number of population number of inhabitants.

3.1.6 Conclusions and Recommendations

The envisaged actions to achieve nearly zero urban settings focused on the following main benefit: the generation of a substantial increase in the real estate value of the buildings through significant energy and architectural transformation (mainly integration of renewable energy sources with new volume additions or new building construction) to go beyond the minimum energy performance and aim at achieving nZEBs.

The strategy may result in an important reduction of the payback time of the interventions, a strengthening of the key investors' confidence, a higher quality and attractiveness of existing buildings and, finally, concrete market acceleration toward energy efficient buildings and nZEBs.

Thus, the comprehensive renovation process presented here, while bringing the existing buildings and urban settings to nearly zero energy, can be linked to the economic, financial, environmental, and social cobenefits of deep renovations, taking into account the payback time and net present value of investments. The strategy presented combines different domains, especially "the creation of new buildings" for the renovation of old ones, providing the economical circle with trust and bankability and ensuring a return of the investment.

The replicable nature of this punctual densification and the important impact on the mobility and the social benefits that could come from the add-on densification strategy demonstrate that this path could be effective toward the major renovation of our cities up to nearly zero energy; furthermore, it could represent a step forward in defining a new paradigm of sustainability and a new role for architecture design and planning. In particular, as illustrated in the beginning of Chapter 2, the possibility of applying punctual densifications appears to be feasible in the different urban contexts of the metropolitan Athens area, due to the peculiar urban gaps derived form the process of substitution and transformation of buildings involving the historical and central parts of the city.[67]

[67] As illustrated, a consistent suburban character is diffusely present in the overall context of the city, embracing the central and peripheral areas as well. Athens cityscape reveals a contradictory, dense and disordered but fascinating structure as a whole. Here, the dramatic city scene seems to define a sort of a conflicting relationship with respect to the historical time: buildings of the 1970s and 1980s overlap the historical urban path, contrasting the old, residual, often abandoned, but still lasting, historical houses of the last century and afar, marking the urban landscape by consecutive shocks and discontinuities. This character is the result of the dramatic grow sped up in the 1980s in the central areas of the cities, destroying and substituting many historical buildings while continuing its spreading at the periphery.

(a) (b)

Figure 3.27 Possible infill option in the center of Athens (a) Existing situation (b).

In Fig. 3.27, the envisaged scenario has left a significant part of the ground floor unbuilt in order to ensure ventilation between the internal urban courtyards and the open public spaces and streets. It is important to note that this contradicts the current rules in the urban regulatory framework for Athens, which suggests avoiding pilotis, considered undesirable for social and health reasons. Nonetheless, a limited use of this architectural solution for densification in dense urban areas could be considered, in the effort to compensate any new construction with the maximum amount of climatic and environmental benefits achievable (Fig. 3.28).

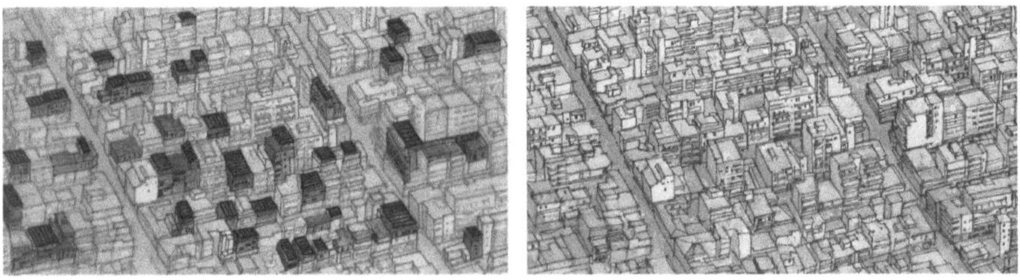

(a) (b)

Figure 3.28 Possible filling and densification at the urban scale in Athens (a) Existing situation Scenario (b).

To expand nZEB practices within the existing built environments, we should work to provide the key market actors with rehabilitation techniques and decision-making tools based on the potential of add-ons, assistant buildings and, in general, densification for promoting major energy renovation. It is also necessary to identify the key enabling conditions to finance deep renovation of buildings through process and organization innovation, capitalizing on the local knowledge. Moreover, at the public level, actions should be promoted to support policies for facilitating the implementation of major renovation strategies through the application of add-ons, assistant buildings, or urban densification. For example, the core elements of binding contracts in a public-private agreement could be represented by such a densification strategy; it should be seen as the delivered bonus in the case of nZEB achievements and environmental compensating actions (as, eg, infrastructural urban adjustments, increasing of permeable surfaces, waste water recycling, etc.). Thus, the nZEB target as the best means of reducing energy consumption should be the threshold target for owners, landlords, and private investors to generate and guarantee a lower-risk and more stable investment climate for public bodies, while promoting a long-term investor confidence market in private sectors and social actions.[68]

To widen the potential actions as designed, it would be necessary to act on norms, since it is through them that the public sphere rules over the private one. Hence, the problem is to neutralize, or, at least, to reduce, the risks in the mutual actions between individuals and organized communities

[68] In this perspective, the designed strategy can help the support of the implementation of European directives to remove regulatory and nonregulatory barriers to energy efficiency, without prejudice to the basic principles of the property and tenancy law of the Member States, in particular as regards: (1) the split of incentives between the owner and the tenant of a building or among owners, with a view to ensure that these parties are not deterred from making energy-efficient investments that they would otherwise have made by the fact that they will not individually obtain the full benefits; (2) costs and benefits sharing between owners/tenants, including national rules and measures regulating decision-making processes in multi-owner properties. At the strategic urban planning level, the strategy here proposed may contribute to implement ambitious, widespread and up-scaled deep energy renovation programs shifting by current public sources of finance toward greater levels of private sector finance provision (EU EED -2012/27/EU- 2013). Contribute to the EU effort toward 'Smart Financing for Smart Buildings' and the related Initiative to be launched by the EC together with the Review of the EPBD in order to make existing buildings more energy-efficient, facilitating the access to existing funding instruments. COM (2015), Communication on a Framework Strategy for a Resilient Energy Union with a Forward-Looking Climate Change Policy, COM (2015) 80 Final, http://ec.europa.eu/priorities/energy-union/docs/energyunion_en.pdf.

toward the adoption of actions to achieve ZEUS. Critical appraisal and comparisons apply to any field of community life, provided it is ruled by norms. One might wonder whether town planning should be considered as a standard for comparison. This also refers to the public behavior, in various countries, in terms of compatibility with the system (environment, market, property, flexibility, etc.). The main public rules, generally applicable throughout the European national territories, define the general principles of the protection of spaces. Every new construction or territorial intervention must conform to those rules that frame the local and special urban regulations.[69]

In the context of possible retrofitting measures to be undertaken toward nearly zero energy environments, the intermediate urban scale (the scale of the real urban settings illustrated), public-private agreements managed at local and municipal levels have to be considered as the first procedural step toward the feasibility of energy-saving and bio-climatic urban design toward nZEBs and ZEUS. Thus, the context of concerted actions and synergies between public and private sectors, including potential sectors of private finance and investors, appears to be the framework that best promotes ZEUS. The presence of a plurality of public-private synergy in the formulation of the central urban laws in many European countries still represents a quite new or innovative formula inside the current tools or instruments provided by urban legislation. The importance of involving citizens (or associations of private bodies that deputize citizens), stakeholders, and private investors is evident in order to define possible medium- or long-term actions that will ensure the feasibility of hypothesized energy-conscious scenarios in the framework of current planning processes and existing legislative issues. The involvement of legislative authorities, private bodies and urban planners in climatic issues is effective in ways that even the best educational projects cannot be.

Concerted management zone plans may be applied to specific sectors or urban areas, and to urban development or restoration at intermediate-scale

[69] The operational urban development uses statutory rules to frame an urban sector-related operation. The possible actions at urban scales are generally case-by case design assuming a series of norms and legislative issues from the national, regional and municipal level. They are subjected to special rules whose decision is mainly public (municipality), with a large set of various and different private involvement all over cities. These operational urban plans represent an opened, flexible and operational framework for a specific urban area. Public communities can act inside them to proceed to the layout and the realization of equipped built or non built areas in the framework of the whole urban plan.

planning decision processes. In many European countries, the urban plan at the intermediate scale leads to the production of very precise plan documents that greatly influence the volumetric features of designed buildings, the designation of spaces, and the main characteristics of the design of public spaces. The layout of this urban tool integrates the third dimension much more than in the other urban tool plans. To deal with urban planning, we have to consider the main urban parameters of the urban built zone[70] on the relevant parameters for urban green zones[71] and consider them in an intercalary dimension, starting from the intermediate urban scale. However, these steps must verify the compatibility with planning processes designed at a larger scale and/or by wider public levels. The additions and densifications as designed in our examples might be not applied or could not be desirable because of other social, political, and economic factors influencing an existing community. We are perfectly aware that the designed scenario cannot solve the problems of market uptake and the social engagement toward ZEUS.

Nonetheless, in the assumption of world population and urban concentration increase as introduced in the first chapter,[72] calibrated, distributed, and locally-based interventions based on densification and add-ons toward

[70] These parameters can be: building density: ratio of the surface building relative to the surface of the urban area considered (impact on porosity, roughness, shading); building continuity: appreciate the buildings arrangement such as isolated buildings, buildings arranges in raw, cluster of buildings (impact on air porosity and roughness); building height: mean height of the buildings (roughness); mineral/vegetal ratio: of the use land (evaporation, filtration, thermal inertia); ground porosity (may be complementary to the precedent); types of buildings and soils (material, heavy, light structure); impact of inertia; colors (of pavement and buildings materials, roofs, walls), impact on reflection; structure and forms of urban spaces (street, courtyards, squares, rectangular grid, predominant exposure); ratio between building height and the relative distance with other buildings or the openness degree to appreciate the openness of public spaces (sky factor); impact on reflective and radiative load; water presence; kind of vegetation; surroundings characteristics (urban, suburbs, rural, park, river, lake).

[71] The relevant parameters for urban green zone can be identified as follows: height of the vegetation, tree density, ratio deciduous/permanent trees; surface, water presence and water required by vegetation; kind and type of vegetation; and the like. As far as it regards the different parameters involved we will mention the following ones: building density, building continuity, building height, mineral/vegetal ratio, type of space, solar orientation, exposure to wind, color, albedo, pavement characteristic, buildings volumetric shape. The typological factors involved in the design interfere with energy, indoor and outdoor comfort, and urban micro-climatic situation in term of porosity, roughness, shading effect.

[72] The percentage of human beings living in cities has increased from 3% in 1800 to 14% in 1900 and is estimated to rise from the current 50–75% in 2050. The figure for Europe is still higher: 83% of the population is expected to live in cities by 2050 (EU Report, 2010).

nearly ZEUS present the a lot of potential for a two-fold benefit. In fact, urban densification schemes may signify a process limiting the negative effects of urban sprawl at the territorial scale while achieving a full-scale energy savings strategy. As shown in the Peristeri case, densification in urban contexts, when combined with a re-organization of open spaces, may even result in the increase of permeable and green surfaces; in this framework, the concept of add-ons or new assistant buildings may even open the way for "adopting" existing buildings in areas where no free space is available, for example, in the historical centers. Thus, we might even envisage a scale-up in this process[73] and hypothesize the exchange of "urban quotes" among different areas of the city, where the centers may be "adopted" by new, renovated, nearly zero energy – or even energy plus – urban peripheries. Thus, starting from the specific limits encountered in the local RUEs, we may utilize the compensative measures given by the possible densification to search for a balance between global costs connected by the initial energy demand and the related retrofitting options to achieve ZEUS (Fig. 3.29).

It is well known that public administrations are called upon optimize energy measures to go well beyond the mere reduction of energy consumption (EU-20-20-20 targets) and achieve the smartest management of energy. In this framework, both historical buildings and urban centers (see point 1, following) and the buildings in peripheral and marginal areas (see point 2, following) should be addressed to optimize the energy management in the whole context of a city. But (1) in historical buildings, energy demand

[73] It is worth mentioning that energy retrofitting interventions like the ones here presented may be assumed as the punctual nodes of a longer term replanning; in fact, small-scale interventions should be seen as key nodes to promote local involvement, rebuild local identity, arouse urban interest and raise human consciousness (energy consumption awareness, cultural and market exchanges, etc.). Further opportunity on research should focus on the feasibility of a system network re-reconnecting the different urban areas, that is testing the potential of the smart areas to generate other smart areas, thus creating a potential network in the city as a whole. This reconnection would be strategic for the construction of urban greenways connecting function also as a catalyst in urban fruition. To do this a large set of possible inconsistencies have to be addressed: new mobility schemes, parking areas creation, enhanced comprehension on the influence on human behavior by spatial layout of buildings and urban places, and so forth. In particular, it is important to understand the possible strategies to foster the creation of local urban communities, involving them in an evidence-based approach to the spatial layouts aiming at producing more safe and vibrant places for social, economic and environmental capital (since poor layouts risk functional failure, loss of investment and social harm). Thus it is necessary to combine social, safety and energy issues, integrating technical and architectural feasibility of zero energy and low carbon areas with the creation of new built environments associated with new forms of urban communities.

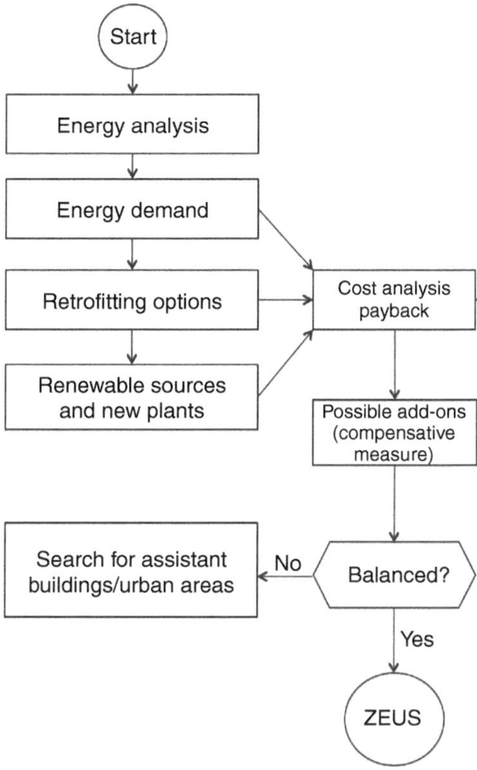

Figure 3.29 ***Proposed procedure for the zero energy urban setting (ZEUS) adoption.***

represents a particular challenge. Here energy-saving solutions via renovation cannot be easily implemented, because historical infrastructures are often connected with and limited by the necessity of preserving authenticity and integrity; thus, they lack a systemic and effective impulse for energy retrofitting.[74] Conversely, (2) major renovation in peripheral areas might

[74] Energy solutions in the field of renovation are not compatible with or adequately adapted for use in historic buildings. In this perspective, a deep renovation of historic buildings appears to be far to be reached by means of mere physical or material processes. Coherently with such constrains, to approach and effectively contribute to solve the problem to reduce the energy consumption and energy costs associated to historical building maintenance and management. Retrofitting actions in these buildings should be based on the human in its relation with the built environment, rather than on usual physical building components (envelope, windows, plants); thus concentrating on the users behavior to reduce temporal and spatial energy consumption both in terms of the period (the time) the users inhabit the building spaces, and in terms of the space dimension (space) they need; the latter is particularly crucial, considering high volumes and surfaces of many historic buildings, often oversized compared to the modern use and habits.

be conceived in a way that can integrate energy requirements and policy measures with the economic and social increase in housing quality by non-energy-related factors, which would contribute to the acceleration of major renovation toward ZEUS. The major challenge of tomorrow's agenda could be the successful integration and combination of issues (1) and (2) into an energy-efficient, holistic strategy for the city of the future, where its outer parts might be conceived to act as the energy suppliers of the historic city core.

Thus, by using the results that have emerged in this work, combined with these final considerations, a first record of proposed steps for policy, legislation, education, and building practices may be outlined. As far as it concerns policy, a carbon zero and zero energy strategy plan, with particular reference to Mediterranean climate, should be oriented at:

- Providing and reinforcing specific target strategies for open areas and public spaces within the urban environment, to promote the use of green and permeable surfaces, thus reducing urban cooling demand at the territorial and urban scale. In this frame, it is most important to promote all forms of passive natural devices according to the potential effects on minimizing the localized pollution problems and to enhance urban microclimate, such as: ensuring pollutant absorption by plants and dispersion by natural ventilation and cooling; improving natural cooling; minimizing solar heat gains both in the open spaces and on the building's facades (allowing protection for southeast and southwest facades by means of shading device systems and natural filters, etc.).
- Incentivizing the use of the most effective and cost-efficient technology mix in relation to local environments and specific sites, considering that passive design features are always cheaper and cost-efficient with respect to active systems. The feasible technology mix should be different according to the type of building, the climate zone, the national energy prices, the availability of know-how and technologies, and cultural and social attitudes.
- Raising capacity at the public level for the adoption of legislative and practicable measures aimed at the deep renovation of existing public buildings up to nearly zero energy buildings in the peripheral or modern areas.
- Setting up special incentives (tax reduction, volumetric building compensation, add-ons, and densification of urban peripheries) to reach fully distributed interventions of zero energy retrofitting of existing buildings with special reference to urban areas, thus achieving win-win

perspectives in relation to urban densification schemes (against urban sprawl) and full-scale energy-saving strategies. This must be combined with a set of specific benchmarks to guarantee the access to the set benefits and incentives. Public bodies should design and request specific measures to be achieved by private owners, landlords, or home managers, with particular reference for the access to the possibility of addition. For example, these can be compulsory measures, binding contracts, required obligations, and the like, set to guarantee the quality and the nearly zero energy target, up to specific technical requirements – example, the respect for cross ventilation between urban zones (see Fig. 3.27), the respect of public network infrastructure, the creation of green surfaces, and limits to increasing impermeable surfaces, and so forth.

- Designing specific public-private agreement tools for the redesign of urban open areas, to assure environmental compensation of densification policies, while providing new layouts of streets and squares to make them more pleasant, accessible, and secure for pedestrians (with specific urban furniture, new pavements, vegetation, and outdoor or covered passages). In general, specific benchmarks and design tools should be provided, and effort should be made to compensate any new construction by achieving the maximum amount of climatic and environmental benefits associable within it and in relation to its boundary conditions.

Thus, as far as it concerns education, care and interest should be focused on:

- Designing education and training aimed at improving the adaptability of buildings, especially within the context of the historical Mediterranean regions, to provide flexibility, adaptability, thermal comfort, and up to nearly zero energy consumption both in the summer and winter seasons.
- Training activities for stakeholders, private owners, and citizens (also within the framework of participative processes) to provide information on the potential of consumers' behavior to reduce cooling demand and save energy. In fact, the minimizing energy use for heating and cooling the buildings could be encouraged by the education of occupants (smart behavior) in addition to the use of free-of cost strategies by intelligent design (using passive architectural strategies).[75]

[75] As for example, training and information on dress behavior and its potential in relation to energy consumption reduction should be encouraged and/or reinforced, since seasonal flexibility in clothing may have huge consequences for energy saving in buildings.

- Training activities and information should also be disseminated to future generations of designers, architects, and engineers to motivate them to embrace the subject of ZEUS. In fact, future designers' attention should shift from fascinating, monumental, and over-imposed actions of new or innovative zero-energy architecture, and instead should stress the importance of dealing with small interventions in the existing environment, showcasing this opportunity as a creative challenge for tomorrow's agenda.

The envisaged actions toward achieving nearly zero urban settings focused on the following main benefit: the generation of a substantial increase of the existing value of the current buildings through significant energy and architectural transformation, in order to go beyond the minimum energy performance and aim at achieving ZEUS. In the search for transitional pathways toward a low-carbon future, in place of large or incremental development plans, we may envision a series of limited, but challenging, nearly zero energy urban settings, which can be conceived as the punctual nodes of a long term re-planning, as they have the potential to generate a new identity for the city as a whole. While on this path, should not refer to the concept of energy performance only in a mechanistic manner.[76] We should rather observe that when the idea of performance is translated into architecture, optimization is embodied by:

- Energy, technical, and constructive efficiency
- Economic and social feasibility
- Multi-functionality
- Aesthetics

These four elements provide added value due to the complex and interconnected system in which the overall worth is grander than the sum of its parts. Following the definition of efficiency as the ratio of the useful energy delivered by a dynamic system to the energy supplied to it, a building could be seen as a system designed to perform in the most efficient way possible. And of course, the most efficient way to perform is the one that

[76] In fact, the notion of performance is normally associated to mechanical elements like engines, and is understood as the manner in which a system reacts to environmental stimuli. A performance analysis can be carried for any object, space, or machinery that has been designed to a specific function. In order to obtain the best performance from a system, it is necessary to provide changes that can make it as functional and effective as possible by minimizing the losses and maximizing results, which is at the base of optimization in mechanics. A system is just a group of interacting units establishing a much more complex and integrated altogether. Given this, a city, a building, an office and even a piece of street furniture can be considered as a system.

can minimize losses and optimize the space required to accomplish the tasks set at the beginning of the design process. An efficient system should also be durable and should require the least maintenance possible; this means that an efficient building also provides economic savings in the long run, when compared to a nonefficient structure.

Incorporating more functions within the same space can be a challenge for designers, but it embodies a variety of positive aspects. In fact, the combination of functions often improves density levels by attracting more people in the same space, preventing neighborhoods from only being used and lived in partially during the day (ie, financial districts versus "dormitory neighborhoods"). The space is filled with new and vibrant energy and is "technically" more efficient and attractive. So much has been written about aesthetics, but we believe that it has always been hard to find a universal definition of it. While other features in a building, such as efficiency and functionality, seem to be more relatable to a general and understandable definition, aesthetics is a very difficult aspect of architecture in all its expressions. So, what makes a building/structure aesthetically appealing? A city, a building, a room, a house, a street is beautiful if it functions properly, if urban dwellers actually use it, if it relates in any way to the built environment in which it resides, and if it creates vitality and identity at the street level or, more in generally, in the city landscape.

Beauty is such a subjective notion, that it is probably not even worth talking about it, but, in the end, we are always attracted to it, by the way order and disorder may combine with each other and generate a vibration. Beauty feeds us, human observers, urban dwellers, and users alike, with peaceful feelings, while at the same time it instills curiosity. Filling the built environments with equality while respecting individuality, in a shared, but animated and exciting vision of the future city, whether in the Mediterranean or elsewhere, is a noble pursuit (Fig. 3.30).

> *Our hope is that (...) we can shift to a culture of sustainability – economic, social and environmental – and that the motto (...) "Better city, better life" can become a reality for the next generation (Baer, EU Report World and European Sustainable Cities, 2010).*

ACKNOWLEDGMENTS

The work presented here is part of the result from the project *Urban Recreation: Energy-efficient Retrofit for Carbon-Zero and Socio-oriented Urban Environments*, project no. 326060 FP7-PEOPLE-2012-IEF, funded by the European Commission, within the framework of the funding scheme Marie Curie Actions— Intra-European Fellowships (IEF).

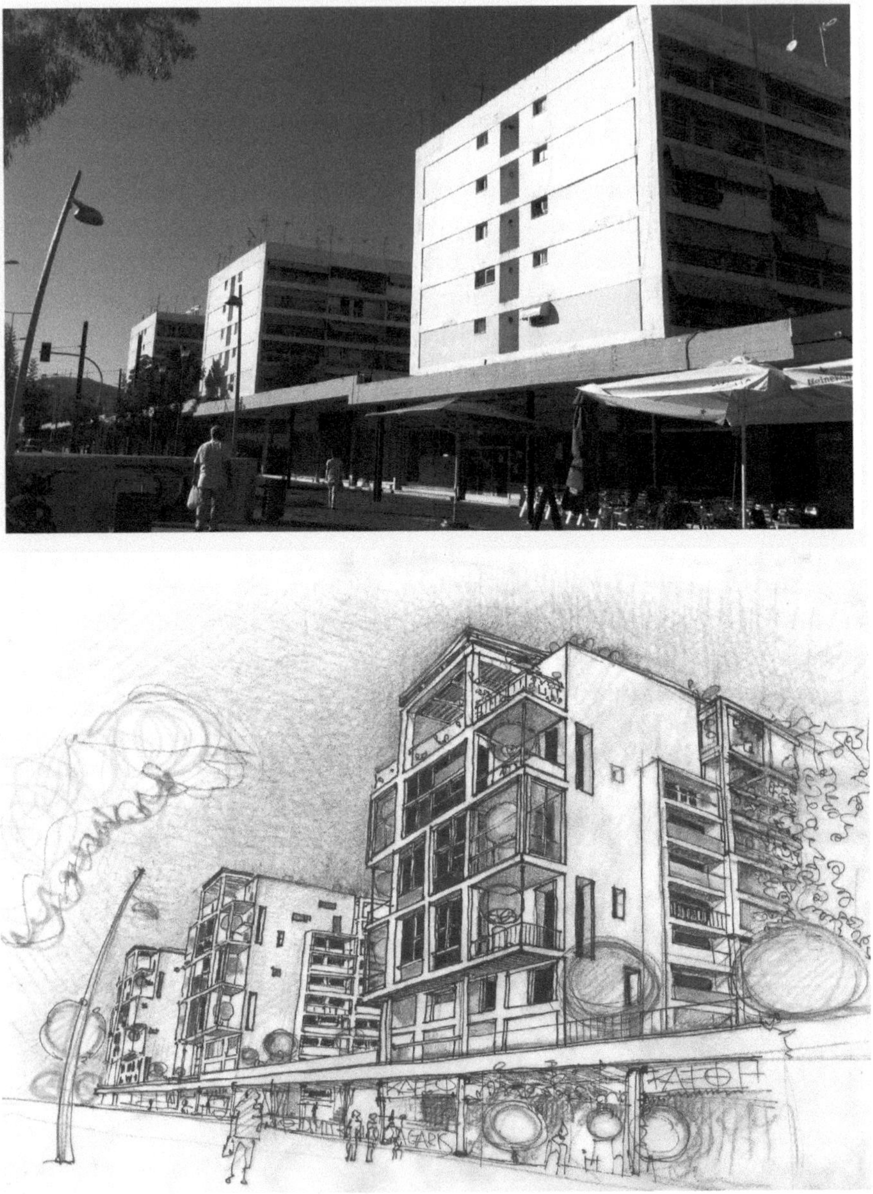

Figure 3.30 ***Zero energy urban setting (ZEUS) through densification in Peristeri.***

This book contains data and images gathered during the Master Theses in Construction Engineering and Architecture, 2012–2013; 2013–2014 by Laura Zuffi, Nicoletta Salvi, Lorenza Cavasino, Rachele Iannone, and Andrea Barbieri. The author acknowledges Laura, Nicoletta, Lorenza, Rachele, and Andrea. Many images have been revised and redesigned by the author herself. Furthermore, data have been elaborated upon under the supervision of the author and in collaboration with colleagues Luca Boiardi and Giovanni Semprini.

Data and calculations have been assessed, revised, and updated by the author and Eng. Massimo Monacelli. Massimo Monacelli also performed the variation of climatic data files for simulating the HI effect and the alternative green scenarios in the outdoor spaces on the energy demand for cooling in a representative building of Peristeri (see Chapter 2).

The author also acknowledges Prof Mat Santamouris and the researchers Theoni Karlessi and Afroditi Symnefa of the Group Building Environmental Research UOA, National University of Athens, Physics Department, for their precious help in the production of physical data. The author is also very thankful to Marina Laskari, Vivian Dorizas and Niki Gaitani for all the wonderful days spent togheter.

The author appreciates and thanks Marisa Luciani for her valuable help and support, and also acknowledges Dimitris Lagaris and Nikos Venetas of the municipality of Peristeri. Without Dimitris, all the work on Peristeri would not have been possible. The Peristeri urban compound will be always my vision of a possible future nearly zero urban setting.

To my beloved friend Beatrice Bettazzi, I owe the lost images of Athens that she gave back to me. She is the author of the picture on the cover of the book. I will forever preserve the wonderful memory of those few beautiful days together in Athens.

To my beloved husband and daughters, I will never find the proper words to express my gratitude for your help, patience, and support. To my parents, brother and family, you always have my thankfulness and my love.

REFERENCES

http://ec.europa.eu/eurostat/statistics-explained/index.php/Housing_statistics.

Bernstein, H.M., 1999. Highlights of the CERF/CIB symposium. Build. Res. Inf. 27 (6), 437–438.

Buildings Performance Institute Europe (BPIE), 2013. A Guide to Developing Strategies for Building Energy Renovation, Delivering Article 4 of the EED, BPIE.

Cecodhas, 2013. Power House, Nearly Zero Energy Challenge, Progress Report Mar. 2013.

COM, 2012. Consultation paper: financial support for energy efficiency in buildings.

COM, 2015. Communication on a framework strategy for a resilient energy union with a forward-looking climate change policy, 80 Final, http://ec.europa.eu/priorities/energy-union/docs/energyunion_en.pdf.

Convertible City, 2006. http://www.convertiblecity.de/CONVERTIBLE%20CITY_english.pdf.

Davis, M., 1981. A wall for all seasons. Riba J. 88 (2), 55–57.

Dimitrijevic, B., Langford, D., MacLeod, I., Maver, T., 2002. An assessment tool for the durability adaptability and energy conservation of buildings, Ninth International Conference on Durability of Building Materials and Components, Paper 140.

Dunster, B., Simmons, C., Gilbert, B., 2009. The ZEDbook. Solutions for a Shrinking World. Taylor and Francis, New York.

Energy Efficiency Financial Institutions Group, 2015. EE, The First Fuel for the EU Economy: How to Drive New Finance for Energy Efficiency Investments, https://ec.europa.eu/energy/sites/ener/files/documents/Final%20Report%20EEFIG%20v%209.1%2024022015%20clean%20FINAL%20sent.pdf. Final report.

Eisenman, P., 1971. Notes on Conceptual Architecture: Towards a Definition, in Casabella, Nov./Dec., 1971.

EU EED, 2013. EU EED (2012/27/EU), Guidebook for Strong Implementation. The Coalition for Energy Savings. Rue de Toulouse 49, 1040, Brussels, Belgium. In: Scheuer, S. (Ed.) accessed at: http://eedguidebook.energycoalition.eu/images/PDF/EED.pdf
EU Report, 2008. Improvement Potential of Residential Buildings, JRC, EU, Institute Prospective Technological Studies, IMPRO-Building, Env. Seville, EU, 2008.
EU Report, 2010. World and European Sustainable Cities. Insights from EU Research, Directorate-General for Research. Directorate L, Science, Economy and Society, Unit L.2. Research in the Economic, Social Sciences and Humanities. European Commission, Bureau SDME 07/34, B-1049 Brussels, 2010. Socio-Economic Sciences and Humanities; EUR 24353 EN.
Ferrante, A., Mochi, G., Gulli, R., Cattani, E., 2011. Retrofitting and adaptability in urban areas. Procedia Eng. 21, 795–804.
Fischer, R. Schwehr, P., 2008. Typenbasierte Evaluation. Chancen für den ganzheitlichen Wohnungsbau, contribution to 15th Swiss Status-Seminar, Energie und Umweltforschung im Bauwesen", Zürich 2008.
Fraunhofer, 2002. Fraunhofer-Institut, für sichere Informationstechnologie, Forschungsprojekt FlexHaus.
Friedman, Y., 1975. Towards a Scientific Architecture. MIT Press, Cambridge, Massachusetts, 169 pp.
Gehl, J., 2010. Cities for People. Island Press, Chicago.
Gertz, C., 1983. Local Knowledge. Basic Books, New York.
Habraken, J., 1999. In: Teicher, Jonathan (Ed.), Supports: An Alternative to Mass Housing. Urban International Press, UK.
Housing Statistics in the European Union, 2010. The Hague: Ministry of the Interior and Kingdom Relations, In: Dol, K., Haffner, M. (Eds.), OTB Research Institute for the Built Environment, Delft University of Technology.
Hutchison, R., 2010. Encyclopedia of Urban Studies. SAGE Publications, California.
IEA, 2001. Assessing the Adaptability of Buildings. International Energy Agency. Annex 31. Energy-Related Environmental Impact of Buildings. Nov. 2001.
JWG, 2013. Towards assisting EU Member States on developing long term strategies for mobilizing investment in building energy renovation (per EU Energy Efficiency Directive Article 4), Composite Document of the Joint Working Group of CA EED, CA EPBD and CA RES, Nov. 2013. (http://www.ca-eed.eu/reports/art-4-guidance-document/eed-article-4-assistance-document).
Klanten, R., Feireiss, L., 2007. Spacecraft. Fleeting Architecture and Hideouts. Die Gestalten Verlag, Berlin.
Kobayakawa, S., Chae, S., Kano N., 2006. Development of an adaptable system for effective utilization of architectural space, Proceedings of the Twenty Third ISARC, Tokyo, Japan, pp. 176–181.
Marini S., 2008. Architettura parassita, strategie di riciclaggio per la città, Architettura Ascoli Piceno, Quodlibet Studio.
MacGuirk, J., 2011. PREVI: the Metabolist utopia, architecture report from Lima. In: Domus, http://www.domusweb.it/en/architecture/2011/04/21/previ-the-metabolist-utopia.html
Nilsson, K., Pauleit, S., Bell, S., Aalbers, C., Thomas, A., Nielsen, S., 2013. Peri-urban Futures: Scenarios and Models for Land Use Change in Europe. Springer, Netherlands.
Peattie, K., 2010. Green consumption: behaviour and norms. Ann. Rev. Environ. Resour. 35, 195–228.
Plagaro, N., Schwehr, P., 2008. Die Typologie der Flexibilität im Hochbau, interact Verlag, Luzern.
Preiser, W.F.E., Vischer, J.C., White, E.T., 1991. Design Intervention: Toward a More Humane Architecture. Van Nostrand Reinhold, New York.
Price, C., 2001. Think the Unthinkable. Exhibition at the Lighthouse, Glasgow.

Santamouris, M., Kapsis, K., Korres, D., Livada, I., Pavlou, C., Assimakopoulos, M.N., 2007. On the relation between the energy and social characteristics of the residential sector. J. Energy Buildings 39 (2007), 893–905.
Schwehr, P., 2010. Evolutionary algorithms in architecture, O&SB2010. Open and Sustainable Building.In: Chica, Elguezabal, Meno, and Amundarain (Eds.), Labein -TECNALIA. ISBN 978-84-88734-06-8.
Shnapp. S., Sitjà, R., Laustsen, J., 2013.What is a Deep Renovation Definition? Global Buildings Performance Network, Technical Report, February 2013, info@gbpn.org, www.gbpn.org, @GBPNetwork.
Vreedenburgh, E., Melet, E., 2012. Rooftop Architecture, Building on an Elevated Surface. Rotterdam: NAi Publishers; New York: Available in North America through D.A.P./ Distributed Art Publishers, 2005.
Webler, T., Tuler, S.P., 2010. Getting the engineering right is not always enough: researching the human dimensions of the new energy technologies. Energy Policy 38, 2690–2691.

APPENDIX

Tables reporting the main thermal properties of the ground floor layers, the roofs, and the glazing components, estimated for each study in the as-built scenario.

Aghia Varvara

Table A1 Main thermal properties of the ground floor layers

Ground floor (U = 4.762 W/m² K)						
Layers	**Materials**			**Properties**		
External layer	Gravel	mm	170	Convective heat transfer coefficient	W/m²K	44.87
				Radiative heat transfer coefficient	W/m²K	5.13
				Surface resistance	m²K/W	0.02
Layer 2	Screed	mm	160			
Layer 3	Reinforced concrete	mm	300			
Layer 4	Screed	mm	80			
Internal layer	Ceramic/marble/tiles	mm	20	Convective heat transfer coefficient	W/m²K	0.342
				Radiative heat transfer coefficient	W/m²K	5.54
				Surface resistance	m²K/W	0.17

Table A2 Main thermal properties of the roof layers' components

Flat roof (U = 1.795 W/m² K)						
Layers	**Materials**			**Properties**		
External layer	Concrete tiles (roof)	mm	20	Convective heat transfer coefficient	W/m²K	44.87
				Radiative heat transfer coefficient	W/m²K	5.13
				Surface resistance	m²K/W	0.02
Layer 2	Screed	mm	80			
Layer 3	EPDM (waterproofing)	mm	20			
Layer 4	Concrete slab	mm	250			

(Continued)

Towards Nearly Zero Energy
http://dx.doi.org/10.1016/B978-0-08-100735-8.00004-0

Table A2 Main thermal properties of the roof layers' components (*cont.*)

Flat roof (U = 1.795 W/m² K)

Layers	**Materials**			**Properties**		
Internal layer	Plaster (lightweight)	mm	15	Convective heat transfer coefficient	W/m²K	4.46
				Radiative heat transfer coefficient	W/m²K	5.54
				Surface resistance	m²K/W	0.1

Table A3 Main thermal properties of the intermediate floor layers

Intermediate floor (U = 1.327 W/m² K)

Layers	**Materials**			**Properties**		
External layer	Ceramic/ porcelain tiles	mm	20	Convective heat transfer coefficient	W/m²K	44.87
				Radiative heat transfer coefficient	W/m²K	5.13
				Surface resistance	m²K/W	0.02
Layer 2	Screed	mm	80			
Layer 3	Concrete slab	mm	180			
Internal layer	Plaster (light)	mm	15	Convective heat transfer coefficient	W/m²K	4.46
				Radiative heat transfer coefficient	W/m²K	5.54
				Surface resistance	m²K/W	0.11

Table A4 Main thermal properties of the building's vertical external components

External wall (U = 1.183 W/m² K)

Layers	**Materials**			**Properties**		
External layer	Plaster (dense)	mm	20	Convective heat transfer coefficient	W/m²K	44.87
				Radiative heat transfer coefficient	W/m²K	5.13
				Surface resistance	m²K/W	0.02
Layer 2	Blocks	mm	220			
Internal layer	Plaster (lightweight)	mm	15	Convective heat transfer coefficient	W/m²K	2.15
				Radiative heat transfer coefficient	W/m²K	5.54
				Surface resistance	m²K/W	0.13

Stratigraphic description of the building envelopes' components. Main components in the as built scenario in Evripidou.

Table A5 Resistance and U-value for the existing building envelope

Type	Exterior walls		
Description	Brick wall		
	Inner surface resistance	m^2K/W	0.130
	External surface resistance	m^2K/W	0.040
	U-Value	W/m^2K	1.493

Layers:

Table A6 Thermo-physical properties for each component layer in the existing building envelope

Material	Thickness (cm)	Conducibility (W/mK)	Conduttance (W/m^2K)	Specific heat (J/kg K)	Density (kg/m^3)	Resistance (m^2K/W)
Mortar	1.00	0.5		1000	1300	0.02
Hollow brick	28.00		2.174	800	1700	0.46
Mortar	1.00	0.5		1000	1300	0.02
Total	30.00					0.50

Table A7 Resistance and U-value for the existing internal walls

Typet	Internal walls		
Description	Internal partition		
	Inner surface resistance	m^2K/W	0.130
	External surface resistance	m^2K/W	0.130
	U-Value	W/m^2K	1.786

Layers:

Table A8 Thermo-physical properties for each component layer in the existing internal walls

Material	Thickness (cm)	Conducibility (W/mK)	Conduttance (W/m^2K)	Specific heat (J/kg K)	Density (kg/m^3)	Resistance (m^2K/W)
Mortar	1.50	0.5		1000	1300	0.03
Hollow brick	12.00		4.167	800	1700	0.24
Mortar	1.50	0.5		1000	1300	0.03
Total	15.00					0.30

Table A9 Resistance and U-value for the existing ground floor

Type	Ground floor		
Description	Brick walls		
	Inner surface resistance	m^2K/W	0.170
	External surface resistance	m^2K/W	0.040
	U-Value	W/m^2K	1.559

Layers:

Table A10 Thermo-physical properties for each component layer in the existing ground floors

Material	Thickness (cm)	Conducibility (W/mK)	Conduttance (W/m^2K)	Specific heat (J/kg K)	Density (kg/m^3)	Resistance (m^2K/W)
Tiles	2	1.30		840	2300	0.015
Screed	7	1.06		840	1200	0.066
Reinforced concrete	20	2.3		1000	2300	0.087
Oversite concrete	10	0.38		1000	1200	0.263
Total	39					0.432

Table A11 Resistance and U-value for the existing internal floors

Type	Internal floor		
Description	Brick walls		
	Inner surface resistance	m^2K/W	0.130
	External surface resistance	m^2K/W	0.040
	U-Value	W/m^2K	1.402

Layers:

Table A12 Thermo-physical properties for each component layer in the existing internal floors

Material	Thickness (cm)	Conducibility (W/mK)	Conduttance (W/m^2K)	Specific heat (J/kg K)	Density (kg/m^3)	Resistance (m^2K/W)
Tiles	2	1.3		840	2300	0.015
Screed	4	1.06		1000	2500	0.038
Hollow brick	20		2.128	800	1700	0.47
Mortar	1	0.5		1000	1300	0.02
Total	27					0.543

Table A13 Resistance and U-value for the existing roof floors

Type	Roof floors		
Description	Roof floor		
	Inner surface resistance	m^2K/W	0.100
	External surface resistance	m^2K/W	0.040
	U-Value	W/m^2K	1.511

Layers:

Table A14 Thermo-physical properties for each component layer in the existing roof floors

Material	Thickness (cm)	Conducibility (W/mK)	Conduttance (W/m^2K)	Specific heat (J/kgK)	Density (kg/m^3)	Resistance (m^2K/W)
Asphalt	1	0.7		1000	2100	0.014
Reinforced concrete	4	2.3		1000	2500	0.017
Hollow brick	20		2.128	800	1700	0.47
Mortar	1	0.5		1000	1300	0.010
Total	26					0.522

Stratigraphic description of the building envelopes' components. Designed scenarios in the three case studies:

Table A15 Resistance and U-value for the designed exterior walls

Type	Exterior walls		
Description	Exterior wall with AAC		
	Inner surface resistance	m^2K/W	0.130
	External surface resistance	m^2K/W	0.040
	U-Value	W/m^2K	0.492

Layers:

Table A16 Thermo-physical properties for each component layer in the designed exterior walls

Material	Thickness (cm)	Conducibility (W/mK)	Conduttance (W/m^2K)	Specific heat (J/kgK)	Density (kg/m^3)	Resistance (m^2K/W)
Mortar	1.00	0.500		1000	1300	0.02
Hollow brick	28.00		2.174	800	1700	0.46
Mortar	1.00	0.500		1000	1300	0.02

(*Continued*)

Table A16 Thermo-physical properties for each component layer in the designed exterior walls (*cont.*)

Material	Thickness (cm)	Conducibility (W/mK)	Conduttance (W/m²K)	Specific heat (J/kgK)	Density (kg/m³)	Resistance (m²K/W)
Gypsum mortar for plastering	1.00	0.180		1000	600	0.056
Autoclaved areated concrete	10.00	0.190		1300	115	1.250
Gypsum mortar for plastering	1.00	0.180		1000	600	0.056
Total	42.00					1.861

Table A17 Resistance and U-value for the designed exterior walls (insulation with polyurethane)

Type	Exterior walls		
Description	Exterior wall with polyurethane		
	Inner surface resistance	m²K/W	0.130
	External surface resistance	m²K/W	0.040
	U-Value	W/m²K	0.219

Layers:

Table A18 Thermo-physical properties for each component layer in the designed exterior walls (insulation with polyurethane)

Material	Thickness (cm)	Conducibility (W/mK)	Conduttance (W/m²K)	Specific heat (J/kg K)	Density (kg/m³)	Resistance (m²K/W)
Mortar	1.00	0.500		1000	1300	0.02
Hollow brick	28.00		2.174	800	1700	0.46
Mortar	1.00	0.500		1000	1300	0.02
Polyurethane	10.00	0.026		1590	35	3.846
Gypsum mortar for plastering	1.00	0.180		1000	600	0.05
Total	41.00					4.402

Table A19 Resistance and U-value for the designed roof floors (insulation with AAC)

Type	Roof floors		
Description	Roof floor with AAC		
	Inner surface resistance	m^2K/W	0.100
	External surface resistance	m^2K/W	0.040
	U-Value	W/m^2K	0.494

Layers:

Table A20 Thermo-physical properties for each component layer in the designed roof floors (insulation with AAC)

Material	Thickness (cm)	Conducibility (W/mK)	Conduttance (W/m^2K)	Specific heat (J/kgK)	Density (kg/m^3)	Resistance (m^2K/W)
Sand and gravel	4.00	2.000		650	2100	0.020
Nonwoven fabric	0.10	0.190		1200	1400	0.005
Autoclaved aerated concrete	10.00	0.080		1300	115	1.250
Sheet of bitumen	0.20	0.23		1000	1100	0.01
Gypsum mortar for plastering	1.00	0.180		1000	600	0.05
Screed	4.00	1.060		896	2800	0.038
Reinforced concrete	4.00	2.3		1000	2500	0.017
Hollow brick	20.00		2.128	800	1700	0.47
Mortar	1.00	0.5		1000	1300	0.02
Total	44.30					1.885

Table A21 Resistance and U-value for the designed roof floors (insulation with polyurethane)

Type	Roof floors		
Description	Roof floor with polyurethane		
	Inner surface resistance	m^2K/W	0.100
	External surface resistance	m^2K/W	0.040
	U-Value	W/m^2K	0.216

Layers:

Table A22 Thermo-physical properties for each component layer in the designed roof floors (insulation with polyurethane)

Material	Thickness (cm)	Conducibility (W/mK)	Conduttance (W/m^2K)	Specific heat (J/kgK)	Density (kg/m^3)	Resistance (m^2K/W)
Sand and gravel	4.00	2.000		650	2100	0.020
Nonwoven fabric	0.10	0.190		1200	1400	0.05
Polyurethane	10.00	0.026		1590	35	3.85
Sheet of bitumen	0.20	0.23		1000	1100	0.01
Gypsum mortar for plastering	1.00	0.180		1000	600	0.05
Screed	4.00	1.060		896	2800	0.05
Reinforced concrete	4.00	2.3		1000	2500	0.017
Hollow brick	20.00		2.128	800	1700	0.47
Mortar	1.00	0.5		1000	1300	0.02
Total	44.00					4.481

Table A23 Resistance and U-value for the designed green roof (insulated with AAC)

Type	Roof floors		
Description	Green roof floor with AAC		
	Inner surface resistance	m^2K/W	0.100
	External surface resistance	m^2K/W	0.040
	U-Value	W/m^2K	0.456

Layers:

Table A24 Thermo-physical properties for each component layer in the designed green roof (insulation with AAC)

Material	Thickness (cm)	Conducibility (W/mK)	Conduttance (W/m²K)	Specific heat (J/kgK)	Density (kg/m³)	Resistance (m²K/W)
Topsoil	4.00	1.50		1550	2000	0.027
Gravel	8.00	1.20		650	2100	0.06
Oversite concrete	4	1.06		840	1200	0.038
Nonwoven fabric	0.10	0.190		1200	1400	0.005
Gypsum mortar for plastering	1.00	0.180		1000	600	0.05
Autoclaved areated concrete	10.00	0.080		1300	115	1.250
Sheet of bitumen	0.20	0.23		1000	1100	0.01
Gypsum mortar for plastering	1.00	0.180		1000	600	0.05
Screed	4.00	1.060		896	2800	0.039
Reinforced concrete	4.00	2.3		1000	2500	0.017
Hollow brick	20.00		2.128	800	1700	0.24
Mortar	1.00	0.5		1000	1300	0.02
Total	56.00					2.051

Table A25 Resistance and U-value for the designed green roof (insulated with polyurethane)

Type	Roof floors		
Description	Green roof floor with polyurethane		
	Inner surface resistance	m²K/W	0.100
	External surface resistance	m²K/W	0.040
	U-Value	W/m²K	0.209

Layers:

Table A26 Thermo-physical properties for each component layer in the designed green roof (insulation with polyurethane)

Material	Thickness (cm)	Conducibility (W/mK)	Conduttance (W/m²K)	Specific heat (J/kgK)	Density (kg/m³)	Resistance (m²K/W)
Topsoil	4.00	1.50		1550	2000	0.027
Gravel	8.00	1.20		650	2100	0.067
Oversite concrete	4	1.206		840	1200	0.038
Nonwoven fabric	0.10	0.190		1200	1400	0.005
Gypsum mortar for plastering	1.00	0.180		1000	600	0.056
Polyurethane	10.00	0.026		1590	35	3.85
Sheet of bitumen	0.20	0.23		1000	1100	0.009
Gypsum mortar for plastering	1.00	0.180		1000	600	0.05
Screed	4.00	1.060		896	2800	0.038
Reinforced concrete	4.00	2.3		1000	2500	0.017
Hollow brick	20.00		2.128	800	1700	0.47
Mortar	1.00	0.5		1000	1300	0.02
Total	57.30					4.647

GLOSSARY

AMA Athens metropolitan area
BAT Best available technologies
EPB Energy performance of buildings
EPBD Energy performance of buildings directive
EU European Union
EPc Energy performance index for cooling
EPh Energy performance index for heating
EE Energy efficient, energy efficiency
EC Energy consumption
ES Energy saving
GHG Green house gas
HI Heat island
HVAC Heating, ventilation air conditioning
nZEBs Nearly zero energy buildings
RES Renewable energy sources
RUEs Real urban environments
UHI Urban heat island
ZEUS Zero energy urban settings

http://dx.doi.org/10.1016/B978-0-08-100735-8.00005-2

SUBJECT INDEX

F

R

V

W

Z

Kolibri 360 GmbH
Ferdinand-Jühlke-Straße 7
99095 Erfurt, Deutschland
produktsicherheit@kolibri360.de

Global Value Chains and Production Networks

Global Value Chains and Production Networks

Case Studies of Siemens and Huawei

Cui Fengru
Party School of Beijing Municipal Materials Co., Ltd., Beijing, P.R. China

Liu Guitang
Beijing Electro-mechanical Engineering Institute, Beijing, P.R. China

ELSEVIER

ACADEMIC PRESS
An imprint of Elsevier

社会科学文献出版社
SOCIAL SCIENCES ACADEMIC PRESS (CHINA)

Academic Press is an imprint of Elsevier
125 London Wall, London EC2Y 5AS, United Kingdom
525 B Street, Suite 1650, San Diego, CA 92101, United States
50 Hampshire Street, 5th Floor, Cambridge, MA 02139, United States
The Boulevard, Langford Lane, Kidlington, Oxford OX5 1GB, United Kingdom

British Library Cataloguing-in-Publication Data
A catalogue record for this book is available from the British Library

Library of Congress Cataloging-in-Publication Data
A catalog record for this book is available from the Library of Congress

ISBN: 978-0-12-814847-1

Publisher: Candice Janco
Acquisition Editor: Scott J. Bentley
Editorial Project Manager: Susan Ikeda
Production Project Manager: Vijayaraj Purushothaman
Cover Designer: Miles Hitchen

Typeset by MPS Limited, Chennai, India

Contents

Section III
GPN Trends in the PostCrisis Era and Conclusions

Foreword

Global production networks (GPNs), the cornerstone of economic globalization, have been transformed in depth, breadth, and complexity amid technological changes and macroeconomic shocks. In particular, the global financial crisis triggered by the US subprime crisis in 2008 changed the landscape of GPNs profoundly, exerted strong impact on the international economic system, and brought about reindustrialization and value chain contraction in developed countries. As the financial crisis declined after reaching its height, the global economy also began to recover slowly, stabilize, and show positive signs of growth. On the whole, it has entered the postcrisis era that is marked by the mixture of recession and recovery as well as moderation and fluctuation. Instability and uncertainty is growing in the global economy. New breakthroughs are coming as the new technological and industrial revolutions unfold. International trade rules are being redesigned. The global economic and industrial landscape is to be reshaped radically. Special attention should be given to the impact of technological changes and replatforming on GPNs and global value chains, as noted by Prof. Andrew Michael Spence, recipient of the 2001 Nobel Memorial Prize in Economic Sciences, when delivering a keynote speech at China Development Forum 2016. While the international environment takes on a new look in the postcrisis era, China has also seen significant changes in the domestic economic environment. The resource, environmental, and factor cost constraints are becoming increasingly tight. Having entered a new normal in economic development, the country faces the double challenge of stabilizing growth and adjusting structure, competition from both developed countries and emerging markets, and the harsh fact that its low cost advantage is diminishing rapidly but new competitive edges are yet to be gained. Apparently there is an uphill battle to fight. A key contributor to the success of China's transformation and upgrading is Chinese enterprises' competitiveness and capacity for sustainable development.

China is now in a period of strategic opportunity when the country's accelerating endeavor to change its growth model meets the new technological and industrial revolutions. Major strategic plans such as the Internet Plus action plan and the Made in China 2025 initiative have been launched one after another and all-round efforts have been made to build China into a manufacturer of quality. As China's dependence on foreign trade keeps growing and the manufacturing sector shifts its focus from size to quality, the open economy in a new era will be largely marked by large-scale "going global" as well as "bringing in" on higher levels. Implementing the "going global" strategy is an important part of the efforts to help Chinese manufacturers grow into international players, build China's manufacturing competitiveness, and advance its Belt and Road Initiative. However, Chinese industrial enterprises remain at the early stage of "going global." Their progress in creating GPNs is undesirable due to the absence of a clear strategy and they still have a long way to go in such aspects as increasing their global presence and capacity to operate globally. Against such a backdrop, it becomes increasingly important to study major organizational innovations in GPNs. Chinese manufacturers seeking to go global may need to pay special attention to some questions. What are GPNs created for, what are their main characteristics, and how do they work? How do multinational corporations (MNCs)—essential organizations in the international economic system today—become successfully embedded in local economies and grow into global companies? How does the financial crisis impact the international division of labor and industrial landscape and in what direction will GPNs move in the postcrisis era? What should China, a country whose dependence on GPNs keeps growing, do in response to the new features of GPNs in the postcrisis era?

The book *Global Value Chains and Production Networks: Case Studies of Siemens and Huawei* builds on microcosmic GPN studies and combines theoretical analysis with case studies. It presents an overview of the research into GPNs, examines the theoretical basis of microcosmic GPN studies, offers the analytical framework of such studies, discusses the Siemens and Huawei cases that are used to illustrate the status of Germany and China in the world economy and the changes in this regard, and describes GPN trends in the postcrisis era. In my view, the author has made a bold attempt to answer the abovementioned questions in this book. It will be a source of inspiration for manufacturing enterprises in China and other parts of the world that aim to make foreign investments safely and efficiently, increase their capacity to operate globally, and reshape their international competitiveness. This is a book of great value to researchers in economics and management, market analysts, business owners, government employees, and those who devise the development strategy for MNCs.

I would like to recommend this book to readers interested in Chinese enterprises going global and expect to see more studies by the author to help Chinese enterprises transform and upgrade themselves and achieve innovation-driven development.

Pei Changhong
Institute of Economics, Chinese Academy of Social Sciences, Beijing, P.R. China

Preface

The framework of the research into global production networks (GPNs) has been developed over time by Dieter Ernst, Jeffrey Henderson, and other authors based on such theories as value chain, value-added chain, global commodity chain, global value chain, networks and embeddedness, and actor-network theory, with traditional division of labor theories serving as meta-theories. GPNs have made it possible to explain, in a wider context, the new changes in international production organizations today and provided a new analytical framework for the research into globalization, regional development, and industrial development. Many scholars have studied GPNs in depth from different perspectives so the framework of GPN research has been further expanded and improved.

There are many deficiencies in previous GPN studies in terms of theoretical framework, research content, attention to new developments, etc. A special feature of today's international economic and trade landscape is the increasingly prominent status of multinational corporations (MNCs). On the one hand, MNCs are players in GPNs on the microlevel and it is the global expansion of MNCs that has facilitated the formation of GPNs. On the other hand, as the international division of labor goes further and businesses become increasingly globalized, the influence of heterogeneous firms on international trade has become a hot topic for researchers in international trade. Previous GPN studies were mostly analyses on regional and industry levels. Very little attention was given to the microlevel enterprises. Therefore, the research content is yet to be enriched.

What is a GPN then? A widely recognized definition is yet to be provided. In this book, a GPN is understood as "a multitier system of diverse networked organizations that allocate resources, connect with value chains, and produce and supply products and services worldwide." This definition underscores three characteristics of GPNs, that is, global spatial configuration, integration of value-added activities, and diversity of network participants.

This book builds on microcosmic GPN studies and combines theoretical exploration with case studies. It presents an overview of the research into GPNs, examines the theoretical basis of microcosmic GPN studies, offers the analytical framework of such studies, discusses the Siemens and Huawei cases, describes GPN trends in the postcrisis era, and makes suggestions on how to transform and upgrade Chinese manufacturers. The research findings presented in this book can help to enrich and improve GPN theories, provide theoretical support for China's strategy of building itself into a manufacturer of quality and Chinese enterprises' "going global" initiative, and inform policymaking in this regard.

The book is divided into eight chapters in three parts, namely "Theoretical Basis and Analytical Framework of Microcosmic GPN Studies," "The Siemens and Huawei Cases," and "GPN Trends in the Postcrisis Era and Conclusions."

Chapter 1, Overview of the Research Into GPNs, defines the concept of the GPN, examines the theoretical origins of GPN research, describes the research framework of the Managerial School and the University of Manchester, summarizes and comments on previous GPN studies, and notes the future direction of GPN research.

Chapter 2, Theoretical Basis of Microcosmic GPN Studies, introduces three theories essential to GPN research, that is, Multinational Corporations Theory, New Classical Trade Theory, and "New" New Trade Theory, as well as their interpretation of GPNs.

Chapter 3, Analytical Framework of Microcosmic GPN Studies, proposes the analytical frameworks of microcosmic GPN studies based on the theories introduced in Chapter 1, Overview of the Research Into GPNs, and Chapter 2, Theoretical Basis of Microcosmic GPN Studies, and the analytical frameworks of Jeffrey Henderson and other authors. There is one framework from the perspective of value with five basic components, that is, Value Objectives, Global Strategy, Networks, Location Selection, and Network Governance, and another framework from the perspective of embeddedness with three areas of focus, that is, Strategy, Dimension, and Region.

Chapter 4, A Study of Siemens's GPN, begins with looking at Siemens's overall GPN as well as global R&D network, global manufacturing network, global supply chain network, and global marketing service network from the perspective of value, and assessing the value added of all networks. Then it analyzes in detail Siemens Healthcare's embeddedness in China in an attempt to figure out how an MNC manages to be embedded locally and how the process evolves.

Chapter 5, A Study of Huawei's GPN, addresses Huawei's overall GPN as well as global R&D network, global production and operation network, and global marketing service network from the perspective of value, and examines the process of Huawei's embeddedness in Europe.

Chapter 6, A Comparison Between GPNs of Huawei and Siemens, compares the GPNs of Huawei and Siemens in general and then from the perspectives of value and embeddedness, and summarizes Huawei's experience.

Chapter 7, GPN Trends in the Postcrisis Era, explores the four driving forces behind the development of GPNs, describes the GPN trends in the postcrisis era as well as the new features of GPNs and their impact on China, and looks at the topological structure of GPNs in the postcrisis era.

Chapter 8, Conclusions and Research Prospects, presents the conclusions drawn from the research and discusses the issues to be further studied.

A highlight of this book is the detailed study of the Siemens and Huawei cases that can help readers better understand GPN theories and the framework of microcosmic GPN studies, and offer Chinese manufacturing enterprises seeking to go global successful experience to draw upon. It will inspire Chinese enterprises striving to raise the level of foreign investment utilization and international cooperation, build a networked manufacturing innovation system, and enhance their capacity to run multinational business and international competitiveness.

Lastly, I would like to thank the Research Institute for Global Value Chains, University of International Business and Economics for its full support for the publication of this book, and my PhD supervisor Prof. Zhao Zhongxiu for helping me get my foot in the door of economics research, offering me careful guidance when I was writing my doctoral dissertation on microcosmic GPN studies, and encouraging me to polish my dissertation for publication after starting my professional career. I also owe much to my parents whose positive influence has made me a tenacious and enterprising person, and to my husband Liu Guitang who always gives me the strength to overcome difficulties when I need it most. Guitang is an expert on IT application in the manufacturing industry as well as the second author of this book. He spent his spare time writing the majority of Chapter 4, A Study of Siemens's GPN, and Chapter 5, A Study of Huawei's GPN, which earns him my eternal gratitude. The publication of the Chinese and English versions of this book coincides with the birth of our daughter Liu Chengcheng so I dedicate this book to her and wish her a life of happiness and fulfillment.

Eight years have passed since I began to write this book. I consulted the works by many authors in China and other countries and documents of Siemens and Huawei and cited their new views, new frameworks, new methods and practical experience in GPN-related fields. I cannot reach everyone who has inspired me but I am thankful to them all. Anyone who finds an improper citation in this book may reach me and revisions will be made when the second edition is published. The views expressed in the book may not be accurate enough, perspectives not broad enough, and case studies not thorough enough. Errors may also be found. Feedback in this regard would be appreciated.

January 2018
Beijing

Introduction

Global production networks (GPNs) emerged and grew in the late 20th century and early 21st century when the international production system went through major changes. Since the 1980s, along with the deepening of economic globalization, trade and investment liberalization has progressed rapidly and technological advances have brought down enterprises' management, transportation, and communication costs. Multinational corporations (MNCs) have started to allocate resources and organize production activities on a global scale. The international division of labor has reached a stage marked by the coexistence of specialization between industries, within industries, and specific to products. Amidst increasingly fierce competition and wide fluctuations in the global market, both the competitive and collaborative ties between countries, regions, industries, and enterprises have become increasingly close. Driven by various forces, the new international division of labor system has been expanded and GPNs have emerged and spread rapidly.

GPNs represent a major organizational innovation in the international production system. In the context of globalization, they have brought development opportunities for the countries and enterprises involved and will exert greater impact on the politics, economy, and culture of these countries. Ever since the concept of GPN[1] was put forth by Dieter Ernst,[2] Jeffrey Henderson,[3] and other authors in 1999, many scholars have studied GPNs on national, regional, and industry levels. For example, the study conducted by Ernst et al. focused on the scope of production, power asymmetry, and knowledge diffusion (mainly flagships) in GPNs, as well as the capacity upgrading of local suppliers and industrial upgrading of developing countries. Henderson et al. paid more attention to the whole social process that involves the production of products and services as well as the reproduction of knowledge, capital, and labor, and stressed that local development could be achieved and social welfare improved amid the interaction between globalization and localization.

---

1. Global production networks are also called international production systems, global production systems, multinational production networks, etc.
2. Dieter Ernst used the term "global production networks" in a conference paper in 1999 for the first time.
3. Jeffrey Henderson and Peter Dicken made the first attempt to devise a detailed GPN framework in a research proposal in 1999.

However, the status of MNCs in today's international economic and trade system becomes increasingly prominent. MNCs in fact act as the main driving force behind economic globalization. As the core organizers of international economic activities today, MNCs not only have greater "industrial sovereignty," but also will gain more and more "economic sovereignty" and even "political say." MNCs are microcosmic participants in GPNs and they have absolute dominance over other participants. In fact, it is the global expansion of MNCs that has contributed to the formation of GPNs. MNCs distribute and configure the various parts and functions of value chains according to the advantages of different countries and regions. The production capacity of all countries is involved. The competitive edges of countries are embodied no longer in industries, but in businesses or one part or several parts of the product value chain. There is vertical, horizontal, or mixed division of labor. Industrial activities become globalized and lay the microfoundation for GPNs. Therefore, microcosmic GPN studies from the perspective of MNCs play a significant role in helping to figure out the main characteristics of GPNs and how they work.

As the international division of labor goes deeper and business activities get increasingly globalized, enterprises stand out at the center of the new international trade landscape and the effects of enterprises' different features on different areas of international trade become a hot subject of study for economists worldwide. Many theories have emerged in the new areas of international trade research, including "new" new trade theories[4] represented by trade models with heterogeneous firms and endogenous boundary model of the firm. Moreover, the MNCs theory has introduced the concept of the network. The new classical trade theory underlines the dilemma between the economies of division of labor and transaction costs as well as the half measure of professional decision-making by individuals in reexamining international trade theories. They all have significant influence on the microcosmic research into international trade. Therefore, the latest international trade theories are introduced in this book to enrich the framework of GPN research and help explain the main characteristics of GPNs and how they work on the microlevel and look at how enterprises use global resources and networks to enhance their international competitiveness and share the fruits of globalization in a better way.

This book features the combination of theoretical exploration and case study so the significance of research is in both theoretical and practical terms. On the front of theoretical exploration, this book offers an overview of the research into GPNs and proposes the analytical framework for microcosmic GPN studies based on the latest international trade theories

4. Some researchers consider endogenous boundary model of the firm as part of trade models with heterogeneous firms. In this book, the two are categorized as "new" new trade theory to highlight their difference in research interest.

such as MNCs theory, new classical trade theory, and "new" new trade theories. In terms of case study, the GPNs of Siemens and Huawei and the embeddedness of their business in specific regions is analyzed in detail from the perspectives of value and embeddedness. There is also dynamic analysis of GPN trends in the postcrisis era.

The research findings presented in this book can help enrich and improve GPN theories. Ever since the concept of GPN was introduced in 1999, its implications have been enriched by many researchers in China and other countries but the theoretical framework is yet to be fully fledged. To be specific, there may be mature theories about GPNs but they only constitute a research framework and there are no methodological tools to put the framework into use. In addition, previous studies are mostly analyses on the macrolevel. Very few studies focus on enterprises, and how GPNs work on the microlevel is yet to be explored. Apart from defining GPNs, this book gives an introduction to the history of GPN research and the theoretical basis of microcosmic GPN studies, provides the analytical framework for such studies, explains what each component of the framework means and how they are categorized, and describes the basic paradigm of the research into each component. In short, this research shows some innovation and has great value in theoretical terms in that it is successful in expanding the framework of GPN research, enriching research methods in this regard, and making microcosmic GPN studies more operable.

The research findings can provide theoretical support for China's strategy of building itself into a manufacturer of quality and Chinese enterprises' "going global" initiative, and inform policymaking in this regard. Ever since its accession to the WTO, China has quickly involved itself in GPNs along with an increasingly open economy. Today, it has the world's largest manufacturing industry and plays an extremely important role in the international production system. However, Chinese manufacturers are still much less competitive than their international counterparts in such aspects as independent innovation capacity, efficiency in resource utilization, industrial structure, IT application, quality, etc. Chinese enterprises have limited presence in other countries and small capacity to operate globally so it is a pressing task for them to transform and upgrade themselves to achieve leapfrogging. The Chinese government has devised the Made in China 2025 plan, listing "increasing the global presence of manufacturing enterprises" as one of the nine strategic tasks. The plan also specifies such priorities as "raising the level of foreign investment utilization and international cooperation," "enhancing the capacity to run multinational business and international competitiveness," and "promoting international cooperation on deeper levels and accelerating the 'going global' of Chinese enterprises." The description of GPNs' main characteristics and how they work on the microlevel, the discussion on how MNCs leverage global resources and networks to enhance their international competitiveness, and the analysis of GPN trends in the postcrisis era in this book can help Chinese enterprises in choosing the right strategy and path in their endeavor to go global. This research can also inspire the Chinese government in devising the right policies in the postcrisis era to help Chinese enterprises become better embedded in GPNs, build a networked manufacturing innovation system with enterprises serving as main players, and enhance the country's innovation capacity and competitive edges.

Section I

Theoretical Basis and Analytical Framework of Microcosmic GPN Studies

Chapter 1

Overview of the Research into GPNs

Chapter Outline

Global production networks (GPNs) represent a major organizational innovation in the international production system. In the context of globalization, they have brought development opportunities for the countries and enterprises actively involved and will exert greater impact on the politics, economy, and culture of these countries. From the perspective of economics, the issue of GPNs boils down to the division of labor, which also provides the essential theoretical basis. However, the development of GPNs has posed challenges to traditional division of labor theory and international division of labor theory. These traditional theories are no longer sufficient to explain the latest changes in the global production system and the driving forces behind them so it is pressing to create a new theoretical framework. Since the 1980s, many new theories have been derived from the meta-theory of division of labor in conjunction with theories in other fields such as economics, management science, sociology, and geography, including value chain,

Global Value Chains and Production Networks. DOI: https://doi.org/10.1016/B978-0-12-814847-1.00001-4

value-added chain, networks and embeddedness, and actor-network theory (ANT), global commodity chain (GCC), and global value chain (GVC).

On the basis of those theories, Dieter Ernst, Jeffrey Henderson, and other authors introduced the concept of GPN in 1999 and further improved its implications in later studies. Now there are two parallel approaches to GPN research, one adopted by the managerial school represented by Dieter Ernst and the other by the University of Manchester represented by Jeffrey Henderson. Comparatively speaking, Dieter Ernst and other like-minded researchers follow the route of GVC research and see three essential characteristics of GPNs, that is, scope, asymmetry, and knowledge diffusion. They put the focus of research on the scope of production, power asymmetry, and knowledge diffusion (mainly flagships) in GPNs, as well as the capacity upgrading of local suppliers and industrial upgrading of developing countries. Jeffrey Henderson and other researchers follow the route of GCC research and underscore three elements of GPNs, that is, value, power, and embeddedness.[1] They pay more attention to the whole social process that involves the production of products and services as well as the reproduction of knowledge, capital, and labor, and stress that local development could be achieved and social welfare improved amid the interaction between globalization and localization. GPNs make it possible to explain, in a wider context, the new changes in international production organizations today and provide a new analytical framework for the research into globalization, regional development, and industrial development. Within the frameworks designed by Dieter Ernst and Jeffrey Henderson, many scholars in China and other countries have studied GPNs in depth from various perspectives, be it theoretical, empirical, or case study. Their remarkable research findings have made the framework of GPN research even better.

1.1 THE CONCEPT OF GLOBAL PRODUCTION NETWORK

1.1.1 Previous Definitions of Global Production Network

Typical definitions of GPN include the following:

Ernst and Kim (2002) consider GPNs a major innovation in the organization of international business and argue that these networks combine concentrated dispersion of the value chain across firm and national boundaries, with a parallel process of integration of hierarchical layers of network participants.

Henderson et al. (2002) propose the GPN as a conceptual framework for grasping the global, regional, and local economic and social dimensions of the processes involved in many (though by no means all) forms of economic

1. There are several Chinese translations of the term *embeddedness*.

globalization. Production networks—the nexus of interconnected functions and operations through which goods and services are produced, distributed, and consumed—have become both organizationally more complex and also increasingly global in their geographic extent.

According to Sturgeon (2002a,b), production networks are the various interfirm relationships that link a group of firms to form larger economic entities. The focus is on the features and level of interfirm relationships. Production networks refer to not only the value chains formed as a result of the vertical division of labor for the same product, but also the integrated relations between firms linked by the production of the same product.

Roper and Grimes (2005) define the concept of GPN proposed by Henderson et al. as "the global network of firms, institutions and other economic agents which shapes, and is shaped by: the fundamental processes of knowledge and wealth creation, enhancement and exploitation; corporate, collective and institutional elements of organizational power; and, spatial and network embeddedness."

Chai (2003) holds that GPNs are international production networks centering around multinational corporations (MNCs) formed as a result of the manufacturing process in which a number of countries are involved in the different stages of a product's production. During such a process, MNCs distribute different stages of production in locations with the highest efficiency and lowest cost and benefit from the differences between locations in cost, resource, logistics, and market.

Liu (2008) defines GPNs as the relationships between highly interdependent production processes distributed in different countries and regions and bonded by value chains, fostered as a result of the division of labor between and integration of the wholly owned subsidiaries and joint ventures of lead firms (MNCs) and external organizations such as independent suppliers, contractors, distributors, and partners in strategic alliances.

1.1.2 The Definition of Global Production Network in This Book

In this book, a GPN is understood as "a multitier system of diverse networked organizations that allocate resources, connect with value chains, and produce and supply products and services worldwide."

This definition underscores the following characteristics of GPNs:

1. *Global spatial configuration.* Geographically, GPNs cross many national boundaries. They are intended to leverage the location advantages of different countries and regions for proper allocation of resources and division of labor along value chains, and to integrate them for maximum overall efficiency worldwide. The adoption of the term *global* instead of *transnational* or *international* is intended to highlight the ever-growing

interdependence between institutions and economic actors of all countries on economic, political, and technological fronts. The terms *international* and *transnational* derive from essentially state-centric discourses. Thus while they incorporate notions of cross-border activity of many sorts, they do not adequately express the way in which nonplace-specific processes penetrate and transform place-specific ones, and vice versa. They do not, therefore, help to deliver the imaginative sensibilities necessary to grasp the dialectics of global–local relations that are now a precondition for the analysis of economic globalization and its asymmetric consequences (Henderson et al., 2002). Internationalization stresses cooperation between entities across national boundaries while globalization focuses on the decreasing importance of national boundaries and increasingly free flows of factors between countries.

2. *Integration of value-added activities*. GPNs integrate dispersed organizations and cover division of labor in all parts of the value chain, including not only production activities but also technology R&D, logistics and purchasing, marketing and sales, after-sale services, and other activities. Value-added activities worldwide are integrated in a global value network system. Here the term *production* is used instead of *commodity*. In contemporary usage the term *commodity* generally connotes standardized products and with that, the fixity of their production in time and space. It clearly has rich social implications (including consumption) but it does not capture product-specific division of labor and flexible production that characterize international production models today. Besides, GPN researchers' preference for a discourse of "production" places the analytic emphasis on the social processes involved in producing goods and services and reproducing knowledge, capital, and labor power. In particular, as long as "production" is couched broadly to include intermediate and final markets and as long as the dynamics of power and knowledge between actors and institutions are understood in a multidirectional and nondeterministic fashion, then the GPN framework allows for far greater complexity and geographical variation in producer–consumer relations than the GCC approach has so far achieved.
3. *Diversity of network participants*. A GPN covers flagships and their subsidiaries, affiliates, and joint ventures as well as suppliers, subcontractors, distribution channels, and value-added resellers (Ernst and Kim, 2002). Network participants are in various hierarchical layers among which power distribution is asymmetric and they are heterogeneous. Flagships (lead firms) are at the center of the networks. There are diverse cooperative relations between the participants such as joint venture, alliance, and outsourcing, based on their complementarity in the division of labor and the principle of mutual benefit. Such relations place emphasis on the models of external governance and shared governance, featuring flexible organizational governance and network connections. The term

network is adopted instead of *chain* because the concept of *chain* stresses the vertical sequence of products or services and other parts of the value chain as well as the arrangement and division of economic activities among different economic actors. Given the unidirectional linear process it indicates, the chain metaphor seems to have difficulties in incorporating due attention to the issues of the reproduction of labor power, etc. and works against the possibility of the individual firms incorporated into a production system having room for autonomous action within that system. The term *network*, however, incorporates the idea of spatial clustering. It stresses how different parts of the value chain are distributed and concentrated worldwide according to the demand for factors. Therefore, *network* can better describe the flows of key factors such as capital, knowledge, and technology among producers, consumers, and intermediaries and underscores that the flows are organized vertically, horizontally, and diagonally in complex and dynamic configurations to form a multidimension, multitier framework of economic activities. Comparatively speaking, a discourse of networks is more inclusive, empirically adequate, and thus more analytically fertile.

1.2 THEORETICAL ORIGINS OF GLOBAL PRODUCTION NETWORK

The GPN framework builds on many theories and has been improving. Quite a few researchers have discussed the theoretical origins of GPN. For example, Hess and Yeung (2006) have identified four major historical precursors of the GPN network: (1) the value chain framework in strategic management since the early 1980s; (2) the networks and embeddedness perspectives in economic and organizational sociology since the mid-1980s; (3) the actor-network analysis in science studies since the mid-1980s; and (4) the global commodity/value chain analysis in economic sociology and development studies since the mid-1990s.

Drawing upon their results, I find that the GPN framework is derived mainly from three sets of theories, that is, division of labor theories, chain theories, and network theories. Fig. 1.1 shows the evolution of GPN research. Division of labor theories mainly includes traditional division of labor theories and traditional international division of labor theories before the 1980s, which are the most fundamental meta-theories of the GPN framework. Chain theories mainly include value chain, value-added chain, GCC, and GVC that emerged in the early 1980s and are the direct precursors of the GPN framework. Network theories mainly include networks, embeddedness, and ANT that provide important theoretical basis for the GPN framework. This chapter focuses on all those theories and addresses their relations with the GPN research.

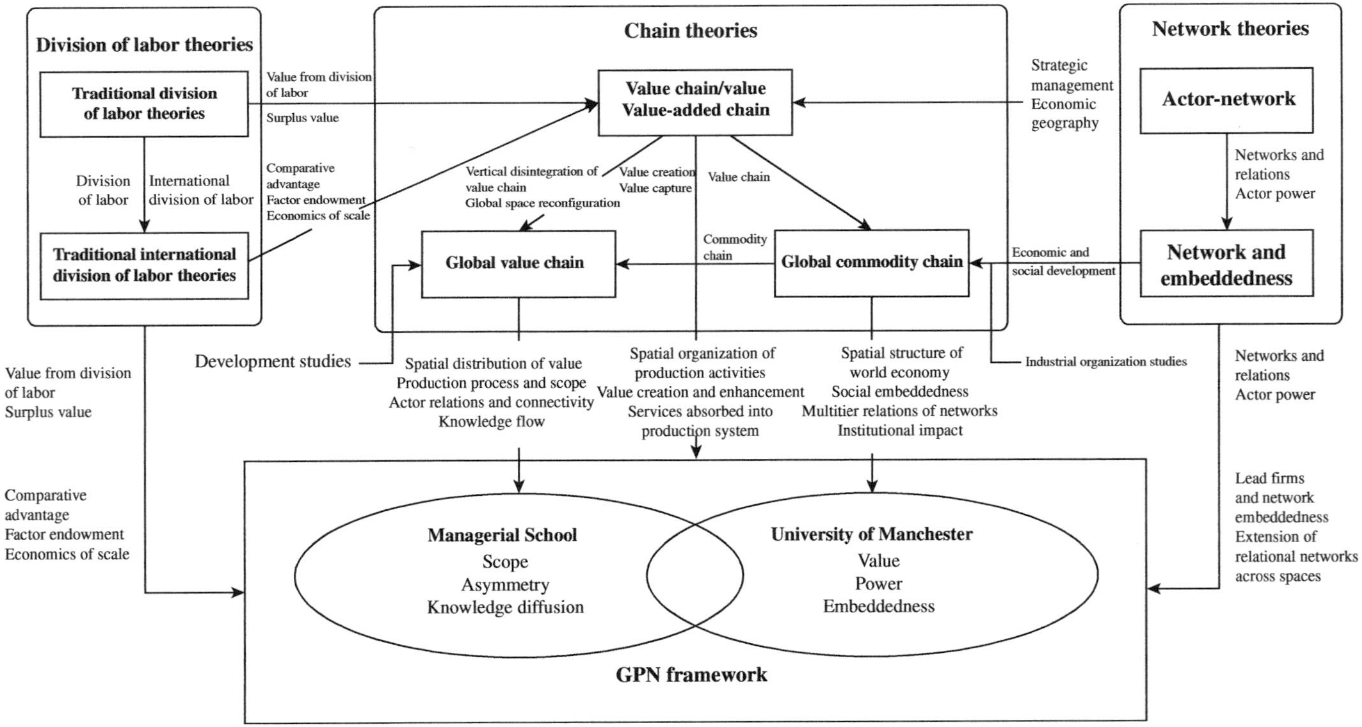

FIGURE 1.1 Overview of GPN research. Source*: Prepared by the author.*

1.2.1 Traditional Division of Labor Theories

The division of labor is the soul of classical economics. It is deemed to be the source of efficiency and productivity or the logical starting point of economic analysis. The issue of GPNs boils down to the division of labor. Traditional division of labor and specialization theories laid the foundation for the GPN framework and its major precursors. The economic efficiency achieved through division of labor is what GPNs are created for. Despite their limitations in explaining practical problems, traditional division of labor theories laid a solid foundation for the GPN research system.

1. Traditional division of labor theories:

   Classical economics, Marxist political economy,[2] and neoclassical economics all recognized the role of division of labor in enhancing labor productivity and driving economic growth. Classical economics is a school of thought in economics that emerged in the transition from agriculture-centered to industry-centered economy and the discussion is centered on the division of labor is the specialization brought by the division of labor. The most representative economist in this regard is Adam Smith. Marxist political economy focuses on the determination of social development by the division of labor and the representative economist is Karl Marx (1867). Neoclassical economics replaces the labor theory of value in classical economics with the marginal utility theory, the discussion is centered on the allocation of resources, and there are analytical concepts about the division of labor and increasing returns to scale. The typical representatives of this school are Alfred Marshall (1890) and Allyn Young (1928).
2. Traditional international division of labor theories:

   Since the division of labor may occur domestically and internationally, there are also many international division of labor theories. The international division of labor has close ties with international trade, which are two inseparable aspects of one process. The international division of labor evolves in the same direction as international trade theories develop. The two are interdependent and promote each other. The international division of labor always holds the attention of economists. Various theories have been proposed, analyzing the causes and benefits of the international division of labor. These theories provide an important basis for the GPN framework and its major precursors as well as good economic interpretation of GPN formation. According to traditional international economics, the two basic types of international trade are interin-

2. It is generally believed that classical economics does not cover Marxist political economy but the latter is influenced by and built on the former.

dustry trade and intraindustry trade. Correspondingly, there are also interindustry specialization and intraindustry specialization.[3]

It is universally agreed that the theories explaining interindustry specialization and interindustry trade are classical and neoclassical trade theories. Typical classical trade theories include the theory of absolute advantage proposed by Adam Smith (1776) and the theory of comparative advantage by David Ricardo (1817). Such theories build on assumptions such as perfectly competitive market and explain interindustry specialization and international trade from the perspective of labor productivity differentials. A typical neoclassical trade theory is the factor endowment theory proposed by E.F. Heckscher and B. Ohlin (Heckscher and Ohlin, 1991). Neoclassical trade theories also build on assumptions such as perfectly competitive market but they address interindustry specialization and international trade from the perspective of factor endowment differences.

The theories explaining intraindustry specialization and intraindustry trade are intraindustry trade theories. A typical example is Raymond Vernon's international product life cycle theory that dynamizes the theory of comparative advantage and the resource endowment theory and combines the product life cycle theory in marketing studies with technological advances in discussing the pattern in international relocation of industries and explaining intraindustry specialization and trade from the perspective of technology differentials. The "new" new trade theory[4] developed by Paul Krugman et al. (Helpman and Krugman, 1985) breaks with traditional assumptions and introduces increasing returns to scale and imperfect competition in explaining intraindustry specialization and trade from the perspectives of horizontal differentiation and homogeneity. The new-factor proportion approach proposed by R.E. Falvey (1981) makes adjustments to the assumptions of H−O model and establishes relationships between various combinations of product features and basic factors such as labor and capital in explaining the causes of intraindustry trade and describing intraindustry specialization and trade from the perspective of vertical differentiation. Of these intraindustry trade theories, the new trade theory of Paul Krugman et al. is the most influential.

3. Limitations of traditional international division of labor theories:

Since the 1980s, global trade and production have witnessed fast development and the implications of the international division of labor and trade have changed significantly. The most important change is that

3. This part mainly refers to "International Trade Theories and Policies" (Zhao and Lv, 2009a,b).

4. In some literature, the intraindustry trade theory is equated with the new trade theory. In this book, the new trade theory in a narrow sense is adopted, meaning that it is a theory that breaks with the theoretical paradigms of classical and neoclassical trade theories and explains the phenomenon of intraindustry trade.

more and more countries use their factor endowment and technology to engage in a production process of certain product through international vertical specialization,[5] resulting in the so-called intraproduct international specialization, and then intermediate trade on a large scale (intraproduct international trade). The global division of labor system is undergoing profound changes, featuring the coexistence of interindustry specialization, intraindustry specialization, and intraproduct specialization. Moreover, intraproduct international specialization and trade grows rapidly and its influence is also growing.

Standard trade theories mostly focus on the specialization and exchange of final products (Arndt, 1997). With nonsubdivisible products as the implicated premise, these theories lack due attention to intraproduct international specialization and have limitations in explaining modern business operation and management and economic development. Many scholars have studied international vertical specialization within the framework of traditional international trade theories. For example, Sanyal and Jones (1982), Hummels (1999), and Deardorff (2001) used the Ricardian model to explain the causes and models of international vertical specialization and trade. Feenstra and Hanson (1996), Arndt (1997), and Deardorff (2001) analyzed international vertical specialization using the H–O model of factor endowment theory. Jones and Kierzkowski (1990, 2001) and Krugman (1991) explained international vertical specialization based on the new trade theory. However, traditional international division of labor theories have failed to systematically address the issue of intraproduct trade, which stands in stark contrast to the importance of intraproduct international trade to the economy.

Since the 1990s, many economists have attempted to establish a new theoretical system and research framework. A typical example is the concept of intraproduct specialization proposed by Prof. Lu Feng at China Center for Economic Research (CCER) in 2004 in *Intra-Product Specialization*. He discussed the new form of division of labor on basic levels, examined the causes, determinants and driving forces of intraproduct specialization, and described the features of products and production processes in the international division of labor today. However, the new theories are yet to be fully developed. There is no sufficient proof of their accuracy or value, no concepts are widely accepted, and the studies based

5. A widely accepted definition of international vertical specialization is yet to be developed. Similar terms include vertical specialization, intraproduct specialization, production fragmentation, disintegration of production, production sharing, outsourcing, slicing up of the value chain, modular production network, etc.

on different notions have different focuses. Therefore, the new theories including intraproduct specialization are yet to become the mainstream of international economics.

1.2.2 Chain Theories

Unlike the new international division of labor theories, the research into contemporary international division of labor and economic development started earlier in the fields where economics overlaps with management studies and sociology, and the research findings have more far-reaching influences. Since the 1980s, an important reflection of economic globalization on the corporate level is the emergence of a series of revolutionary management and operation methods and the introduction of concepts in strategic competition and other related fields of management studies in the context of deepening intraproduct specialization (Lu, 2004). Chain theories such as value chain, value-added chain, GCC, and GVC go beyond the assumption that one country finishes the whole process of a product's production, proceed from the international specialization and global combinations of production blocks, and offer deeper interpretation of contemporary international division of labor and economic development. They are the direct precursors of the GPN framework.

1.2.2.1 Value Chain

In 1985, Michael Porter proposed the analytical framework of value chain in the book *Competitive Advantage*, noting that the advantage of the individual firm in competition is the advantage in certain parts of the value chain with strategic value (Porter, 1985). The basis of value chain analysis is "value," which is derived from the value from division of labor in classical economics and surplus value in Marxism. The theory of competitive advantage incorporates comparative advantage, factor endowment, economies of scale, and other traditional international trade theories that are summarized and improved.

The analytical framework of value chain provides a key concept of space for analysis and reveals how value is created, enhanced, and captured in different spatial structures. It also suggests the inseparability of the manufacturing and service sectors that constitute the main body of economic and production activities. The service sector's importance in a modern economy means that this will be an important task of GPN research.

1.2.2.2 Value-Added Chain

In the book *Designing Global Strategies: Comparative and Competitive Value-Added Chains*, Kogut (1985a,b) developed the concept of value-added chain to analyze the advantage in international strategy. The value-added chain was used to analyze the competitive position of the firm in a global industry and for structuring the strategic allocation decision. Therefore, the concept of value chain is extended to countries and regions. Compared with the value chain perspective, value-added chain can better reflect the relationship between the vertical disintegration of value chain and global space reconfiguration. Hence its crucial importance to the formation of the GVC perspective.

From the perspective of GPN research, the theories of value chain and value-added chain have a limited utility because they pay no attention to issues of corporate power, the institutional contexts of—and influences upon—firm-based activities, or to the territorial arrangements (and their profound economic and social asymmetries) in which the chains are embedded. As a consequence, they have little relevance for the study of economic development (Henderson et al., 2002).

1.2.2.3 Global Commodity Chain

Gereffi and Korzeniewicz (1994) proposed the framework of GCC on the basis of combining value chain analysis and industrial organization research. They discussed internal structure of GCCs and how lead companies create and control GCCs and argued that technological, institutional, and organizational innovations and regulatory environment determine and change the structure of industries and the power of lead companies in the industries, thus determining the evolution of the governance structure of GCCs. Gereffi (1995) stresses four tiers of GCC research: (1) input–output structure, (2) territoriality, (3) governance structure, and (4) institutional framework.

The concept of GCC incorporates most of the elements relevant to the organization of firm and interfirm networks and their relation to the possibilities for economic and social development, with the exception of trade unions and other NGOs, and the focus of research is on the dynamics of global organization of production. Gereffi's contribution was an explicit attempt to operationalize some of the world-systems categories for the empirical study of cross-border, firm-based transactions and their relation to development, as well as the structure of corporate power embedded in the intra- and interfirm networks that circle the globe. It transcends the limitations of state-centric conceptualizations in the study of business management and organization as well as the impact of MNCs on economic and social development, thus extending the scope of GPN research.

1.2.2.4 Global Value Chain

As global production and trade networks became more complicated and the studies of such networks go deeper, Gereffi et al. (2001) made one more step forward by proposing the framework of GVC, connecting the concept of value chain with developing countries to make up for the shortcomings of the GCC framework, and laying the foundation for GPN research. The GVC framework places more emphasis on the input–output process of production, spatial distribution of value, and lead firms' dominance over chain governance, focusing on the hierarchical relationships of production organizations. It reveals the dynamics of the global industry and examines where value is and who creates and distributes it.

A milestone in GVC research is the publication of a special issue of *IDS Bulletin* ("The Value of Value Chains") in 2001. Gereffi and other researchers in this field analyzed the process of economic globalization from the perspective of value chains. They understood the trade in goods and services as a governance system and argued that understanding the significance of value chain operation is particularly important to enterprises and policymakers in developing countries. That is because the process of a GVC's formation is one in which enterprises worldwide become integrated in value chains and obtain necessary technological capacity and service support. In the special issue of the magazine, many scholars discussed the governance, evolution, upgrading, and other dimensions of GVCs, making clear the basic concepts and theoretical framework.

The GVC perspective's broader analysis of global economic integration can be seen in four aspects. First, the focus of research is shifted from manufacturing to other parts of the goods and labor supply chain, including distribution and marketing. Second, it addresses the relationships between different actors in the chains and their implications for development. Third, the idea of interorganizational connectivity makes it easier to analyze the relationship between formal/informal jobs and workers, especially for developing countries. Fourth, the attention to all parts and processes (not only production) in the chains and related activities helps to determine how to improve returns through imperfect competition and fragmentation in the market.

1.2.2.5 Limitations of Chain Theories

The value chain, value-added chain, GCC, and GVC theories all employ a "chain" metaphor in analyzing organizations and connectivity in the process of product production. A chain maps the vertical sequence of events leading to the delivery, consumption, and maintenance of goods and services—recognizing that various value chains often share common economic actors and are dynamic in that they are reused and reconfigured on an ongoing basis—while a network highlights the nature and extent of the interfirm relationships that bind sets of firms into larger economic groupings (Sturgeon, 2001). A major weakness of the "chain" approach is its conceptualization of production and distribution processes as being essentially vertical and linear. In fact, such processes are better conceptualized as being highly complex network structures in which there are intricate links—horizontal, diagonal, as well as vertical—forming multidimensional, multilayered lattices of economic activity. For that reason, an explicitly relational, network-focused approach promises to offer a better understanding of production systems (Henderson et al., 2002).

1.2.3 Theories of Network, Embeddedness, and Actor-Network

1.2.3.1 Networks and Embeddedness

In the 1920s and the 1930s, social network analysis began to catch the interest of sociologists, focusing on social interaction as the microfoundation of society. In the mid-1980s, Granovetter (1985) started to embed economic activities in social network for study, noting that the core of social fabric is people's social network and the network mechanism of embeddedness is trust that is derived from and embedded in the social network. After that, a large number of researchers in economic sociology, organization studies, and strategic management began to pay attention to networks and embeddedness but the concept's diffusion into economic geography's lexicon remained relatively slow until the early 1990s. In particular, Dicken and Thrift (1992) made a strong case for economic geographers to take networks and embeddedness very seriously in the geographical analysis of firms and their productive activities. By the late 1990s, the concept of networks and embeddedness had become an analytical cornerstone of the GPN framework in economic geography.

1.2.3.2 Actor-Network Theory

What might seemingly be missing in an embeddedness framework that relies on the structural analysis of network relations, however, is the role of geographical agents such as firms (Hess and Yeung, 2006). This has compelled economic geographers to understand better the nature and properties of networks and their constituents. The work of actor-network analysis (Callon, 1986; Latour, 1987, 1999; Law, 1987, 1999) in science and technology studies since the mid-1980s becomes highly useful here. The ANT emphasizes the relationality of both objects and agency in heterogeneous networks, pointing out that entities in networks are shaped by their relations and connectivity to other entities (Law and Hassard, 1999). For the study of GPNs, this means that space and distance have to be seen not in absolute terms, but as "spatial fields" and relational scopes of influence, power, and connectivity. The ANT rejects dualism and conceptualizes an actor-network as a heterogeneous network consisting of both humans and nonhumans, thus allowing the consideration of the important technological elements that underlie and influence economic activities. The conceptualization of actors and their relations breaks with the previous focus on economic connections between firms in economic geography and ushers in the consideration of business organizations and their spatial effect on a larger social scale.

1.2.3.3 Limitations of Networks, Embeddedness, and Actor-Network Theory

While the ANT offers an interesting methodology that has been adopted already for the study of globalization and production networks, its contribution to the analysis of economic development is constrained by the fact that it lacks an appreciation of the structural preconditions and power relations that inevitably shape production networks (Dicken et al., 2001). Actors are theorized in the GPN framework devised by Dicken and coworkers not as individual agents per se, but as a constitutive part of the wider network through which emergent power and effects are realized over space (Hess and Yeung, 2006).

1.3 FRAMEWORK OF THE RESEARCH INTO GLOBAL PRODUCTION NETWORKS

Both the Managerial School and the Manchester School conceptualize GPNs as a governance model for global production organizations that organizes the value chain across firm and national boundaries according to some formal rules (contract), with a parallel process of integration of hierarchical layers of network participants. Compared with the GVC/GCC framework, the GPN research integrates different value chain perspectives to form a single analytical framework, directs attention to the differences between different parts of the value chain in value creation, and addresses their spatial implications. Greater emphasis is put on localization in the process of globalization as well as the global and regional mechanisms and their effects. The GPN framework offers better understanding of and theoretical support for production organizations worldwide.

The two schools of thought adopt two parallel approaches to GPN research and their research content differs. Comparatively speaking, Ernst follows a route close to that of the GVC framework and shows strong influence of management studies. Ernst et al. hold that GPNs have three essential characteristics, that is, scope, asymmetry, and knowledge diffusion. Their approach focuses on the relations between business organizations assuming different functions and their production scope, and underscores the central role of flagships, the interdependence in networks, and power asymmetry. They hold that the power distribution among network actors is dynamic, and that R&D and technology transfer play a part in driving GPN growth. Special attention is given to flagship-led knowledge diffusion, the flow and sharing of knowledge, and its impact on local suppliers' capacity upgrading and local industries' upgrading and development.

Henderson et al. recognize the implications of the GCC perspective and their studies are much influenced by sociology. They underscore three elements of GPNs, that is, value, power, and embeddedness; address the role of institutions and value enhancement and distribution; pay more attention to the whole social process that involves the production of products and services as well as the reproduction of knowledge, capital, and labor; and stress that local development could be achieved and social welfare improved amid the interaction between globalization and localization. They argue that the control and coordination in production networks is accomplished through the hierarchical control mechanism of firms, MNCs in particular, under the functioning of the market mechanism so greater emphasis is placed on hierarchical control relations. It is also believed that local development is the result of the global functional configuration of lead firms and their suppliers so local response to globalization is highlighted.

1.3.1 The Managerial School

The GPN studies by Ernst et al. are based on electronics and information technology industries. Ernst and Kim (2002) propose a conceptual framework for discussing the relations between the evolution of GPNs and the role of network flagships in knowledge transfer and diffusion, and local suppliers' capacity. According to Ernst et al., GPNs represent a major innovation in the organization of international business and these networks combine concentrated dispersion of the value chain across firm and national boundaries, with a parallel process of integration of hierarchical layers of network participants. A GPN covers flagships and their subsidiaries, affiliates, and joint ventures as well as suppliers, subcontractors, distribution channels, value-added resellers, and R&D alliances and a series of cooperative agreements such as standards consortia.

1.3.1.1 Flagships

GPN actors fall into two major categories, that is, flagships that are in a dominant position and local suppliers. There are two types of global flagships: (1) "brand leaders" (BL), like Cisco, GE, IBM, and Dell, that allocate resources worldwide to create their own GPNs for cost reduction, product differentiation, and faster speed-to-market as well as horizontal specialization; and (2) "contract manufacturers" (CMs), like Solectron and Flextronics, that establish their own GPNs through vertical specialization to provide global supply chain services for the BLs. At the heart of GPNs, flagships provide strategic and organizational leadership beyond the resources that lie directly under their management control such as technology and brands. They control strategic value-added chains and activities and thus have absolute influence and control over other firms in the networks.

1.3.1.2 Local Suppliers

Local suppliers are divided into higher-tier suppliers and lower-tier suppliers. Higher-tier suppliers, such as Taiwan's Acer Group, play an intermediary role between global flagships and local suppliers. They deal directly with global flagships that determine core GPN-related R&D activities as well as core business activities such as the formulation of overall marketing strategy. They possess valuable proprietary assets (technology and channels) and have developed their own mini-GPNs so higher-tier suppliers are at the heart of these mini-networks. Lower-tier suppliers' main competitive advantages are low cost, speed, and flexibility of delivery. They are subordinate to local higher-tier suppliers in business and rarely deal directly with global flagships.

The conceptualization of Ernst et al. highlights two important factors: (1) the transition from intrafirm to interfirm transactions and forms of coordination and (2) an ever-increasing scope of networks. GPNs link flagships' subsidiaries, affiliates, and joint ventures with subcontractors, suppliers, service providers, and partners in strategic alliances. Lead firms (flagships) are at the heart of networks and have strategic and organizational leadership, with other members under their direct management and control. The driving force of firms seeking globalization comes from the interplay of three determinants, that is, technology, institutions, and competition. The main purpose of the networks is to provide global flagships with quick and low-cost access to resources, capabilities, and knowledge that are complementary to their core competencies.

Ernst et al. hold that GPNs have three essential characteristics. The first is scope, that is, a GPN covers all stages of the value chain, not just the production part. The second is asymmetry, that is, flagships dominate and control resources and decision-making in the networks. The third is knowledge diffusion, that is, the sharing of knowledge is the adhesive that maintains network growth (Ernst, 2003a,b). Their analytical framework directs attention to the behavior of flagships: allocation of resources and innovation. For that, they place the focus of research on the scope of production, power asymmetry, knowledge diffusion (mainly flagships) in GPNs, as well as the capacity upgrading of local suppliers and industrial upgrading of developing countries. They have considerable research findings in this regard.

1.3.2 The Manchester School

Henderson et al. (2002) comment on the GPN studies by Ernst. Apart from noting his shortcomings, they also recognize the helpfulness of his work in that he highlights a number of key problems that have hindered previous research in this area. First, he criticizes the tendency to focus narrowly on the role of key flagships within GPNs at the expense of attention to network suppliers. This means failure to reveal the source of success in the GPNs in both theoretical and practical terms. Second, he notes that in mapping the dispersion of production units, research has often overlooked the wide range of service functions (from design to marketing and beyond) that are crucial to the viability of GPNs. Third, Ernst notes a preoccupation with formal

R&D and technology transfers, which may preclude an appreciation of the importance of diffusion of less codified forms of knowledge.

Henderson et al. are committed to provide a more rigorous and comprehensive analytical framework than that of Ernst. The GPN proposed is a conceptual framework that is capable of grasping the global, regional, and local economic and social dimensions of the processes involved in many (though by no means all) forms of economic globalization. Production networks—the nexus of interconnected functions and operations through which goods and services are produced, distributed, and consumed—have become both organizationally more complex and also increasingly global in their geographic extent. Such networks not only integrate firms into structures that blur traditional organizational boundaries, but also integrate national economies in ways that have enormous implications for their well-being. They propose three conceptual categories, that is, value, power, and embeddedness, and four conceptual dimensions, that is, firms, sectors, networks, and institutions. That is basically the framework for GPN analysis proposed by Henderson et al.

1.3.2.1 Value

It means both Marxian notions of surplus value and more orthodox ones associated with economic rent. The creation, enhancement, and capture of value stands at the center of GPNs.

1. *Value creation.* The significant issues here include the conditions under which labor power is converted into actual labor through the labor process; and the possibilities for generating various forms of rent.
2. *Value enhancement.* The issues involved here include: (1) the nature and extent of technology transfers both from within and without the given production network; (2) the extent to which lead and other major firms within the network engage with supplier and subcontractors to improve the quality and technological sophistication of their products; (3) as a consequence, whether demands for skill in given labor processes increase over time; (4) whether local firms can begin to create organizational, relational, and brand rents of their own. In all of these cases, the national institutional influences to which the firms are subject (government agencies, trade unions, employer associations, for instance) may be decisive for the possibilities of value enhancement.
3. *Value capture.* The pertinent issues here partly involve: (1) the nature of property rights and thus laws governing ownership structures and the repatriation of profits; (2) the extent to which firms are totally foreign owned, totally domestically owned, or shared equity as in joint-venture arrangements; and (3) the extent to which corporate governance is founded on stakeholder principles having important consequences for whether value generated in a given location is retained there and indeed used to the benefit of the commonweal.

1.3.2.2 Power

The source of power within the GPNs and the ways in which it is exercised is decisive for value enhancement and capture and thus for the prospects for development and prosperity. There are three forms of power that are significant here, that is, corporate power, institutional power, and collective power.

1. *Corporate power*. The lead firms in GPNs have the capacity to influence decisions and resource allocations decisively and consistently in their own interests. While power is usually asymmetrically distributed in production networks, lesser firms sometimes (and for contingent reasons) have sufficient autonomy to develop and exercise their own strategies for upgrading their operations. Additionally, and at least in principle, lesser firms incorporated into networks have the possibility of combining with other lesser firms to improve their collective situation within the GPNs.
2. *Institutional power*. Institutions in the GPNs include national and local institutions, international organizations (e.g., EU, ASEAN, NAFTA, etc.), world organizations (e.g., IMF, World Bank, WTO, etc.), UN agencies (particularly ILO), etc. Different types of institutions have different power to exercise and play different roles in the GPNs.
3. *Collective power*. There are collective agents who seek to influence companies at particular locations in GPNs. Examples of such collective agents include trade unions, employers associations, NGOs concerned with human rights, environmental and other issues, etc.

1.3.2.3 Embeddedness

GPNs do not only connect firms functionally and territorially but also they connect aspects of the social and spatial arrangements in which those firms are embedded and which influence their strategies and the values, priorities, and expectations of managers, workers, and communities alike. There are two forms of embeddedness of interest here, that is, territorial embeddedness and network embeddedness.

1. *Territorial embeddedness*. GPNs do not merely locate in particular places. They may become embedded there in the sense that they absorb, and in some cases become constrained by, the economic activities and social dynamics that already exist in those places. For example, the GPNs of particular lead firms may take advantage of clusters of small and medium enterprises (with their decisively important social networks and local labor markets) that predate the establishment of subcontracting or subsidiary operations by such firms. Moreover, the location of lead firms in particular places might generate a new local or regional network of economic and social relations, involving existing firms as well as attracting new ones. Embeddedness, then, becomes a key element in regional economic growth and in capturing global opportunities.

Similarly, national and local government policies (training programs, tax advantages, etc.) may influence embeddedness. The mode of territorial embeddedness, or the degree of a GPN firm's commitment to a particular location, is an important factor for value creation, enhancement, and capture.

2. *Network embeddedness*. GPNs are characterized not only by their territorial embeddedness, but also by the connections between network members regardless of their country of origin. It is most notably the "architecture," durability, and stability of these relations, both formal and informal, which determine the agents' individual network embeddedness as well as the structure and evolution of the GPN as a whole. Network embeddedness can be regarded as the product of a process of trust building between network agents, which is important for successful and stable relationships.

Methodologically, the GPN perspective of Henderson et al. directs attention to (1) the networks of firms involved in R&D, design, production, and marketing of a given product, and how these are organized globally and regionally; (2) the distribution of corporate power within those networks, and changes therein; (3) the significance of labor and the processes of value creation and transfer; (4) government agencies, trade unions, employer associations, and NGOs that influence firm strategy in the particular locations absorbed into the production chain; and (5) the implications of all of these for technological upgrading, value adding and capturing, economic prosperity, etc. for the various firms and societies absorbed into the chains. In terms of research content, Henderson et al. focus more on the whole social process that involves the production of products and services as well as the reproduction of knowledge, capital, and labor.

1.4 GLOBAL PRODUCTION NETWORK STUDIES WORLDWIDE

Compared with the Managerial School, the Manchester School has proposed a more comprehensive GPN framework that offers more applicable and broader perspectives for the research into economic globalization. It is particularly true with China, which is experiencing the transition from a planned economy to a market economy. The influence of institutional factors and the inertia of traditional systems remains significant and enterprises are confronted by political, social, cultural, and other problems in the production process. Therefore, the GPN studies, be they theoretical, empirical, or case study, conducted by other scholars in China and other countries[6] are summarized in this section, with the main thread of the Manchester School's

6. Some research results of Ernst, Henderson, and their collaborators have been described in previous sections.

framework (value, power, and embeddedness) running through together with the Managerial School's perspectives on networks of firms and local upgrading in the flow of knowledge.

1.4.1 Global Production Network Studies in Other Countries

In terms of value chain and industrial upgrading, Humphrey and Schmitz (2002) describe four types of industrial upgrading: process upgrading, product upgrading, function upgrading, and cross-industry upgrading. Taudes et al. (2002) studied the organizational learning mechanisms of industrial networks. Bair and Gereffi (2003), Palpacuer and Parisotto (2003), Coe et al. (2004), and Liu et al. (2006) carry out studies about the multitier process of value creation and industrial upgrading worldwide. Gereffi and Memedovic (2003) discuss the prospects for industrial upgrading in developing countries taking the garment industry as an example. Schmitz (2004) and Gereffi et al. (2005) use the GVC framework to analyze the processes of industrial upgrading, value creation, and value adding and capture. Parthasarathy and Aoyama (2006) attempts to prove that studying informal GPNs helps to understand local upgrading in the global economy, through a case study of the development of Bangalore's software industry.

As for organizational structure and power distribution, Levy (2005) examines the hegemony of global factories in GPNs. Roper and Grimes (2005) look at the characteristics of the development of information and communication technologies in different cities in the context of GPN. Hess and Coe (2006) note that power and embeddedness play a central role in shaping the organization and spatiality of production networks, taking the telecommunications industry as an example. Sally (2006) conducted a study on GPN embeddedness in Fiji's garment industry and reinforced the idea of power asymmetry by analyzing the dynamic process of production networks' development and transformation.

Regarding territorial and network embeddedness, Smith et al. (2002) and Bair (2005) note that the cultural, political, and institutional environments of GPN firms' operation should be understood anew. Depner and Bathlt (2005) explain how a German company overcomes difficulties against an unfamiliar institutional and cultural background based on the study of the supply chain in Shanghai's automobile cluster. Liu and Dicken (2006) attempt to reveal how the Chinese government uses institutional power to get foreign investors passively embedded in the country's economic system, taking China's automobile industry as an example. Hess and Rodrigue (2006) note the need to study the role of logistics and transportation in GPNs and give a brief introduction to previous studies done by members of American Association of Geographers (AAG). Yeung (2009) addresses the relations between regional development and the changes in GPN competitiveness, taking East Asia as an example.

1.4.2 Global Production Network Studies in China

GPNs remain a new area of study for Chinese scholars but many research results can be found, especially in terms of regional industrial upgrading. Many Chinese scholars have adopted the approach of empirical analysis or case study to explore regional status upgrading and related industrial upgrading in different regions and industries.

On the front of value chain and industrial upgrading, Liu et al. (2006) look at the transformation and upgrading of Chinese enterprises engaging in processing trade in GPNs. Pu and Zhang (2008) analyze the features of certain industries in GPNs, taking the automobile industry as an example. Tang and Zhang (2009) analyze empirically the status of Chinese manufacturing sector in the new international division of labor system and East Asia's production networks and the enhancement of value chains. Zhao and Gu (2010) discuss the relationship between wage level and industrial upgrading in the GPNs based on a survey into China's textiles and garment industries.

In terms of organizational structure and power distribution, Su et al. (2005) examine the governance mechanism of GPNs and Bu (2007) looks at the impact of GPN governance model on processing trade. Li et al. (2008) discuss the general organizational framework of GPNs as well as the value distribution and spatial competition of networks, taking the computer industry as an example. Zhu and Sun (2008) select different industries for the analysis of MNCs' globalized production model. Zhuo (2011) compares the governance models of GVCs.

As for territorial and network embeddedness, Gao et al. (2011) explore how international trade barriers affect the organizational model and spatial features of GPNs, taking China–Canada trade in bicycles as an example. Li and Ning (2011) summarize theories related to the GPN framework and propose a framework of local development based on globalization. Liu (2011) discusses China's path options for opening up to the outside world in the context of GPN.

1.5 SUMMARY

The research literature shows that the past two decades have seen a large number of theoretical, empirical, and case studies done by scholars in China and other countries from various perspectives. All these have helped us to acquire a profound understanding of the GPN framework and its components. However, the research results also show that there are still many improvements to make, which should be addressed in future GPN studies.

First, the theoretical framework is yet to be fully developed. There may be mature theories about GPNs but they only constitute a research framework and there are no methodological tools to put the framework into use. GPN research still faces challenges posed by firms' organization and management and integration on social, institutional, and cultural levels, that is, the integration of and balance between the Managerial School and the Manchester School. More new international trade theories should be included, such as new classical trade theory and "new" new trade theory. This will allow for better interpretation of the world's economic phenomena.

Second, the research content is yet to be enriched. There are very few detailed GPN studies on the microlevel. Previous studies are mostly analyses on national, regional, and industry levels, with little support of corporate data and analysis of business configuration and network integration across regions and industries from the perspective of firms. Too much emphasis is placed on chains and basic activities therein (especially production activities). Although the value chain in the GPN framework covers all stages, very little attention is given to supporting activities, especially producer services, in empirical studies. Besides, existing studies focus on a specific value chain or commodity chain. The various chains should be examined together and the implications of networks are not stressed sufficiently. The scope and depth of industry studies should be expanded. Now most studies focus on automobiles, electronics/communication, software, garments, toys, etc. Very little attention is given to emerging, high-end, and key basic industries, such as medical equipment, new energy, aerospace, aviation, shipbuilding, locomotives, CNC machine tools, etc. Moreover, the research on finished products is much more than that into core parts such as engines and controllers. Regional research should also be promoted in scope and depth. Now studies in this regard focus on global–local relations and the relations between developing countries and developed countries. More studies should be carried out to compare China with emerging markets, especially with BRICS countries. In addition, regional differences are not fully addressed in current studies. Given China's vast territory and regional imbalances in development, different parts of China surely have different ways of getting embedded in the GPN.

Lastly, due attention should be given to the dynamics in the operating environment of GPNs. The "system" or "world" composed by material and economic societies constitutes the basic environment for the formation, operation, and evolution of GPNs. Now most studies are carried out on the assumption that the basic environment is relatively static and trade rules are relatively stable. The US financial crisis, the European debt crisis, Japan's nuclear accident, Thailand's floods, political turmoil in North Africa and the Middle East, and other major events in recent years are bound to exert far-reaching impact on the global political, economic, and trade climate. The postcrisis era is marked by the mixture of recession and recovery as well as moderation and fluctuation. Instability and uncertainty is growing in the global economy. International trade disputes are intensifying, and international trade policies are full of uncertainty. Will new changes occur to the regional and industrial configurations of GPNs? Are there new configurations emerging in the existing framework? Can the existing research results stand the test of GPN changes under the new circumstances? All these questions are yet to be addressed.

Since its accession to the WTO, China has quickly involved itself in GPNs and its dependence on GPNs has been growing. In light of the findings and shortcomings in current GPN studies, I would suggest continuing to improve the GPN framework, extending the scope of research, and conducting more dynamic analysis of GPNs. This is of great theoretical and practical value to the implementation of China's overarching policy of stabilizing growth, stimulating domestic demand, adjusting structure, and promoting reform, and improvement of China's international competitiveness.

Chapter 2

Theoretical Basis of Microcosmic GPN Studies

Chapter Outline

As it is noted in Chapter 1, Overview of the Research Into GPNs, there are many shortcomings in the current global production network (GPN) studies in terms of theoretical framework, research subject, dynamic analysis, etc. In particular, the research into international trade has been conducted on corporate level while there are very few GPN studies on the microlevel and a more comprehensive approach is yet to be adopted. This chapter and Chapters 3, Analytical Framework of Microcosmic GPN Studies, explore the theoretical basis and analytical framework of GPN studies on the microlevel, building on the GPN framework proposed by Henderson et al. The basic theories concerning GPNs introduced in Chapter 1, Overview of the Research Into GPNs, including traditional division of labor and international division of labor theories, value chain, value-added chain, global commodity chain (GCC), global value chain (GVC), networks and embeddedness, and actor-network theory, provide the theoretical basis for microcosmic GPN studies. This chapter focuses on three important theories in this regard, that is, multinational corporations (MNCs) theory, new classical trade theory, and "new" new trade theory, as well as their interpretations of GPNs.

Global Value Chains and Production Networks. DOI: https://doi.org/10.1016/B978-0-12-814847-1.00002-6

2.1 MULTINATIONAL CORPORATIONS THEORY

2.1.1 Traditional Theories of Multinational Corporations

Early theories of MNCs are mainly built on the transaction cost theory and address three core issues, that is, why is it necessary to make foreign direct investment (FDI), what makes successful FDI possible, and where should FDI be located. Typical examples in this regard include monopolistic advantage theory (Hymer, 1960; Kindleberger, 1969), product life cycle theory (Vernon, 1966), internalization theory (Buckley and Casson, 1976), eclectic paradigm (Dunning, 1977, 1979, 1998), etc. These are FDI-related theories developed in the early stage of MNCs and mainly targeted at MNCs in developed countries so they can be considered traditional theories of MNCs. These theories can explain the internal mechanism and conditions of MNCs' integrated expansion in GPNs, the formation and change of production locations in GPNs, integrated production and hierarchy within MNCs, etc.

2.1.2 Strategic Management Theory

The word *strategy* derives from military activities and it evolved into an approach to business management after World War II, especially in the United States. Igor Ansoff's *Corporate Strategy*, the world's first book about strategy, was published in 1965, marking the beginning of corporate strategy research in modern history. Economic globalization has been an overwhelming trend since the 1980s. MNCs have been looking at the global market and drastic changes have occurred in the competition landscape, driving people to study strategic management in depth. Businesses' competitive strategies were at the forefront of economic and management studies so competitive strategy theories were further developed and three major models emerged, including industry structure, core competencies, and strategic resources. Michael Porter introduced the theory of industry structure and also proposed the value chain framework, based on which the concepts of value-added chain, GCC, and GVC, direct precursors of the GPN framework, were developed. The theory of core competencies was introduced by C.K. Prahalad and Gary Hamel (1990) and can be used to explain the scale and form of GPNs and the boundary of the firm. The theory of strategic resources proposed by David J. Collins and Cynthia A. Montgomery (Collins and Montgomery, 1995, 2005) can be used to explain the dynamic governance of GPNs.

2.1.3 Network of Multinational Corporations

Since the 1990s, economic globalization has been gathering momentum, and GPNs have emerged and been expanding, effecting dramatic changes to the competition and cooperation among MNCs. The competition between individual firms has been gradually replaced by that between consortia of firms.

As GPNs undergo further changes, modular value chains, relational value chains, captive relationships, and other new types of governance grow in popularity and MNCs no longer make FDI through internalization but form production networks through contract manufacturing, alliances, and other nonequity-based means. Traditional theories of MNCs focus on developed countries, parent firms, hierarchy, and economy but pay little attention to developing countries, subsidiaries, networks, and lack of economy. The strategic management theory addresses competition and competitive advantages much more than the alliances of firms in various forms. Therefore, both face serious challenges in explaining the formation and development of GPNs.

Against such a backdrop, the theory about the network of MNCs has drawn increasing attention and includes such concepts as internal network, external network, strategic linkage, and local embeddedness, as summarized by Ye (2008). This theory responds to the changes in organizations and can explain the various factors that cause the formation and evolution of GPNs. It also examines the various network relationships that can be induced or exist inside or outside firms and their structures. The theory offers good explanations about why GPNs exist, network participants and their relations, what network resources firms have access to, what competitive advantages they have, the spatial characteristics of networks, and their relations with regional economy. Its all-round and in-depth examination of how GPNs operate helps us understand the significance of GPNs to the organizational form and performance of MNCs and to local industries and economic development. Therefore, the theory about the network of MNCs is an important theoretical foundation of GPN research.

2.2 NEW CLASSICAL TRADE THEORY

2.2.1 New Classical Economics

The division of labor and the economies of scale have been hot topics to economists. The division of labor is the soul of classical economics. It is considered the source of efficiency and productivity, and used by some economists as the logical starting point. Advocated by Dixit, Stiglitz, and Krugman, the economies of scale became a mainstream concept in the field of international economics after the 1970s. However, the theory based on this concept faces some dilemmas: first, the prediction that economic growth and other phenomena will occur if and only if the average size of firms increases is a departure from reality; second, the model adopts a black box approach to firms and fails to address why firms exit and the economic implications of the enterprise system itself so it has difficulty in explaining many modern economic issues; and third, transaction costs do not have substantive meanings in this model since no transaction costs are involved in the increase of firm size and other phenomena.

After the 1980s, economists represented by Rosen, Becker, Yang, Borland, and Ng used inframarginal analysis to turn brilliant ideas about division of labor and specialization in classical economics into decision-making and equilibrium models for explaining all economic activities. This broke down the barriers between traditional macroeconomic and microeconomic models, and started the trend of using modern analytical tools to revive classical economics. They used the endogenous decision-making model for personal choice of specialization level developed based on inframarginal analysis to examine how the market and pricing determine the level of division of labor across all sectors. They interpreted Adam Smith's division of labor theory and views on the causes of international trade based on the personal decision-making for specialization level and changes in the equilibrium level of division of labor. This new school of thought, known as "new classical economics," is advocated by economists like Yang Xiaokai. Studies in this regard, proceeding from the evolution of the division of labor, are intended to find the micromechanism of economic growth and establish the microeconomic model of macroeconomic growth. By formalizing the thoughts on division of labor and specialization in classical economics, they change the subject matter of economic research from the optimal allocation of resources under a given structure of economic organizations to the interactions between technology and economic organizations and how such interactions evolve.

New classical economics has the following characteristics:

First, every decision-maker is both a consumer and a producer and there is no absolute separation between pure consumers and firms. The separation between the two means that the theoretical basis of domestic trade differs from that of international trade. In the case of domestic trade, pure consumers would starve without trading because they do not produce things; in the case of international trade, however, every country is both a consumer and a producer so there are comparative advantages, economies of scale, and preference differences. The assumption that every decision-maker is both a consumer and a producer, according to new classical economics, is closer to the reality and means that the optimal decision representing self-interested behavior is always the corner solution. Within the framework of consumer–producer unity, the model of new classical economics introduces endogenous division of labor, specialization level, and market integration. It is believed that the emergence of firms is endogenous rather than exogenously given and can explain why domestic trade is expanded to international trade. This is a model of endogenous trade.

Second, every person favors diversity as a consumer and achieves the economies of specialization as a producer. The economies of specialization have something to do with the scope of a person's production activities; they are not the economic effect of an increase in firm size. All people's economies of specialization combined are the economies of division of labor, which is the effect of social networks. New classical economics replaces the economies of scale with the economies of specialization and introduces the concept of transaction costs, thus resulting in the dilemma between the economies of specialization and transaction costs. To be specific, on the one hand, specialization enhances productivity and gives decision-makers greater production capacity; on the other hand, due to the diversity in consumer preferences, specialization also means that every decision-maker needs to buy more goods from other specialized decision-makers, which incurs higher transaction costs. Specialization and trade are the basic components of the economy and a main thread that runs through the course of economic development is the dilemma and trade-off between specialization and transaction costs.

Third, the analysis of supply and demand is based on corner solution while the marginal analysis based on corner solution is not used. New classical economists still adopt a mathematical method to describe their theory so that the abovementioned views are more scientific. The method they use is inframarginal analysis, that is, to make a marginal analysis of every corner and a cost–benefit analysis between corners to find an optimal approach to decision-making. The marginal analysis of every corner is to address the issue of resource allocation with a given division of labor structure and it determines the structure of relative demand for and supply of different products when the total demand is given. The cost–benefit analysis between corners determines the specialization level and model (structure of economic organizations) and all people's decision-making in this regard determines the level of division of labor that determines market capacity and total demand.

2.2.2 New Classical Trade Theory's Interpretation of Global Production Network

New classical economics offers a brand new analytical framework, and reorganizes modern economic theories. It excludes the assumption that producers and consumers are absolutely separated, replaces the concept of economies of scale with that of economies of specialization, and considers the significance of various transaction costs in general equilibrium. A shared framework is established and the new ideas that challenge classical new economics, such as transaction cost economics, property rights economics, new trade theory, endogenous growth theory, evolutionary economics, information economics, and game theory, can be integrated into a new mainstream school of thought. In new classical economics, there is no longer a distinction between microeconomics and macroeconomics or independent development economics, economics of international trade, growth theory, and macroeconomics. There is no absolute separation between pure consumers and producers; the principles of international trade are the same as those of domestic trade. New classical economics unifies the isolated branches of economics with an internally consistent core theory, increasing the explanatory power of economics considerably. A new classical trade theory is created using its analytical framework, theoretical paradigm, and analytical approaches. According to the model of new classical trade theory, the level of division of labor and market capacity are both determined by transaction efficiency and rise at the same time as transaction efficiency is enhanced. It means that market capacity and the level of division of labor are the two sides of the same coin (Yang and Zhang, 2003). The basic proposition is that the division of labor will evolve as transaction efficiency is enhanced and economic development, trade, and market structure changes are the different facets of this evolution process. As the division of labor evolves, every person's specialization level will rise along with productivity, dependence on foreign trade, level of commercialization, endogenous comparative advantage, production concentration, degree of market integration, diversity in economic structure, variety of tradable goods, and the number of related markets while self-sufficiency rate goes down. If transaction efficiency is extremely high, the whole world will be an integrated market due to the full division of labor (Yang and Zhang, 2003).

New classical economics applies the analytical approach to the division of labor and trade among individuals to the international division of labor and international trade, reexamines the international trade theory using the individual specialization decision-making model in the context of the dilemma and trade-off between the economies of division of labor and transaction costs, and offers new explanations about the basic issues of trade theory with a model of division of labor evolution. The new classical trade theory with division of labor and specialization at the core can analyze the economic reality that GPNs as a new way of organizing production is growing in popularity from a more rigorous and standard perspective of economics. It explores technology transfer and spillover, trade in intermediate goods and inputs, trade and economic growth, industrial upgrading, increase in employment, MNCs' industrial organization, and gains from trade on the microlevel. This will lead to different explanations about key issues such as the causes, models, and outcomes of trade from those offered by traditional international trade theory, and it can be used to analyze corresponding economic policies. The new classical trade theory is intended to find out the micromechanism of economic growth using a scientific approach and build a micromodel of macroeconomic growth so it is a basic theory of crucial importance to microcosmic GPN studies.

According to new classical economics, the creation of GPNs as a new type of international production organization is caused by the endogenous evolution of division of labor. GPNs in which MNCs are the main players are in essence a new model of international division of labor and the direct outcome of individual specialization decision-making and division of labor across sectors. The cause is the endogenous advantage brought and reinforced by division of labor and specialization. GPNs make it possible to put different stages of the production process in different countries so that the benefits of specialization can be fully exploited, that is, helping to promote knowledge building and learning outcomes among all parties involved in the division of labor, improve productivity, achieve increasing returns to scale, and eventually reduce production costs, build the competitiveness of specialized firms, and bring about an increase in economic growth and social welfare. The advancement of trade and investment liberalization worldwide provides the institutional support for the formation of GPNs while technological advances play an important role in driving the development of GPNs. These technical and institutional factors have reduced barriers to the division of labor worldwide significantly, and improved transaction efficiency. Therefore, MNCs are active in creating and expanding GPNs. Moreover, the development of GPNs and increase in the number of stages in the cross-border division of labor will result in a rise in transaction costs. To benefit as much as possible from the division of labor, the government has to keep improving the economic system to lower various transaction costs, and works to increase specialization and improve the scale of every stage in the division of labor, which in turn leads to a further decrease in transaction costs. The benign interactions between the two brings about dynamic effects of learning and sustained growth.

According to new classical economics, both domestic trade and international trade are the results of a trade-off between the economies of specialization and the need to reduce transaction costs; international trade is the outcome of domestic trade development. This can also explain the evolution of MNCs' efforts to create GPNs. When transaction efficiency is low enough, self-sufficiency rate is in general equilibrium and firms tend to be internally integrated. As transaction efficiency increases, domestic division of labor advances and some lead firms begin to create national production networks and introduce cooperation with other firms at home. When transaction efficiency continues to rise, the division of labor goes deeper and beyond the domestic market, international division of labor and trade emerges, and lead firms begin to create GPNs and engage in cross-border cooperation. The consumer–producer assumption applies perfectly to a country so individual decision-making process has explanatory power in the analysis of countries' economic behavior. As long as the benefits brought by division of labor among countries exceed the transaction costs incurred by international trade, countries will choose specialization and trade with other countries, resulting in the formation and expansion of GPNs. Besides, a bigger market provides more space for the trade-off between the benefits brought by division of labor and transaction costs. GPNs emerge after national production networks because international trade incurs additional transaction costs compared with domestic trade. According to new classical economics, the degree of a country's involvement in GPNs can also be explained by the level of division of labor and transaction efficiency. Developed countries have higher transaction efficiency and equilibrium level of division of labor than developing countries so they need international trade to benefit fully from the high level of division of labor. This is also why developed countries tend to adopt unilateral free trade and actively expand GPNs. In contrast, developing countries have lower transaction efficiency and equilibrium level of division of labor and benefit little from international trade so they tend to change the relative prices of tradable goods through tariffs and other means so as to make more gains from trade. Therefore, the key to developing a country's economy leveraging GPNs is to improve transaction efficiency and create endogenous comparative advantages through the division of labor rather than simply rely on exogenous comparative advantages.

Transaction cost or transaction efficiency is a core concept in the theoretical framework of new classical economics. It is a key factor that determines the division of labor and specialization in economies as well as the types of economic organization. The scale, scope, and depth of the division of labor within GPNs are determined by the comparison between the marginal benefit and marginal transaction cost of the division of labor. If the former is greater than the latter, intranetwork division of labor will go deeper and GPNs will continue to develop; if the former is lesser than the latter, intranetwork division of labor and GPNs will tend to shrink. New classical economics draws distinctions between different types of transaction cost from the perspective of division of labor. It distinguishes between endogenous and exogenous transaction costs and endogenous transaction costs are divided into transaction costs in narrow and broad senses. A new classical property rights theory is also developed based on this. New classical economics focuses the research into transaction costs. The significance of the research is that making the degree of monopoly and externality endogenous according to the dilemma helps to study moral hazard and transaction costs incurred by information asymmetry and the game theory can be used to examine interactions between strategic moves. According to new classical economics, the government should be committed to lowering transaction costs through means in two aspects so that more firms are involved in GPNs and sustained economic growth can be achieved. On the front of exogenous transaction costs, the government should encourage technological advances and bring down hard or tangible costs such as transportation and communication costs. On the front of endogenous transaction costs, the government should keep working on institutional innovation within important frameworks such as WTO rules and achieve multilateral free trade through negotiations, reduce tariff and nontariff barriers, improve the legal framework, relax administrative regulation, enhance management efficiency, and lower soft or intangible costs to drive the division of labor to go wider and deeper and bring greater productivity gains. The latter is particularly important to developing countries including China.[1,2]

2.3 "NEW" NEW TRADE THEORY

Since the beginning of the 21st century, a new international trade system with firms at the core has emerged as the international division of labor and globalization of business activities goes further. The effects of firms' different characteristics on international trade in different fields have become a hot research topic of international economics. "New" new trade theory[3] represented by trade models with heterogeneous firms and endogenous boundary model of the firm has emerged as the latest theory of international trade. "New" new trade theory breaks with the firm homogeneity assumption in traditional trade theory and new trade theory and includes heterogeneity in the analytical framework for firms so that the variables of analysis are changed from countries and industries on the macrolevel to firms on the microlevel (Fan, 2007). The analysis of firms' trade and investment for internationalization and outsourcing and integration for global organization of production marks a new area of international trade research.

The concept of "new" new trade theory was first proposed by Baldwin and Nicoud (2004) but the earliest studies were conducted by Melitz (2003), Antras (2003), and Bernard et al. (2003). The theory has two branches. One of them concerns the choice of internationalization, based on the study by Melitz (2003), is also known as trade models with heterogeneous firms, and it explains why in reality only some firms choose export and FDI. The other concerns the choice of global organization of production, based on the study by Antras (2003), and is also known as endogenous boundary model of the firm. It integrates the concepts of industry organization and contract into the trade model, well explains intrafirm trade, and introduces a theoretical innovation in the research into firms' global production.[4]

---

1. The concept of "local embeddedness" was proposed very early but Yang and Xia (2005) were the first to suggest replacing the traditional concept of "location advantage" with "local embeddedness."
2. Based on related studies, including Yang (1997 and 1998), Yang and Huang (1999, 2003), and Yang and Zhang (2003).
3. In some literature, the endogenous boundary model of the firm is considered the extension of trade models with heterogeneous firms. In this book, the concept of "new" new trade theory is adopted to stress their difference in the area of research.
4. Based on related studies, including Melitz (2003), Helpman et al. (2004), Helpman (2006) and Li (2010).

2.3.1 Trade Models with Heterogeneous Firms

In the early 1980s, new trade theory included heterogeneity in the trade model to explain the phenomenon of interindustry trade but the heterogeneity was mainly reflected in product differences and monopolistic competition and no attention was paid to the differences between firms in productivity. Therefore, it is assumed that firms in the same industry are symmetrical and at the same technological level, all firms have the same productivity level, and all other firms will also export if one firm exports. However, more and more empirical studies have proved that the symmetry assumption of new trade theory has major limitations. In the 1990s, a large number of empirical studies on firms showed that only some firms export products to other countries and exporting firms are better than nonexporting firms in terms of size and productivity.

Melitz (2003) combines the monopolistic competition model of new trade theory with the firm heterogeneity assumption and constructs a model of intraindustry dynamics with heterogeneous firms. The model builds the general-equilibrium monopolistic competition model with industry dynamics of Hopenhayn (1992), expands the trade model of Krugman (1980), and introduces firms' differences in productivity. Melitz examines the relationship between international trade and intraindustry resource allocation and proves that firms with higher productivity take the initiative to enter the export market while firms with lower productivity are forced to exit from the market so that the productivity level of the entire industry is raised and that trade brings development opportunities for some firms and great challenges to others. The predictions made based on this model basically go with those of empirical studies so it is widely acknowledged that the research into firm heterogeneity and the basic framework of international trade and investment has great significance of theoretical foundation.[5,6]

Many theoretical models introduced after Melitz (2003) can well explain the relationship between heterogeneous firms and their internationalization moves. Building on the Melitz model, Helpman et al. (2004) include heterogeneous firms, export and FDI in the same analytical framework, construct a multicountry, multisector, general-equilibrium model to examine firms' export and FDI activities, and prove that firms' productivity difference is an important factor affecting their export and FDI. Yeaple (2005) attempts to explain the systematic differences between exporting and nonexporting firms

---

5. For individual firms, A can be considered an exogenous variable since the market size can be deemed infinitely great. E stands for total expenditure and J for the variety of differentiated goods.

6. According to the market clearing condition, it can be concluded that the output of the firm is "$A(a/\alpha)A/\alpha-\varepsilon$," total income "$A(a/\alpha)A/\alpha_1-\varepsilon$," and variable costs "$\alpha A(a/\alpha)A/\alpha_1-\varepsilon$."

and effectively explains why skill premium has been growing by connecting trade costs with firms' decision-making in four aspects, that is, entry, technology, export, and worker type. Melitz and Ottaviano (2008) construct a variable markup model in analyzing the relations among market size, productivity, and trade, and prove that market size and trade will affect the intensity of competition and heterogeneous firms' production decision. Bernard et al. (2007) successfully explain the causes of intraindustry trade and find factors that influence firms' entry into the export market by introducing firm heterogeneity in a standard trade model. Helpman et al. (2007) create a theoretical model for analyzing MNCs' choice of integration strategy by combining firm heterogeneity with two types of FDI (vertical FDI and horizontal FDI). Manova (2008) integrates credit constraint in the model of Melitz (2003) and finds that firms with higher productivity enjoy advantages in winning export credit support and firms in financially developed countries have easier access to the export market and export more products, particularly in sectors that rely on external financing.

2.3.2 Endogenous Boundary Model of the Firm

There are two basic models of firm boundary. One is to apply the transaction cost theory of Coase and Williamson to the study of business internationalization; the other is to adopt the property right analysis method of Grossman et al. (2003) combines the Grossman–Hart–Moore firm theory with the Helpman–Krugman new trade theory in the same analytical framework and proposes a model of incomplete contract and property right model concerning firm boundary to analyze MNCs' positioning and control decision, which marks the starting point of the endogenous boundary model of the firm. The model defines MNCs' boundary and the international position of production and can be used to predict the type of intrafirm trade. Inspection by variables show that the qualitative and quantitative characteristics of the model and data are consistent. Antras and Helpman (2004) combine the trade models with heterogeneous firms proposed by Melitz (2003) and the endogenous boundary model of the firm proposed by Antras (2003) to create a new theoretical model that analyzes firms' behavior in international business from their differences in organizational structure and concludes that the decisions on outsourcing or integration, operating at home or abroad, etc. are all firms' endogenous organizational choices. Antras (2005) creates a dynamic general-equilibrium model of North–South trade to explain the appearance of product life cycle caused by the incompleteness of international contracts. Antras and Helpman (2006) generalize the international production organization model with heterogeneous firms proposed by Antras and Helpman (2004), introduce contractual frictions in the model, and allows for their varying degrees according to inputs and country. This model suggests that firms' productivity level would influence their strategic choice of organizational form.[7,8]

In general, the endogenous boundary model of the firm proceeds from the organizational choice of individual firms and combines international trade theory with firm theory in the same framework. It marks a further development in the trade theory with heterogeneous firms and offers a brand new perspective for the research into firms' globalization and industrial organization. The endogenous boundary model of the firm is typically started by the studies of Antras (2003) and Antras and Helpman (2004), examines firm heterogeneity's influence on the strategic choice of outsourcing and insourcing, and explores how firms' organizational form affects trade model.

2.3.3 Interpretation of Global Production Networks in "New" New Trade Theory

The "new" new trade theory is highly consistent with microcosmic GPN studies. The inclusion of the theory in the GPN framework provides a new microbasis and new perspectives for the research into GPNs. The theory can explain the heterogeneity and power asymmetry of firms in GPNs, including the behavioral pattern and inner traits of firms and productivity, technology, and workers with heterogeneous skills as the sources of heterogeneity. In conjunction with the fixed costs of trade, the theory can explain firms' productivity differences, and be used to analyze the influence of firms' organizational form on international trade and GPNs and in particular, the features and causes of international trade within MNCs, thus enriching the research into GPNs. The "new" new trade theory mainly explains the following issues about GPNs: what kind of firms would create GPNs to serve international markets; what organizational forms would they choose (export or FDI, horizontal or vertical FDI, creation of new business or cross-border M&A, and outsourcing or integration); how do they choose locations (domestic or international and South or North); can GPNs improve firms' performance and competitiveness. Therefore, the "new" new trade theory offers significant support to microcosmic GPN studies.

The "new" new trade theory proceeds from firm heterogeneity in explaining the benefits of GPNs. Countries and regions that lag behind should be actively involved in the international division of labor and raise the level of openness to the outside world, which will help enhance industry productivity and promote local economic development. The theory finds a new way of improving productivity. Under the condition that the productivity of

7. This does not include the export of foreign-funded enterprises since only the behavior of domestic enterprises is the subject of study here.

8. According to the market clearing condition, it can be concluded that the output of the firm is "$A(a/\alpha)A/\alpha-\varepsilon$," total income "$A(a/\alpha)A/\alpha_1-\varepsilon$," and variable costs "$\alpha A(a/\alpha)A/\alpha_1-\varepsilon$."

individual firms stays unchanged, a country can still raise the level of productivity in an industry and even across all sectors through trade and opening up. It also suggests that GPNs may have negative impacts on less developed regions. Opening the market may plunge industries with lower efficiency in less developed regions into recession and industries entering less developed regions may be industries with low productivity and high environmental costs. Business relocation may lead to a reversion in small countries' productivity gains while helping increase big countries' productivity gains. GPNs lead to the reallocation of resources and the shift of profit and market share to firms with high productivity, which may cause excessive resource monopoly and thus loss of efficiency in the entire market.

2.4 SUMMARY

This chapter focuses on three basic theories important to microcosmic GPN studies, including MNCs theory, new classical trade theory, and "new" new trade theory, as well as their interpretation of GPNs. The MNCs theory can explain such issues as the formation and evolution of GPNs; network participants and their relationships; internal mechanism and conditions of MNCs' integrated expansion; operation mechanism and dynamic governance of networks; spatial characteristics of networks and their relations with regional economy, scale, and form of GPNs; and boundary of the firm. The new classical trade theory with the division of labor and specialization at the core can analyze the economic reality that GPNs as a new way of organizing production is growing in popularity from a more rigorous and standard perspective of economics. It explores technology transfer and spillover, trade in intermediate goods and inputs, trade and economic growth, industrial upgrading, increase in employment, MNCs' industrial organization and gains from trade on the microlevel. This will lead to different explanations about key issues such as the causes, models, and outcomes of trade from those offered by traditional international trade theory. The "new" new trade theory is highly consistent with microcosmic GPN studies, and can explain the heterogeneity and power asymmetry of firms in GPNs, and firms' productivity differences. It can be used to analyze the influence of firms' organizational form on international trade and GPNs and in particular, the features and causes of international trade within MNCs, thus enriching the research into GPNs. In short, these basic theories are of crucial importance to microcosmic GPN studies.

Chapter 3

Analytical Framework of Microcosmic GPN Studies

Chapter Outline

This chapter proposes the analytical frameworks of microcosmic GPN studies based on the theories introduced in Chapter 1, Overview of the Research into GPNs, and Chapter 2, Theoretical Basis of Microcosmic GPN Studies, and the GPN framework proposed by Jeffrey Henderson and other authors. On the microlevel, global production networks are in essence global value networks and the agents in networks with the same value objectives form an ecosystem of symbiosis. Value is a driving force behind the evolution of GPNs. Therefore, value as one of the three elements of the GPN framework proposed by Henderson et al. (2002) stands at the center of microcosmic GPN studies. Embeddedness connects MNCs closely with local economy and contributes to their interactions and common development so it also plays an important role in microcosmic GPN studies. Power is largely influenced by firms' contribution to network value and how they are embedded in the networks so it is not considered a key subject of microcosmic GPN studies in this book. The analytical framework of microcosmic studies proposed in this chapter comprises one framework from the perspective of value and another from the perspective of embeddedness.

Global Value Chains and Production Networks. DOI: https://doi.org/10.1016/B978-0-12-814847-1.00003-8

3.1 FRAMEWORK OF MICROCOSMIC GPN STUDIES

The GPN framework proposed by Henderson et al. (2002) has three elements, that is, value, power, and embeddedness. Multinational corporations (MNCs) are the participants of GPNs on the microlevel. For any firms including MNCs, core competencies are a prerequisite for gaining sustained competitive edge and fat profits. The most important attribute of firms' core competencies is value. Firms can make profits only by satisfying market demand and creating value for customers. The competitive edge of core competencies will be bigger if market demand is better met and greater value is created for customers. Therefore, from the perspective of value, the process of making profit is one of creating value and profit is just a result of the complex value creation process.

With the further development of GPNs, value creation depends not simply on the integration of firms' internal resources but more and more on the coordination between all participants of GPNs. As it is noted in *World Investment Report 2011*, "Foremost among the core competencies of a TNC is its ability to coordinate activities within a global value chain... In a typical value chain, a TNC oversees a sequence of activities from procurement of inputs, through manufacturing operations to distribution, sales and after-sales services. In addition, firms undertake activities—such as IT functions or R&D—which support all parts of the value chain." According to Porter's competitive advantage theory, the advantages of firms in competition are those they have in these specific parts of strategic value and firms that have gained competitive advantages usually organize the important activities of strategic value in better and cheaper ways than their competitors. MNCs not only need to integrate resources of GPNs to create greater value for themselves but also need to think about how to create value for other network participants and achieve win–win outcomes. Therefore, MNCs and all other stakeholders form a network in which they create value together. The characteristics of all participants in the network and their relationships will affect the creation, capture, and enhancement of network value. Every participant's contribution to network value determines its position and power in the network and influences the organizational form of the GPN. Therefore, on the microlevel, value is the core of GPNs and drives the evolution of GPNs. The value-centered framework has the greatest importance to microcosmic GPN studies.

Another element of the GPN framework proposed by Henderson et al. (2002) is embeddedness. GPNs do not only connect firms functionally and territorially but also they connect aspects of the social and spatial arrangements in which those firms are embedded and which influence their strategies and the values, priorities, and expectations of managers, workers, and communities alike. Embeddedness makes MNCs play a key role in the global and local economy. The dialectic relationship between MNCs and local economy is that the firm grows out of and also contributes to the local economy. The type of embeddedness determines a firm's position and power in the GPN and has significant influence on value creation, enhancement, and capture in the network. Therefore, embeddedness is essential to regional economic development and access to global opportunities. The embeddedness-centered framework is also important to microcosmic GPN studies.

Power is also an element of the GPN framework proposed by Henderson et al. (2002). There are three forms of power in GPNs, that is, corporate power, institutional power, and collective power. Corporate power should be the focus of microcosmic GPN studies but it is not what firms have inherently or does not stay unchanged. Firms' power in the network is influenced by their contribution to network value and how they are embedded in the network. The change in firms' value contribution or type of embeddedness determines the dynamic change of corporate power. Therefore, power-centered framework is not considered important to microcosmic GPN studies in this book.

To sum up, the framework of microcosmic GPN studies in this book is proposed from the perspectives of value and embeddedness based on the GPN framework proposed by Henderson et al. as well as the theoretical basis of microcosmic studies (Fig. 3.1). The value-centered framework has five basic elements, namely value objectives, global strategy, networks, location selection, and network governance. The embeddedness-centered framework has three areas of focus, that is, strategy, dimension, and region.

3.2 FRAMEWORK OF MICROCOSMIC GPN STUDIES FROM THE PERSPECTIVE OF VALUE

The framework of microcosmic GPN studies from the perspective of value (Fig. 3.2) has five basic elements, that is, value objectives, global strategy, networks, location selection, and network governance, as well as four basic components, that is, global R&D network, global manufacturing network, global supply chain network, and global sales and service network.

The overall goal of GPNs is to create value in a sustainable way so value objectives are the cornerstone of network organizations. MNCs have to make a tradeoff between the overall goal of the network and their value objectives. Global strategy is a whole package of action plans that help MNCs to reach their value objectives as well as a program of action for the entire network. In addition, the effective implementation of global strategy will direct more of MNCs' attention to localization strategy. GPN participants are connected by the division of labor as network nodes. Every participant's position and role in the network depend on the contribution of its core resources and key technologies to value creation. The organizational forms of network include the internal and external networks of firms. While expanding globally, MNCs would choose the best location in every part of the value chain and

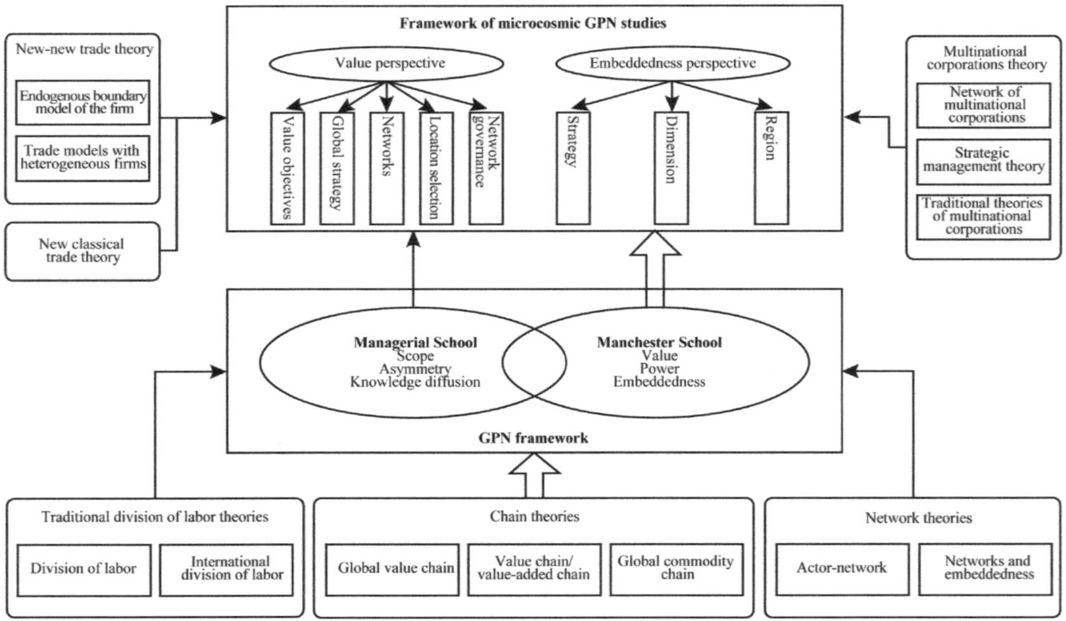

FIGURE 3.1 Framework of microcosmic GPN studies. Source: *Prepared by the author.*

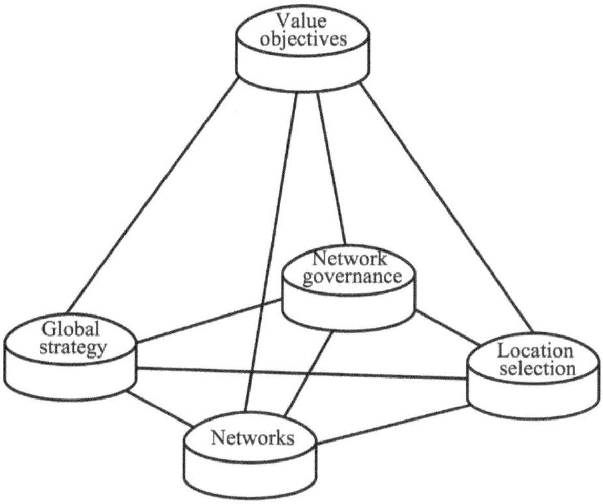

FIGURE 3.2 Framework of microcosmic GPN studies from the perspective of value. Source: *Prepared by the author.*

establish connections to firms at the location so as to form new networks of regional economic and social connections. Therefore, MNCs' location choice has significant impact on value creation and regional economic development. Network governance shows how network participants coordinate their actions and how the network operates. Global production networks feature external governance and cogovernance but MNCs usually have dominance in the network so the governance type of firms affects the power and distribution of other firms in the network and make themselves more competitive.

R&D, manufacturing, purchasing, sales, and aftersales service are usually considered the most important and strategic value creation activities of MNCs and these geographically scattered activities are also closely connected and form the global R&D network, global manufacturing network, global supply chain network, and global sales and service network. These networks are the basic components of MNCs' global production networks.

3.2.1 Value Objectives

3.2.1.1 The Basic Meaning of Value Objectives

Black's Law Dictionary defines "value objectives" as the overall evaluation and views about the significance, importance, worthiness to obtain or utility of people, things, and objects. The overall goal of GPNs is to create value in as sustainable way. Value objectives are the cornerstone of network organizations and point the way for MNCs in network building, network governance, and location selection. Shared value objectives ensure high levels of consistency and agreement and thus cooperation among network participants.

For MNCs, the overall goal of the network and their value objectives are connected but are not completely consistent. To reach their value objectives, they have to make a tradeoff between the two. Therefore, value objectives are the soul of the entire network architecture.

The concept of value carries great importance in the value of division of labor theory in classical economics, labor theory of value, and theory of surplus value in Marxism, and the competitive strategy theories about multinational corporations. In competitive strategy theories, the model of core competencies describes value as an important attribute of core competencies; the model of strategic resources describes value as an important attribute of strategic resources; the model of industry structure introduces the concepts of value chain and value system, based on which the theories of value-added chain, global commodity chain, and global value chain are developed. Brandenburger and Nalebuff (1996), Kathandaraman and Wilson (2001), Slywotzky (2002), Han (2010), and Zhou and Cheng (2011) put forward the concept of "value net" and argue that the value net incorporates strategy, structure, management, and culture, is a customer-centered value creation system, and has four components, namely customers, suppliers, competitors, and complementors. The structure of value net may undergo steady, gradual changes or out-of-order violent changes. The competition among MNCs has evolved from value chain competition to the competition over global value system and even global value network.

In this book, a global production network is seen as a net of one or multiple complex value chains, or a value network. Greater emphasis is placed on the cooperation and competition between firms. In a win–win market, network participants form an organic network of symbiosis, which creates new room for value creation. The characteristics of network participants and their relations will influence value creation and capture and firms in the network can increase overall value by working together extensively, thus influencing the organizational form of GPNs. In brief, value creation stands at the center of global production networks and drives their evolution.

3.2.1.2 Evaluation of Value Objectives

Microcosmic GPN studies mainly address corporate valuation. Since the mid-20th century, finance researchers have been committed to exploring scientific approaches to corporate valuation and have worked out some mainstream methods, including relative value (P/E, EV/EBIT, and EV/EBITDA), free cash flow to firm (FCFF), free cash flow to equity (FCFE), and economic value added (EVA) (Chi, 2008).

EVA is a corporate performance and value management system introduced by American scholar Stewart and registered by Stern Stewart & Co. It was fully published on the *Fortune* magazine in September 1993. EVA has marked advantages in measuring the returns to shareholders and corporate value compared with other indicators of accounting earnings. It suggests that shareholders need to obtain at least the average returns to venture investment in the market to ensure that the firm's value is not undermined by its previous operations. The founder of EVA describes the formula for its calculation as:

$$EVA = NOPAT - TC \times WACC \tag{3.1}$$

where NOPAT stands for net operating profit after taxes, TC for total capital, and WACC for weighted average cost of capital.

EVA can be simply described as the firm's net operating profit after taxes minus liability and equity costs so it is a kind of residual income after all capital charges are deducted. It is based on and better than the traditional indicator of accounting profit. The biggest difference between the two is in the handling of capital charges. In traditional accounting methods, interest charges are used to reflect the debt financing cost but the cost of equity capital is not considered so the minimum return under risk constraint required by shareholders cannot be embodied. In the calculation of EVA, however, the cost of equity capital is included in the total capital charges as the opportunity cost of investors' capital so the returns required by all sources of financing can be reflected.

The important significance of EVA is to give managers and shareholders a clear concept of "value creation" and thus help managers realize the value management of the company. To date, EVA has been applied to more than 400 companies worldwide, including Coca-Cola and Siemens, and EVA maximization has replaced profit maximization as these companies' management objective (Han, 2011). With the rapid growth of China's economy, the government attaches more and more importance to economic sustainability. In 2003, the State-owned Assets Supervision and Administration Commission of the State Council (SASAC) began to try to use EVA to assess the performance of managers in central SOEs. On December 28, 2009, the *Interim Measures for Assessing the Performance of Central SOE Leaders* was issued and EVA carries the greatest weight among the three basic measures. It marked the beginning of conducting EVA assessment among central SOEs.

3.2.2 Global Strategy

3.2.2.1 The Basic Meaning of Global Strategy

Global strategy, also known as globalization strategy, means that MNCs allocate different parts of the value chain in a way to leverage the comparative advantages of different countries and regions, undertake integrated operations, and strive to reduce operating costs so as to gain long-term, stable global competitive advantages and maximize global efficiency. The global strategy of MNCs is a whole package of action plans intended to help them reach their value objectives as well as a program of action for the entire network architecture. As the global strategy is implemented on deeper levels, MNCs will gradually create global R&D, supply chain, manufacturing, and sales and service networks to achieve optimal allocation of resources worldwide. In the meantime, these networks become increasingly integrated in the economy of the host country and the market demand in the host country becomes increasingly important. MNCs will give more and more attention to localization strategy. Localization means that the overseas subsidiaries of MNCs try to be as local as possible in such aspects as staffing, funding, sources of spare parts, and technology development as an effort to rapidly adapt to the economic, cultural, and political environments of the host country. It is actually a process of integrating MNCs in the host country's economy in all aspects, including production, marketing, management, and personnel as well as a process of fulfilling their corporate responsibility in the host country and incorporating corporate culture into local culture. Therefore, striking a balance between globalization and localization is one of the keys to MNCs' success in operating globally.

3.2.2.2 Formulation of Global Strategies by MNCs

According to Wild et al. (2009), MNCs' global strategies are formulated on three levels, that is, corporate level, business level, and departmental level.

Corporate-level strategies specify the national markets and industries the company will enter, overall objectives for the company's business units, and every department's role in reaching these objectives. Corporate strategies fall into four categories, including growth, retrenchment, stability, and combination strategies. Combination strategy is intended to apply growth, entrenchment, and stability strategies to different business units of the company. It is widespread among MNCs since very few companies formulate the same strategy for different business units.

Business-level strategies are Porter's generic strategies, namely cost leadership, differentiation, and focus. Cost leadership strategy is to use the economies of scale to get the lowest cost among all competitors in the same industry and suits companies that produce standardized products and have integrated marketing strategy. Differentiation strategy is intended to make buyers think that the products are unique in the industry. Differentiation can be achieved by improving the reputation of product quality, building a different brand image, and designing products differently. Focus strategy is formulated for a specific market segment and the company may achieve the goal by becoming a cost leader, producing differentiated products, or combining the two.

Departmental-level strategy addresses the specific activity of turning resources into products and supports corporate and business strategies. Porter's value chain analysis is adopted in formulating departmental-level strategies. On the basis of examining the role of every department's activities, including primary and support activities, in the process of creating value for customers, strategies are devised to ensure that the company has advantages in value creation.

3.2.2.3 International Market Entry Strategy

There are many ways for MNCs to enter international markets to implement their global strategies, including trade (export), FDI (wholly owned enterprise, joint venture, etc.), nonequity-based arrangements (alliance, outsourcing, licensing, cooperation agreement, contract manufacturing, etc.). In the case that property rights cannot be separated (e.g., to avoid knowledge or technology diffusion, excessively high supervision cost, etc.), equity-based means is the first choice; in the case that property rights can be separated, nonequity-based means are important ways to create global production networks. Three typical means in this regard are FDI, outsourcing, and alliance.

Buckley and Casson (1998) expanded their internationalization model created in 1976 and created a detailed systematic model of MNCs' international market entry strategy, including where to locate production, whether production and distribution facilities are owned by the entrant, whether sole ownership or shared ownership is achieved through joint venture, and whether ownership is gained through greenfield investment or acquisition, to examine how MNCs carry out their global strategies.

"New" new trade theory looks at MNCs' internationalization path and boundary of the firm from a microperspective. It mainly explains the following issues: what kind of firms would choose to enter international markets; how would they enter international markets (export or FDI, horizontal or vertical FDI, creation of new business or cross-border M&A, and outsourcing or integration); how do they choose locations (domestic or international and South or North). Related discussions can be found in Chapter 2, Theoretical Basis of Microcosmic GPN Studies.

3.2.3 Networks

3.2.3.1 The Basic Meaning of Networks

A network is composed of nodes and resources connected in particular ways. The nodes are network participants on different levels. They are the primary targets of network organization and implement global strategies and network governance. The nodes provide core resources and key technologies for the network and their contribution to network value determines their position and role in the network. A network is not a simple superposition of nodes but a topological structure in which the nodes are connected through certain ways of communication and interaction. The process of establishing connections is one of network members building trust in each other and such connections include both internal and external ties of firms. The networks of nodes connected according to the division of labor result from competition.

3.2.3.2 *Participants and Drivers of GPNs*

A global production network (GPN) is a type of organization in between firms and the market and its participants fall into two categories, that is, firms and institutions. The GPN framework proposed by Ernst et al. offers a more detailed analysis of firms' relational network. It suggests that a GPN covers flagships and their subsidiaries, affiliates, and joint ventures as well as suppliers, subcontractors, distribution channels, value-added resellers, and R&D alliances and a series of cooperative agreements. Institutions in the GPNs include national and local institutions (see related parts in Chapter 1: Overview of the Research into GPNs); there is apparent asymmetry and heterogeneity among firms and their cooperative relations are diverse. The GPN framework proposed by Henderson et al. provides a more detailed analysis of institutions in the network. The institutions in a GPN include national and local institutions, international organizations (e.g., EU, ASEAN, NAFTA, etc.), world organizations (e.g., IMF, World Bank, WTO, etc.), UN agencies (particularly ILO), etc. as well as trade unions, employers' associations, NGOs concerned with human rights, environmental and other issues, etc. (see related parts in Chapter 1: Overview of the Research into GPNs). Institutional arrangements influence the development of GPNs.

In a GPN, lead firms usually occupy the strategic parts of the network and control the entire production network and industry, assuming the role of an initiator, organizer, or core driver of the network. Drawing upon the classification of global commodity chain by Gereffi (1999a,b), we divide global production networks into buyer-driven and producer-driven ones. The structure of network differs widely according to the driver type and the form of a GPN depends largely on the driver type.

3.2.3.2.1 Buyer-Driven GPNs

In a buyer-driven GPN, branded marketers and retailers are the lead firms of the industry. They establish production networks on a global scale leveraging their outstanding advantages in design, brand, and marketing. Production activities are well organized by subcontractors in developing countries and finished products are provided as required by overseas buyers. Buyer-driven GPNs are characterized by intense competition, local ownership and globally distributed production system. Lead firms in the network manage and operate production and trade networks, make sure that the business activities of all individual firms can be fully integrated, and serve as intermediaries for integrating overseas factories of the global consumer goods market. They do not control GPNs through direct ownership but place more emphasis on the building of software environments. To consolidate their position in the network, lead firms usually work on developing channels or building brands to increase their influence on consumers.

3.2.3.2.2 Producer-Driven GPNs

In a producer-driven GPN, manufacturers in the industry, international oligopolists, are the lead firms of the network. They can exercise control over backward linkage of supplying raw materials and spare parts and forward linkage of entering distribution and retail, and coordinate the production networks. Producer-driven GPNs are common in capital- and technology-intensive industries. The lead firms always achieve the economies of scale through vertical integration and focus more on infrastructure and other parts of hardware. Production and sales across the network and national strategic alliances among international competitors are often seen in producer-driven GPNs.

3.2.3.2.3 Buyer-Driven and Producer-Driven GPNs Compared

From the perspective of economics, in a global production network, the parts where there are entry barriers to new firms are where more profits are yielded. Such high entry thresholds are usually the special assets of lead firms that bring them high returns like economic rents. The lead firms in producer-driven GPNs depend mainly on technology and organizational rents while buyer-driven GPNs mainly involve brand, marketing, relational, and trade policy rents. In both buyer-driven and producer-driven GPNs, actors take part in activities in a number of aspects to gradually obtain or develop special assets and thus gain various economic rents.

3.2.3.3 Division of Labor and Organizational Form of GPNs

The organizational form of a GPN depends largely on the governance type and incentive mechanism of the MNC that creates the network in its home country. In other words, GPNs can be considered the cross-border extension of domestic competition to some extent. The cultural endowments of different countries/nations are crucial to the organizational form of GPNs. Zysman et al. (1997) believe that network structure types can be distinguished from two dimensions. One is verticality/horizontality in the division of labor between firms, which describes the durability of cooperative relations between firms in the network and their balance of power; the other is the openness/closedness of the network that reflects the entry difficulty for firms outside the network. Therefore, GPNs have four types of organizational form, that is, vertically closed network, vertically open network, horizontally closed network, and horizontally open network. The focus of their discussion is on the emergence of Asian manufacturers and creation of Asian production networks, and they associate national features with the types of production networks.

First, vertically closed network. Japanese and South Korean firms usually keep production stages of high value added at home but transfer those of low value added such as assembling to other countries like China. The spare parts that have higher technological requirements are usually provided by follow-up suppliers in the home country or purchased from the home country. The parent company maintains a high level of control over its overseas affiliates. Therefore, production networks dominated by Japanese and South Korean firms show a model of hierarchical vertical integration. The networks built by lead firms are actually the cross-border extension of their domestic production networks and are highly closed. Local firms and other foreign firms in the host country find it hard to enter the core system of the network. These are the features of typical vertically closed networks.

Second, vertically open network. American firms tend to outsource and give greater managerial autonomy and more manufacturing functions of high value added to partner companies and more decision-making power and innovation opportunities to their overseas subsidiaries. Parent companies only engage in product development, system integration, brand promotion, and other activities of high value added. Therefore, the production networks dominated by American firms are more flexible, open, and adaptive. American firms are technical standard formulators and brand leaders.

Third, horizontally open network. Taiwan's firms are mostly small and medium-sized enterprises and they focus on one or two types of products. The production network among suppliers is complex, short-term, and changing cooperative relationship based on "responsiveness to the market." Such temporary cooperative relations start when an order is placed and end when an order is fulfilled. Therefore, the networks dominated by Taiwan's firms are usually quite open. Network members are usually innovative and there is a good balance of power between them. With flexible and complicated relations, they often change their partners.

Fourth, horizontally closed network. In the networks of Chinese firms in Hong Kong and Taiwan of China, Singapore, and other regions, the Chinese firms are independent but the same culture and language and interpersonal relations make it very difficult for non-Chinese firms to be involved. Therefore, the networks established by overseas Chinese firms are usually horizontally closed and their operation is mainly based on interpersonal relations instead of market transactions. This facilitates the cooperation and coordination between firms and also creates high entry barriers to firms outside the network.

Zysman et al. believe that the organizational form of networks changes as the environment changes. The openness or closedness of networks is important to regional economic development. Comparatively speaking, open networks are more beneficial to industrial upgrading in developing countries.

3.2.4 Location Selection

3.2.4.1 The Basic Meaning of Location Selection

Location selection is a process of finding the best locations of branches or paths to reach value objectives according to prescribed standards and based on spatial analysis. It involves location motives, location decisions, plant locations, etc. Location choices directly serve MNCs' global strategy and are influenced by existing economic activities and social changes in the region. While expanding globally, MNCs would choose the best location in every part of the value chain and establish connections to firms at the location so as to form new networks of regional economic and social connections. The global networks of MNCs and regional industrial networks eventually lead to the formation of global production networks. The global networks of MNCs are connected to multiple locations, cover various value chains, and involve the cross-border flows and diffusion of knowledge so MNCs' location choices have significant impact on their value creation and regional economic development.

3.2.4.2 Determinants of MNCs' Location Choices

Location theory is a core theory and a common area of study in economic geography and regional economics. It originated from classical economics and can be traced back to the classical location theory of Germany. Economists found that locations and places are related to economic processes very early. Von Thunen (1826) proposed the agricultural location theory for the first time in *The Isolated State*; Weber (1909) put forward the industrial location theory; Christaller (1933) introduced the central place theory; Losch (1944) proposed the economic location theory. After World War II, location theory made the transition from classic to modern stage. Isard (1956) proposed regional science in the book *Location and Space-Economy*. Moses (1958) introduced a new classical theory that integrates location theory and production theory. Smith (1971) extends the cost idea in Weber's classical industrial location theory to the space–cost curve and uses the curve and the spatial boundary analysis introduced by Losch to find the "optimal location," "quasi-optimal location,", or "suboptimal location." Since the 1970s, the scope of spatial location research has become wide thanks to the successful application of control theory and game theory. The spatial pricing and output theory of Holaham and Bechman, spatial competition theory of Lancaster, and location choice theory and spatial equilibrium theory of Beckman are the basic components of the modern location theory system.

According to the eclectic paradigm proposed by Dunning (1993), location-specific advantage is a determinant of MNCs' location choice for investment. Location advantages include advantages arising from the host country's factor endowments, such as natural resources, labor, geographical location, etc., as well as advantages in the investment environment, such as political and economic systems, laws, regulations and policies, infrastructure, market capacity, external economies of scale, etc. The two types of factors play an important role in the decision-making on FDI locations. The eclectic paradigm holds that FDI is related to the host country's location advantages and is consistent with MNCs' strategic objectives.

3.2.4.3 Factors Influencing the Spatial Agglomeration of GPNs

Since the 1990s, the importance of economic activities' spatial location to economic development and international economic relations has drawn much attention from academia worldwide. New economic geography has made outstanding contributions in this regard. Some famous economists, such as Paul Krugman, Fujita, Michael Porter, Robert Barro, W. Brian Arthur, and Anthony Venables, have all made active efforts to promote the integration of economics and geography. New economic geography includes traditional thoughts about spatial economics such as urban economics, regional science, and economic location theory; introduces such concepts as region, location, and distance into economics; and uses increasing returns to scale, external economy, imperfect competition, and spatial agglomeration to explain the competitive advantages of national and regional economic development and describes them with mathematical models (Liu and Yin, 2006). There are two subjects of research. One is the spatial agglomeration of economic activities; the other is the dynamics of regional growth and agglomeration. The study of industrial location is an important component.

The core–periphery model of Krugman (1991) is the most typical general-equilibrium location model with two regions and two sectors in new economic geography. This model shows how the symmetry (same factor endowment) of two regions turns into an asymmetric structure of core–periphery under the impact of increasing returns to scale, free flows of labor, and transportation costs. It suggests that the evolution of industries' structure and spatial configuration is endogenous, which makes the model more effective than new trade theory in explaining location choices about international trade and FDI. Since the introduction of this model, many scholars have studied firms' location choices worldwide (Table 3.1).

The basic thinking and research methods of new economic geography show that its international model has offered new perspectives for the study of MNCs' location choices. Its strengths can be seen in three aspects. First, the model provides a general-equilibrium approach based on the choices of a single transnational firm, introduces such parameters as transportation costs, degree of economies of scale, and manufacturing consumption share, and examines the influence of changes in production costs on firms' location choices. Second, the model can be used to study the influence of existing agglomeration of economic entities on equilibrium results. Third, the model

TABLE 3.1 Representative NEG Literature on MNCs' Location Choice

Category	Researcher	Description
Basic model (core–periphery model)	Krugman (1991)	Examines the influence of firms' spatial distribution choices on the agglomeration of businesses or labor in different regions in the equilibrium state and the changes in trade flows caused thereof, leveraging the Dixit–Stiglitz monopolistic competition model to conclude that transportation costs are endogenous
Single-plant corporation model with the headquarters and plant in the same country	Krugman and Venables (1995)	Examines the impact of changes in trade costs on MNCs' locational choice and how such choice affects the agglomeration of manufacturing and trade flows in a country
Single-plant vertical MNC model with the headquarters and plant in two countries	Ting Gao (1999)	Introduces the ownership structure of MNCs and examines the influences of trade costs and factor differences on the locations of MNCs' headquarters and plant
Horizontal MNC model with plants in two regions	Raybaudi-Massili (2000)	Examines the influences of transportation costs, plant costs, and products' elasticity of substitution on horizontal MNCs
Multiregional vertical and horizontal MNC model	Ekholm and Forlid (2001)	Examines the influence of MNCs' locational choice on spatial agglomeration of an industry based on a multiregional assumption
	Markusen (2005)	Shifts the focus of analysis to the effect of reduced investment barriers on MNCs' locational choice and examines the two basic models of MNCs

Prepared by the authors based on the study of Li and Ding (2007).

can be used to analyze how institutional, policy, infrastructure, communications, and other changes in the country attracting FDIs affect the centripetal and centrifugal forces of agglomeration and thus influence the location choices of transnational firms (Li and Ding, 2007). The theoretical system of new economic geography can be applied to study the following issues concerning location selection in GPNs: overall spatial configuration of MNCs in global networks and the possible causes of such configuration; factors influencing MNCs' choice of specific countries for making FDI; influence of MNCs' investment behavior on income levels of countries; influence of MNCs' investment behavior on industrial agglomeration and coordinated development between regions in the host country; how the host country influences the regional distribution of MNCs by establishing institutions, implementing policies, improving infrastructure, and creating an enabling environment.

3.2.5 Network Governance

3.2.5.1 The Basic Meaning of Network Governance

Network governance shows how network participants coordinate their actions and how the network operates. The effective operation of network organizations requires not only proper form of the network and location choices but also the establishment of trust, coordination, power, decision-making, incentive, and constraint mechanisms. Network rules are the code of conduct for all participants, serving as laws for network organizations and laying the foundation for their operation. Participants of GPNs are connected by complementary division of labor and stable bilateral transaction contract (relational contract) for mutual benefit so that the organization and transactions of GPNs are more efficient. GPNs feature external governance and cogovernance and require corresponding arrangements concerning decision-making power and control. To some extent, all stakeholders have a role to play in decision-making and governance. However, MNCs have dominance in the network and control the entire network by seizing strategic parts of global value chains. They play a central role in network governance as the lead firms.

3.2.5.2 Main Types of Network Governance

Many scholars have studied in depth the governance types of GPNs. Sturgeon (2002a,b) explains three types of network governance according to the different roles of lead firms and the relation between lead firms and suppliers and taking the electronics industry as an example: captive production networks represented by the networks dominated by Japanese and South Korean firms; relational production networks represented by networks dominated by overseas Chinese firms in Germany, Italy, and Southeast Asia; modular production networks represented by networks dominated by American firms.

On the basis of transaction cost economics, network of firms, technology and learning ability of firms, and other theories, Gereffi et al. (2005) put forward three factors that determine the power relations between firms in the network, namely complexity of transactions, ability to codify transactions, and capabilities in the supply base and five governance types according to the power differences among organizations. In the ascending order of the degree of coordination and power asymmetry, they are markets, modular networks, relational networks, captive networks, and hierarchy. Markets and hierarchy are two extreme governance types while modular, relational, and captive networks are similar to the three types proposed by Sturgeon (2002a,b).

3.2.5.2.1 Market

In this governance model, transactions are easily codified, product specifications are simple, and suppliers have strong capabilities. Asset specificity will fail to accumulate and network operation is based on market exchange. There is little explicit coordination between lead firms and suppliers. The core mechanism of market governance is the traditional price mechanism and the arms-length exchange relations are based on market contract. Such a governance model usually occurs in the early stage of value chain development in developed and developing countries or products in which developing countries have core technologies and competitive advantages. Its occurrence is caused by developing countries' obtaining technological and market prowess for some product or unique endowment in some production factor.

3.2.5.2.2 Modular

This governance model involves complex and modularized products. Specifications or related knowledge are easy to codify and technical standards are introduced. Suppliers have the competence to supply full packages or some modules. Hard-to-codify information is internalized so lead firms need to regulate and control. Standardized contract can reduce transaction costs quite well. Most typical modular production networks are created by American firms. Suppliers provide turnkey services for lead firms and there is basically no need for support or input from lead firms except design. Modular networks feature market responsiveness and have many of market governance's strengths but are not based on price. The frequency and intensity of exchanges and interdependence between firms are lower compared with relational networks. There is no need for spatial proximity and cultural similarity and the cost of switching to new partners is low.

3.2.5.2.3 Relational

When product specifications cannot be codified, transactions are complex, and supplier capabilities are high, relational network governance can be expected. The exchange of knowledge between suppliers and lead firms is most often accomplished by frequent face-to-face exchange of information and there is strong mutual dependence. The explicit coordination of lead firms plays a big role in this process. In the meantime, highly competent raw material and spare part suppliers provide a strong motivation for lead firms to outsource to gain access to complementary competencies. Relational production networks are similar to market-based and modular networks. There is a balance of power, high asset specificity, and very complex relations. Relational networks are widely seen among in overseas Chinese firms in Germany, Italy, and other countries. The governance depends mainly on the social connections among network participants rather than authority of lead firms. Transaction costs are low, interfirm relations are very flexible, and firms are highly competent to adapt to the market.

3.2.5.2.4 Captive

When the complexity of product specifications and the ability to codify are both high but supplier capabilities are low, network governance will tend toward the captive type because it requires a great deal of intervention and control on the part of lead firms. In the process, lead firms will invest lots of resources and make great efforts to achieve governance through strict control and regulation of suppliers, and provide various support to ensure suppliers are willing to maintain the partnership. Therefore, lead firms seek to lock-in suppliers to exclude others from reaping the benefits of their efforts. The suppliers face significant switching costs and are frequently confined to a narrow range of tasks (e.g., simple assembly) and are dependent on the lead firm, becoming "captive suppliers." Despite the power asymmetry, captive production networks control opportunism of both sides and provide suppliers with market access. Typical captive networks are dominated by Japanese and South Korean firms. The famous lean production system of Japan is based on such a governance type and it had significant impact on automobiles and other industries.

3.2.5.2.5 Hierarchy

When products are complex, product specifications are hard to codify, suppliers' capabilities are low, transaction costs are high, and it is hard to control opportunism through contract, lead firms have to adopt vertical integration for intrafirm governance. This also satisfies the need to exchange tacit knowledge within the network as well as the need to effectively manage complex production networks and to control resources, especially intellectual property. Lead firms are involved in many explicit coordination activities and the power of lead firms and suppliers is extremely asymmetric. The biggest feature of this governance type is vertical integration. At the center of operation is management and control and the network is based on employment relations so it is also known as integration governance. This governance type is more often seen in Japanese and South Korean firms. The parent company establishes subsidiaries in developing countries through FDI and exercises tight control over overseas branches through vertically integrated hierarchy.

3.2.5.2.6 Five Governance Types Compared

The governance type determines the degree of coordination and power asymmetry among network participants, behavioral pattern of lead firms, participation of firms in the periphery, and possibility of knowledge transfer, and has far-reaching impact on the upgrading speed and approach of participating firms. In general, relational networks enable suppliers to acquire more complex knowledge, followed by modular and captive networks. Market and hierarchy types make it most difficult for suppliers in the network to acquire knowledge. Consequently, the different governance types foretell suppliers' different corporate learning and potential to achieve industrial upgrading. For lead firms, the choice of governance type is a choice made on the basis of weighing the benefits and costs of different organizational forms. As mentioned above, the three key factors that influence network governance are complexity of transactions, ability to codify transactions, and capabilities in the supply base (Table 3.2).

As shown in Table 3.2, each of the five governance types is a combination of risks and returns brought by the international division of labor. From market to modular and hierarchy, the degree of explicit coordination and power asymmetry is on the rise. They also differ in transaction, conflict resolution, degree of elasticity, amount of entrustment among economic entities, atmosphere in the organization, and actors' behavioral choice (Table 3.3).

It should be noted that the five governance types are not static but mutually transformational. In the process of GPN formation, assets of specificity in strategic links play an essential role. Core asset is a dynamic concept and changes with time. The variables that determine the governance type are all changing so the GPN governance type in a specific industry is evolving all

TABLE 3.2 Determinants of GPN Governance

Governance Type	Complexity of Transactions	Ability to Codify Transactions	Capabilities in the Supply Base	Degree of Explicit Coordination and Power Asymmetry
Market	Low	High	High	Low
Modular	High	High	High	↓
Relational	High	Low	High	
Captive	High	High	Low	
Hierarchy	High	Low	Low	High

Prepared by the authors based on the study of Gereffi et al. (2005).

TABLE 3.3 Five GPN Governance Types Compared

Governance Type	Market	Modular	Relational	Captive	Hierarchy
General basis	Contract–property right	Complementary division of labor	Family ties, etc.	Market forces	Employment relations
Transaction	Price	Network relations	Network relations	Network relations	Company rules
Conflict resolution	Haggling over price–mandatory enforcement	Mutual benefit	Reputation	Conflict resolution led by big enterprises	Administrative order–supervision
Degree of elasticity	High	Medium	Medium	Medium	Low
Amount of entrustment among economic entities	Low	Medium to high	Medium to high	Medium to high	Medium to high
Atmosphere in the organization	Distrust; bargaining	Mutual benefit	Lots of leeway	Big enterprises having the upper hand	Bureaucracy; doing everything by the book
Actors' behavioral choice	Independence; low conversion costs	Interdependence	Interdependence	Small and medium-sized enterprises depending on big ones	Subordination to and dependence on higher authorities

Prepared by the authors based on the study of Zhuo (2011).

the time. It is amid the dynamics between strategic and nonstrategic links and core and noncore assets that the structural arrangements of GPNs are marginally adjusted and corrected to ensure dynamic governance. This dynamic governance process actually allows for the possibility that latecomer processing firms take part in GPNs to get upgraded.

Moreover, the governance model of GPNs has diversity and what type of governance a lead firm decides to adopt depends on its weighing the benefits and risks of outsourcing. Lead firms decide on the proper degree of network connection according to transaction type and changes in the transaction environment and make different governance arrangements for different transactions to minimize transaction costs. Therefore, there may be a mixture of governance types in a single network while the governance type of lead firms influence the power and distribution of other firms in the network and makes them more competitive.

3.3 FRAMEWORK OF MICROCOSMIC GPN STUDIES FROM THE PERSPECTIVE OF EMBEDDEDNESS

From the perspective of embeddedness, microcosmic GPN studies mainly address issues in three aspects: MNCs' embeddedness strategy including strategic linkage, strategic embeddedness, and strategic coupling; MNCs' embeddedness dimensions, including economic, technical, social, cultural and institutional embeddedness; embeddedness regions on macro-, meso-, and microlevels. Fig. 3.3 shows the embeddedness-centered framework of microcosmic GPN studies.

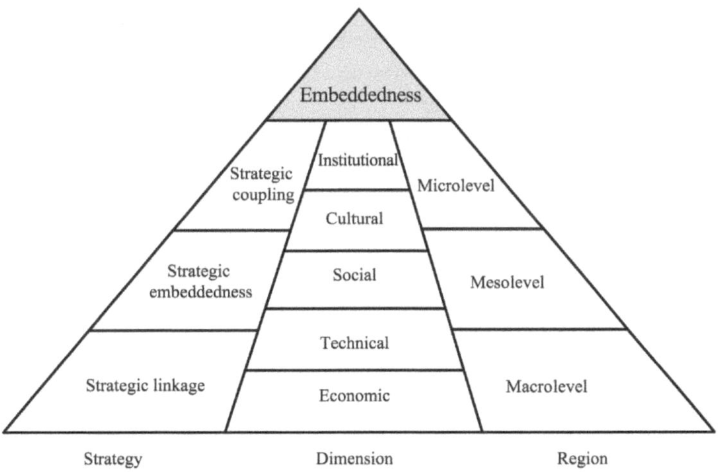

FIGURE 3.3 Framework of microcosmic GPN studies from the perspective of embeddedness. Source: *Prepared by the author.*

3.3.1 The Basic Meaning of Embeddedness

3.3.1.1 Studies on Embeddedness

The term *embeddedness* was introduced by Polanyi (1944) from the perspective of traditional political economy. He believed that economic motives are embedded in social relations and economic behavior is part of social activities. Since then, a large number of scholars have studied embeddedness from the economic and societal subsystem perspectives. In particular, the classic embeddedness study by the representative of economic sociology Granovetter (1985) has been widely accepted and understood. In his view, the social fabric is the social network in people's life and the network mechanism of embeddedness is trust that arises from and is embedded in the social network. Granovetter distinguished between relational embeddedness and structural embeddedness and divided relationships into strong linkages and weak linkages on the basis of introducing the concept of relationship strength.

Economic geographers have contributed a lot to the development of the embeddedness concept. Zukin and DiMaggio (1990) followed the route of economic sociology and held that economic behavior is affected not only by social structure and relations but also by other noneconomic factors so they took one step further and divided embeddedness into four types, namely structural, cultural, cognitive, and political embeddedness. Based on the empirical analyses of specific regions by Dicken and Thrift (1992), economic geographers began to do network and embeddedness studies in conducting geospatial analysis of firms and their production activities and connected the concept of embeddedness closely with the ideas of cultural shift and institutional shift in economic geography, further developing the concept of embeddedness. At the end of the 1990s, the concepts of network and embeddedness in economic geography became an analytical cornerstone of the GPN framework. In the GPN framework proposed by Henderson et al. (2002), embeddedness, power, and value are the three elements of a global production network, and territorial embeddedness and network embeddedness are the core components of GPN research. With the advent of the network concept, scholars in all disciplines (e.g., sociology, management studies, economics, etc.) have started to use the concept of embeddedness.

The concept of embeddedness is also expanded significantly in the fields of organizational and business management studies. For example, Halinen and Tomroos (1998) divided embeddedness into horizontal and vertical embeddedness as well as temporal, spatial, social, political, market, and technical embeddedness when studying the evolution of business networks. As MNCs' importance grows in the context of economic globalization, researchers have also directed attention to the embeddedness of MNCs. Since the mid-1990s, Andersson and Forsgrcn (Andersson, 1996; Andersson et al., 2001, 2002, 2005) have done a series of studies on the network embeddedness of MNCs' subsidiaries. In 1996, they divided their network embeddedness into internal embeddedness (parent group embeddedness) and external embeddedness (local embeddedness in the host country). In 2001, they introduced the concept of technical embeddedness. In 2002, they divided local embeddedness further into business embeddedness and technical embeddedness. In 2005, they studied the influence of the parent company's control mechanism on subsidiaries' external embeddedness. Humphrey and Schmitz (2002) proposed four types of embeddedness in global value chains, namely hierarchical embeddedness, market embeddedness, market embeddedness, and quasi-hierarchical embeddedness when studying the evolution of industry cluster embeddedness. Hochtberger et al. (2003) find that the competitiveness of MNCs' subsidiaries is influenced not only by their local embeddedness but also by their connections to external and internal networks. Whitel (2003) studied MNCs in Ireland's software industry as a typical case and concluded that the degree of local embeddedness is higher when subsidiaries have wider and closer connections and grow better locally.

Chinese scholars have also done some research into the local embeddedness of MNCs. Zhao (2004) divides embeddedness into economic, social, and institutional embeddedness and explores the relationship between embeddedness and industry cluster competitiveness. Wen et al. (2007) examine the relationship between embeddedness and FDI-driven industry clusters from three dimensions, namely cultural, network, and territorial embeddedness. Qiu and Chen (2010) propose three levels (macro, meso, and micro) and four dimensions (economic, social, technical, and institutional) of MNCs' embeddedness. Ye (2008) studied the process and mechanism of MNCs' local embeddedness and proposed four drivers of MNCs' local embeddedness, that is, MNCs' overall strategic planning, subsidiaries' voluntary actions, local governments' push, and local firms' growth.

3.3.1.2 Definition of Embeddedness in This Book

The term *embeddedness* appears frequently in the literature of all disciplines but it is understood differently. To clarify the meaning of embeddedness, at least three elements should be defined, namely who is embedded in what and how. On the basis of previous literature, I have defined the concept from the perspective of microcosmic GPN studies as follows: "firms" in global production networks are embedded locally; the types of embeddedness include economic embeddedness, technical embeddedness, social embeddedness, cultural embeddedness, and institutional embeddedness; the intensity levels of embeddedness include connection (weak linkage), embeddedness (balanced connection), and coupling (strong linkage); and there is both active and passive embeddedness as well as one-way and two-way embeddedness. In the framework of microcosmic GPN studies, the focus of research is on the form, intensity, and structure of MNCs' linkage with local economy and their relationships with firms' organizational management and behavioral performance; it is stressed that the embeddedness of firms depends on the social capital and tacit knowledge of specific regions, and that the affiliation and dependence of firms on local economy and the embedded networks have significant influence on the sustained development of firms and GPNs.

It is easy to confuse embeddedness with localization. Localization means that MNCs integrate their production, marketing, managerial, staffing, and all other operations into the economy of the host country. To put it simply, it is to do as the Romans do. It is more significant to study the embeddedness of MNCs than to study their localization for a number of reasons. First, embeddedness means to be locally integrated while localization means to do as local firms do. Second, embeddedness focuses on the interactions and linkages between MNCs and local economy while localization focuses on the adaptation of MNCs. Third, embeddedness involves economic, social, technical, institutional, and many other factors while localization focuses on economic and technical aspects. Fourth, embeddedness concerns MNCs' strategic management while localization is MNCs' strategy. Localization is an integral part of embeddedness and embeddedness is far beyond the scope of localization in terms of both implications and applications. Embeddedness makes MNCs play a key role in the global and local economy. The dialectic relationship between firms and local economy is that firms grow out of and also contribute to local economy. Therefore, for both MNCs and local governments, it is more practical to stress the embeddedness of MNCs than to simply advocate localization (Qiu and Chen, 2010).

3.3.2 Embeddedness Studies by Region

Embeddedness studies by region are to examine where and how MNCs are embedded and the influence of embeddedness on MNCs and local development. It has something in common with location selection analysis but the latter focuses more on the spatial distribution of MNCs' subsidiaries, interactions between the parent company and subsidiaries, as well as the relationships between location choices and MNCs' global strategy while the former adopts a local perspective in analyzing the spatial configuration of MNCs, interactions and linkages between MNCs and local economy, as well as the influence of embeddedness on local development and improvement of social welfare.

Embeddedness studies by region on the macrolevel consider a country as an organic whole and look at how MNCs are embedded in the country and its impact. The studies on the mesolevel divide a country into different administrative or economic regions and examine how MNCs are embedded in different regions and its impact, such as the studies on the issue of embeddedness in eastern, central and western China, or the Yangtze River Delta Economic Zone and Pearl River Delta Economic Zone, or some provinces. The studies on the microlevel deal with cities or even more specific locations, such as the business districts and development areas of cities or towns and villages under their jurisdiction.

3.3.3 Embeddedness Studies by Dimension

Embeddedness studies by dimension look at how MNCs are embedded locally, including economic, technical, social, cultural, and institutional embeddedness. Embeddedness type is an embodiment of embeddedness strategy, influences the regions of embeddedness and local firms' position and power in GPNs, and thus exerts significant impact on value creation, enhancement, and capture.

3.3.3.1 Economic Embeddedness

Economic embeddedness refers to the process in which MNCs keep making local investments and establish sustained and stable industry linkages, including the purchasing of raw materials, localization of upstream and downstream product supply, interfaces with local industry, etc. Economic embeddedness can be divided into forward and backward embeddedness. Forward embeddedness refers to MNCs' linkages with local distributors, agents, and consumers while backward embeddedness refers to MNCs' linkages with upstream suppliers, especially the utilization of local raw materials, machines, and labor. Economic embeddedness provides local firms with market opportunities and also helps to improve the competences of local firms, which is good for the coordination of various entities in the local and global networks and thus makes MNCs and local networks more competitive. The economic embeddedness of MNCs can be seen in the following aspects: investments for serving different purposes, investments in line with local strategies and geographically agglomerated, upgrading investment structure, cemented economic ties between MNCs and local firms, and MNCs' efforts to develop local business partners and suppliers.

3.3.3.2 Technical Embeddedness

Technical embeddedness refers to the technology linkages between MNCs and local firms, including the technology transfer and spillover effects on local firms and promotion of local industrial structure upgrading. Technology spillover effects can be brought about by demonstration and imitation, flow of people, and the forward and backward linkages of the industry chain. Technology spillover occurs as a result of MNCs' unintentional behavior. As a reflection of externalities, it can help improve local firms' technology and productivity, promote the growth of related firms and the creation of new firms, and thus drive local industrial development. The technical embeddedness of MNCs is embodied in the following aspects: local agglomeration of R&D institutions, sustained growth in R&D funding, marked technology introduction and spillover effects, and significantly enhanced product grade and technological level.

3.3.3.3 Social Embeddedness

Social embeddedness is a broad concept in many studies. It covers all aspects of noneconomic embeddedness and may be equivalent to embeddedness. In this book, social embeddedness is defined as a narrow concept. It refers to the closeness of interpersonal relationships and social ties MNCs have established with local communities and also involves the trust and reputation they have gained. It lays the social foundation for MNCs' local operations. Social embeddedness helps MNCs to increase their local influence and visibility, build a friendly and reliable image, improve their relations with stakeholders, and achieve sustainable development in the host country. The social embeddedness of MNCs is reflected by their efforts to be locally rooted, serve the local economy, and fulfill their social responsibilities as corporate citizens.

3.3.3.4 Cultural Embeddedness

Cultural embeddedness includes both MNCs' adaptation to and compliance with local cultural traditions, customs, and other informal norms and local people's adaptation to MNCs' corporate culture and acceptance of their values. It is not a purely economic motive but it offers a binding framework for entities' action and stresses the role of shared cultural understanding in shaping economic strategies and objectives. The corporate culture of MNCs is deeply influenced by the culture of their home countries so wherever they are, their economic behavior is branded by the culture of their home countries, and it is easy to build synergy and trust with other firms with the same cultural background. Therefore, the cultural embeddedness of MNCs helps to foster a new culture and motivate local firms to reform their corporate system and culture but conflicts with local culture may arise and should be well addressed. The cultural embeddedness of MNCs is embodied in their efforts to adapt to local culture and have cultural exchanges and cooperation with local firms, develop localized networks based on identity and cultural proximity, promote the local application of the parent company's culture, foster cultural diversity and create mechanisms for cross-cultural communication and management, and implement personnel and cultural localization strategies.

3.3.3.5 Institutional Embeddedness

Institutional embeddedness refers the linkages of MNCs with other institutions including government agencies, associations, banks, universities, research institutes, training system, etc.; the degree of influence on such institutions; and the influence on the change of local systems. The government and research institutes play a very important role in this regard. Many problems cannot be solved without diverse institutional arrangements. Institutional embeddedness cements the ties between local government, associations, and firms; helps to create synergies among network members; and raises the levels of firms and various institutions in the network. Such embeddedness is embodied in the efforts of MNCs to establish ties with other institutions, help put in place a sound modern administrative system, and promote the development of intermediaries in the network.

3.3.4 Studies on Embeddedness Strategy

Embeddedness strategy refers to a whole package of action plans that MNCs devise to become locally embedded and the studies in this regard look at the strength of MNCs' linkages with local networks. Embeddedness strategy is also an important component of MNCs' global strategy.

3.3.4.1 Strategic Linkage

The concept of strategic linkage was introduced by Nohria and Garcia-Pont (1991) and it means that firms obtain resources through strategic linkage activities. They may gain access to needed resources by establishing linkages with firms that have complementary competencies or share resources with firms that have similar competencies to consolidate their competencies and achieve the economies of scale and scope, reduce risks, and boost competitiveness. According to the theory of strategic linkages, embeddedness has at least two aspects of significance to MNCs: global utilization of their resources, and access to local resources in the host country. It is possible that both aspects are satisfied. Strategic linkages are mainly intended to get resources so the initial strategy of MNCs is to establish weak linkages with the local economy. By region, embeddedness usually occurs in developed regions and first- and second-tier cities. By dimension, it begins with economic embeddedness and gradually moves to social, cultural, and institutional embeddedness but technical embeddedness is rather weak. Except economic embeddedness, other types are all one-way embeddedness.

3.3.4.2 Strategic Embeddedness

Strategic embeddedness means that MNCs gradually incorporate the local economy in their global strategy in light of local markets' long-term potential and their development strategy, and makes strategic "global–local" arrangements with a view to growing together with local firms and local communities in the long run. In the case of strategic embeddedness, there are balanced connections between MNCs and local economy. MNCs focus more on the local market's potential and competitive human resources. The investments they make are mainly for market development and they see the investment destinations as important markets to sell their products. In making investment, MNCs tend to establish wholly owned enterprises or enterprises in which they have a controlling stake. Most investments go to capital and technology intensive industries and industries with wide and in-depth linkages, expanding from manufacturing activities to the service sector and to research institutes and regional headquarters. Local investments are included in the division of labor system of MNCs' global networks and show a distinctive feature of spatial agglomeration. By region, the embeddedness spreads from developed regions and first- and second-tier cities to less developed regions and third-tier cities. By dimension, two-way economic embeddedness is the basis and technical, social, cultural, and institutional embeddedness gradually gains strength, which is particular true with technical embeddedness.

3.3.4.3 Strategic Coupling

Strategic coupling refers to the interaction and integration between MNCs' embeddedness and local industrial networks, which results in the creation of inseparable connections between the two and a complete network of industrial ecosystem. In the case of strategic coupling, there are strong linkages between MNCs and local economy, which show three features. First, business cluster. MNCs and a large number of firms and supporting institutions are spatially agglomerated and maintain a long-term, nonspecific cooperative relationship marked by complementarity, thus gaining strong and sustained competitive advantages. Second, high-end industry chain. The interactions between MNCs and local industrial networks eventually lead local industries to the high end of the value chain. Third, symbiotic cooperation. MNCs and local industries together form an industrial value network in which member firms have shared interests and evolve at the same time for win—win outcomes.

3.4 SUMMARY

This chapter proposes the analytical frameworks of microcosmic GPN studies from the perspectives of value and embeddedness, based on the theories introduced in Chapter 1, Overview of the Research into GPNs, and Chapter 2, Theoretical Basis of Microcosmic GPN Studies, and the GPN framework proposed by Jeffrey Henderson and other authors. The value-centered framework has five basic elements, namely value objectives, global strategy, networks, location selection, and network governance. The embeddedness-centered framework has three areas of focus, that is, strategy, dimension, and region. All the concepts are defined and how to achieve or assess them are also explained. Therefore, both the analytical framework and specific research methods of microcosmic GPN studies are provided in this chapter.

Section II

The Siemens and Huawei Cases

Chapter 4

A Study of Siemens's GPN

Chapter Outline

The microlevel participants of a global production network (GPN) are multinational corporations (MNCs) and they assume overwhelming dominance in the network. In fact, it is their global expansion that has promoted the formation of GPNs and shaped the new landscape of international trade centering around MNCs, which is a manifestation of economic globalization on the microlevel. International trade research has also gone deeper from the national and industry levels to the corporate level, with new areas of research explored. In contrast, however, existing GPN studies are mostly analyses on the national, regional, and industry levels; very few microlevel studies are carried out and data support on the corporate level is insufficient. Therefore, thorough GPN research is yet to be conducted. Since different industries have different features, there might be sharp differences between the GPNs of the same company in different industries. The research into MNCs' business performance in specific industries or regions is yet to go deeper. In addition, given the different cultural, political, and institutional environments

Global Value Chains and Production Networks. DOI: https://doi.org/10.1016/B978-0-12-814847-1.00004-X

in different overseas markets, it is necessary to look at how MNCs overcome all the challenges to get "embedded" in local economic systems and grow into global businesses and how they impact local economic and social development.

Within the framework of microcosmic GPN studies proposed in Chapter 3, Analytical Framework of Microcosmic GPN Studies, this chapter examines the global networks of Siemens as well as related strategic framework, location choices, and governance models (Fig. 4.1) from the perspective of value, and discusses how this company builds its global research and development (R&D), manufacturing, supply chain, and marketing service networks centering around its value objectives to configure its business across regions and across industries, and how the company uses the power of networks to integrate value-creating activities, enhances its core competencies, and achieve sustainable revenue growth. Then the embeddedness of Siemens's healthcare business in China by strategy, dimension, and region (Fig. 4.22) is analyzed in detail to help figure out how MNCs are embedded locally and the pattern in its evolution.

4.1 OVERVIEW OF SIEMENS'S GLOBAL PRODUCTION NETWORK

4.1.1 About Siemens AG

Siemens AG is one of the world's largest electronics and electrical engineering companies headquartered in Berlin and Munich and with a history of more than 160 years. It works to develop and manufacture products, design and install complex systems and projects, and tailor a wide range of solutions

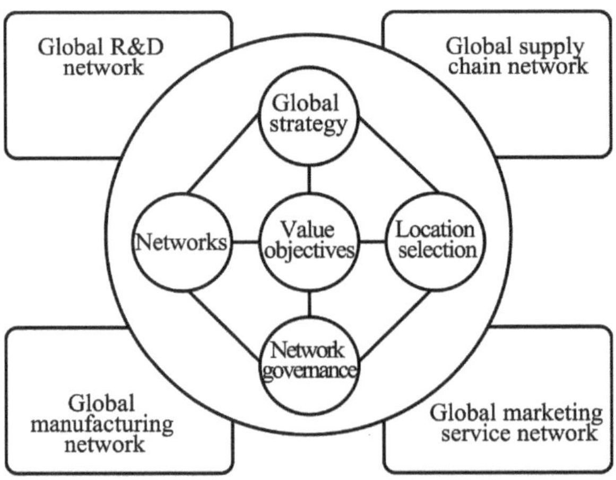

FIGURE 4.1 Siemens's GPN framework from the perspective of value. Source: *Prepared by the author.*

for individual customer requirements, as a market and technology leader in Energy, Healthcare, Industry, and Infrastructure & Cities. In fiscal year 2014, which ended on September 30, 2014,[1] Siemens generated revenue from continuing operations[2] of EUR 71.9 billion and net income of EUR 5.5 billion (Siemens AG, 2014). The company has around 360,000 active employees in more than 200 countries, about 160 R&D centers in about 30 countries/regions, over 285 manufacturing plants across the globe, as well as office buildings, warehouses, R&D facilities or sales offices in almost every country/region. Siemens has around 680,000 shareholders in more than 100 countries, making it one of the world's largest listed companies. About 60% of the company's stock is held outside Germany. Siemens began to list its shares on the New York Stock Exchange (NYSE) in March 2001, making it a company listed on both the Frankfurt Stock Exchange and the NYSE.

4.1.2 Global Expansion and Transnationality Index

Founded by Werner von Siemens in Berlin in 1847, Siemens has experienced eight stages of development, namely founding and initial expansion (1847–65), triumph of international projects (1865–90), growth through consolidation and partnerships (1890–1918), returning to the global market (1918–33), National Socialism and the war economy (1933–45), rebuilding and rise to a worldwide corporation (1945–66), new markets and business lines (1966–89), and globalization (1989–). It has witnessed the first wave of globalization from the mid-19th century to the 1920s, interruptions to globalization from World War I to the Cold War, and the progressing of the second wave of globalization since the 1990s.

Siemens began to pursue international operations even in its early years. Before World War I, the company had established institutions in Russia, Britain, Austria, France, Belgium, Spain, China, and Brazil. In fiscal year 1914, it had 82,000 employees, about one-fourth of whom were from foreign countries. After the two world wars, the company began to gradually resume overseas business in the mid-1950s, and its exports exceeded DM 1 billion for the first time in 1961. In the 1970s, the company's overseas operations were mainly in Western Europe and North America, and a few were in Japan, Latin America, India, South Africa, the Middle East, and some island countries in the South Pacific Ocean. In 1985, Japan, France, Finland, Mexico, Brazil, Singapore, and Malaysia gained higher positions in Siemens' global system. Siemens launched its global strategy in 1989. By 2000, the company's global expansion had become more scattered, with more operations introduced in countries where it had already had presence and more

1. Data in this chapter are mostly from the website of Siemens (www.siemens.com) and its annual reports.

2. The data from Siemens's annual reports concern its continuing operations.

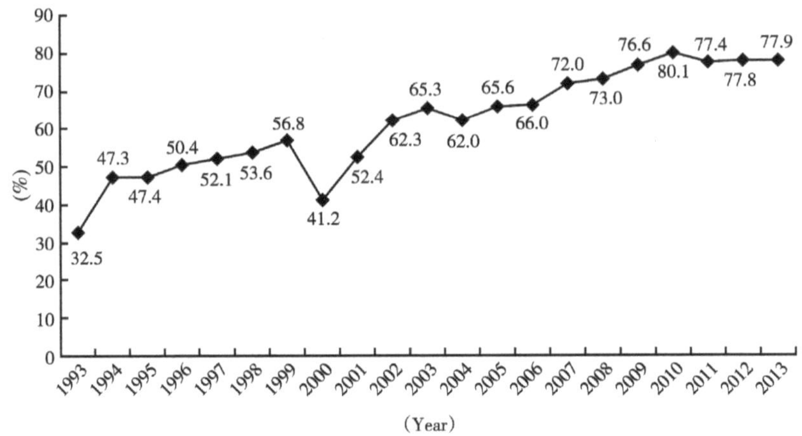

FIGURE 4.2 Siemens's TNI, 1993–2013. Source*: Prepared by the author according to the* World Investment Report *published by the UNCTAD and related data provided in Siemens's annual reports.*

countries like China, Australia, and South Korea included in its global system. In fiscal year 2014, Siemens was divided into three regions and 14 regional clusters on a geographic basis, covering a total of 94 countries and regions.

The United Nations Conference on Trade and Development (UNCTAD) publishes the *World Investment Report* annually, in which it ranks nonfinancial MNCs by foreign assets and estimates their Transnationality Index (TNI). TNI is calculated as the arithmetic mean of the following three ratios: the ratio of foreign assets to total assets, the ratio of foreign sales to total sales, and the ratio of foreign employment to total employment. Siemens had a TNI of 32.5% and 77.8% in 1993 and 2013, ranking 73rd and 25th, respectively. Fig. 4.2[3] shows Siemens's TNI from 1993 to 2013. Apparently, the company's TNI was on the rise from 1993 to 1999, fell in 2000 due to the global economic recession, exceeded the 1999 level in 2002, kept increasing after that and reached the peak level of 80.1% in 2010, and stayed around 77% in the following 3 years. It suggests that Siemens's global networks have been expanding since its global strategy was adopted.

4.1.3 Value Objectives and Strategic Framework

Since the 1990s, in response to the new challenges and pressure from shareholders, Siemens has gradually become a competitor in the global market based on its strategic pillars of productivity, innovation, and growth, which

3. The figures for 2001 and 2009 are the results of the author's calculation based on Siemens's annual reports.

necessitate the company's formulating a corresponding global strategy and business sectors. In brief, the global strategy is intended to create value and global networks are built to enhance the company's core competencies and help it gain and maintain the market and technology leadership in all sectors and establish sustainable competitive edges in a global sense so as to achieve sustainable, profitable growth and continually increase company value. One Siemens is the company's the global strategy framework for turning its vision into reality and reaching its goals (Fig. 4.3). Since Siemens' business covers multiple sectors and many regions/countries, it is necessary to formulate corresponding strategies on sector, regional/national, and division levels within the One Siemens framework to support the company's overall global strategy.

Siemens's vision is to be a pioneer in energy efficiency, industrial productivity, affordable and personalized healthcare, and intelligent infrastructure solutions. The vision is reflected in its company strategy, which guides the company in turning its vision into reality based on its values—to be responsible, excellent, and innovative. The strategy comprises three strategic directions and three concrete focus areas along each of the three strategic

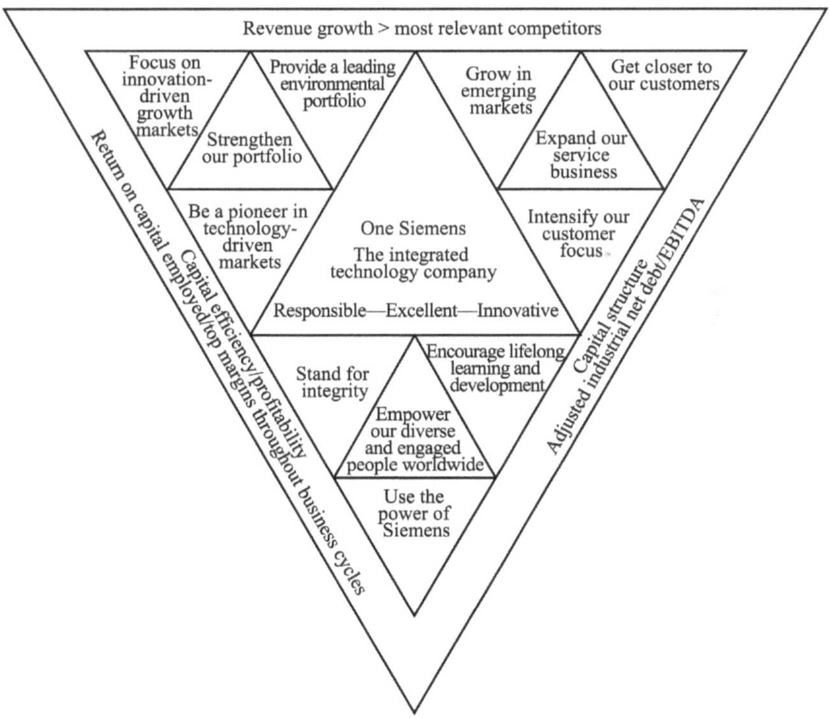

FIGURE 4.3 Siemens's global strategy framework. Source: *Siemens,* Siemens Annual Report 2011.

directions have been defined to exploit the full potential of the company. In the strategic direction of focusing on innovation-driven growth markets, the three focus areas are to be a pioneer in technology-driven markets, strengthen their portfolio, and provide a leading environmental port. In the second strategic direction of getting closer to customers, the three focus areas are to grow in emerging markets, expand service business, and intensify customer focus. In the third strategic direction of using the power of Siemens, the three focus areas are to encourage lifelong learning and development, empowering diverse and engaged people worldwide, and standing for integrity.

One Siemens includes a financial target system for capital-efficient growth and the goal of continuous improvement relative to the market and its competitors. The financial target system defines indicators for revenue growth, for capital efficiency and profitability, and for the optimization of capital structure. The interaction of the target indicators creates a balanced system that drives the sustainable growth of company value. The most important driver for increasing the value of Siemens over the long term is profitable revenue growth. The company's goal is to grow faster than its most relevant competitors. It tracks its performance by comparing its revenue growth with the average growth of the most relevant competitors. And it sets hurdle rates for acquisitions. Siemens's aim is to remain profitable and to work as efficiently as possible with the capital of its shareholders and lenders so capital efficiency is included in the company's financial reporting, and return on capital employed is defined as the indicator for measuring capital efficiency. Sustainable profit and revenue growth can only be achieved on the basis of a healthy capital structure. In addition to operating indicators, the company has also defined an indicator focused on the management of capital structure and strives to optimize the ratio. In addition, Siemens ensures unrestricted access to debt financing instruments in the capital markets and its ability to meet scheduled debt service obligations.

4.1.4 Networks and Governance Model

Siemens uses the One Siemens framework to connect organizations within and outside the company and create global R&D, manufacturing, supply chain, and marketing service networks. Strategically, it has the ultimate ownership of GPNs and the right to control and configure them so as to achieve sustainable value creation and enhancement. As is shown in Fig. 4.4, the nodes of Siemens's global networks include not only the Supervisory Board, Managing Board, Corporate Functions, Sectors, Cross-sector Services, Regional Clusters, and other organizations within the company, but also organizations outside the company such as universities and research institutes, commercial companies, government organizations and public funds, suppliers at all levels, joint venture and affiliated manufacturers, contract

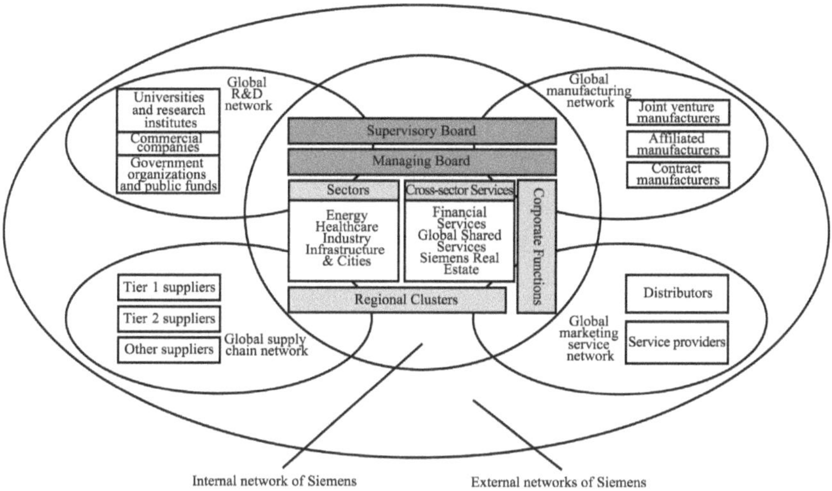

FIGURE 4.4 Siemens's global networks. Source*: Prepared by the author according to related information about Siemens.*

manufacturers, distributors, and service providers. It has created the networks through direct investment (wholly owned enterprise, joint venture, M&A, etc.), trade, nonequity-based arrangements (alliance, outsourcing, cooperation agreement, contract manufacturing, etc.), and many other means. Since Siemens operates in technology-intensive industries, its global networks are producer-driven and its forward linkages (with sellers and service providers) and backward linkages (with suppliers) are both quite close. In terms of the division of labor, there is both horizontal specialization, namely producing the same products in different countries, and horizontal specialization, namely arranging different production activities in different countries. As far as governance model is concerned, listed on both the Frankfurt Stock Exchange and the NYSE, Siemens adopts a two-tier board structure, that is, Managing Board and Supervisory Board, as well as an integrated hierarchical model featuring Corporate Functions, Sectors, Cross-sector Services, and Regional Clusters. Relational governance is adopted for members of the company's external networks. In particular, the governance of tier 1 and tier 2 suppliers and the production network in Germany depends mainly on the social relations (e.g., reputation and trust) between network participants. There are also market relations between Siemens and other enterprises in the networks but modular and captive governance is rarely seen.

As a company based in Germany, Siemens complies with the German Corporate Governance Code and adopts a dual board system comprising the Managing Board and Supervisory Board. Since it is listed on the NYSE, Siemens also needs to observe related US securities laws and regulations, including the Sarbanes–Oxley Act of 2002 (SOX) as well as the rules and regulations of the US Securities and Exchange Commission and NYSE. Despite some differences between German and US governance rules, Siemens still works to improve corporate governance according to the US legislation so it has transformed itself from a German public company to a global listed company that has the features of both US and British governance structure and German and Japanese governance structure, setting a great example of successfully combining typical European and US governance rules. The dual board system remains a defining feature of Siemens's governance structure.

The Supervisory Committee advises and supervises the Managing Board on business management. To improve efficiency and meet the both German and US corporate governance standards, Siemens has established six committees for the Supervisory Board, including Chairman's Committee, Audit Committee, Compliance Committee, Nominating Committee, Mediation Committee, and Finance and Investment Committee. The Supervisory Board of Siemens assumes similar responsibilities to those of the Board of Directors in the US or British model. In terms of composition, however, the Supervisory Board has more members and half of them are employees while the Board of Directors is mainly composed of independent directors.

The Managing Board is the top management body of the company committed to serving the interests of the company and achieving sustainable growth in company value. Members of the Managing Board include CEO, CFO, leaders of Corporate Functions, CEOs of Sectors, etc. According to the German Company Law, the Managing Board can elect one president who is not given any special power. Now the President of Siemens is more like the role of CEO in the United States or the President also serves as the CEO. In recent years, the Managing Board and CEO of Siemens have been gaining greater power and importance; the compensation and incentive stock options for the Managing Board, CEO in particular, have been rising. Therefore, from the perspective of function and power, the Managing Board is more like the management in a US or British governance model.

Siemens reports to its shareholders the company's business development, financial position, and revenue every 4 years and the Annual Shareholders' Meeting is usually held in the fourth month of every fiscal year. The Annual Shareholders' Meeting decides on the distribution of net income, approves proposals of the Managing Board and Supervisory Board, and appoints the independent auditor. The amendments to Articles of Association and measures for changing the company's equity can only be approved at the Annual Shareholders' Meeting and carried out by the Managing Board. The general meeting of shareholders is hosted by Chairman of the Supervisory Board and the election of new shareholder supervisors is on the agenda every 5 years at the general meeting.

Siemens operates under the leadership of the Managing Board. The organizational structure of the company features Sectors, Cross-sector Services,

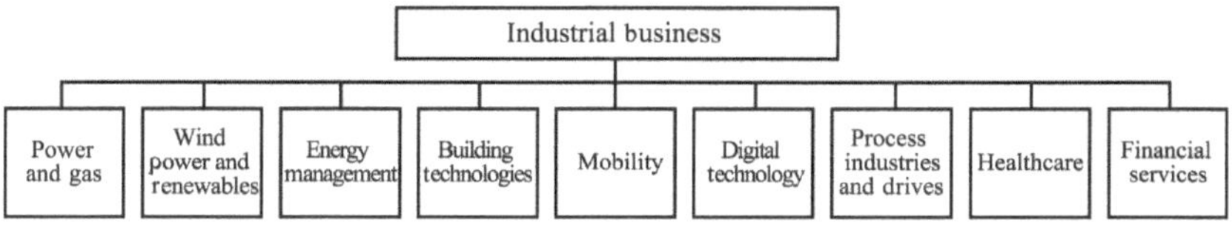

FIGURE 4.5 Siemens' Industrial Business since fiscal year 2015. Source: *Siemens,* Siemens Annual Report 2014.

Corporate Functions, and Regional Clusters. Since fiscal year 2012, Siemens has been developing business in four sectors, that is, Energy, Healthcare, Industry, and Infrastructure & Cities. The Sectors comprise divisions that are further divided into units. Cross-sector Services include Financial Services, Global Shared Services, Siemens Real Estate, etc. Corporate Functions include corporate development, finance and control, human resources, legal affairs and compliance, research institute, supply chain management, communication and governance affairs, information technology, management consulting, data protection, security, sustainability, etc. Moreover, major equity investment enterprises include BSH Home appliances, Nokia Siemens Networks, and Siemens Enterprise Networks.

In May 2014, Siemens launched its Vision 2020, announcing its plan to focus on growth fields along the electrification, automation, and digitalization and improve its portfolio considerably. According to Vision 2020, Siemens will have flatter structures and become more customer-oriented. Executed measures include the removal of additional Sectors since October 1, 2014 and the reduction of divisions from 16 to 9, including Power and Gas, Wind Power and Renewables, Power Generation Services, Energy Management, Building Technologies, Mobility, Digital Factory, Process Industries and Drives, and separately managed healthcare. Siemens will combine and streamline the functional departments serving the entire company and accelerate decision-making by reducing middle layers and delegating power to Regions and Divisions. All the measures are expected to bring about a EUR 1 billion in functional cost reduction, most of which will materialize before the end of 2016. Fig. 4.5 shows Siemens's industrial business (excluding Power Generation Services) since fiscal year 2015.

4.1.5 Global Presence and Performance

In fiscal year 2014, Siemens was divided into three Regions and 14 Regional Clusters on a geographic basis. The three Regions described in the company's financial statements are Europe, CIS, Africa, Middle East; Americas; and Asia, Australia. Sometimes financial information about specific countries, such as Germany, the United States, China, and India, is disclosed. The region Europe, CIS, Africa, Middle East includes seven clusters, that is, Germany, Northwest Europe, Southwest Europe, Central and Eastern

Europe, Russia/Central Asia, the Middle East, and Africa. The Americas region includes three clusters, that is, North America, Brazil, and Southern Andean Region. The Asia, Australia region includes four clusters, that is, Northeast Asia, South Asia, Japan, and Asia-Pacific. All these clusters cover a total of 94 countries and regions, as shown in Table 4.1.

TABLE 4.1 Siemens's Regional Clusters and Related Countries/Regions

Region/ Cluster	Country/Region	Number
Europe, CIS, Africa, Middle East		63
Germany	Germany	1
Northwest Europe	Britain, Denmark, Ireland, Netherlands, Norway, Finland, Poland, Sweden	8
Southwest Europe	Belgium, Switzerland, Spain, France, Greece, Italy, Portugal	7
Central and Eastern Europe	Armenia, Austria, Azerbaijan, Bosnia and Herzegovina, Bulgaria, Czech Republic, Georgia, Croatia, Hungary, Israel, Moldova, Montenegro, Romania, Serbia, Slovenia, Slovakia, Turkey, Ukraine, Kosovo	19
Russia/Central Asia	Russian Federation, Kazakhstan, Belarus, Uzbekistan, Tajikistan, Kyrgyzstan, Turkmenistan	7
Middle East	UAE, Egypt, Iran, Kuwait, Pakistan, Saudi Arabia, Bahrain, Iraq, Jordan, Lebanon, Libya, Oman, Qatar, Syria	14
Africa	South Africa, Ethiopia, Nigeria, Angola, Algeria, Tunisia, Morocco	7
Americas		18
North America	United States, Mexico, Costa Rica, Dominican Republic, Guatemala, Panama, El Salvador, Nicaragua, Honduras	9
Brazil	Brazil	1
Southern Andean Region	Argentina, Bolivia, Chile, Colombia, Ecuador, Peru, Uruguay, Venezuela	8
Asia, Australia		13
Northeast Asia	Chinese Mainland, Hong Kong of China, Taiwan of China	3
South Asia	India, Bangladesh	2

(Continued)

TABLE 4.1 (Continued)

Region/ Cluster	Country/Region	Number
Japan	Japan	1
Asia-Pacific	Indonesia, Malaysia, Philippines, Singapore, Thailand, Vietnam, South Korea	7
Total		94

Prepared by the authors according to related information on the website of Siemens (www.siemens.com).

In fiscal year 2014, Siemens earned EUR 10.9 billion of revenue in Germany and had 115,000 employees and 69 major production facilities in Germany, accounting for about 15%, 33%, and 24% of the total worldwide, respectively. In the region Europe (excluding Germany), CIS, Africa, Middle East, revenue was EUR 27.9 billion, and there were 96,000 employees and 70 major production facilities, accounting for about 38%, 29%, and 25% of the total worldwide, respectively. In the Americas region, revenue reached EUR 18.8 billion and there were 70,000 employees and 76 major production facilities, making up about 26%, 20%, and 26% of the total worldwide, respectively. In the Asia, Australia region, revenue was EUR 14.4 billion and there were 62,000 employees and 74 plants, making up about 20%, 18%, and 26% of the total, respectively. The revenue produced in Europe, Americas, and Asia, three major economic regions of the world, all contributed to more than 20% of the total worldwide (Fig. 4.6) so Siemens can be considered a global company in its true sense (Buckley, 2009).[4]

4.2 SIEMENS'S GLOBAL PRODUCTION NETWORK FROM THE PERSPECTIVE OF VALUE

4.2.1 Siemens's Global R&D Network

4.2.1.1 Value Objectives and Global Innovation Strategy

Siemens's global R&D network is mainly intended to ensure long-term future viability, enhance technological competitiveness, and optimize the

4. Alan M. Rugman and Alain Verbrke noted in the article "Regional Multinationals: the New Research Agenda" that only multinationals whose sales in the world's three major economic areas (EU, North America, and Asia) account for 20% of the total sales or more can be considered global companies. The article is Chapter 7 of the book *What is International Business* edited by Peter J. Buckley.

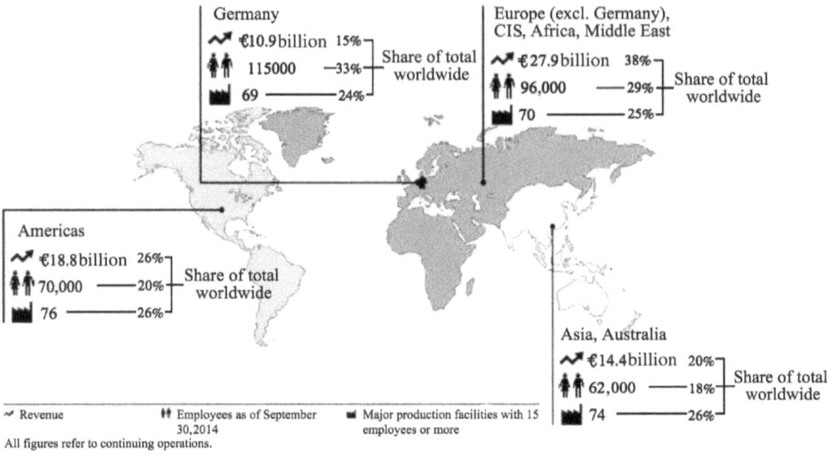

FIGURE 4.6 Siemens's business performance worldwide in fiscal year 2014. Source*: Siemens,* Siemens Company Presentation 2015.

allocation of R&D resources. The company has been considering innovation its lifeblood and through smart innovation management it has developed key technologies and brought important innovations for the market.

Siemens wants to involve more experts and idea providers in innovation activities so as to speed up the innovation process. This requires the creation of a powerful collaborative innovation network covering all sectors and regions as well as internal and external participants. In fiscal year 2009, Siemens introduced a new approach to innovation—Open Innovation (OI)—as the key to future sustainability. The principles of openness, collaboration, and entrepreneurship make the company's innovation process more efficient, powerful, and dynamic. OI means that the time when laboratories are closed is gone and it is intended to bring together the best people of the world in science and business. Laboratories used to be the whole world to Siemens but today the whole world is its laboratory.

The OI initiative of Siemens has four elements. First, the Corporate Technology (CT) department acts as the hub of the innovation network and the overall innovation capacity of Siemens is used to track major goals and ensure the company's technological competitiveness in the long run. Second, Siemens has established trust-based partnerships with 600 universities and many research institutes in about 70 countries. It works with universities and research partners in small teams on research projects. Third, its Technology-to-Business Centers (TTBs) in Berkeley and Shanghai use the ideas from university teams to support the development of new technologies and business models. Such a spin-in model has contributed to the launch of over 30 innovative products and services. Fourth, Siemens Technology Accelerator adopts a spinout model, opening-up possibilities for highly innovative technologies developed by Siemens and yet to be applied. As an alternative, they can be invested in newly founded companies.

Fig. 4.7 shows Siemens's global R&D network including both internal and external global networks.

4.2.1.2 Internal R&D Network

Siemens's R&D work stopped for some time after World War II. In 1955, the company managed to reconnect with the international community of scientific research and set up a laboratory in Munich. In 1965, it established the Erlangen Research Center, research laboratories in Berlin and Princeton, and important research bases in the United States, Austria, and Britain.

In 1969, Siemens adjusted the structure of its R&D facilities and established the Central Technology Department. In 1996, the company combined the Central Technology Department and the Development Center and named it CT. In 2006, Siemens Corporate Technology China was set up in Beijing. Today, Siemens's internal R&D network is composed of CT and the R&D teams in Sectors. In its continuing operations, Siemens had around 28,800 R&D employees at 160 R&D centers in about 30 countries in fiscal year 2014, including about 13,200 in Germany and about 15,600 in other countries.

CT consists of the CTO's Office, Corporate Research Center, Corporate Development Center, Corporate Intellectual Property, and other departments, with more than 6400 employees. CT locations feature proximity to customers, operation centers, and human resource centers. CT is led by the CTO of Siemens, who is also a member of the Managing Board. In Siemens's global R&D network, CT is a hub that leverages the innovation capacity of the entire company to reach major innovation goals and continually attract the world's most talented people to participate in joint research projects. Siemens has more than 1000 research partnerships every year, about half of which are initiated by CT, so it acts as a catalyst in the company's global R&D network.

The Corporate Research Center focuses on key technologies and cross-sector technologies with strategic significance to more than one business unit, works with the research teams of business units to develop new solutions in many application-oriented projects, and maintains close ties with customers and top universities. Now the Corporate Research Center has 13 locations worldwide (Fig. 4.8), including Munich, Erlangen, and Berlin in Germany; Vienna in Austria; Moscow and St. Petersburg in Russia; Princeton and Berkeley in the United States; Beijing and Shanghai in China; Tokyo in Japan; Singapore; and Bangalore in India.

The Corporate Development Center has software development resources for all business sectors and develops and runs software for new products and services as partners of all sectors. Now the Center has 24 locations in the

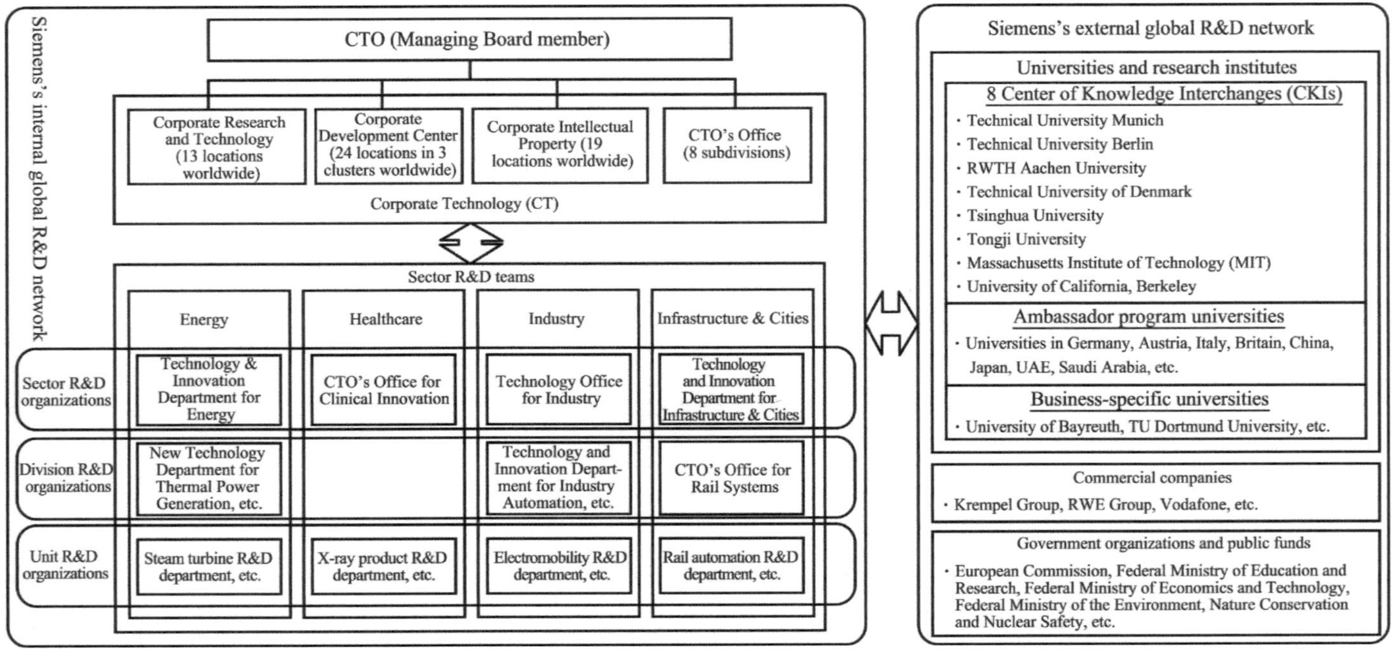

FIGURE 4.7 Siemens's global R&D network. Source: *Prepared by the author according to related information about Siemens.*

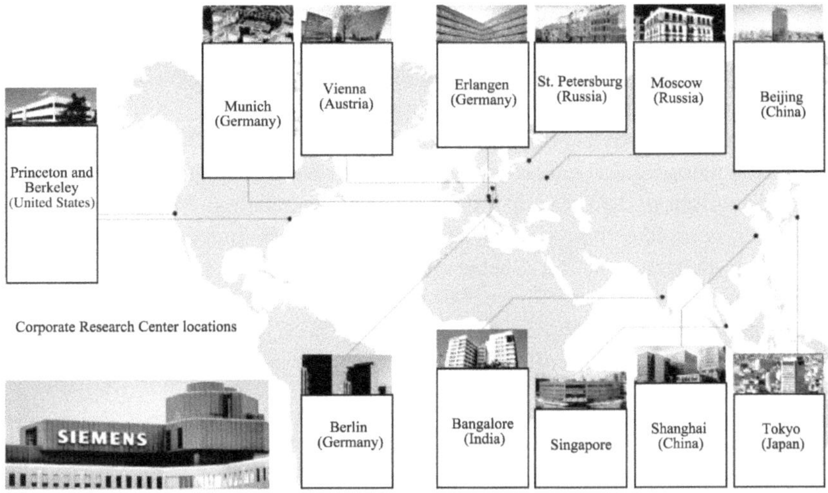

FIGURE 4.8 Siemens's Corporate Research Center locations. Source: *Website of Siemens (www.siemens.com).*

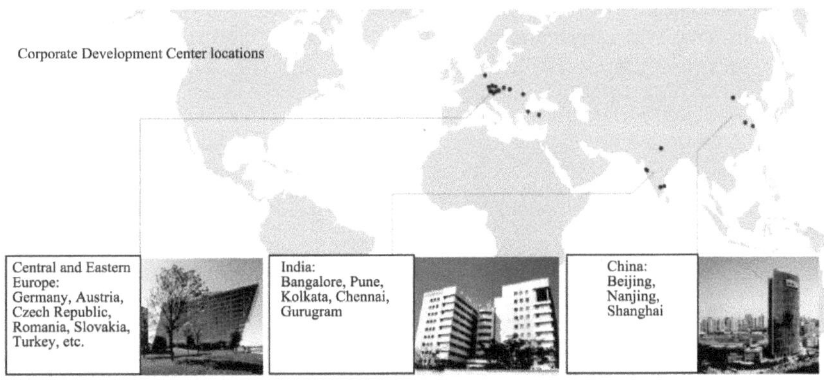

FIGURE 4.9 Siemens's Corporate Development Center locations. Source: *Website of Siemens (www.siemens.com).*

three clusters of Central and Eastern Europe, China, and India (Fig. 4.9), including 16 locations in Germany, Austria, Czech Republic, Romania, Slovakia, Turkey, etc., three in Beijing, Nanjing and Shanghai, and five in Bangalore, Pune, Kolkata, Chennai, and Gurugram.

The Corporate Intellectual Property department protects Siemens's intellectual property rights, using patents, brand, standards, and information about the market, the company, and technologies, and provides support for the company's innovation process. More than 400 experts work at 19 locations worldwide, supporting the registration, execution, and development of the company's industrial property rights and trademark rights. The department is a representative of Siemens's practice of setting up committees in accordance with international regulations and standards. It provides business sectors and regional units with technology- and market-related information and manages the company's patents worldwide.

Siemens has about 21,400 R&D employees in nearly 30 countries around the world, whose R&D efforts are focused on their next generation of products and solutions to prepare for their successful listing. Most of the R&D employees complete their work within business units, but there are also corresponding departments on Sector and Division levels for innovation and cooperation management (Fig. 4.7), that work closely with CT to promote the exchange of knowledge and experience. The R&D teams on the Sector level may also choose to collaborate with external partners.

4.2.1.3 External R&D Network

The cooperation with the world's top universities and research institutes is an indispensable part of Siemens's R&D activities so OI is crucial. Siemens establishes or maintains over 1000 research partnerships with universities, research institutes, and commercial companies every year (Fig. 4.10). To reinforce the innovation process and ensure that its business and research experts are involved in advanced scientific networks, Siemens has developed two long-term strategic partnership programs for the world's outstanding

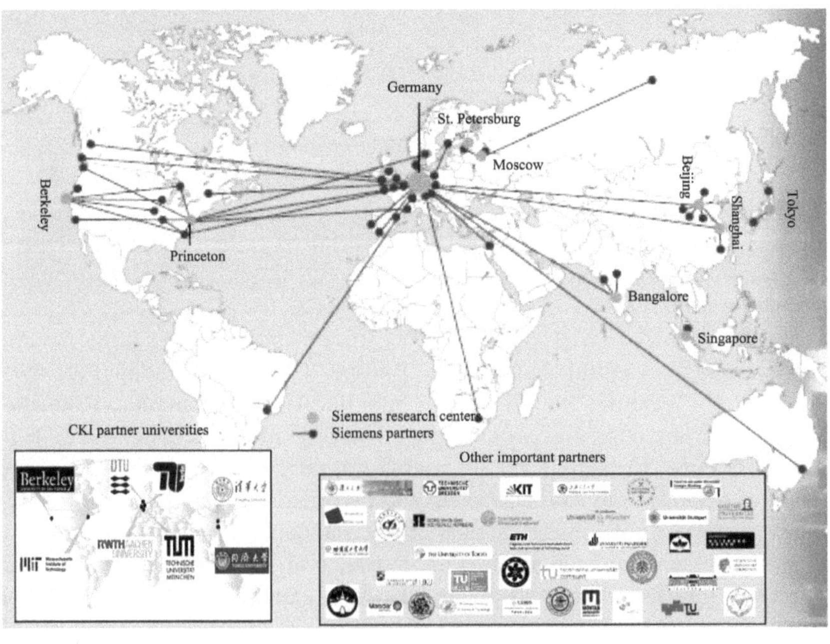

FIGURE 4.10 Siemens's global partners. Source*: Website of Siemens (www.siemens.com).*

universities and research institutes, that is, Center of Knowledge Interchange (CKI) and Ambassador Program. The company also has diverse collaborative partnerships with universities for specific projects.

The CKI program is what Siemens uses to collaborate with universities on the deepest level. The company chooses to establish partnerships with universities that can satisfy its needs for making innovations in different research fields so as to develop in-depth transfer of knowledge. Siemens sets up an office in every CKI university to act as a point of contact for managing CKI activities and the CKI partnership is based on a framework agreement on collaborative research. Now Siemens has eight CKIs, including Technical University Munich, Technical University Berlin, and RWTH Aachen University in Germany, Technical University of Denmark, Tsinghua University and Tongji University in China, as well as Massachusetts Institute of Technology and University of California, Berkeley in the United States.

The Ambassador Program strengthens the long-term cooperation between Siemens and strategically important academic institutions. Compared with CKI, the program focuses more on practicability with a simple organizational structure and no formal processes so joint activities are mainly driven by the program's specific interest. It offers a collaboration environment open to the scope of research, field of technology, specific projects, recruitment activities, etc. The related managers of Siemens are called "University Ambassadors" who work together with corresponding people in universities to drive the collaboration of the two organizations. The ambassadors are usually the CEOs of Sectors, Divisions, Units, or country-specific business. The CTO's Office helps ambassadors to identify partner departments of Siemens, enter the network of university connections, and advise on sustainable cooperation models. Many universities have participated in the Ambassador Program, including TU Dresden, Karlsruhe Institute of Technology (KIT), and TU Dortmund in Germany; Johannes Kepler University Linz, TU Wien, and TU Graz in Austria; Polytechnic University of Milan in Italy; Lancaster University Management School (LUMS) in Britain; Fudan University, Shanghai Jiao Tong University, Beihang University, Peking University, Chinese Academy of Sciences, Harbin Institute of Technology, and Zhejiang University in China; the University of Tokyo in Japan; Masdar Institute of Science and Technology in UAE; and King Abdullah University of Science and Technology (KAUST).

In addition to the CKI and Ambassador programs, Siemens collaborates with universities for specific business and diverse approaches are introduced according to the theme of cooperation (Table 4.2), including contract research, joint research, PhD fellowship, expert lectures, etc. Siemens offers a legal framework, range of costs, and guidelines for each approach. The approaches depending on specific universities and choices do not have to protect all the details and there is no integrity requirement. Since specific

TABLE 4.2 Approaches to Cooperation With Universities Based on Specific Business

Established approaches	Research projects (contract research)	Joint research projects	Joint research projects of public funds	Joint research laboratories	PhD fellowships
	Workshops/seminars/ thematic meetings	Lectures	Visits to Siemens laboratories	Siemens university conferences	
New approaches	Creative competitions	Open research funds	Innovation agents		University competence tours
	Job-based donations	Expert integration		Scientist leasing	

Website of Siemens (www.siemens.com).

projects always involve special considerations, the guideline is not a substitute for negotiations with partner universities.

Moreover, Siemens participates in the publicly funded research programs sponsored by the European Commission, Germany's Federal Ministry of Education and Research, Federal Ministry of Economics and Technology, and Federal Ministry of the Environment, Nature Conservation and Nuclear Safety, and other organizations. Since 1984, researchers at Siemens have taken part in hundreds of EU-sponsored projects. Siemens is considered to be one of the most successful industrial companies involved in the European Commission's research programs. Important areas of research include sustainable technologies, network of machines, new materials, as well as diagnosis and treatment innovations in the public health system. The participation in these projects ensures that the company keeps abreast of the latest R&D activities in different areas of innovation and research and acquires a lot of new knowledge from international research partners. Siemens and the world's best research institutes, customers, and competitors, together with exceptionally talented and enterprising young researchers, form a special global R&D value network.

4.2.2 Siemens's Global Supply Chain Network

4.2.2.1 Value Objectives and Global Supply Chain Strategy

Siemens manages its global supply chain network with a view to fully developing potential value within the global supply chain and making continued contributions to business sectors and clusters through significantly sustainable and safe ways of supply and overall design of supply chain network in the principle of professional ethics and responsibility, thus securing the company's competitive advantages and business success. The primary goal of all supply chain management activities is to make sure that Siemens provides customers with the products they need and ensure the quality, availability, and prices of services. Therefore, Siemens needs a supply chain network where global balance and localization are closely combined. It should provide the process standards, basic guidelines, methods, and tools to make full use of the supply chain's innovation strength, and focus on the excellent execution of global operations.

Supply chain network management is directly responsible for the material costs that account for about 50% of the revenue and makes crucial contributions to the economic success of Siemens. Therefore, the management strategy in this regard must be systematic and has a close focus on the One Siemens strategy and the strategies of Sectors and Divisions, and external best practices should be employed as useful supplements. In 2009, Siemens launched the Supply Chain Management Initiative, focusing on supply chain process, standards, and material pooling in the beginning. In 2010, Siemens

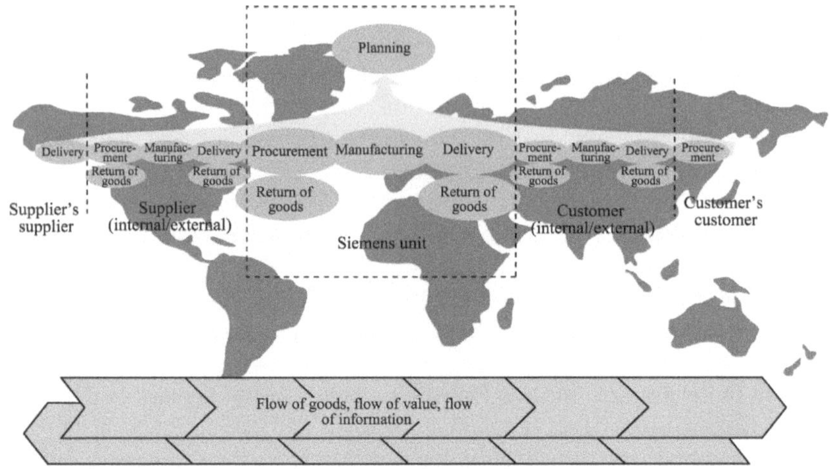

FIGURE 4.11 Supply chain network integrated into Siemens's global value network. Source: *Prepared by the author according to related information about Siemens.*

created an organization—corporate supply chain management (CSCM)—that goes beyond previous projects to begin addressing value creation. Since 2011, Siemens has renewed its supply chain management measure, fully integrating the supply chain network into its value creation network (Fig. 4.11).

Siemens's supply chain management strategy has four core elements that ensure the global supply chain network is made full use of to create value and build competitive advantages: first, using the global supplier network to promote innovation at Siemens; second, supporting business growth worldwide with a sustainable and globally balanced supply chain network; third, achieving world-class excellent execution to reach the highest standards; and fourth, attracting and developing the best employees. With these four elements, supply chain network management as a main means to create value supports the attainment of company goals in a sustainable way. Siemens has devised a supply chain strategy implementation route for each Sector. Supply chain departments, business and cross-functional departments are brought together to ensure that key employees are involved in the early stage; navigation projects are carried out on the Business Unit level to drive the supply chain actions on Sector and Cross-sector levels and support value creation in sectors.

4.2.2.2 Organizational Form of Supply Chain Management Network

Siemens's supply chain network is a global network of businesses and organizations created based customer needs in terms of value, cost, and time and offering products and services. All supply chain management departments of the company collaborate and work in a global network (Fig. 4.12).

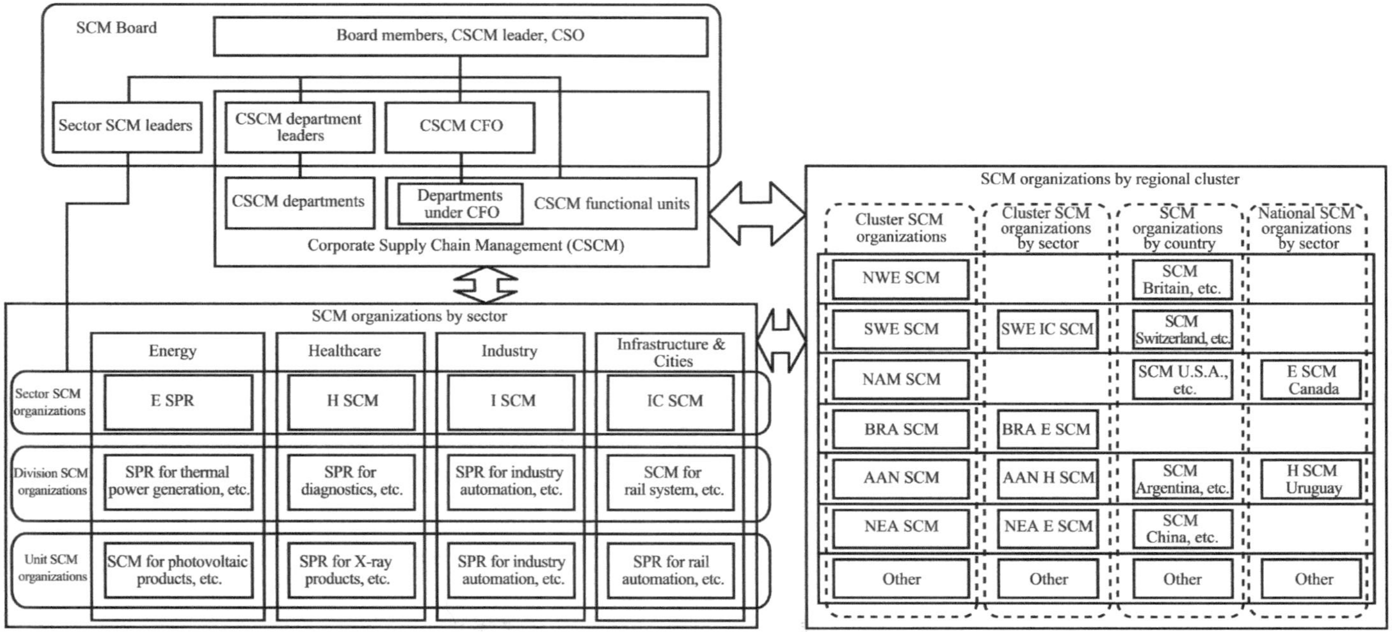

FIGURE 4.12 Siemens's global supply chain management network. Source: *Prepared by the author according to related information about Siemens.*

The network comprises CSCM, supply chain management teams on Sector, Division, and Unit levels, and the supply chain management teams in Regional Clusters. There is also a Supply Chain Management Board (SCM Board) that coordinates operations in the supply chain management network.

Chairman of the SCM Board is a member of Siemens's Managing Board responsible for managing supply chains as well as leader of CSCM. Members of the SCM Board include CSCM leader, CSCM CFO, Sector-level SCM leaders, leaders of CSCM departments, etc. CSCM is a company-level functional unit responsible for addressing issues related to global supply chain, such as control and the environment, methods/processes/systems, finance/control, people development, etc., playing a leading role in Siemens's global supply chain network. SCM teams are put together on Sector, Division, and Unit levels to purchase direct materials worldwide and drive and coordinate all procurement activities within their respective business scopes. The SCM teams on the Regional Cluster level drive and coordinate all procurement activities within the scope of cluster. In Regional Clusters, SCM organizations are also established on the Sector level to take care of direct materials for specific business in the region concerned. In some Regional Clusters, there are also national-level and Sector-level SCM organizations.

4.2.2.3 Governance System of Supply Chain Network

4.2.2.3.1 Supplier Management Process

At Siemens, supply chain management is an integrated system based on which the company collaborates with suppliers and achieves sustained value chain improvement. Supplier management process includes supplier selection and qualification recognition, supplier strategy and forward procurement list, procurement decision-making, supplier assessment and classification, and supplier development and elimination (Fig. 4.13). Supplier assessment is compulsory company-wide and comprises performance, strategic, and risk assessment. According to the assessment results, suppliers are classified into four types, namely Preferred, Advanced, Approved, and Substandard. Siemens provides company-level rewards for suppliers with excellent performance in breakthroughs and innovations, global strategic procurement, sustainability, and best overall. Rewards may also be provided for suppliers in a particular business unit of a Sector or a particular field of materials.

4.2.2.3.2 Global Value Sourcing

A main element of Siemens's SCM moves is to create a competitive and globally balanced supply chain network and one of the key factors for creating balance is to increase the share of procurement from emerging markets, which is achieved through the Global Value Sourcing (GVS) program. The key performance indicator of the program is GVS share, namely the share of goods and services procured from defined GVS countries. Siemens influences the growth in GVS procurement through three key means (Fig. 4.14): (1) increasing GVS domestic procurement by expanding operations in GVS countries, namely Local for Local (L4L); (2) reducing imports from non-GVS countries for operations in GVS countries, namely Localization; and (3) increasing exports from GVS countries to non-GVS countries, namely Local for Global (L4G). To implement the GVS program, Siemens has devised a standard process for implementing the GVS program, including objectives and strategies, landmarks and structure, action implementation, action execution, and monitoring.

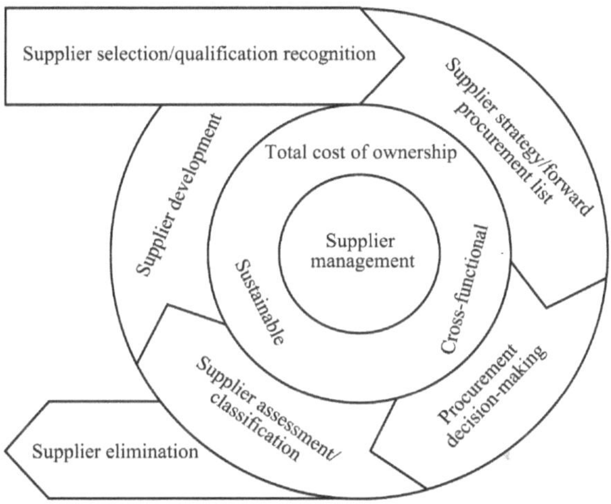

FIGURE 4.13 Siemens's supplier management process. Source: *Website of Siemens (www.siemens.com).*

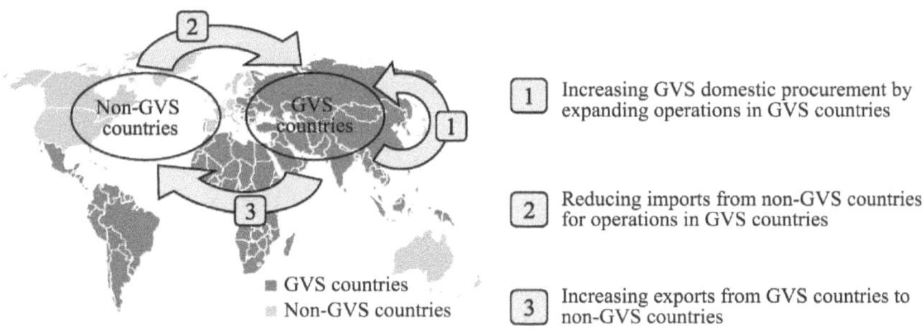

FIGURE 4.14 Siemens's key means for improving GVS. Source: *Prepared by the author according to related information about Siemens.*

4.2.2.3.3 Other Measures for Supply Chain Management

Another important theme of supply chain management is the sustainability of supply chain. All suppliers of Siemens are required to comply with the *Code of Conduct for Siemens Suppliers* and the company promotes the execution of the Code in its supply chain. Siemens also uses external experts to launch onsite sustainable audits to ensure the execution of the Code and promote the sustainable operations of supply chains worldwide.

In addition, the Energy Efficiency Program for Siemens (EEP4S) was launched in 2010 to create a green supply chain up to the highest sustainability standard so that the company's principle of protecting global climate is followed in every part of the value chain. This is of great significance to the improvement of suppliers' energy efficiency and the sustainable competitiveness of the entire supply chain. Under the program, energy and environmental assessments on different levels will be carried out among 19,000 suppliers worldwide according to the quantity supplied and energy consumption in their respective industries. This helps to improve suppliers' energy efficiency and cost structure and eventually enhances the long-term competitiveness of both sides.

4.2.2.4 Location Choices Concerning Global Supply Chain

According to Siemens, since fiscal year 2006, the company's annual purchasing volume for continuing operations accounts for about 50% of its revenue. Based on the analysis of procurement from major countries, and taking into account the GDP growth from 2008 to 2014, and external market analysis data, Siemens selected 58 countries for market analysis, and classified them into four industrial types according to product, industry, and manufacturing (Table 4.3). In 2010, about 87% of Siemens's global purchasing volume was from the 58 countries.

With the macroeconomic data about these countries for 1995–2020 collected from the World Bank's Worldwide Development Indicators (WDI) and Global Development Finance (GDF) as well as Siemens's data, the Current Market Index (CMI) of these countries from 1995 to 2010 and their Market Dynamic Index (MDI) from 2011 to 2020 are calculated and the sum of the two is the Total Index of a Country Model. Based on an analysis of the 58 countries' Total Index of a Country Model, Siemens divides all sourcing countries into non-GVS and GVS countries. There are 20 major non-GVS countries, including Australia, Austria, Belgium, Canada, Denmark, Finland, France, Germany, Ireland, Italy, Japan, Luxembourg, the Netherlands, New Zealand, Norway, Spain, Sweden, Switzerland, Britain, and the United States. There are also five countries with a small purchasing volume, including Andorra, the Vatican, Liechtenstein, Monaco, and San Marino.

TABLE 4.3 Siemens's Main Sourcing Countries and Corresponding Industrial Types

Classification	Type 1: Primary-Stage Economy	Type 2: Agroindustrial Economy	Type 3: Industrialized Economy	Type 4: Advanced Industrialized Economy
Products	Raw materials, semifinished products	Consumer goods, light industry goods	Light and heavy industry goods	Complex machinery, high-value capital goods, service outsourcing
Industry	Foreign-funded enterprises, trade in handicraft	State-owned enterprises, very few capital-intensive	State-owned enterprises, many capital-intensive	Companies operating globally
Manufacturing	Primary-stage production and agriculture	Labor-intensive production	Modern and old-fashioned industrial production	High-level automation, high-tech production, dominance in the service sector
Corresponding countries	Algeria, Iran, Nigeria, Paraguay, Peru, Saudi Arabia, Uruguay	Argentina, Bangladesh, Belarus, Bulgaria, Colombia, Egypt, Greece, Indonesia, Morocco, Pakistan, Romania, Russia, Serbia, Slovenia, Tunisia, Ukraine, Venezuela, Vietnam	Australia, Brazil, China, Croatia, Czech Republic, Hungary, Iceland, India, Lithuania, Malaysia, Mexico, Norway, Philippines, Poland, Portugal, Singapore, Slovakia, South Africa, Spain, Thailand, Turkey	Austria, Canada, France, Germany, Israel, Italy, Japan, South Korea, Sweden, Switzerland, Britain, United States
Number of countries	7	18	21	12
Procurement	Local for Local (L4L)Local for Global (L4G)		L4L and L4G	

Website of Siemens (www.siemens.com).

In addition to 25 non-GVS countries, Siemens considers nearly 170 source countries as GVS countries. According to the Total Index of a Country Model, Siemens has defined the top 20 GVS countries by 2016, including China, India, Brazil, Russia, Mexico, South Korea, Turkey, Indonesia, Malaysia, Singapore, Saudi Arabia, Thailand, South Africa, Hungary, Iran, Czech Republic, Poland, Ukraine, Pakistan, and Egypt. These 20 GVS countries have well-developed infrastructure and enormous potential for economic growth. In fiscal year 2010, 16% of Siemens's purchasing volume was from these 20 countries, and the largest sourcing country was China that contributed to more than 5% of the company's total purchasing volume. Through the GVC program, Siemens can effectively reduce procurement costs and ensure an efficient global sourcing network.

4.2.3 Siemens's Global Manufacturing Network

4.2.3.1 Value Objectives and Global Manufacturing Strategy

Independent production is one of the most important elements of Siemens. Siemens operates more than 289 major manufacturing and production plants in more than 40 countries worldwide, including facilities for joint ventures and affiliated enterprises. The core element of Siemens's global manufacturing strategy is to achieve continuous cost improvement in terms of products and solutions, as well as new market development. The increasing complexity of processes and technologies requires Siemens's deep understanding of the needs of customers and regional markets, which means that maintaining large-scale, independent production is critical for Siemens. It is also important to increase the value added of production through technological advances. Another key element of Siemens's global manufacturing strategy is to increase productivity. At Siemens, about two-thirds of productivity gains are obtained through process improvement, and Siemens is implementing a process improvement initiative based on lean principles—the Siemens Production System (SPS). The remaining one-third of productivity gains are achieved through structural changes, such as global plant configuration and optimization. Siemens adopts a unified and systematic approach, the Global Manufacturing Footprint (GMF). In addition, Siemens considers high quality as a key factor in meeting customer needs and achieving the company's success so it is committed to maintaining a quality culture of continuous improvement and high transparency, and conducting comprehensive quality management.

4.2.3.2 Governance System of Global Manufacturing

The goal of SPS is to continuously improve Siemens's global production processes. As a structured approach, SPS is about designing and operating the Siemens production business in accordance with lean principles, and reducing business activities that do not increase customer value through different methods and principle, thus directly generating customer value, achieving the continuous improvement of Siemens's global production processes, and helping Siemens meet growing customer needs and enhance cost advantages over competitors. SPS is intended to achieve lean management and involves five principles. First, Managers are familiar with the principles of lean production and demonstrate them by their example every day. Second, they systematically apply the current best approach in every activity until they find a better one. Third, they produce exactly what is needed at just the right time. Fourth, they detect problems immediately at the point of origin and systematically rectify them. Fifth, every one of them knows what to do and is challenged.

SPS provides a standardized framework (Fig. 4.15) to optimize all necessary areas of activity, including guiding principles/production strategies, product lifecycle management, supply chain management, customer relations management, goal and performance management, and personnel and continuous improvement. Production plants act as a basic business nodes that connect Siemens's global R&D, supply chain, manufacturing, and marketing service networks so that the company's global networks are made full use of to achieve continuous improvement of its manufacturing business. To date, SPS has been successfully implemented in more than 150 major plants of the company in the world.

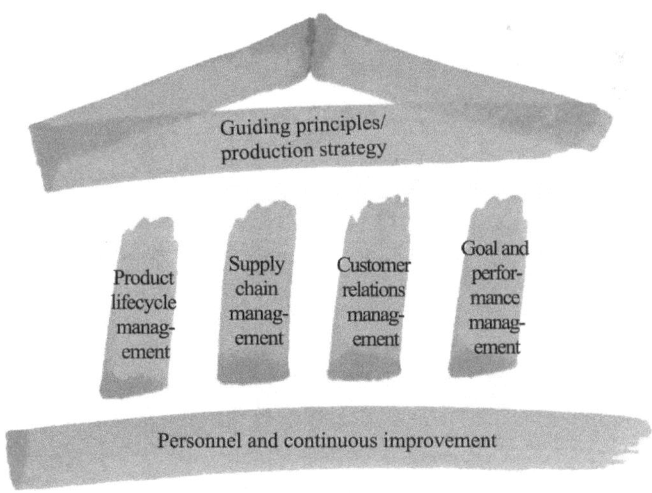

FIGURE 4.15 Framework of Siemens Production System. Source: *Website of Siemens (www.siemens.com).*

4.2.3.3 *Plant Configuration of Global Manufacturing*

Based on GMF, Siemens optimizes plant configuration worldwide and enhances productivity to ensure its leadership in global manufacturing management, and provide further support for profitable growth. This is an important factor in gaining competitive advantages over other companies. GMF involves four main principles. First, customer satisfaction is the top priority. Siemens systematically analyzes customers' final buying decisions, namely key success factors (KSFs), and works with the designed global manufacturing network to achieve the best customer satisfaction. Second, Siemens designs its global manufacturing network based on KSFs to improve the total cost of ownership (including production costs, procurement costs, transportation costs, etc.) so it is not just optimization on the functional unit or regional cluster level. Third, value and information flows are optimized to avoid waste (nonvalue-added processes) through localization. "Green configuration" is also achieved through localization. For example, local R&D and production activities can help connect procurement, manufacturing, and sales to avoid long-distance transportation, communication, travel, etc. Fourth, the quality of existing GMF is reviewed on a regular basis and a continuous improvement process is also run in this regard.

The GMF framework mainly includes GMF design and GMF implementation, and provides specific guidance and reference flow, support tools, capabilities, and expertise building measurements for the global manufacturing network. GMF design supports strategic planning for business units while GMF implementation supports the operation configuration of business units. Therefore, the GMF framework (Fig. 4.16) mainly includes the following

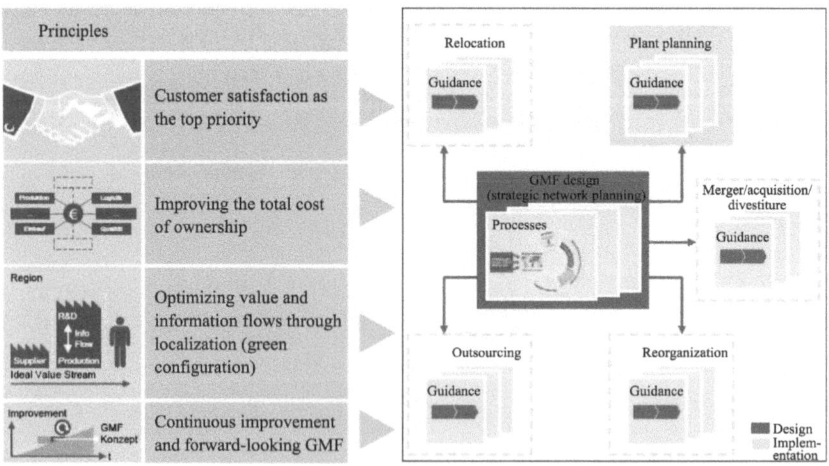

FIGURE 4.16 Framework of Global Manufacturing Footprint. Source: *Website of Siemens (www.siemens.com).*

activities: strategic network planning, plant reorganization, outsourcing, merger/acquisition/divestiture, plant relocation, and plant planning.

4.2.3.4 Analysis of Global Manufacturing Network by Region

In fiscal year 2014, Siemens had over 289 manufacturing plants in nearly 40 countries, including about 139 in the Europe, CIS, Africa, Middle East region, and about 76 in the "Americas" region, and about 74 in the "Asia, Australia" region. In China, one of Siemens's largest markets, there were 45 major production and manufacturing plants. Table 4.4 shows the shares of Siemens' plants and revenue by region from fiscal year 2010 to fiscal year 2014.

Despite the deviation in the statistics resulting from the differences between fiscal years in the scope of continuing operations and the scope of region, it can be seen in the table that the shares of both plants and revenue changed very little in all of the three regions, and the gap between the two was very small. It suggests that Siemens has balanced global manufacturing and global sales. Europe (including Germany) is the main manufacturing base and market of Siemens; the plants and revenue here account for almost 50% of the total worldwide but the share of plants is slightly smaller than that of revenue. The share of plants shows a slight downward trend in the entire Europe, CIS, Africa, Middle East region. The shares of plants and revenue in the Americas region are both more than 25% and the gap is very small, and the two are already basically the same in the last two fiscal years. The share of plants in the Asia, Australia region is slightly higher than that of revenue, which suggests to some extent that manufacturing exports from Asia have increased.

4.2.4 Siemens's Global Marketing Service Network

For Siemens, it is important to build sustained, customer-focused strategic partnerships. As a result, Siemens has a structured approach to key account management that allows for a structure tailored to the size and location of the customer. At Siemens, a Regional Account Manager is responsible for managing cross-sector customers in certain countries and regions. A Global Account Manager is responsible for managing business activities concerning customers worldwide in a particular Division. A Corporate Account Manager is responsible for managing business activities concerning key customers across geographical areas and business divisions.

An important customer-oriented driver of Siemens is the Executive Relationship Program. The program focuses on Siemens's key customers and is intended to build long-term relationships with executives from around 100 customers. This program is run on the Managing Board level. All members of the Siemens Managing Board maintain direct contact and regular

TABLE 4.4 Shares of Siemens's Plants and Revenue by Region, fiscal year (FY) 2010–14

	FY2010		FY2011		FY2012		FY2013		FY2014	
Plants	320		285		290		290		289	
Revenue (EUR 1 billion)	75,978		73,515		78,296		75,882		71,920	
Region	Plants as a Share of Total (%)	Revenue as a Share of Total (%)	Plants as a Share of Total (%)	Revenue as a Share of Total (%)	Plants as a Share of Total (%)	Revenue as a Share of Total (%)	Plants as a Share of Total (%)	Revenue as a Share of Total (%)	Plants as a Share of Total (%)	Revenue as a Share of Total (%)
Europe, CIS, Africa, Middle East	50	55	49	53	48	51	48	53	48	54
Americas	25	27	26	28	28	29	28	28	26	26
Asia, Australia	25	18	25	19	24	20	24	20	26	20

Prepared by the authors according to data in annual reports of Siemens.

conversations with key customers, and listen to their needs in person. In addition, the program has been introduced into the national-level organizations in which senior management representatives from Siemens regional companies are involved.

The business sectors, divisions, and units of Siemens are responsible for the global sales of their own business, and can support customers worldwide directly from their respective headquarters, especially for large contracts and projects. However, most of Siemens's customers are small and medium-sized enterprises and organizations that need local support. To work with them locally, Siemens's regional companies offer guidance to global sales and they are responsible for sales in their respective countries, cross-sector fields, and divisions. This enables Siemens to maintain close ties with customers around the world, and provide them with local partnerships and faster, tailor-made solutions that meet their business needs. Siemens is active in about 200 countries and is a competitive market leader. It is usually considered to be a local citizen due to its continuing local performance for a long time. An essential component of Siemens's global marketing and sales activities is compliance with applicable laws, and international rules and regulations.

4.2.5 Value Added of Siemens's Global Networks

4.2.5.1 Global R&D Network

Siemens considers inventions and patents a key factor of its business strategy. They have created great value added for the company's business activities and enhanced its core competences, acting as a significant measure of global R&D network's value added. Fig. 4.17 shows how many inventions

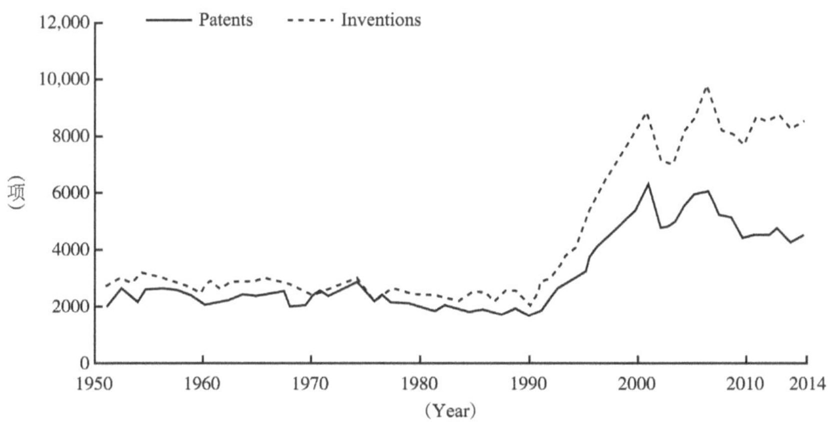

FIGURE 4.17 Siemens's registered inventions and patent applications, fiscal years 1950–2014. Source: *Prepared by the author according to data in Siemens's annual reports.*

Siemens registered and how many patent applications it filed annually during the 65 years from fiscal 1950 to fiscal 2014.

The numbers of Siemens's registered inventions and patent applications changed slightly during the period from fiscal year 1950 to fiscal year 1990 but increased rapidly during the period fiscal years 1990–2001 thanks to the implementation of the company's global strategy in the 1990s. In fiscal year 2001, 6330 new patent applications were filed, reaching a record high in the company's history, and 9060 inventions were registered, second only to the 2006 level. Siemens saw its annual numbers of inventions and patents fall due to the impact of the global recession between fiscal 2002 and fiscal 2003 but they soon returned to a continuing upward trend. In fiscal year 2006, more than 10,000 inventions were registered, reaching a record high, and 6000 new patent applications were filed, second only to the 2001 level. Similarly, Siemens's registered inventions and patent applications declined in number between fiscal 2007 and fiscal 2009 due to the global economic crisis, but began to rise sharply in fiscal year 2009. In fiscal year 2014, a total of 8600 inventions were registered for Siemens's continuing operations, equivalent to an average of 39 inventions per workday, and the number of inventions per researcher doubled compared with the level in fiscal year 2000. In fiscal year 2014, Siemens filed 4300 patent applications, equivalent to an average of 20 applications per workday.

At present, Siemens holds around 56,100 patents worldwide for its continuing operations. Siemens ranked first in the patent statistics of the European Patent Office (EPO) for the first time in 2010. In 2013, the company's patent ranking was number two according to EPO, number four in Germany, and number 13 in the United States. Table 4.5 lists Siemens's patent rankings for the period 2009–13, and the rankings show that Siemens is one of the most innovative companies in the world.

The global R&D network helps Siemens increase its innovation capacity continually, and the company has won many awards worldwide for its OI program. For example, in fiscal year 2011, Siemens stood out against 41 competitors and won the European Most Admired Knowledge Enterprises (MAKE) award. The MAKE awards, initiated and sponsored by Teleos—a

TABLE 4.5 Siemens's Patent Rankings, 2009–13

Authority	2009	2010	2011	2012	2013
DPMA	3	3	3	3	4
EPO	2	1	1	2	2
US PTO	13	9	10	11	13

Prepared by the authors according to data in annual reports of Siemens.

renowned knowledge management and intellectual capital research firm in Britain—together with Know Network in 1998, are presented annually in recognition of institutions or enterprises that have successfully transformed corporate knowledge and intellectual capital into outstanding products, services, and solutions, and created value for shareholders. In 2011, Siemens Corporate Technology China was named the Best Enterprise Institute of the Year by the Chinese business magazine *Global Entrepreneur* for the third time.

4.2.5.2 Global Supply Chain Network

Siemens uses supply chain's contribution to net income (CNI) as an important measure of the value added of its global supply chain network. Since its global supply chain strategy was launched, the supply chain network's CNI has been on the rise, growing 124% in fiscal year 2010 over 2008.

Siemens has also introduced three indicators for assessing value creation in supply chain management, including the share of centrally managed company-level sourcing, the share of GVC, and the number of suppliers (purchasing volume larger than EUR 10,000). The gains brought by the company's global supply chain management from fiscal year 2008 to fiscal year 2010 can be seen in three aspects (Fig. 4.18).

First, increasing share of centrally managed company-level sourcing. The share was 29% in fiscal 2008, 44% in fiscal 2009, and 47% in fiscal 2010, which was up 18 percentage points compared with 2008. It means that the goal set for fiscal year 2010 was reached.

Second, increasing share of GVC. The share was around 20% in fiscal 2008, about 21% in fiscal 2009, and about 24% in fiscal 2010, which was up four percentage points compared with 2008. Siemens set the midterm goal for the share of GVC as 25%, which was reached in fiscal year 2011 when the share of purchasing volume from emerging markets was up to 27%.

Third, decreasing number of suppliers (purchasing volume larger than EUR 10,000). The number was about 113,000 in fiscal 2008, about 97,000

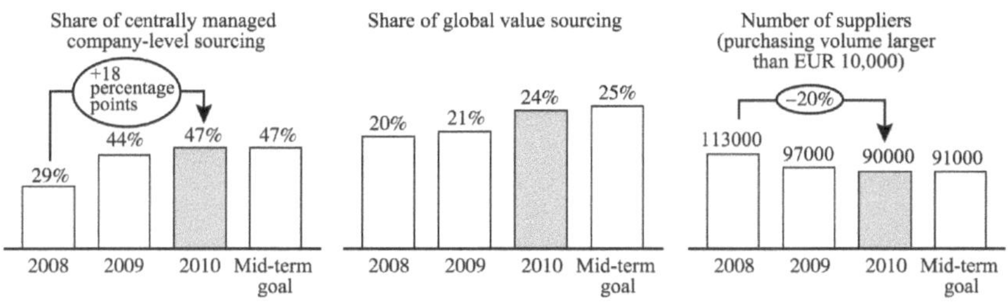

FIGURE 4.18 Important indicators of Siemens global supply chain network's value added. Source: *Website of Siemens (www.siemens.com).*

in fiscal 2009, and about 90,000 in fiscal 2010, which was down by about 20% compared with 2008. Siemens set the midterm goal for the number of such suppliers as 91,000 which was attained in fiscal 2010.

4.2.5.3 Global Manufacturing Network

Through GMF and SPS, Siemens has created a sound global manufacturing network to help lower its total production costs, improve local value and information flows, and obtain shorter delivery time and higher quality. In recent years, Siemens's average annual productivity has increased by around 8%. As a result, the performance of its manufacturing business has improved significantly. Since data about its manufacturing processes are not available, Siemens's economic value added (EVA) is used to assess the value added of the global manufacturing network.

Siemens uses EVA to assess the value creation and economic success of its business. EVA is the net profit less the user cost of capital. Therefore, a business creates value only when it can at least make up for the user cost of capital and achieve the EVA that meets the needs of the capital market. For Siemens, a company that considers manufacturing one of the most important elements, EVA can partly reflect the value added of its manufacturing business. Fig. 4.19 shows Siemens's overall EVA between fiscal 2001 and fiscal 2007, which was apparently on the rise. It suggests that global manufacturing helps to lower its operating costs and increase the value added of its business and decentralized production helps to reduce the regional economic risks incurred by natural disasters.

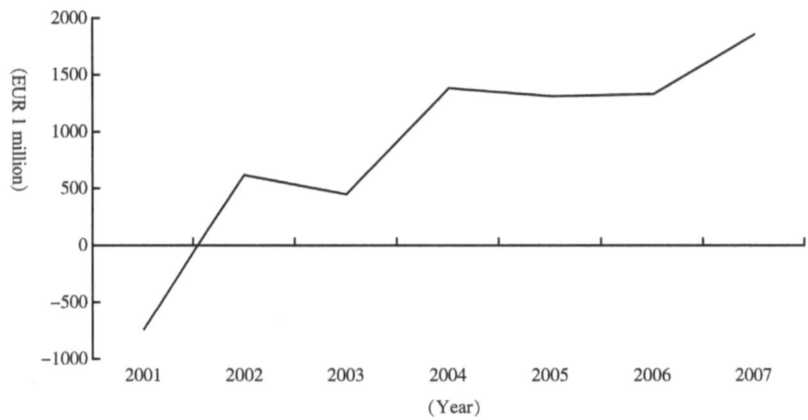

FIGURE 4.19 Siemens's economic value added, fiscal years 2001–07. Source: *Prepared by the author according to the data provided in annual reports of Siemens.*

4.2.5.4 Global Marketing Service Network

Siemens's global revenue has increased sharply since the 1990s when the global strategy was introduced. Fig. 4.20 lists the revenue shares of Siemens regions in fiscal years 1990, 2000, 2010, and 2014.

The revenues for fiscal years 1990 and 2000 are calculated in accordance with the *German Corporate Governance Code* while the figures for fiscal years 2010 and 2014 concern the continuing operations of Siemens. Some adjustments were also made to the scope of each region during this period so there may be some deviations in the data. Despite that, it is apparent that Siemens's total revenue more than doubled from fiscal 1990 to fiscal 2000, growing from EUR 32 billion to EUR 78 billion. The revenue in Germany increased from EUR 14 billion to EUR 19 billion but its share of the total fell from 44% to 24%. The revenue in Europe (excl. Germany), CIS, Africa, and the Middle East grew by around 1.5 times from EUR 11 billion to EUR 28 billion and its share of the total increased from 34% to 36%. The revenue in the Americas increased by over 3 times from EUR 5 billion to EUR 21 billion and its share of the total went up from 16% to 27%. In the Asia, Australia region, revenue rose by 4 times from EUR 2 billion to EUR 10 billion and its share of the total climbed from 6% to 13%. The global marketing service network helps to increase the company's sales overseas, especially in newly explored markets, and thus contributes to sales growth in the global market.

After the bribery scandal, the US financial crisis, and the European debt crisis, the revenue from Siemens's continuing operations in fiscal years 2010 and 2014 dropped slightly compared with that in fiscal 2000. Specifically, the revenue in Germany fell to EUR 11 billion, even lower than the level in fiscal 1990, and its share of the total dropped to around 15%. The revenue in

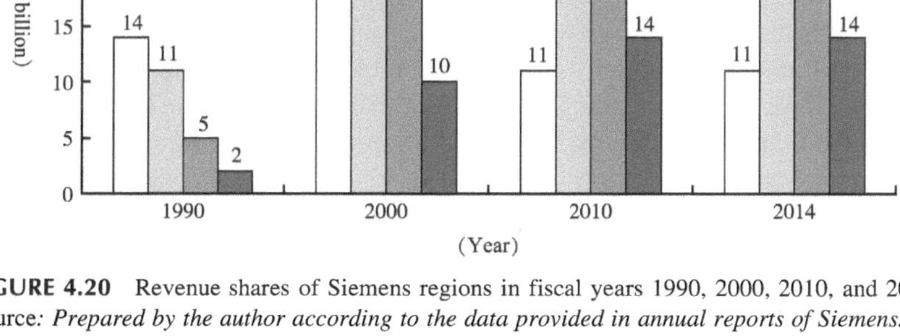

FIGURE 4.20 Revenue shares of Siemens regions in fiscal years 1990, 2000, 2010, and 2014. Source: *Prepared by the author according to the data provided in annual reports of Siemens.*

Europe (excluding Germany), CIS, Africa, and the Middle East increased slightly in fiscal 2010 and the level in fiscal 2014 was basically the same as that in fiscal 2000 but its share of the total rose to around 39%. In the Americas, the revenue in fiscal 2010 was basically at the same level as that in fiscal 2000 and that in 2014 was slightly lower than the 2000 level but its share of the total stayed around 26%. In the Asia, Australia region, the revenue rose to EUR 14 billion in fiscal years 2010 and 2014, up about 40% over fiscal 2000, and its share of the total went up to around 20%. Apparently, by creating a global sales network, Siemens has boosted sales in emerging markets and reduced sales fluctuations and stabilized income flow worldwide.

4.3 SIEMENS'S HEALTHCARE NETWORK FROM THE PERSPECTIVE OF EMBEDDEDNESS

4.3.1 Global Production Network of Medical Devices

The medical device market is one of the fastest growing and most active international trade markets in the world economy. According to the Committee on Medical Devices under the European Commission, sales in the global medical device market rose from USD 187 billion in 2001 to USD 435.3 billion in 2011, registering a compound annual growth rate of 8.35%, higher than the average annual growth rate of around 5% in the global pharmaceutical market during the same period, and about twice the growth rate of the Chinese economy. Even in 2008 and 2009 that witnessed the global recession, medical device sales still achieved a growth rate of 6.99% and 7.02%, respectively. The medical device market will continue to grow as the economy recovers and middle-income consumers demand for healthcare services grows in emerging markets.

The global high-end medical device market is largely dominated by products from the United States, Germany, and Japan, and other European companies have a certain advantage in some specialized fields. The United States is the world's largest medical device producer and consumer, and its consumption contributes to more than 40% of the global total, compared to more than 30% in Europe and about 10% in Japan (www.askci.com, 2012). The competition in medical devices is global. As a focus field of competition among high-tech companies worldwide, this market is marked by a growing trend of oligopoly dominance. According to the European Medical Device Industry Association, the top 25 medical device companies (70% of which are based in the United States) contribute to more than 60% of the global medical device sales, suggesting that the level of market concentration has been raised significantly. US biomedical media FierceBiotech ranked the top 10 medical device companies by R&D spending.[5] According to its data, despite

5. The rankings are Johnson & Johnson, Abbott, Siemens Healthineers, Medtronic, GE Healthcare, Philips Healthcare, Baxter International, Boston Scientific, Ovidien, and BD.

the global economic downturn, the world's medical device giants invested heavily in R&D, and R&D spending registered a growth in the medical device industry.

The global medical device industry will feature two major networks in the future. One is the network of high-tech new medical device industries dominated by the United States, Europe, and Japan. Developed countries have monopolized the vast majority of the global market and 70% of China's high-end medical devices are imported. The other is the network of labor-intensive and low-tech traditional medical device industries, in which product production is gradually transferred from the developed countries in Europe and the United States to developing countries with lower labor costs, especially China, India, and other Asian countries. China is becoming the world's new medical device manufacturing center, and ranks first in the world by the production of a variety of basic products. The two networks are taking shape as the global medical device industry grows rapidly.

4.3.2 Evolution of Siemens Healthineers' Embeddedness in China

Siemens Healthineers is one of the world's largest medical solution suppliers as well as a leading manufacturer and service provider in the areas of medical imaging, laboratory diagnostics, healthcare IT, and hearing instruments, with a history of more than 130 years. With its global headquarters in Erlangen, Germany, Siemens Healthineers has offices and about 40 major manufacturing sites in more than 138 countries. In fiscal year 2014 (as of September 30, 2014), Siemens Healthineers brought in EUR 12.429 billion in revenue, making up 17% of Siemens's total revenue; it had around 51,000 employees, accounting for 14% of the company's total. It earned EUR 2.027 billion profits, with a margin of 16.3%, and spent over EUR 1.3 billion on R&D, accounting for 9.6% of its revenue. It ranked first among all Siemens sectors by profit margin and the ratio of R&D spending to revenue. Its revenue in the Europe, CIS, Africa, Middle East region registered EUR 4.391 billion, accounting for 35% of the company's total. Its revenue in Germany was EUR 880 million, making up 7% of the total; that in the Americas was EUR 4.729 billion, making up 38% of the total; and that in the Asia, Australia region was EUR 3.281 billion, making up 26% of the total.

Siemens Healthineers has operated in China for 117 years, ever since 1899, when Siemens provided China with the first X-ray unit. Now Siemens Healthineers has six operating companies, 33 representative offices, 60 service stations, and over 3000 employees in China, providing the Chinese, Asian, and global markets with many medical products including CT scanners, MRI scanners, X-ray units, ultrasound machines, hearing instruments, and laboratory diagnostics equipment. Its China headquarters is in Shanghai International Medical Zone, North China headquarters in Beijing, and South

China headquarters in Guangzhou. Leveraging its strengths in areas of innovation, Siemens strives to support China's endeavor to enhance its healthcare services, and satisfy the needs of Chinese hospitals in different regions and on different levels, including the diverse needs of leading medical research institutes and hospitals, large general hospitals with huge influxes of patients, community hospitals in cities, and small and medium-sized township hospitals.

Based on the embeddedness-centered framework of microcosmic GPN studies proposed in Chapter 3, Analytical Framework of Microcosmic GPN Studies, this section looks at how Siemens Healthineers gets embedded in China in terms of strategy, dimension, and region (Fig. 4.21).

Since its representative office was opened in China in 1982, Siemens Healthineers has been implementing its "deep embedding" strategy to become integrated into China's economic and social networks and included China in its global strategy to enable interactions with the Chinese healthcare industry and establish a complete industry network featuring two-way embeddedness. From the perspective of strategy, Siemens Healthineers' embeddedness in China can be divided into four phases after the introduction of China's reform and opening-up policy, namely exploratory embeddedness phase (1982–91), strategic linkage phase (1992–2000), strategic embeddedness phase (2001–05), and strategic coupling phase (2006–).

Siemens Healthineers' embeddedness strategy is in line with China's policy of reform and opening-up and healthcare reform. It is also forward-looking in making the gradual transition from "footloose embedding" to

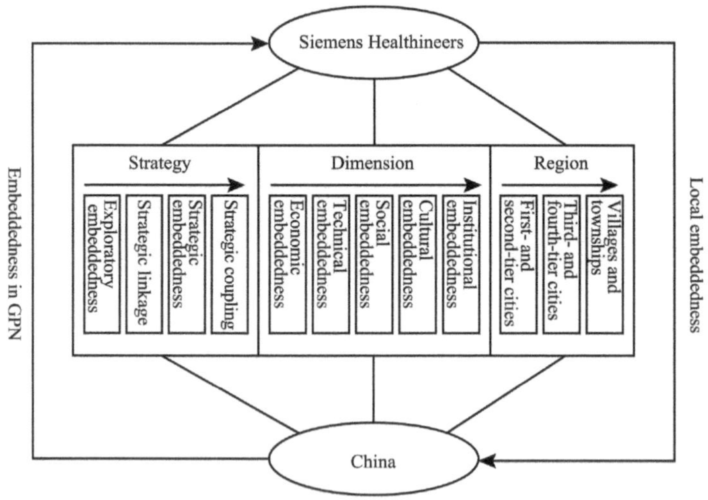

FIGURE 4.21 Analytical framework of Siemens healthcare sector's embeddedness in China. Source: *Prepared by the author.*

"deep embedding." In terms of region, Siemens Healthineers' embeddedness starts in the coastal areas of southeast China and then expands to central and western areas, spreading from first- and second-tier cities to third- and fourth-tier cities and even villages and townships. From the perspective of dimension, it begins with economic embeddedness and then expands to technical, social, cultural, and institutional embeddedness, developing from one-way to two-way embeddedness. As far as the division of labor is concerned, embeddedness starts with sales and services and then expands to local production, procurement, R&D, and brand management, spreading from sales and production activities to R&D centers, service centers, and regional operation headquarters. Local sales embeddedness begins with high-end market and expands to low- and mid-end markets. Local production embeddedness starts with labor-intensive activities and expands to capital- and technology-intensive activities. Local R&D embeddedness begins with low value-added activities and expands to high-value-added activities.

Fig. 4.22 shows the process of Siemens Healthineers' embeddedness in China and major events in this process.

4.3.2.1 Exploratory Embeddedness Phase (1982–91)

Siemens made an attempt to reenter China when the policy of reform and opening up was introduced. In the 1980s, China's healthcare reform was focused on promoting market-based operations and Siemens started to show interest in the Chinese market. In 1984, Siemens created sales channels in China mainly by exporting medical devices to the Chinese market through trade. In 1989, Siemens Technology Development Co., Ltd. (STDC) was established in Beijing to provide service support for Siemens's medical devices in China. During this period, Siemens Healthineers' investment in China was small and there was only one-way economic embeddedness. Operations focused on sale of imported products and provision of technical support. High-end products were sold only in a few big cities and big hospitals in coastal areas of southeast China. In the exploratory embeddedness phase, consumers and end users in China got to know about Siemens and its medical device products while Siemens also gained a better understanding of China's healthcare market, including opportunities, risks, and operations, laying a solid foundation for its embeddedness across the board.

4.3.2.2 Strategic Linkage Phase (1992–2000)

In 1992, Deng Xiaoping's speech in his second South China Tour brought the country's drive to open up and attract foreign investments to a new climax. Siemens's investment in China began to increase significantly. The State Council announced the beginning of healthcare reform in 1994 and issued the *Regulations on the Supervision and Administration of Medical*

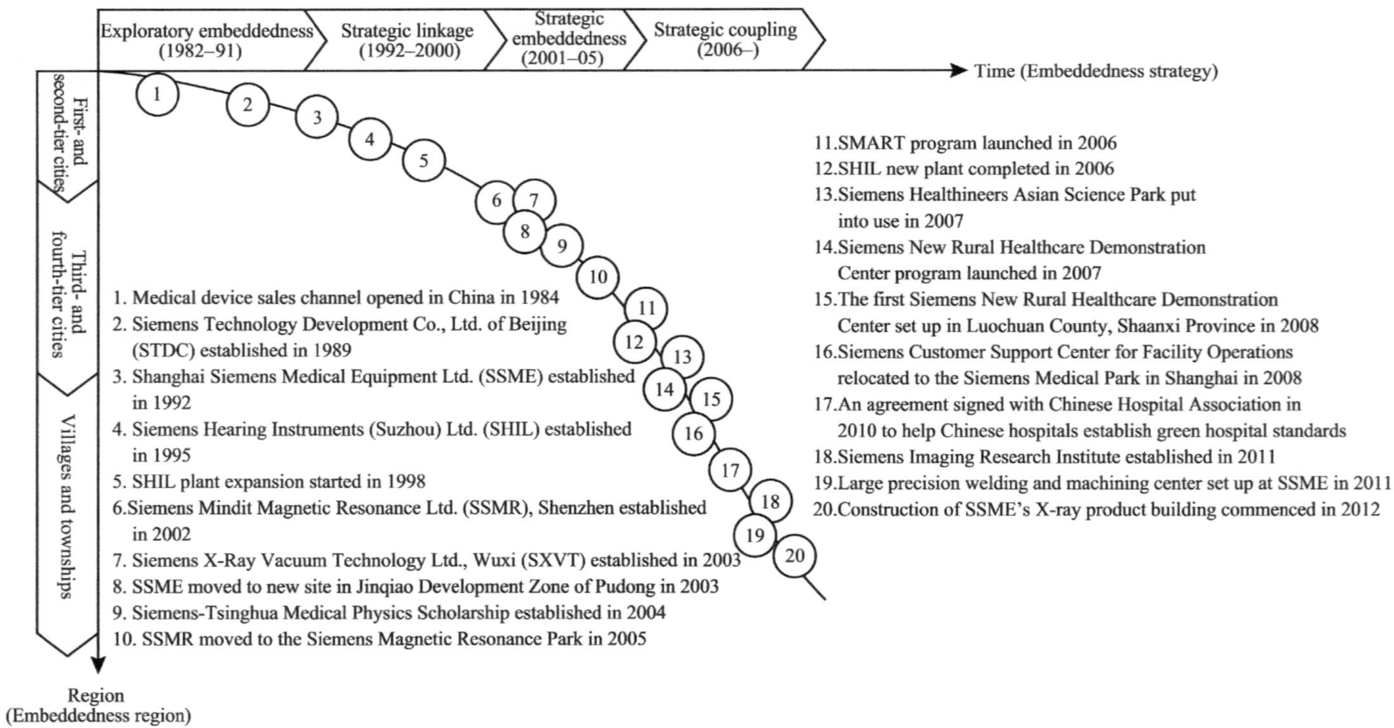

FIGURE 4.22 Siemens Healthineers' embeddedness process in China and major events. Source: *Prepared by the author according to related information about Siemens.*

Devices in 1998, marking the initial establishment of a sound medical device management system. Based on its knowledge of and experience in the Chinese market, Siemens Healthineers became aware of the value in this market and thus increased its investment in China for strategic considerations, officially entering the stage of embeddedness in China.

Shanghai Siemens Medical Equipment Ltd. (SSME) established in 1992 is the first CT scanner production base that Siemens established outside Germany. Siemens Hearing Instruments (Suzhou) Ltd. (SHIL) was established in 1995 as the first ever wholly foreign-funded enterprise brought into operation in the China–Singapore Suzhou Industrial Park. Plant expansion started in 1998. The company was relocated and a new plant opened in 2000.

In the strategic linkage phase, Siemens Healthineers began to invest more in China considering its market and low cost advantages. It expanded from coastal areas in southeast China to the first- and second-tier cities in central and western China but still focused on selling high-end products. In addition to selling imported products and providing technical support, Siemens Healthineers started to establishing joint ventures or wholly owned companies to manufacture medical devices and sell them locally. Its products were supplied in both domestic and international markets, the proportion of locally purchased components was increased through local production, and local supporting enterprises were developed. Through all these strategic linkage activities, the company obtained all the resources it needed to achieve the economies of scale and scope, reduce investment risks, and enhance competitiveness. In this phase, Siemens Healthineers' three lines of business went in parallel, including importing products and selling them in China, producing and selling products locally, and producing products locally and exporting them. It progressed from one-way to two-way economic embeddedness and then from economic to social, cultural, and institutional embeddedness but technical embeddedness remained on a low level. Except for economic embeddedness, the embeddedness in other dimensions is mainly one-way.

4.3.2.3 Strategic Embeddedness Phase (2001–05)

With China's accession to the WTO in 2001, Siemens paid more attention to the Chinese market and began to promote operations in the country across the board, seeking long-term development. China's healthcare system was gradually improved thanks to its accession to the WTO. The average tariff of medical devices declined from 11% to 5%–6% between 2000 and 2003 and the import of large-sized medical equipment no longer required the approval of related authority starting from 2003. The outbreak of SARS led to the government's encouragement of medical service providers' faster replacement of medical devices. The demand for high-end medical equipment generated thereof became a sustained driving force behind the development of the medical devices industry. In 2005, China became the third biggest medical device market, following the United States and Japan. Eight of the world's top ten medical device companies established production bases on China, completing their strategic configuration of operations in China. As a result, competition in the medical device market became increasingly fierce and low cost of manufacturing gradually lost its role as a competitive edge. With its strategic focus shifted from Asia-Pacific to China and growing investment in the country, Siemens Healthineers entered the strategic embeddedness phase in China.

In August 2001, Siemens produced the first SOMATOM Smile CT scanner, the world's most structurally compact and cost-effective CT scanner jointly developed by Chinese researchers and Siemens's R&D centers in Germany and the United States and targeted at both Chinese and international markets. It marked that Siemens's local R&D activities in China arrived at a new stage. In 2002, Siemens and Shenzhen Maidite Instrument Co., Ltd. established a joint venture—Siemens Maidite Shenzhen Magnetic Resonance Ltd., which is known as Siemens Shenzhen Magnetic Resonance Ltd. (SSMR) today. The company is the only R&D and manufacturing base of Siemens for MRI systems and components outside its headquarters in Germany. In 2005, Siemens Magnetic Resonance Park, located in the Shenzhen High-Tech Industrial Park, was completed and SSMR moved to the new site, which increased China's strategic importance in Siemens's global MRI business. In 2003, Siemens X-Ray Vacuum Technology Ltd. (SXVT) was established in Wuxi New Area to engage in the R&D, manufacturing, and sale of various X-ray tubes and assemblies. SXVT is the only manufacturer of X-ray tubes and assemblies for medical use under Siemens outside Germany. In 2003, Siemens Healthineers increased its investment in SSME and moved it to the new site in the Jinqiao Development Zone of Pudong to increase its capacity for developing and manufacturing X-ray and CT scan devices, X-ray units, and medical device components, thus achieving local manufacturing and global supply. In 2005, Siemens Hearing Instruments (Suzhou) Ltd. started its plant expansion project. The new plant was planned to be 3 times larger than the old one and one of the world's top hearing aid plants that would incorporate manufacturing, sales service, and technical support.

In the strategic embeddedness phase, Siemens focused more on the Chinese market's long-term potential and competitive human resources apart from the preferential policies and low labor costs in the country. On the basis of further expanding the market and scaling up local production, Siemens gradually included China in its global strategy and raised China's status in its global business. China was not only an important manufacturing base but also an important sales market for Siemens. It extended the scope of target customers from hospitals in coastal areas in southeast China to those in third- and fourth-tier cities in central and western China, offering customized solutions to hospitals at all levels. In this way, it boosted the sale of medical devices and achieved high-speed growth. In terms of dimension, the company started with two-way economic embeddedness, based on which technical, social, cultural, and institutional embeddedness gradually gained strength. In particular, local R&D and international cooperation were boosted significantly and its R&D capacity kept growing. Siemens wanted to growth together with Chinese enterprises and local communities so as to attain its goal of sustainable development and produce win–win outcomes for both the company and China.

4.3.2.4 Strategic Coupling Phase (2006–)

In 2006, the Chinese government officially launched the new healthcare reform. In 2007, the State Food and Drug Administration revised the *Regulations on the Supervision and Administration of Medical Devices* and the Ministry of Health issued the *Notice on Further Strengthening the Management of the Centralized Procurement of Medical Devices*. In the new healthcare plan introduced in 2009, one of the major tasks was to "improve the public health system, medical service system, medical support system, and drug supply system to build a basic healthcare system covering both urban and rural residents." The government also underscored basic healthcare in the 12th Five-Year Plan. In 2006, Siemens Healthineers entered the strategic coupling phase in China, which was in line with China's healthcare reform and Siemens's embeddedness strategy in China.

In 2006, Siemens announced its plan to invest another RMB 10 billion in China, most of which would be in its healthcare business. The new plant of Siemens Hearing Instruments (Suzhou) Ltd. was completed in 2006; its representative office in Xi'an was set up in 2007 and its R&D center was established in 2009. In 2007, the Siemens Healthineers Asian Science Park located in Shanghai International Medical Zone, involving a total investment of RMB 300 million, was put into use. The Northeast Asia headquarters and China headquarters of Siemens Healthineers are both located in the park and SSME moved to the park. In 2008, its UPTIME Service Center (USC) was relocated to the Siemens Medical Park in Shanghai. In 2010, Siemens Laboratory Diagnostics moved to the medical park in Shanghai. In 2011, a large precision welding and machining center was set up at SSME and Siemens Imaging Research Institute was established. In 2012, the construction of SSME's X-ray product building commenced and the building, once completed, would provide a floor area of 32,000 m^2 for R&D and manufacturing activities.

In 2011, the first year of the 12th Five-Year Plan period, Siemens signed a memorandum of understanding with the National Development and Reform Commission (NDRC) of China to renew its comprehensive cooperation, continuing the historic partnership established in 1985, and paving the way for both sides to achieve sustainable development in the future. Siemens entered a new era of all-round cooperation with China. During the 12th Five-Year Plan period, Siemens Healthineers continued to adjust its R&D, production, sales, and service strategies in response to China's policy orientation. In particular, it strived to boost business collaboration, strengthen R&D, and increase mergers and acquisitions efforts. While providing economically reliable and state-of-the-art technologies, Siemens also paid special attention to basic and regional healthcare, focused on promoting local-for-local solutions, launched the "Healthy China" program, and committed itself to developing products suitable for community and rural hospitals so as to provide people with affordable medical services.

In the strategic coupling phase, the collaborations and interactions of Siemens Healthineers with China's medical device industry have fostered a complete industrial ecosystem with three characteristics of strategic coupling. First, clustering of companies. Siemens and a large number of enterprises are spatially concentrated, and it maintains long-term, nonspecific partnerships with them, such as Nanhui International Medical Science Park in Shanghai. Second, high-end value chain. Siemens Healthineers has changed the role of China from simply a sales and manufacturing base to a regional R&D center, service center, and operations headquarters. Third, synergistic symbiosis. Siemens Healthineers has created in China R&D, manufacturing, supply chain, sales, and service networks featuring "global–local" interactions. Together with local enterprises, it has built an industrial value system for long-term cooperation, common development, and win–win results. The system has also reinforced Siemens's strategy in China, namely gaining a foothold in China, serving China, and facing the world.

Geographically, in the strategic coupling stage, Siemens's embeddedness gradually moves from cities to villages and townships. By dimension, the two-way economic, technical, social, cultural, and institutional embeddedness keeps deepening. Through strategic coupling, on the one hand, Siemens Healthineers has achieved ultrahigh-speed growth in China. From 2010 to 2012, it was ranked the most competitive medical device company in the field of radiology for 3 consecutive years. On the other hand, it has contributed to the rapid development of China's medical device industry and the fast growth in medical device exports. At present, about 70% of SSME's products are exported to foreign countries; 85% of SHIL's products are sold all over the world; about 90% of the Magnetom Essenza MRI systems produced by SSMR are exported to international markets. In addition, 80% of the components of Siemens's medical devices manufactured in China are from China, and the growing number of key components developed and manufactured in China serves as a perfect illustration of how deeply Siemens Healthineers is embedded in China.

4.3.3 Analysis of Siemens Healthineers' Embeddedness in China by Dimension

The process of Siemens's development from exploratory embeddedness to strategic linkage, strategic embeddedness, and strategic coupling is also one of transition from economic to technical, social, cultural, and institutional embeddedness and from one-way to two-way embeddedness.

4.3.3.1 Economic Embeddedness

Since the beginning of reform and opening up, Siemens Healthineers has been increasingly embedded in the Chinese economy. It has achieved two-way, in-depth economic embeddedness in China by building a complete value chain that covers all aspects including product management, R&D, procurement, production, and service delivery.

4.3.3.1.1 Scaling Up Investment in China

Siemens Healthineers' major investments in China have been explained in the section about its embedding process in China. Table 4.6 lists the major operating companies of Siemens Healthineers in China. Besides, Siemens acquired Bayer HealthCare's Diagnostics Division in 2006 and established Beijing Siemens Medical Diagnostics Equipment Ltd. in 2007. It also acquired Diagnostic Products Corporation (DPC) in 2006, including DPC's wholly owned subsidiary—DPC (Tianjin) Co., Ltd.

Siemens has set a new example to other multinational healthcare companies and plans to continue increasing investment in the future. The configuration of Siemens Healthineers' operations in China shows the importance of Shanghai, Beijing, Guangdong, and Jiangsu in China's healthcare industry and has driven the development of medical device industry in these regions and even across the country.

4.3.3.1.2 Extending the Scope of Business and Upgrading Investment Structure

Siemens Healthineers has been extending the scope of business in China, which can be seen in two aspects. One is the rich variety of its product lines. Currently, the products that Siemens Healthineers develops and manufactures in China include CT scanners, diagnostic X-ray units, ultrasonic devices, hearing instruments, MRI systems, X-ray tubes, as well as related components. The other is the increasing number of functional departments that make Siemens capable of providing customers with a full range of diagnostics products and solutions. Siemens Healthineers has expanded business from installation and assembling in the beginning to manufacturing center, R&D center, service center, and regional operations headquarters, with its

TABLE 4.6 Operating Companies of Siemens Healthineers in China

Operating Company	Year of Establishment	Location	Main Line of Business	Employees
Siemens Technology Development Co., Ltd., Beijing	1989	Beijing	Service and technical support	
Shanghai Siemens Medical Equipment Ltd.	1992	Shanghai	R&D, manufacturing and services concerning CT scanners, X-ray units and components	1000 +
Rexton (Suzhou) Hearing Systems Co., Ltd.	1995	Suzhou	R&D, manufacturing and services concerning hearing aids	600 +
Siemens Shenzhen Magnetic Resonance Ltd.	2002	Shenzhen	R&D, manufacturing and services concerning MRI systems	About 500
Siemens X-Ray Vacuum Technology Ltd.	2003	Wuxi	R&D and manufacturing of X-ray tubes and assemblies for medical use	100 +

Prepared by the authors according to related information about Siemens.

investment structure upgraded all the way. Its Northeast Asia Headquarters and China Headquarters are both located in the Siemens Medical Park in Shanghai that was put into use in 2007. The medical park features highly integrated medical devices and a service center and incorporates R&D, manufacturing, marketing, sales, and service functions, serving as an activity hub of Siemens Healthineers in China. In 2008, Siemens's USC was established in the medical park in Shanghai. Based on the facility operations monitoring center, technical support center, resource mobilization center, and customer experience center, the USC has fulfilled the vision of "prospective" customer service in China.

4.3.3.1.3 Deepening Local Economic Linkages

Siemens has seen its forward and backward economic embeddedness in China go deeper and deeper and achieved sustainable development by striving to promote localization. The former is embodied in the ever-improving sales network, service quality, and marketing channels, and especially the penetration of communities through the SMART program. The latter is reflected in the fact that the GVS program has helped the company to increase the share of procurement in China, develop local suppliers, and localize support services. At present, 80% of the components for Siemens Healthineers' medical devices are from China and the proportion of key components developed and manufactured by China has been on the rise. In fiscal year 2012, about 90% of Siemens's raw materials and semifinished products were procured in China. In comparison, the share of procurement in China remains to be increased for Siemens Healthineers.

In response to the ever-changing Chinese market, Siemens Healthineers has established sound sales network and service system in China. Now it has 33 representative offices and 60 service stations across the country to better meet users' inquiry, purchasing, and repair needs. To ensure timely service, Siemens Healthineers has especially set up a Spare Parts Center that operates around the clock and provide 80% of the key spare parts. It has also established an USC in Beijing and Shanghai to provide hospitals across China with 24-hour distance diagnostic services and introduced a 800 free after-sales service hotline to ensure timely trouble shooting for customers. The USC of Siemens took the lead in China to fulfill the vision of prospective customer service in 2008 and was the first to adopt a PDA mobile service system in 2009, further improving its customer service in China.

To bring quality and affordable medical products and services to more patients, Siemens Healthineers has been building channels and closer ties have been established with more than 165 distributors across the country. On the one hand, more quality products such as Magnetom Essenza MRI system and the latest floor-mounted cardiology system Artis Zee are put on the distribution catalogue to promote related sales. Siemens would help distributors improve efficiency so that more of its products can benefit patients. On the other hand, Siemens seeks to achieve greater synergies among product groups, encourages outstanding distributors to shift the focus from a single product group to multiple product groups, and works with distributors to bring more quality products to the market in a faster way.

4.3.3.1.4 Integrating China Business Into Global Production Network

Siemens Healthineers invests in China to serve the Chinese market. More importantly, it considers China an important node of its GPN, makes full use of China's competitive edges to configure and improve its value chain worldwide, and strives to lower the costs of production and operations to support its global strategy. Therefore, while Siemens Healthineers becomes increasingly embedded in the Chinese economy, it has also promoted its global business, boosted the imports and exports of China's healthcare industry, and helped China better integrate into global networks of the healthcare industry.

SSME is the medical imaging device manufacturing base of Siemens Healthineers that supplies products for the global market. It is also one of the largest high-tech companies engaging in the development and manufacturing of medical devices in Shanghai. About 70% of its products are exported to other countries. In 2010, it ranked among the top ten medical device manufacturers by exports and first among medical device companies exporting to the German market. In 2011, it ranked among the top five healthcare companies in Shanghai by total revenue. Of every two Siemens's CT products sold globally, one is manufactured by SSME. SHIL is the world's largest hearing aids manufacturing base that contributes to about one-fourth of the global output. It has an over 50% market share in China and 85% of its products are marketed worldwide. In 2010, it ranked among the top 20 exporters of hospital diagnostic and treatment equipment. SSMR, as an Asian MRI center, is an integral part of Siemens's global MRI strategy and about 90% of its Magnetom Essenza MRI systems are exported to international markets. In 2010, it ranked among the top 10 medical device manufacturers by exports and the top three medical device exporters in Shenzhen. In 2012, the 1000th Magnetom Essenza MRI system came off the production line in the Siemens MRI Park in Shenzhen and was sold to Japan, further underlining China's role as the strategic center of Siemens's global healthcare network.

Ever since Siemens Healthineers invested in the Suzhou Industrial Park in 1995, the world's top five hearing aids manufacturers that contribute to around 80% of the industry's output have established subsidiaries and manufacturing bases in China one after another. Three of them including Siemens have set up manufacturing bases in Suzhou. Swiss company Phonak established Unitron Hearing (Suzhou) Co., Ltd. in Suzhou as one of its largest manufacturing bases. Starkey established Starkey Hearing Technologies Co., Ltd. in Suzhou as its manufacturing and technical service center. Suzhou is therefore the hearing aids manufacturing center of China and even the world, which fully shows that Siemens Healthineers plays a part in driving regional industrial agglomeration.

4.3.3.2 Technical Embeddedness

Compared with economic embeddedness, Siemens Healthineers' technical embeddedness can better show how important it considers China business and its role in promoting the structural upgrading of China's healthcare industry. Just as a Siemens employee put it, "We may be more Chinese than

Chinese companies since many of them place their technical centers outside China but our original designs are right here."

4.3.3.2.1 Many R&D Facilities and High Spending on R&D

Siemens Healthineers considers the R&D investment in China very important. Its operating companies have established R&D institutions, many of which are their only or largest R&D facilities outside Germany. About one-third of their employees are R&D staff. R&D institutions and employees play an important role in Siemens Healthineers' global innovation network. SSME is the only CT R&D and manufacturing center of Siemens Healthineers outside Germany that has grown into a global R&D center for basic X-ray products. It is now the largest R&D and manufacturing center of Siemens Healthineers in the Asia-Pacific region. SHIL set an ITE laboratory in 1996 and a R&D center in 2009. SXVT is the only X-ray tube R&D and manufacturing of Siemens Healthineers outside Germany and it established a laboratory in 2010. SSMR is the largest MRI R&D and manufacturing base of Siemens Healthineers outside Germany.

4.3.3.2.2 Extensive International Cooperation and High Quality of R&D Work

As an important component of Siemens Healthineers' global innovation network, Chinese R&D employees play a strategic role, work closely with the company's global R&D team, fully leverage the global R&D network to develop innovation, and strive to meet the same high standards as those at headquarters to ensure Chinese R&D work with German quality and increase the Chinese team's R&D capacity. For example, SSME's CT software development team works closely with Siemens Healthineers' R&D center in Germany to develop software for advanced CT systems. SSMR collaborates closely with the Erlangen headquarters and Siemens Magnet Technology in Oxford to develop the best MRI products. In addition, CT plays a leading role in Siemens's innovation activities and offers new solutions regarding medical information systems and image processing. Siemens Corporate Technology China is a direct and strong independent innovation partner of Siemens Healthineers in China.

R&D localization has driven the improvement of Chinese team's R&D abilities. For example, SSME launched in the global market its brand new 6-slice and 16-slice CT scanners in 2007, the upgraded 6-slice and 16-slice CT scanners in 2010, and the new-generation 128-slice spiral CT scanners in November 2011. In the meantime, more and more key components are designed and manufactured by the company itself, such as various mechanical structural parts and control units for radiology, gastroenterography, and angiography; diagnostic beds for mid-end MRI and CT scanners; high-voltage X-ray generators; etc. SSME has grown rapidly into a central player in locally designing high-quality products and improving costs for Siemens.

4.3.3.2.3 Advancing the SMART Program and Moving From Being Local to Global in R&D

The SMART program launched in 2006 is the most forceful measure that Siemens Healthineers has introduced for technical embeddedness in China. Under this program, the Chinese R&D team focuses on developing low- and mid-end products suitable for China's basic healthcare market, and promoting SMART products to other emerging markets, helping Chinese products go global.

In 2006, Siemens officially announced its SMART program and it began with the healthcare business. In 2007, Siemens Healthineers produced the first CT scanner under the SMART program. The cost is reduced significantly since 80% of the components were procured in China. It is sold only at RMB 2 million to RMB 3 million, far lower than the price of a high-end four-dimensional spiral CT scanner. With good cost performance, this product is popular in the mid- and low-end market and it has the largest sales volume among all of Siemens's CT products. Now SMART products contribute to 20% of the total sales. Siemens expects the proportion to rise to around 40% and plans to launch SMART products in other fields and win more customers in second- and third-tier cities. Therefore, Siemens Corporate Technology China is more active in channeling human resources into the development of SMART products. Now 50% of the research projects are carried out under the SMART program while the other 50% are Siemens's traditional projects of high technology development.

Since the idea of SMART was introduced, Siemens has developed product series in the basic healthcare market, which are then launched in international markets as total solutions. For example, SOMATOM Emotion 16 CT scanner, a product developed, manufactured, and managed by the Chinese team, is now 70% exported to the global market. Another product Magnetom Essenza MRI system, also developed, manufactured, and managed by the Chinese team, is 90% exported to the global market.

4.3.3.2.4 Introducing Open Innovation to Promote Cooperation With Other Companies, Hospitals, Academic Groups, and Research Institutes

Through OI, Siemens aims to bring together the world's best people in science and business to make the innovation process more efficient, stronger, and more dynamic. Since the OI strategy is also implemented in China, Siemens Healthineers works closely with other companies, hospitals, academic groups, and research institutes in China. Siemens also plans to combine the essence of traditional Chinese medicine with Western science and

TABLE 4.7 Some Partners of Siemens Healthineers in China

Partner	Starting Year	Location	Collaboration
Fuwai Hospital	2004	Beijing	Established Fuwai Hospital–Siemens Global Heart Center
Chinese PLA General Hospital	2005	Beijing	Established a modern Molecular Imaging Center
Huashan Hospital	2005	Shanghai	Developed diagnostics technology that integrates in vivo and in vitro diagnostic solutions
The First Affiliated Hospital of Guangxi Medical University	2005	Nanning	Established medical image storage and transmission system training center
Haifu	2005	Chongqing	Developed MRI-guided high intensity focused ultrasound therapeutic systems
Dirui	2005	Changchun	Developed the CS1300 biochemical analyzer
Chinese medical imaging community	2011	Beijing	Established the Siemens Imaging Institute with an academic board of 36 members and an advisory board of 17 experts

Prepared by the authors according to related information about Huawei.

technology. Therefore, it is now working hard to develop new therapeutic methods together with its Chinese partners. Table 4.7 lists some of its partners in this regard.

4.3.3.3 Social Embeddedness

One of Siemens's core values is commitment to being an outstanding corporate citizen, which builds on the idea that the company as an employer, a client, and an outstanding corporate citizen constitutes an integral part of the Chinese economy and society. Therefore, throughout the long history of its development in China, Siemens has been committed to promoting the public good. Its corporate citizenship activities can be traced back to 1937 when John Rabe, Business Representative of Nanjing and Regional President for International Security, provided humanitarian aid for more than 250,000

Chinese refugees during the Nanjing Massacre, and used the premises of Siemens Nanjing to provide shelter for them.

Over the years, Siemens has made positive contributions to building a harmonious society in China. Its corporate citizenship activities focus on three areas, namely environmental protection, education on science and technology, and humanitarian aid. In 2012, Siemens officially established the Employee Volunteers' Association to provide a platform for employees, the company, and the general public to provide volunteer service and have cross-sector cooperation. It is worth noting that the association is intended to grow into an organization completely composed of and managed by employees.

4.3.3.3.1 Environmental Protection

As a socially responsible company Siemens works to satisfy commercial needs in an eco- and community-friendly way. It has been running the Green Building Program since 2007 to improve energy efficiency significantly at all of its production sites and offices. In September 2010, it launched a global supplier energy efficiency program to create a green value chain up to the highest sustainability standard. Siemens Healthineers has launched a "green + hospital" program comprising products and solutions in various fields of application, including energy production and allocation, building automation, IT and communications infrastructure, green IT, and medical technologies. This is to help hospitals meet ecological, economic, and social requirements for sustainable development. In 2010, Chinese Hospital Association (CHA) and Siemens signed a strategic cooperation agreement to promote the building of green hospitals.

4.3.3.3.2 Education on Science and Technology

Siemens supports preschool, primary, secondary, and tertiary education and helps to develop future scientists and engineers for China through various means, such as launching the I-Green Education Program and I-Green Dream Laboratory, signing strategic cooperation agreements with famous universities in China, establishing strategic CKI partnerships with Tsinghua University and Tongji University, setting up centers for postdoctoral studies. Siemens Healthineers also organizes many science and technology education activities. For example, in 2004, it signed an agreement with the International Exchange and Cooperation Center under the Ministry of Health on the creation of Siemens Medical Scholarship program to provide excellent young Chinese physicians with more training opportunities in other countries. In the same year, Siemens Healthineers signed an agreement with Tsinghua University on the establishment of Siemens–Tsinghua University Medical Physics Scholarship, the first of its kind in China. This collaboration was intended to improve Chinese medical physicists' theoretical and technical competence.

4.3.3.3.3 Humanitarian Aid

Siemens fulfills its social responsibilities in China and is committed to helping more people meet their basic needs, including basic healthcare, social assistance to vulnerable groups, and timely technical and humanitarian aid in the event of natural disasters. In the area of disaster relief, Siemens Healthineers provided rapid and effective assistance in many emergency cases. When the earthquake hit Sichuan in 2008, Siemens Healthineers, together with West China Hospital, Sichuan University, Huaxi Public Medical Information Service Company, and China Telecom Chengdu, established a remote imaging and diagnostic system on May 18 to achieve remote connection with hard-hit Mianzhu, a city 70 km away from Chengdu. It represents an innovation in the model of emergency rescue in the earthquake-stricken areas of Sichuan. Siemens also provides timely technical and humanitarian aid for other areas hit by natural disasters, such as Yushu of Qinghai Province hit by an earthquake and Jilin Province hit by a flood in 2010. Siemens Healthineers considers supporting China's healthcare industry and donating advanced medical devices an important contribution. For example, in 2005, it donated a Multimobil 2.5 X-ray machine to a hospital in Fuping County, Hebei Province. In 2006, it donated hearing aids worth more than RMB 10 million to China Foundation for Disabled Persons (CFDP), following the donation of a thousand hearing aids in 2002. In 2008, it donated a mobile intensive ultrafiltration water treatment system worth about RMB 1 million to the Red Cross Society of China Wuxi Branch. In 2011, it donated devices worth about RMB 1.9 million to the Jinggangshan Hospital of TCM and eighteen township hospitals in the Jinggangshan, and worked with Huashan Hospital, Fudan University to provide all-round training for the Jinggangshan Hospital of TCM.

4.3.3.4 Cultural Embeddedness

As a global company active in more than 140 countries and with about 400,000 employees, Siemens considers cultural diversity very important and underscores the necessity to allow diverse cultures and attract people from across the world. It carries out surveys among employees in 40 languages on a regular basis so as to improve the diversity management process. As an important part of Siemens's China business, Siemens Healthineers gets culturally embedded in this country in the same way as that adopted by Siemens AG as a whole, which can be seen in two aspects. On the one hand, Siemens Healthineers integrates the corporate culture of Siemens into its China business within the One Siemens framework, such as introducing employee stock ownership plan and annuity plan, publishing *Siemens World* for global employees and *Voice of Siemens* for Chinese employees, offering three career development paths (general manager, project manager, and technical expert), establishing Siemens Management School for training, providing overseas training, and job opportunities, etc. On the other hand, Siemens Healthineers fully recognizes and respects Chinese culture and considers itself a Chinese company while doing business in the country and promoting the integration of diversity and localization. Talent localization is a core part of Siemens's localization strategy. At present, more than 99% of Siemens employees in China are Chinese people. Asian faces make up the majority of Siemens's Northeast Asia management team. Therefore, Siemens has set a good example in talent localization.

4.3.3.5 Institutional Embeddedness

Siemens considers itself a Chinese company and strictly observes Chinese laws and regulations. After the 2006 bribery scandal, Siemens took all-round measures immediately and established a global compliance organization to ensure that operations at all levels are carried out ethically and lawfully. The business organizations across the country and the positions of regional general manager and provincial general manager stand as clear testimony to the company's institutional embeddedness in China. The institutional embeddedness of Siemens Healthineers is also reflected in the connections of its operation system with other social organizations and its influence on them, its impact on the institutional changes in China, its role in driving regional industrial agglomeration, etc.

Some advanced concepts and systems of Siemens Healthineers are taking root in China and will effect changes in institutions and business model in China's healthcare industry. Siemens has brought medical device finance leasing to China, and is the only qualified foreign company in this regard in China. Since the onset of the financial crisis, many countries have begun to pay more attention to more efficient equipment financing, such as leasing, since large sums of money cannot be used if medical devices are purchased in the traditional form of making full payment. In China, medical equipment leasing will emerge as a viable way of financing as the Chinese government has promised to increase spending on basic medical devices in the new healthcare reform. To make this a reality, the financial management and financing supply chain in the healthcare system must be reformed. Siemens Finance and Leasing Ltd. (SFLL) in China is a part of Siemens Financial Services (SFS) based in Munich, Germany, and has opened 18 offices in China to provide customized, flexible and long-term financing solutions for local customers in China. The Affiliated Hospital of Jining Medical University was the first medical institution to use Siemens's financial leasing services. To date, the hospital has procured equipment worth a total of RMB 282 million. Now 20% of Siemens Healthineers' operations in China are driven by financial leasing. According to Siemens, the ratio of leasing penetration in the Chinese market to the total spending on medical devices will rise rapidly, and is expected to continue growing during the second decade of the 21st century, effectively easing the sunk cost burden of China's healthcare system.

4.4 SUMMARY

In the context of economic globalization and intensifying market competition since the 1990s, Siemens has to formulate a corresponding global strategy and business sectors. Aiming to achieve sustainable value creation, Siemens operates within the framework of One Siemens and mobilizes global resources and technical personnel within and outside the company to build global R&D, manufacturing, supply chain, and sales and service networks so as to configure operations across regions and industries and globally integrate value-creating activities. This helps the company to expand market opportunities, lower operational costs, increase innovation capacity, and raise the level of customer satisfaction, thus enhancing the company's core competencies and helping it establish a sustainable competitive edge in a global sense, to achieve sustainable, profitable growth and continually increase company value.

Based on the framework of microcosmic GPN studies proposed in Chapter 3, Analytical Framework of Microcosmic GPN Studies, this chapter examined Siemens's global networks on the whole and its global R&D, manufacturing, supply chain, and sales and service networks from the value perspective in terms of value objectives, global strategy, networks, location selection, and network governance and assesses the value added of all the networks. It also explains the GPN of medical devices and the trade structure of China's medical device industry from the perspective of embeddedness, and describes how Siemens Healthineers gets embedded in China on the levels of strategy, region, and dimension. The case study of Siemens helps to better understand the micro-GPN framework and explain the drivers and benefits of GPNs as well as how MNCs operate based on GPNs on the microlevel, and figure out how MNCs get embedded in China and the pattern in the evolution of the process, thus contributing to development of both China's economy and MNCs.

Chapter 5

A Study of Huawei's GPN

Chapter Outline

Since the Chinese government adopted the "going global" strategy in 2001, quite a number of Chinese enterprises have been out on the global market. China's overseas investments have seen a rapid growth in the past decade or so. According to the Ministry of Commerce of China, Chinese foreign investments have increased for 13 consecutive years. China is now among the top three in the world in terms of foreign investment flow. China's foreign investments reached up to USD 140 billion in 2014, USD 20 billion more than its use of foreign capital, indicating China's turn from an economic and trade power to an economic and trade superpower with the positive net capital outflow as a new normal. The 2008 Global Financial Crisis hit hard. The global economy is yet to be rebalanced, slow in recovery in the crisis' aftermath, and riddled with uncertainties. Chinese manufacturing enterprises must be facilitated with a stronger competitive edge in going global. Despite increasing advantages in foreign investments and cooperation, Chinese manufacturing enterprises are yet to be perfect in capital operation, management, brand operation, risk control, talents for multinational

Global Value Chains and Production Networks. DOI: https://doi.org/10.1016/B978-0-12-814847-1.00005-1

business management, and good operation practice. In addition, Chinese manufacturing enterprises have been heavily restrained by market uncertainties and protectionist policies of other countries. Thus, the "going global" strategy has seen on the whole unsatisfactory effects in Chinese manufacturing enterprises.

Huawei Investment & Holding Co., Ltd. (Huawei) has been successfully globalized in the "going global" tides. In the past two decades or so, riding on the tides of reform and opening up and of fast development in information and communications technology (ICT), Huawei has been staying customer-centered and employee-oriented and won respect and trust from its customers with ongoing customer-tailored innovations. Initially a private enterprise with a startup capital of RMB 21,000, Huawei has solidly grown into a Fortune 500 company boasting annual sales of around RMB 240 billion. Its sales revenue surpassed that of Ericsson in 2013. In 2014, Huawei saw a stronger leading edge, surpassing Ericsson in carrier business for the first time and becoming the industry leader in every sense. Huawei has set an example of independent innovation and global development of Chinese companies by integrating Western management science with Chinese ethos. It has something of significance for "going global" Chinese enterprises.

Based on the framework of microcosmic global production network (GPN) studies proposed in Chapter 3, Analytical Framework of Microcosmic GPN Studies, this chapter analyzes, from the perspectives of value, value objectives, global strategy, networks, location selection and governance model of Huawei and its global research and development (R&D), production and operation, and marketing service networks; and details, from the perspective of embeddedness, Huawei's embeddedness in Europe by economic, technical, social, cultural, and institutional dimensions. The chapter makes a comparison between GPNs of Siemens and Huawei and summarizes Huawei's experience, hoping to help more Chinese enterprises take a foothold in reality and rise in globalization with stronger international competitiveness by getting embedded in GPNs.

5.1 OVERVIEW OF HUAWEI'S GPN

5.1.1 About Huawei

Huawei is a leading global provider of information and communication technology (ICT) infrastructure and smart devices. With integrated solutions across four key domains—telecom networks, IT, smart devices, and cloud services—Huawei is committed to bringing digital to every person, home, and organization for a fully connected, intelligent world. Huawei's end-to-end portfolio of products, solutions, and services are both competitive and secure. Through open collaboration with ecosystem partners, Huawei creates lasting value for the customers, working to empower people, enrich home life, and inspire innovation in organizations of all shapes and sizes. According to the data published in Huawei's 2017 Annual Report, its annual revenue in 2017 totaled CNY 603,621 million (US$92,549 million based on the year-end exchange rate), and the net profit grew to CNY47,455 million (Huawei, 2017).

5.1.2 Global Expansion and Transnationality Index

Huawei is a successful example of "going global" Chinese enterprises. Established in Shenzhen in 1987, with 30 years' development, Huawei is now a globalized multinational company in every sense. At present, Huawei works in more than 170 countries and regions, serving over one-third of the world's population. Among 180,000 employees, there are more than 160 different nationalities with almost 70% localization rate. The revenue outside China reached up to RMB 298,529 million in 2017. The overseas market contributed approximately 50% of the company's revenue. Huawei has made explorations in its globalization, basically in three stages, namely "going global," internationalization, and globalization.

5.1.2.1 "Going Global": 1996–2003

Huawei adopted the "going global" strategy in the 1990s when the domestic market was sluggish, "to live on." Following the strategy of "entering from the remote outside" it adopted in the domestic market, Huawei bypassed restrictions set in various standards in developed countries and first entered big developing countries with its price advantage. And from there, the company expanded little by little. Huawei was in cooperation with Hutchison Telecommunications, the second largest telecom carrier in Hong Kong, in 1996, and entered the market of the Commonwealth of Independent States (CIS) in the same year. In 1997, the company established a joint venture in Russia and entered the Latin American market. It went to India and Sub-Saharan Africa in 1998. However, Huawei's overseas performance was not that satisfactory during that period.

Bubbles in the IT industry burst in 2000. Equipment suppliers began to shorten their fronts and withdraw from some fringe markets. But carriers' demand for equipment remained. Huawei then took the opportunity and gained a foothold on international markets. It entered the Middle East and Africa in 2000 and saw USD 100 million generated from international markets. Huawei made a fast expansion in 2001 into over 40 countries and regions in Southeast Asia and Europe. Its international market sales reached USD 552 million in 2002. In the meantime, Huawei established R&D centers in Bangalore, India (1999), Stockholm, Sweden (2000), and the United States (2001), to enhance the company's core competence by making full use of regional superior resources around the world. In 2003, 4000 employees were engaged in the company's international system. Its international market sales surpassed USD 1.05 billion. During that period, Huawei worked hard on cooperating with telecom carriers in developing countries and found its international markets largely in Southeast Asia, Russia and Eastern Europe, Latin America, and Africa. Nonetheless, the company basically remained outside of developed countries in Europe and America where the level of technology and threshold were rather high.

5.1.2.2 Internationalization: 2004–08

Huawei saw both opportunities and challenges in 2004. Even though Cisco dropped its suit against Huawei, the latter still lost its market in the United States for a while. However, Huawei consequently became known more in Europe and the United States. Meanwhile, Huawei found it a must to engage in international cooperation and began to focus on the European market. With cooperation with reputed local agents, Huawei entered developed countries and regions, including Germany, France, Spain, and Britain. Its operation got fully internationalized. In 2008, Huawei saw 75% of its revenue from overseas markets and the year's contract sales reached USD 23.3 billion.

Huawei made the breakthrough on the European market by acquiring British Telecom's (BT) certification. From then on, the company has followed the international regulations and standards on international markets and grown into a major supplier on the international telecom market. At the end of 2003, Huawei and Siemens sealed a cooperation agreement, with which Siemens would increase year by year the sales volume of Huawei's data communication products on the European market. In 2004, Huawei achieved its first significant contract win in Europe valued at over USD 25 million with Dutch carrier, Telfort. Huawei signed in 2005 with Vodafone a Global Framework Agreement, as Vodafone's preferred telecom equipment supplier. In the same year, Huawei was selected as a preferred 21 Century Network (21CN) supplier by British Telecom and saw its international contract orders exceed domestic sales for the first time. In 2006, Huawei signed with Vodafone a strategic cooperation agreement for 3G handsets and established a Shanghai-based joint R&D center with Motorola. In 2007, Huawei established a joint venture with Symantec and another with Global Marine, and its core communication equipment was adopted by all the major carriers in Europe. At the end of 2007, Huawei became a partner to all the top carriers in Europe.

Nonetheless, Huawei experienced setbacks in North America. Entering the United States as early as in 2002, Huawei had 2.5% of its contract sales from North America in 2005. But the figure kept declining to less than 1% in the next few years. Huawei never gave in. In early 2007, Huawei had its first cooperation with the US mobile carrier, Leap Wireless, deploying CDMA 3G network for the latter. In July 2007, Huawei won the CDMA2000 network contract from Leap Wireless. In 2008, Huawei sealed the contracts with TELUS and Bell Canada for HSPA deployment. Huawei was selected in 2009 by the US telecom carrier, CoxCommunications as the provider of end-to-end

CDMA mobile network solutions. In the same year, Clearwire of the United States, the world's largest WiMax carrier selected Huawei as one of its equipment suppliers. Despite its ongoing efforts on expansion in North America, Huawei met one barrier after another there. In 2008, Bain Capital and Huawei submit an application for a buyout of 3Com. But the deal collapsed due to being unable to come to agreement with the Committee on Foreign Investment in the United States (CFIUS). Huawei bought certain 3Leaf assets for USD 2 million in May 2010. But with the CFIUS's disapproval, Huawei dropped the takeover bid in 2011. In August 2010, Republican senators showed concern over Huawei-Sprint telecom equipment deal, stating that the deal may threaten the US national security, as Huawei had deals with Iranian and Chinese armies. Thus, Huawei's bid hit a hurdle. Huawei bid for Motorola's wireless business and failed in 2010, despite an offer USD 1.2 billion higher than its rival's. The United States was the last and most difficult market for Huawei's international expansion strategy.

5.1.2.3 Globalization: From 2009 Onward

In 2009, Huawei aimed at optimized regional organizations, operational coordination, global approach, well-organized and coordinated implementation, and flexible deployment of its global resources, to serve customers better. The international telecom manufacturing saw a slide generally due to the financial crisis. But Huawei bucked up, ranking No. 2 in global market share of radio access equipment and successfully delivering the world's first LTE/EPC commercial network, with the largest number of LTE commercial contracts. In 2010, Huawei deployed worldwide over 80 SingleRAN networks and established its Cyber Security Evaluation Center in Britain

The ICT industry went to a new point in 2011 at the moment of which Huawei needed a transformation. It aimed at an accelerated pace of globalization, a global management structure, convergence of global superior resources, and a global business layout, to build a solid foundation for growth in globalization. Huawei also changed the single platform operation into multiservice centers (carrier networks, enterprise service, and consumer service), and grew from a telecom equipment manufacturer into a world-class provider of information and communications solutions, to build itself into a globalized Chinese company in every sense. In 2012, the company continuously promoted its globalized operations. It stepped up investments in Europe, particularly in Britain, and established a new R&D center in Finland. Also in that year, Huawei set up local boards of directors (BoDs) and advisory boards in France and Britain. In 2013, Huawei set up the Financial Risk Control Center (FRCC) in London to manage global financial risks and ensure that its financial operations remain efficient, secure, and standard-compliant. Huawei's European Logistics Center was put into official operation in Hungary, covering countries throughout Europe, Central Asia, the Middle East, and Africa. During the same period, Huawei made small M&As. In 2011, Huawei acquired Symantec's shares in Huawei Symantec at USD 530 million, holding 49% of Huawei Symantec's shares, to consolidate its enterprise business. In addition, Huawei made M&As in Europe, Israel, and the Middle East.

Huawei has made full use of its global resources in a globalized layout, to fully support the company's globalized operation. With its competence and cost-efficiency superiorities in finance, service, consulting, and human resources in different regions around the globe, Huawei has set up over 40 competence centers and more than 30 shared service centers. It has seen an enhanced integrated capacity thanks to its localized operation and global partnership. Also, local competence has been introduced to other places around the world along Huawei's global value chain and created more values. In this process, job opportunities have been created for locals and local industrial development has been substantially boosted. Huawei has worked on building HR centers in its major localities around the world and run local business with local advantages. Over 70% of its employees working outside the Chinese mainland are local people. Huawei is thus a globalized company with a globalized staff. Huawei today is an integral part of the global economy.

In 2011, China Enterprise Confederation (CEC) and China Enterprise Directors Association (CEDA), following the common practice of international organizations, launched for the first time the Top 100 Multinational Corporations of China and Transnationality Index (TNI), ranked by their foreign assets. The TNI is calculated as the arithmetic mean of the following three ratios:

- the ratio of foreign sales to total sales (a);
- the ratio of foreign assets to total assets (b); and
- the ratio of foreign employment to total employment (c).

The formula is written as: $(a + b + c)/3 \times 100\%$. According to Huawei's data published in 2011, the company had as of 2010 foreign assets worth RMB 66.67 billion, foreign sales of RMB 120.4 billion and 21,700 foreign employees, with a TNI of 42.08%. Huawei thus ranked 10th by foreign assets and sixth by the TNI. Though much higher than the average TNI for top 100 multinational corporations of China at 13.37% in 2011, Huawei's TNI was only slightly higher than the average TNI for top 100 multinational corporations of developing countries at 40.13% in that year, and much lower than the average TNI for the world's top 100 multinational corporations at 60.78% of the year.

5.1.3 Value Propositions and Global Strategy

For the past 30 years Huawei has maintained an unwavering focus, rejecting shortcuts and easy opportunities that do not align with its core business.

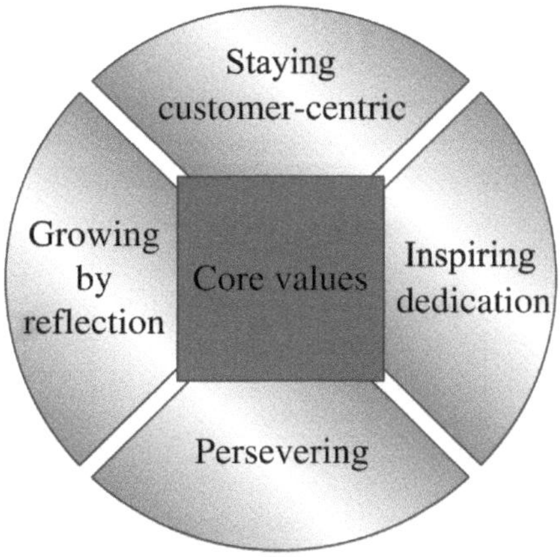

FIGURE 5.1 Huawei's core values. Source*: The authors made it based on related materials on Huawei website (https://www.huawei.com/en/).*

With a practical approach to everything it does, Huawei concentrates its efforts and invests patiently to drive technological breakthroughs. This strategic focus is a reflection of Huawei's core values (Fig. 5.1): staying customer-centric, inspiring dedication, persevering, and growing by reflection.

Huawei's vision and mission is to bring digital to every person, home, and organization for a fully connected, intelligent world. The digital world is decentralized and virtual, whereas the intelligent world is cohesive and based in a physical space. Huawei serves as a bridge between the two worlds, actively cultivating a fertile environment where all things can be connected. By giving every person, home, and organization access to an immersive virtual space, Huawei is building something greater than before—a new, intelligent world. Huawei's strategy is to enable digital transformation with ICT infrastructure and intelligent devices (Fig. 5.2) focused on building ubiquitous connectivity, create a better experience with broadband, develop open and trusted cloud platforms, and grow an experience-centric device ecosystem.

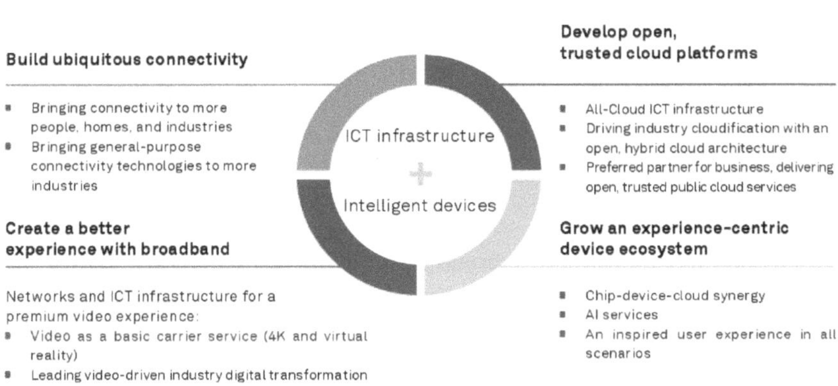

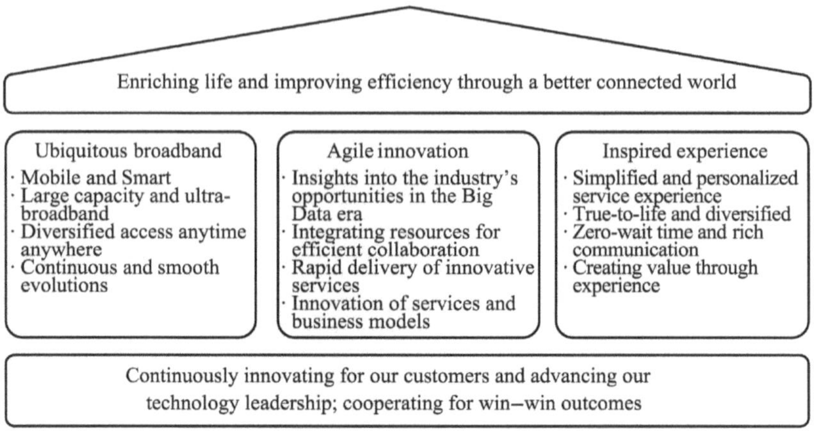

FIGURE 5.2 Huawei's vision, mission, and strategy. Source*: Huawei, Huawei Investment & Holding Co., Ltd. 2017 Annual Report (https://www.huawei.com/en/).*

In the context of globalization, Huawei believes that open, cooperative, and free trade policies are guarantees for improving the competitiveness of the ICT industry and promoting the development of the digital economy. With a global view in mind, Huawei integrates the best resources with comparative advantages and manages these resources consistently across the globe. The company is working hard to ensure a global industry chain grounded in win–win approaches and reciprocal obligations to help build a sound business ecosystem for the long-term development of the industry. As a responsible corporate citizen, Huawei always contributes significantly to the local markets in which it operates. Huawei has grown into a globalized company by getting globalized and localized, following international and market rules, and adhering to openness, collaboration, and innovation (Fig. 5.3).

1. *Globalization and localization*. Huawei adopts a "glocalization" strategy: a global perspective of the trends, opportunities and risks in the industry and of the company's strategic deployment; a global layout for the optimal resource integration and enhanced efficiency and capacity; a global management system of recognized Western concepts, models, and processes for the company's global strategy; a local operation supported by local talents; incentives to a local business team to be in full charge of local business, with a proper authorized power regardless of nationality; an optimized operation and management system for overseas subsidiaries to guarantee compliance operations; and an industry ecosystem builder and corporate citizen that substantially contributes to the local society.
2. *Following international and market rules*. On its way to being a globalized company, Huawei has seen the significance in following international rules and maintaining market orders. The company hires Chinese legal experts. In addition, all the representative offices around the globe

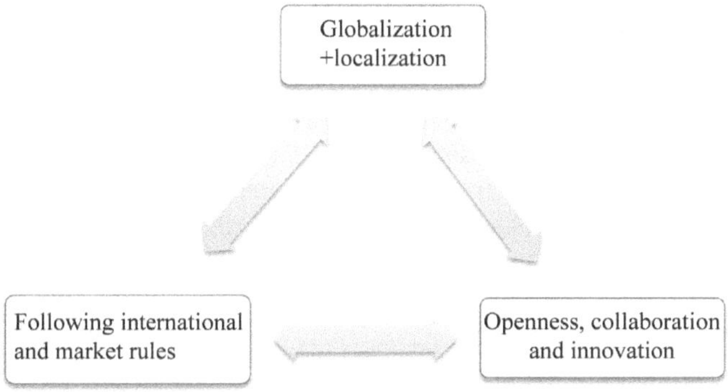

FIGURE 5.3 Huawei's path to globalization. Source*: The authors made it based on related materials on Huawei website (http://www.huawei.com/cn/).*

invite local legal experts to join them as senior employees of the Legal Affairs Department. Huawei also seeks legal consultancy from local senior experts. Chinese enterprises have been perplexed with intellectual property right (IPR) issues in the process of "going global." Huawei is no exception. The company now gives the first priority to IPRs in its innovation, inputting enormously in patent R&D and paying IPR fees. Huawei costs USD 300 million on international patent fees. Moreover, Huawei is actively engaged in negotiations for international trade frictions. It believes that better communication and collaboration is the best way to solve problems on international markets and guarantee long-term benefits and shared success.

3. *Openness, collaboration, and innovation.* Huawei stays customer-centered and adheres to openness, collaboration, and innovation in a globalized context. It has coestablished with its customers over 30 joint innovation centers around the globe. Customer-based research institutes make quick response to customers' needs and satisfy those needs. Huawei is dedicated to customers' success by creating long-term value for customers. In this way, Huawei makes its own success.

5.1.4 Networks and Governance Model

Huawei's GPN features product development-based (breakthroughs at "points") integrated product development (IPD) (readiness on "lines"), modular value chains (expansion in "areas") and world-oriented value networks (operations in "networks") (Li et al., 2010). Huawei's learning decision model with sales of proxy switches gradually developed into a leading decision model that focused on the production of hi-tech flagship products. With IBM's IPD, Huawei turned from a technology-driven company to a market-driven one and its R&D began to keep up with international advanced manufacturers. With the IPD, Huawei has built a complete value chain and created, on the basis of market management, business process reengineering and product management, marketing service, production and operation, and R&D modules. The company globally outsources modules with lower added values and focuses on improving the added values of core modules, so as to stand on the upper part of the value chain. Based on its specific needs in modules, Huawei seeks partners and establishes strategic partnerships with Chinese and foreign scientific research institutes, universities and colleges, high-tech companies, multinational corporations, suppliers, distributors, and customers by way of alliance-building, substantial profit release, product compatibility, capital fusion, and stock-holding. In addition, the company has kept close contact with ministries and local governments. In this way, Huawei has built a global value chain of major interested parties (Fig. 5.4), by making full use of internal and external resources.

By staying customer-centric and inspiring dedication, Huawei has sustained long-term growth through continuous improvement of our corporate governance structure (Fig. 5.5), organizations, processes, and appraisal systems.

The Shareholders' Meeting is the company's authoritative body, making decisions on major issues such as company's capital increase, profit distribution, and selection of the members of the Board of Directors/Supervisory Board.

The Board of Directors (BOD) is the highest body responsible for corporate strategy, operations management, and customer satisfaction. The BOD's mission is to lead the company forward. It exercises decision-making authority for corporate strategy and operations management, and ensures the protection of customer and shareholder interests.

The BOD and its Executive Committee is led by rotating chairmen, who serve as the foremost leader of the company during their respective terms.

The key responsibilities of the Supervisory Board include overseeing the responsibility fulfillment of BOD members and senior management,

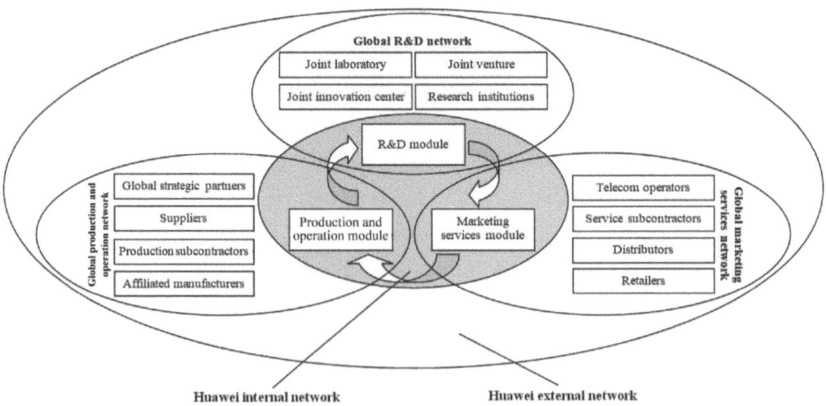

FIGURE 5.4 Huawei's global networks. Source*: The authors made it based on related materials on Huawei website (http://www.huawei.com/cn/).*

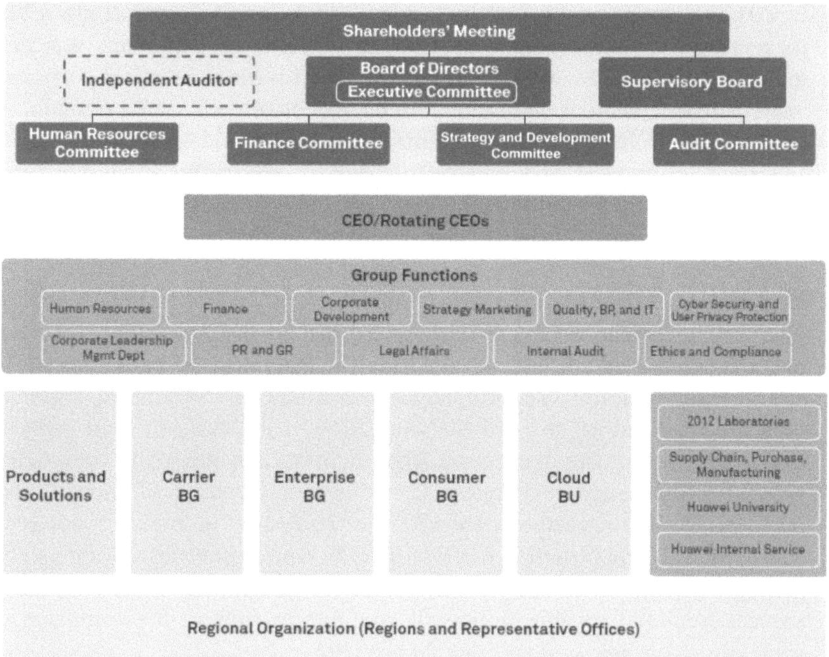

FIGURE 5.5 Huawei corporate governance structure. Source: *Huawei (https://www.huawei.com/en).*

monitoring the company's operational and financial status, and supervising internal control and legal compliance.

KPMG has been Huawei's independent auditor since 2000 and is responsible for auditing the company's annual financial statements. In accordance with applicable accounting standards and audit procedures, this independent auditor expresses an opinion as to whether the financial statements are true and fair.

The company has established a business structure that focuses on three dimensions: customers, products, and regions. All organizations jointly create value for customers, and are responsible for the company's financial results, market competitiveness, and customer satisfaction.

- The Carrier BG and the Enterprise BG manage and support solution marketing, sales, and services that target carrier customers and enterprise/industry customers, respectively. The two BGs provide innovative, differentiated, and advanced solutions based on the business characteristics and operational patterns of different customers while continuously improving the company's industry competitiveness and customer satisfaction. The Consumer BG focuses on serving device consumers and deals with all the aspects of consumer domain. This BG is responsible for business performance, risk controls, market competitiveness, and customer satisfaction in the consumer business.

- In 2017, Huawei established the Cloud BU, a business unit responsible for end-to-end management of Huawei's offers to the cloud services industry. It is tasked with enhancing competitiveness in cloud services, and ensuring customer satisfaction and business success in this domain.
- Products & Solutions is an organization that provides integrated ICT solutions to carriers and enterprise/industry customers. In addition to product planning, development, and delivery, this organization is also responsible for developing product competitiveness in order to deliver a better user experience and to support the company's business success.
- Regional organizations are the company's regional operations centers. They are responsible for developing and effectively leveraging regional resources and capabilities, and also for the execution of corporate strategy in their regions. The company has continuously optimized regional organizations and accelerated the delegation of authority to field offices. Command and on-site decision-making authority has gradually been delegated to representative offices. Currently, to improve efficiency and responsiveness to customer needs, the company is piloting contract approval at the representative office level in some countries. While establishing closer partnerships with customers to help them achieve business success, regional organizations will continue to support the company in achieving profitable and sustainable growth.
- Group Functions provide business support, services, and supervision. They are positioned to offer accurate, timely, and effective services to field offices and strengthen supervision while delegating sufficient authority to them.

Huawei's global management system enables it to promote its corporate culture company-wide and effectively manage its business. The aims are to:

- Remain customer-centric; continue to innovate in order to address customer needs and maintain Huawei's technological leadership; and build an ecosystem for shared success;
- Control risks and ensure business continuity; and
- Pursue corporate social responsibility (CSR) initiatives to promote sustainable social development.

Huawei has successfully passed comprehensive audits, regular assessments, and stringent reviews conducted by 31 of the world's top 50 carriers as well as by major enterprise and industry customers. Audited domains include financial robustness, quality management, risk management, delivery and services, supply chain management, knowledge management, project management, cyber security, information security, EHS, CSR, sustainability, and business continuity management. Huawei received full and extensive recognition from its customers in these key domains, as evidenced by their choice of Huawei as a strategic partner for their future-oriented transformations.

Over the last year, Huawei added a corporate vision and mission management module to Develop Strategy to Execute (DSTE) management system. The aim is to use this vision to help guide effective strategy management of the company and its business units, and further strengthen strategic communication inside and outside of Huawei. These strategies serve as drivers for annual business planning, budgeting, and performance appraisals. This management system ensures that the strategic objectives of the company and business units are incorporated into annual plans and budgets, so that all business units are well coordinated. With well-managed corporate investment, the company has been able to achieve its strategic and business objectives.

Huawei's management transformation focused on simple, efficient, and secure transactions with customers, as well as improvement of internal efficiency and effectiveness. Huawei has started to explore how to treat each of its 100-plus representative offices as a smaller form of Huawei and build process management systems for representative offices based on real-world scenarios.

5.1.5 Global Presence and Performance

According to Huawei's 2017 Annual Report, the company ran its global businesses in four regions, namely China, Americas, Asia Pacific, and Europe/Middle East/Africa (EMEA). Americas Region covers the United States and Canada in North America; and Argentina, Bolivia and Brazil, Mesoamerica and the Caribbean, Chile, Colombia, Costa Rica, Ecuador, Guatemala, Mexico, Peru, Paraguay, Uruguay, and Venezuela in Latin America. Asia Pacific Region (excluding the Chinese mainland) includes Australia, New Zealand, Bangladesh, Cambodia, Hong Kong of China, Taiwan of China, India, Indonesia, Japan, Kazakhstan, South Korea, Malaysia, Myanmar, the Philippines, Singapore, Sri Lanka, Thailand, Turkey, Uzbekistan, and Vietnam. EMEA Region covers Austria, Belarus, Czech, Finland, France, Germany, Greece, Italy, the Netherlands, Norway, Poland, Portugal, Romania, Russia, Spain, Sweden, and Britain in Europe; Qatar, Saudi Arabia, and the UAE in the Middle East; and Algeria, Cameroon, the Congo, Egypt, Ghana, Kenya, Morocco, Nigeria, South Africa, Sudan, Tanzania, Uganda, Zambia, and Zimbabwe in Africa.

In 2017, Huawei's well-balanced, worldwide presence helped the company achieve stable and healthy growth in China, Asia Pacific, and EMEA (Fig. 5.6). The annual revenue was CNY603,621 million, up 15.7% year-on-year. These are listed as follows:

- Based on ongoing 4G network rollout, sustained growth in the smartphone sector, and our growing capabilities in enterprise and industry

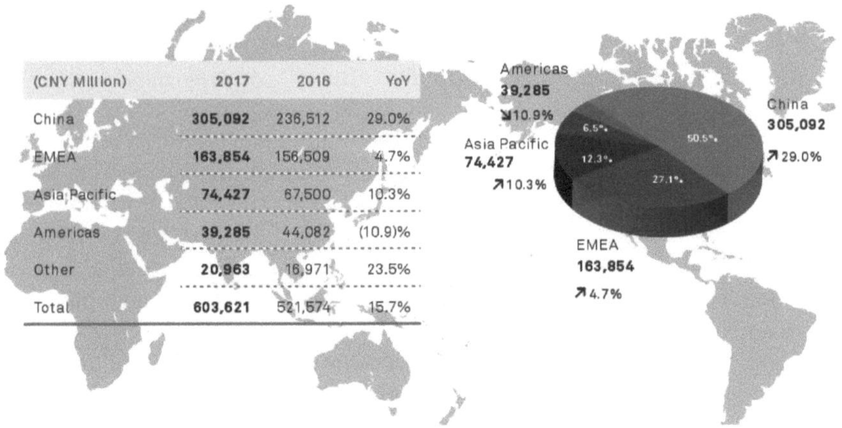

FIGURE 5.6 Huawei's global performance, 2017. Source: *Huawei, Huawei Investment & Holding Co., Ltd. 2017 Annual Report (https://www.huawei.com/en/).*

solutions, Huawei earned CNY305,092 million (about 50.5% of total) in revenue from the Chinese market, up 29.0% year-on-year.

- Due to the accelerated pace of digital transformation amongst enterprises and a growing share of the smartphone market, Huawei earned CNY163,854 million (about 27.1% of total) in revenue from Europe, the Middle East, and Africa (EMEA), up 4.7% year-on-year.
- Due to the accelerated pace of digital transformation amongst enterprises and a growing share of the smartphone market, Huawei maintained its growth momentum in the Asia-Pacific Region and achieved CNY74,427 million (about 12.3% of total) in revenue, up 10.3% year-on-year.
- In the United States, affected by fluctuations in telco investment cycles in Latin America, Huawei's revenue from this region decreased to CNY39,285 million (about 6.5% of total), down 10.9% over 2016.

5.2 HUAWEI'S GLOBAL PRODUCTION NETWORK FROM THE PERSPECTIVE OF VALUE

Huawei basically relied on the "wolf blood" for its development before 1998; that is, it grasped market needs with an acute sense and quickly launched products. But the excessively subjective way ignored market changes and technological developments of its rivals. Huawei was engaged in R&D "with eyes shut." The company paid a terrible price for this. Pricked by the enormous loss resulted from errors, Mr. Ren Zhengfei decided to introduce IBM's integrated production development (IPD) R&D process. The market-oriented IPD stresses product innovation based upon market needs and competition analysis and effectively manages a product from project analysis through to marketing. With IPD, Huawei has established a complete set of R&D, production and operations, and marketing service modules, which have supported the operations of the three BGs for carrier networks, enterprise service and customers and the businesses in China, Americas, Asia Pacific, and EMEA. SBGs and Group Functions provide support, services, and supervision for BGs. The matrix-type business structure has enabled Huawei to make the best use of internal and external resources and build a global value network encompassing manufacturing and services and suppliers and consumers, so that the company has well adapted itself to the new trend in which multinational corporations around the world are advancing to high-end services. As of December 31, 2013, Huawei had over 150,000 employees in its different services around the world, 45% of whom worked in R&D, around 8% in manufacturing and supply chain, and about 34% in marketing, sales, delivery, and service.

5.2.1 Global Research and Development Networks

5.2.1.1 Value Objectives and Global Strategy

Innovation has been in Huawei's blood since the founding day of the company. Technological innovation is the company's lifeline and efforts on it have never stopped. As stipulated in *Huawei's Basic Law* published in 1998, at 10% of annual sales revenue is used for R&D. The company has been firm on this. And 10% of the R&D input is used for forward research, with continuous research and tracking of new technologies and in new areas. The company's R&D expenses totaled RMB 40.8 billion (USD 6.6 billion), 14.2% of the year's sales revenue. In the previous decade, Huawei had input over RMB 151 billion in its R&D. Huawei's independent innovation is well guaranteed with ample financial and human resources on the company's path to globalization. Huawei has now boasted numerous core technologies and products, thanks to its firm stand of technological innovation.

Huawei has continuously internationalized its R&D, to well integrate and make good use of global technological resources and talents, supporting the company's internationalized businesses. With decades of exploration and practice, Huawei has formed a new set of paths and approaches of open innovation adapting to economic and technological globalization, and has quickly grown into a world-class high-tech multinational corporation. As Huawei's president Ren Zhengfei put it, a company runs on two wheels, namely business model and technological innovation. Huawei's open innovation system well combines the two wheels and is a pillar to the company's leapfrog development. "A closed system is doomed to die and openness brings vitality." "Being not innovative generates the biggest risk." Such ideas have been Huawei's philosophy fully incorporated in its operations and management.

1. An international market-oriented culture of innovation that focuses on creating value for customers.

   Huawei has upheld that "Huawei exists only for its customers and customers' needs are the primary driving force for the company's development." The company improves its technologies, services, and products around customers' needs. Huawei creates a market-oriented culture of innovation that focuses on creating value for customers and establishes its values at providing stable service for customers in the long run. The company sees its core competence at winning an ever-higher customer satisfaction, turning from a technology-driven enterprise to a market-driven one. Huawei's R&D is market-oriented. Technicians are responsible for success in R&D and market. Accordingly, the company sets up customers' needs research, to draw roadmaps for product development by reflecting customers' needs around the world to the R&D sector. Each year, the company sends 5% of its R&D staff to marketing and 5% of its marketing personnel to R&D. Huawei has coestablished joint innovation centers with world-known telecom carriers and set up laboratories for individual and enterprise customers. In this market-oriented innovation system, passive technological innovations have been made on initiative, to adapt to market needs by grasping opportunities, and lead market by creating opportunities. By staying customer-centered, Huawei has made success in domestic and international markets.
2. Setting world-class enterprises as the benchmark and creating an open innovation system facilitated by management and service innovation.

   Huawei has long been positioned to be a world-class enterprise in line with international norms and practice. But the company was well aware of the huge gaps between it and foreign telecom giants, particularly in terms of high technology and management. Thus, Huawei learned hard from outstanding domestic and foreign companies, standing on the shoulders of giants. The company learned how to compete by diving into competition. In its early days, Huawei already sought management consultancy from world-known enterprises, reengineering its management and business processes. Following leaders of the industry such as Siemens, Alcatel, and Ericsson, Huawei has set up internationally recognized management and operations systems (Wang, 2013), including HAY Group's Job Evaluation and Salary System, IBM's IPD and Integrated Supply Chain (ICS), and the UK's National Vocational Qualification (NVQ) as the company's vocational qualification system. Huawei also invites Fraunhofer–Gesellschaft (FhG) of Germany as the company's quality management consultant, Pricewaterhouse & Coopers (PwC) as its financial consultant, and KPMG as its audit consultant. Huawei facilitates its open technological innovation system with management and service innovation. Technologically, Huawei adopts a follow-up innovation mode: Huawei will not be fully engaged in product development until an industry leader has developed a new technology or product that proves to be commercially valued. Huawei inputs at least 90% of its R&D power in applied technologies, with only 10% in basic research. The company grabs international markets first with its superiority in cost in domestic operations before "overtaking at a corner" with its accumulated strength in technology and market. Huawei is now in the world caliber and even slightly goes ahead of the others in product development. It is turning from a follower to a leader.

3. Integrating a take-in principle with independent R&D and realizing technological and market expansion via integrated innovation.

Basically, Huawei has integrated a "take-in" principle with independent R&D in its innovation. It adopts integrated innovation as an international practice. Initially, as a startup, Huawei was weak in technology, while many patents for basic technologies and applications were already filed in the field of telecoms. Thus, it was not wise to make zero-based independent R&D. Thus, Huawei followed the take-in principle for R&D, admitting its weakness in technology and absorbing the latest achievements in electronic information industry worldwide. In the meantime, the company relies on its independent R&D and develops advanced core technologies by open cooperation. Mr. Ren Zhengfei used to say, "When the rate of the newly developed is higher than 30%, it's not innovation. It's waste." Huawei encourages its R&D people to reduce the portion of invention when they develop a new product, and to make use of existing technological achievements and cooperate with or purchase from the outside. The take-in principle is followed in various ways. It is a common practice to enter international markets by negotiating and paying reasonable patent royalties. The company has realized fast technological accumulation by way of technological cooperation, strategic alliances, cross-border M&As, and investment partnerships. Huawei has realized its technological and market expansion that thrives on share success, through establishing investment partnerships and technological alliances with Texas Instruments, HP, IBM, Motorola, 3Com, Siemens, Infineon, Cisco, and SAP. Huawei has made reinnovation and integrated innovation on the advanced technologies acquired in various ways, around customers' needs. In this way, Huawei has kept up with the world's advanced technologies and enhanced its core competence, realizing its objectives to seize the domestic market and take shares in international market service.

4. Integrating systematic tracking with breakthroughs and focusing on core technologies in fields of strength.

Turning from a telecom equipment manufacturer to an ICT solution provider, Huawei has always followed the law of specialization-based division of labor and the principle of focus, focusing on its major businesses and core technologies in fields of strength. Facing the pressure from well-capitalized, technologically monopolizing multinational corporations and government-supported large state-owned enterprises, Huawei has not made hasty chase in technology. Instead, the company has made systematic tracking of international communications technology and advanced around core network technologies. Once it targets at a technology, it directs resources and development strength on it. By applying this "pressure tactic" on the key part of software and hardware in core networks, Huawei has achieved its own core technologies. In the late 1990s when 2G was in its heyday, Huawei began to focus on 3G. With years' efforts on 3G, during which period Huawei withdrew from the market of Little Smart (personal handy-phone system, PHS) terminals, Huawei finally put its world-class 3G technology into commercial operation at the end of 2003, laying a solid foundation for the company's entering international markets in future. Huawei has incorporated mainstream international standards into the industry and been in close cooperation with major telecom carriers in the world. By the end of 2013, Huawei had joined more than 170 industry standards and open source organizations hold 185 positions in these organizations. Huawei focuses investment in key technologies, architectures, and standards in the ICT field with the aim of providing broader, smarter, and more energy-efficient pipes that require zero wait time and create a better experience for users. It is committed to continuous innovations, and has made significant achievements in the fields of future 5G communications, network architecture, computing, and storage. Huawei has always focused on its specialized fields. With this "pressure and focus principle" in technological innovation, the company has been in the lead, continuously consolidating and enhancing its competitive edge in the field of telecom equipment.

5. Setting IPR management as the company's core strategy and establishing an effective mechanism for continuous innovation.

   Huawei has created a strong IPR group for its innovation management system. The group consists of over 300 IPR-related technical experts, patent engineers, and lawyers in copyright, trademark, and license. The company also keeps optimizing its international IPR management systems and process and encourages employees to apply for international patents. IPR strategy is one of Huawei's core strategies, with which the company has created an innovation-driving and failure-tolerant culture of innovation and formed an effective mechanism for continuous innovation. Huawei's IPR strategy outlines four key aspects: focusing on the core fields of strength and strengthening the company's core competence, IPRs, and brand strength, to expand the company's map of patents around the world and enhance its global competence; getting actively engaged in constitution of international standards, trying to having all the technical solutions included in international standards and accumulating basic patents; learning, abiding by, and applying international IPR rules with

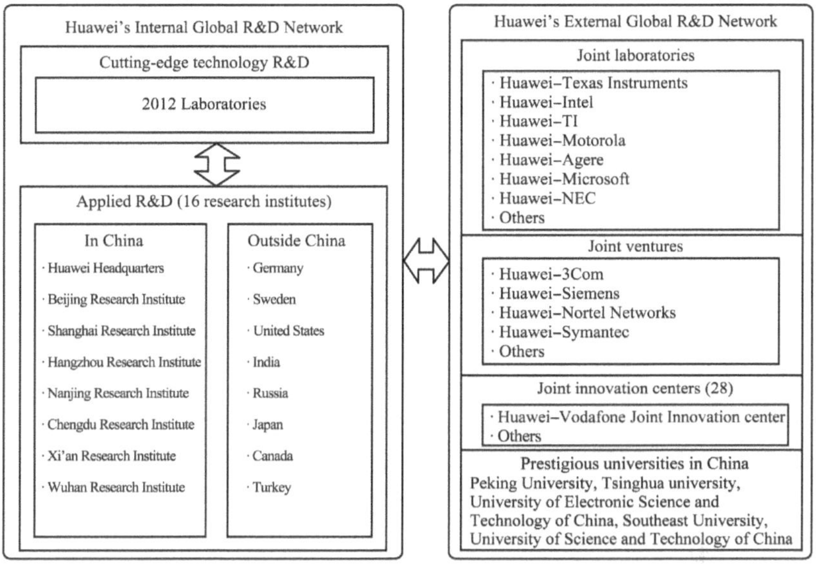

FIGURE 5.7 Huawei's global R&D networks. Source*: Prepared by the author according to related information about Huawei.*

an open mind and settling disputes concerning foreign patents via negotiations and cooperation for products; and exchanging patents for technologies, to realize integrated utilization of innovations worldwide by way of cross licensing. Huawei spends tens of millions of dollars on patent applications around the world. With cross licensing, the company saves hundreds of millions of or even billions of dollars on patent royalties each year. Single-purpose patent applications may generate no profits for a company, but when the company makes cross licenses of its patents IPRs and technologies with leading multinational corporations, it saves significantly on cost of one-way permits and sees wide application of its patented technologies, which generates enormous profits.

Huawei's global R&D networks include internal and external ones, as shown in Fig. 5.7.

5.2.1.2 Huawei's Internal Research and Development Networks

Huawei boasts internationalized technological innovation. The company is well positioned in the domestic market and accurately oriented on the international telecom equipment market, thanks to its worldwide R&D networks. Huawei has continuously internationalized its R&D via its open innovation system, to well integrate and make good use of global technological resources and talents, supporting the company's internationalized businesses. Huawei has now established 16 research institutes in Germany, Sweden, the

United States, India, Russia, Japan, Canada, Turkey, and China. About 76,000 employees (45% of the company's total employment) are engaged in product and solution R&D. Huawei has attracted technologies, expertise, and talents from all over the world to its internationalized worldwide synchronized R&D system, and in the meantime, worked closely with partners from industry, academia, and research institutes, enabling it to take the lead in researching, innovating, and implementing future networks. The company also set up 28 joint innovation centers with leading carriers to translate leading technologies into competitive edges and business success for customers.

Huawei's global R&D consists of two parts, namely cutting-edge technology and applied R&D. Cutting-edge technology R&D focuses on new technologies that will be in full application in 3–5 years. Product R&D is included in this part. Huawei's 2012 Laboratories are engaged in cutting-edge technology R&D, with its funds mainly from the company's budget. Product R&D is conducted in R&D offices in all business units, to develop new products for different products and markets. Huawei inputs 15% of its R&D funds in cutting-edge technology R&D and the remaining 85% in applied R&D. The company's 16 research institutes have formed a multi-functional global R&D system that effectively serves the company and is capable of sustainable development. Huawei has integrated IPD with capability maturity model (CMM) that focuses on process management, forming Huawei's branded IPD–CMM that is a key management tool of Huawei's global R&D system. IPD–CMM serves as the standards for Huawei's software developers and the universal language of Huawei's R&D staff. A universal development management process and tool enables cross-cultural teamwork between different research institutes for worldwide synchronized R&D. Huawei is thus able to effectively manage and coordinate its research institutes around the globe.

5.2.1.2.1 2012 Laboratories

2012 Laboratories is one of Huawei's departments engaged in basic innovation research. Namely after the film 2012, the Laboratories is committed to innovation, research and platform technology development. An as end-to-end responsibility center that provides BGs with support and services, 2012 Laboratories works on laying the cornerstone for Huawei's future technology and R&D capability. It is Huawei's decision and action made to the huge transition in ICT industry. IT and IC are integrated at a faster pace, thanks to smart terminals and cloud computing. Data floods into people's lives in an era of broadband. The era of ICT is filled with enormous opportunities and challenges. Huawei has to strengthen its cutting-edge technology research and solves the problem it faces after turning from a follower to a leader in the industry, fulfilling its role as an industry leader and seizing the commanding heights in technological innovation. Huawei sets up 2012

Laboratories for a better future development, focusing on continuous input in key technologies, frameworks, and standards in ICT, building broader, smarter, and more effective pipes, and working closely with its partners in industrial, academic, and research circles, to take the lead in research and innovation for future networks.

5.2.1.2.2 Research Institutes

Huawei has established 16 research institutes in China and abroad. Established at different times and in different places, the research institutes complement each other in functions and position and form a multifunctional global R&D system. Huawei's research institutes in China include Huawei Headquarters and research institutes in Beijing, Shanghai, Hangzhou, Nanjing, Chengdu, Xi'an, and Wuhan (Table 5.1). Its overseas research institutes are set up in Germany, Sweden, the United States, India, Russia, Japan, Canada, and Turkey (Table 5.2). The research institutes are mostly located in

TABLE 5.1 Huawei's Research Institutes in China

Institute	Location	Major Business	Note
Headquarters	Bantian, Longgang District, Shenzhen	Major R&D center	
Beijing	Zhongguancun Environmental Protection Park, Haidian District, Beijing	R&D related to carrier network business	
Shanghai	Jinqiao, Pudong District, Shanghai	R&D related to carrier network and consumer business	
Hangzhou	Binjiang District, Hangzhou	R&D related to carrier network and enterprise business	
Nanjing	Yuhua District, Nanjing	R&D related to carrier network business	
Chengdu	Chengdu Hi-Tech Industrial Development Zone, Chengdu	R&D related to carrier network and enterprise business	
Xi'an	Xi'an Hi-Tech Industries Development Zone, Xi'an	R&D related to carrier network business	
Wuhan	Future City, Wuhan East Lake High-tech Development Zone, Wuhan	R&D related to carrier network and consumer business	

Prepared by the author according to related information about Huawei.

TABLE 5.2 Huawei's Overseas Research Institutes

Institute	Time of Founding	Location	Major Business	Note
India	1999	Bangalore, India	Platform and middleware development, covering APPs and functions for Huawei's most products; developing networks and enterprise solution software for other businesses	Huawei's largest overseas research institute and Huawei's first solely-run overseas research institute
Russia	1999	Moscow, Russia	Technology in wireless networks and optical fiber communications, multimedia and cloud computing	
Sweden	2000	Stockholm, Sweden	R&D of GSM, 3G and LTE Ultra 3G systems	
United States	2001	Dallas and Silicon Valley, United States	Next-generation communications solutions targeting American users	A new research institute established in Silicon Valley in 2012
Europe	2008	Munich and Nuremberg, Germany; Milan, Italy; London, Britain; Brussels, Belgium	"Overseas brain" for basic research and core technology	Headquartered in Munich, Germany and five offices in four countries
Turkey	2010	Istanbul, Turkey	Research concerning value-added services, pricing and charging, call center and customer relationship management	

(*Continued*)

TABLE 5.2 (Continued)

Institute	Time of Founding	Location	Major Business	Note
Canada	2010	Ottawa, Canada	Development of strategic products of wireless, cable, optical fiber and IP networks; a major role in 5G R&D	A new R&D center established in Ottawa in 2014
Japan	2010	Tokyo and Yokohama, Japan	Consolidating the cooperation with current suppliers and expanding cooperation with new suppliers, to codevelop new-generation technologies for Huawei's worldwide terminal and network products	Two R&D centers in Tokyo relocated to Yokohama and reorganized in 2013

Prepared by the author according to related information about Huawei.

areas with developed electronic information and software industry, advanced and mature technology, and a fully developed market. Huawei sets up a research institute either in a technology-intensive developed country, that is, a country with certain leading technologies; or in a developing country rich in human resources and with satisfactory R&D environment, so as to find strong support for its technological innovation from internationally advanced technologies and talents. Huawei's India, Nanjing, and Shanghai research institutes and Central Software Department have been certified with CMM5, indicating the company's advanced software process management and quality control in the industry.

Europe is becoming Huawei's second homeland. The company aims to double its European employment in coming years. From 2010 onward, Huawei has set up R&D centers in Belgium, Italy's Milan, Finland's Helsinki, Ireland's Cork City and Dublin, Britain's Ipswich Town (in the east) and Bristol (in the southwest), and France's Paris and Sofia Antipolis in Alpes-Maritimes. Huawei now has 17 R&D institutes in eight European countries, namely Belgium, Finland, France, Germany, Ireland, Italy, Sweden, and Britain.

Huawei has always seen talent as the core power of innovation and tries to attract as many talents, so as to seize the commanding heights in high-tech development. It is necessary to take a foothold in a local market, communicating with the local market and understanding the market needs, to run business in overseas markets. That is the superiority overseas local employees have. Thus, Huawei has accelerated its pace in internationalizing its employment, as the company keeps expanding its international businesses. The company recruits local talents in localized operations, and introduces advanced talents fostering mechanism and international assessment standards, to encourage a local business team to be in full charge of local business, with a proper authorized power regardless of nationality. According to Huawei, its overseas employment has kept increasing by at least 15% annually. As of the end of 2013, Huawei had over 30,000 foreign employees from 150 countries. More than 70% of its employees outside China were locals, most of whom were engaged in R&D. Huawei has focused on human resources center building in the major markets it runs. The company has set up 36 training centers to foster local technicians. Huawei has been significantly facilitated in its expansion in European and other markets, thanks to its ever-enlarging overseas employment, particularly senior foreign employees.

5.2.1.3 Huawei's External Research and Development Networks

With long-established strategic partnerships, Huawei has been able to focus on R&D innovation in fields of strength. Also, it has established stable long-term partnerships with Chinese and foreign suppliers and cocreated R&D centers with multinational magnates, forming an open innovation system. Huawei has integrated resources of innovation in global value network and materialized its goals of mutual benefits and shared success. Huawei makes global strategic collaboration in three ways, namely joint laboratories coestablished with world-class enterprises, joint ventures, and joint innovation centers coestablished with leading carriers. In addition, Huawei has made brilliant achievements via collaboration with prestigious universities in China and abroad.

Huawei coestablishes joint laboratories with world-class enterprises. Huawei–Texas Instruments DSP Joint Laboratory codevelops digital signal processing (DSP) hardware and software for communications products, enhancing the capability of developing and applying DSP chips of Huawei's development engineers and fast applying Huawei's new technologies in multimedia. Huawei–Intel Joint Laboratory focuses on communications solutions based upon Intel's IX framework. In addition, Huawei has set up joint laboratories with other world-class enterprises, including TI, Motorola, Agere, ALTERA, SUN Microsoft, and NEC, to codevelop new technologies (Jiang and Li, 2011).

Huawei establishes joint ventures with its partners. In 2003, Huawei-3Com Co., Ltd was created by Huawei and 3Com, a magnate in the world's data communications. Huawei and Siemens set up a joint venture in 2004, focusing on the development, production, sales, and service of TD-SCDMA technology and products. The joint venture replaced Cisco as the manufacturer of Siemens's data products. A joint venture of Huawei and the telecom magnate Nortel Networks was established in 2006, to develop ultrabroadband access solutions targeting the global market. Huawei and Symantec coestablished in 2007 a joint venture engaged in secure storage systems and devices for telecommunications and enterprises. It was collaboration between a hardware manufacturer and a software company. Huawei has for years selected partners to its own strategic objectives. For instance, Huawei got benefited from Siemens's advanced 3G technology in its collaboration with Siemens. Nortel Networks helped a lot in Huawei's expansion in North American markets. And thanks to its cooperation with Symantec, Huawei made a success in the IP field. By establishing joint ventures, Huawei has had access to advance technologies and wider markets and has financially benefited for its technological development and market expansion.

Strategically, Huawei also coestablishes joint innovation centers together with leading carriers. Mutually benefiting from such collaboration, Huawei and its carrier partners make continuous exploration of business models and CT innovation, so that users and carriers are both benefited. In October 2006, Huawei and Vodafone established in Spain the first joint innovation center MIC. By 2013, Huawei already set up 28 joint innovation centers with over 10 carriers around the world, including important telecom carriers in China, Europe, North America, the Middle East, and Southeast Asia, and made success in more than 100 key innovation projects. Joint innovation centers have been a key way to continuously enhance Huawei's strategic partnerships with its partners.

Huawei has since 1996 been in cooperation with prestigious universities in China for R&D. The IS95-based narrow band CDMA codeveloped with Peking University laid a foundation for Huawei's later-day CDMA technology, and a group of technological talents grew out of this program. Huawei attributed its success in SDH optical networking technology to its cooperation with the Radio and Electronics Department, Tsinghua University. In addition, Huawei has been in years' collaboration with University of Electronic Science and Technology of China, Southeast University, and University of Science and Technology of China.

5.2.2 Global Production and Operation Network

5.2.2.1 Value Objectives and Global Strategy

Huawei's global production and operation network comprises plan, production, procurement, and logistics, and designs, plans, and controls information, goods, and capital flows in the supply chain, via ICS management, to realize the two key objectives of the supply chain, namely elevated customer satisfaction and lowered cost of the supply chain. Huawei has upgraded its ICS to the global supply chain network, building an advanced manufacturing model for the global value network, "to focus on market needs, create a coordinated operation of global supply chains that is quick in response, high-quality, low-cost, and flexible, enhance the company's integrated competence and ensure perfect contract deliveries." In this way, Huawei has successfully enhanced its competence.

Huawei has seen its customer groups grow fast, expanding from carriers and enterprises to individual consumers. The supply chain keeps extending, as the company grows with globalized businesses. Huawei has fully adopted a globalized integrated procurement-supply chain, to deal with ever-fiercer competition on domestic and international markets and satisfy diversified customer needs. Huawei holds that the robust supply chain remains the key to the sustainable development of the entire industrial chain. A company can only realize the sustainable development that thrives on shared success by integrating industrial resources with an open mind, attaching significance to the cooperation with enterprises on its supply chain, and creating a platform for integrating global resources. Customers, Huawei, suppliers, and subsuppliers share the same business ecosystem that remains sound when all parties in it are healthy. As a globalized company that runs businesses in more than 170 countries and regions, Huawei has adopted a "glocalization" strategy, sharing all values generated on the value chain with its customers around the globe. Huawei runs localized operations in the regions and countries where it runs businesses, to boost local employment and taxes, contributing to local development. With cooperation with excellent local enterprises based on division of labor, Huawei has fully integrated its strength in global value chain with local innovation power, helping local innovation gain global value. Huawei has worked hard on its own sustainable development and on sustainable development management and capability building of suppliers, advancing together with its suppliers.

In 2013, apart from extensively adopting customer-oriented methods for managing supplier CSR risks, Huawei also transformed its risk management approach into one that focuses on efficiency management. It has incorporated sustainability requirements into its supplier management process, to set the trend for sustainability initiatives across the industry chain. Huawei's supply chain sustainable development management focuses on the following four aspects. (1) *Risk management*: Huawei stays focused on key issues and incorporates CSR risk management into procurement process and supplier lifecycle with a hierarchical management system. The company uses procurement quotas to continuously drive supplier improvements, and takes viable measures to monitor and control risks, to achieve the best practice. (2) *Efficiency management*: Huawei goes farther than mere compliance and being customer-driven and tries to find more opportunities in CSR, to enhance suppliers' capability, improve the efficiency of procurement, optimize business process, reduce waste and cost, and enhance the company's business competence. (3) *Business innovation*: Huawei takes forward thinking and in-depth coordination. With CSR, the company searches for new opportunities for business innovation, develops new products, opens up new markets, and explores new business models, fully incorporating CSR into its business strategy and brand. (4) *Collaboration*: Huawei pays attention to system problems in the industry and conducts crossover dialogs and collaboration on representative CSR issues. The company is actively engaged in working out industry rules and sets industry benchmarks, leading the development initiatives across the industry and highlighting the company's mission and brand.

5.2.2.2 Networks

In its early days, Huawei worked all across the production chain. It manufactured all parts of a product in different workshops and assembled into the final product. The product's value chain was completed within the company's workshops. However, this production mode could not keep up with the company's development. Experts from IBM found in 1999 Huawei's core competence in the company's R&D and market-leading capability. So it would be perfect that Huawei has a firm control of its core competence in production and operation management and outsources all the others, to form an effective supply chain. IBM tailored the integrated supply chain (ISC) management for Huawei. In ISC, each link in the company's operations, be it an internal issue or an external partner, is seen as an integral part of the supply chain and put under effective management, to improve the operational and economic efficiency of the supply chain. With ISC, Huawei made an organizational restructuring, merging its Production Dept., Plan Dept., Procurement Dept., Import & Export Dept., Qualification Dept., Outsourcing Cooperation Dept., Dispatching Dept., and Warehousing Dept. into the Supply Chain Management Dept. in charge of the integrated management of supply chain. Supply chain management has been an effective means to lower costs and inventory and improve supply quality, rate of capital turnover, supply speed, and project quality.

Since 2000, Huawei has outsourced its manufacturing, assembling, packaging, dispatching, and logistics to specialized companies. In Shenzhen alone, there are over 100 subcontractors working for Huawei. Production and service outsourcing of noncore products features a comparative advantage in costs. Huawei is searching for quality suppliers worldwide and increasingly outsourcing its parts manufacturing, assembling and intermediates. The company establishes a long and stable purchaser–supplier relationship with multinational magnates in high-end areas. The noncore services are widely outsourced, including information service, software, training, data recovery, management consultancy, equipment testing and maintenance, and office support. There are thousands of enterprises upstream and downstream on Huawei's supply chain, including numerous renowned Chinese and foreign companies. Huawei has created a dynamic cobweb-shaped global production and operations network (Fig. 5.8). In this production mode, functions of a product's value chain are delegated to enterprises in different countries and the product is a cowork of numerous enterprises around the globe. Huawei stands at the core, focusing on product design, core module manufacturing, and production of tailored products. Subcontractors work around and for Huawei, with their superiorities in different divisions of labor. As of December 31, 2013, Huawei had an employment of over 150,000 employees around the globe, about 8% of whom were engaged in manufacturing and supply chain.

Huawei's ICS consists of four stages (Yi et al., 2011).

1. *Foundation.* Huawei analyzes and sums up the status quo based upon the existing supply chain, positioning internal obstacles to and strengths in the supply chain management and analyzing the external market environment for its features and uncertainties, to work out the company's strategy and direction of supply chain.
2. *Functions integrated.* Huawei focuses in this stage on its internal operations, launching integrated management of its core functions and reengineering its business processes, for an optimal integration of functions. A cross-department team is created for the project planning and executing, to enhance interdepartmental collaboration.
3. *Internal supply chains integrated.* Huawei has the supply chains under its direct control integrated in this stage, integrating its internal supply

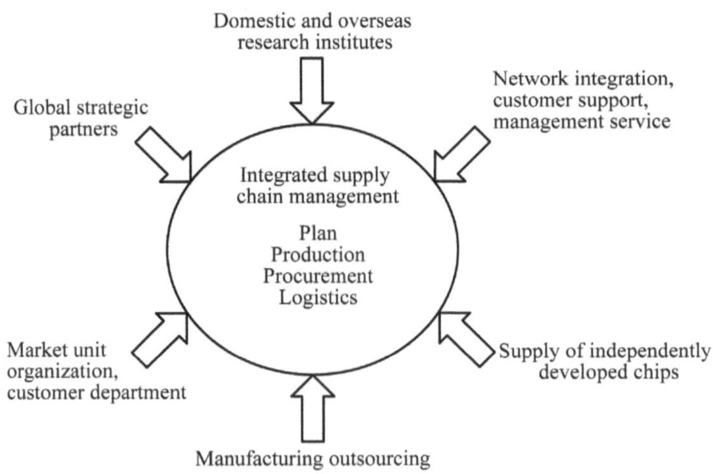

FIGURE 5.8 Huawei's integrated supply chain. Source*: Prepared by the author according to related information about Huawei.*

chains with supplier and user management in the supply chain, to form an internal ISC.

4. *External supply chains integrated.* This is the key stage of successful ISC. The company's internal supply chains are integrated with external suppliers and customers, to form an ISC, and the company establishes good partnerships with its suppliers and customers.

With the above four stages completed, Huawei establishes its ISC that integrates finance, information, and management models. This fast-reconstructed dynamic global production and operations network is supplier- and user-oriented, instead of being product-oriented, and ideal for integration of and cooperation with customers, suppliers, and service providers. ISC well meets market needs, thanks to its improved adaptation to market changes, flexibility, speed, innovation, and knowledge.

5.2.2.3 Governance Models

Huawei aims to "enhance customer satisfaction and reduce the total cost of supply chain" and makes substantial effort on its ISC management, dedicated to building a transparent and open business cooperation environment and helping its suppliers understand and be well-informed of Huawei's global procurement policies and requirements, to value and protect the common interests of suppliers and Huawei. ISC never drags the company behind. Rather, it has successfully turned into the company's competence on market and been the company's "invisible wings" of fast development, bringing Huawei from China to the world (Tang, 2014).

5.2.2.3.1 Procurement Model of Strategic Significance

Huawei makes detailed classification of its material families and suppliers. It establishes well-organized technical and business qualification systems. With its centralized qualification and decentralized procurement, Huawei makes flexible decisions and quick responses to market. Huawei's Material Expert Panel is in charge of the procurement of materials required by business departments or local markets and selects suppliers for each material family and assesses each shortlisted supplier. Suppliers are classified as potential, qualified, core, and strategic ones, in accordance with their qualification and assessment results. Following a collective decision, a procurement team launches decentralized procurement with the "80/20" principle, so that the supplier group for each material family is rationally structured. The procurement model is of strategic significance. The centralized management and control improves the benefits and efficiency of procurement, to fulfill local needs, so that Huawei is always able to find the best suppliers in terms of price, quality, and supply speed. This procurement model facilitates Huawei to provide more innovative, differentiated, and competitive products and services.

5.2.2.3.2 Effective Supplier Management

Huawei establishes "harsh" requirements for its suppliers. A supplier is required to have thorough stock plans. It is expected to deliver fast, despite any changes to product specifications, and to deliver at the preagreed price, despite any changes to technical conditions. Huawei launched a well-organized supplier assessment system. The assessment models include technical performance, quality, responsiveness, delivery, costs, and contract fulfillment, as well as environmental impact and social responsibility. A quantitative assessment is conducted with detailed indicators. The final assessment report is formed for management decision and regularly sent to suppliers. Huawei attaches great importance to the two-way communication with it suppliers. Expert Panels and Technical Qualification Center keep close contact with suppliers. Each Expert Panel has a liaison officer in charge of communication with suppliers. Huawei sets up the Office of Suppliers' Feedback, to deal with issues rising in Huawei–supplier collaboration. Huawei's effective supplier management features "harsh" requirements, scientific assessment, and two-way communication. With this model, Huawei creates a fair-play platform and equal partnership for its suppliers and helps them go to the world. Huawei and its suppliers thus enjoy an open, effective, and mutually beneficial partnership.

5.2.2.3.3 Ongoing Quality Improvement

Huawei has ongoing technical optimization in its independent production. In addition, the company established a strict quality control system. Huawei is qualified with TL9000 and ISO9001:2000. It has seen the DPMO (defect per million opportunity) decreasing in the past 5 years, and the FTY (first throughput yield) increasing over the same period of time, up approximately 30%. In the meantime, quality improvement is never absent from Huawei's ISC. With Six Sigma, Lean Production, employees' independent feedback, QCC (Quality Control Circle), and other quality improvement measures, quality improvement has been rooted in Huawei's employment. Huawei adopts an advanced barcode system, in which a barcode is like an ID that plays a role in the accurate management of dispatching and transport and enables a backward pass of any product defects. In this way, Huawei is confident of the high-quality error-free delivery of its products.

5.2.2.3.4 All-Inclusive Risk Management and Sustainability

The fulfillment of the supply chain usually fails due to strikes, natural disasters, or port congestion around the globe. The supply chain's continuity is much more guaranteed, if such risks could be identified beforehand and alternatives made ready. Huawei's supply chain risk management follows the general risk management procedures, identifying, analyzing, evaluating, and dealing with any risks. In particular, a production continuity system is created, to identify and restrict crisis points that are likely to break up production continuity. Alternatives are worked out according to emergency measures, to fully guarantee supply continuity.

Huawei has to deal with a huge number of diversified suppliers around the globe. Sustainability is part of Huawei's procurement process and supplier lifecycle. The company uses procurement quotas to continuously drive supplier improvements. Huawei audits suppliers and assigns them one of three priority levels: high, medium, and low. The factors considered during the audits are supplier location; product category; business volume and relationships; sustainability performance; potential environmental risks; and risk management systems and capabilities. Huawei conducts onsite audits on suppliers with high-level potential risks; sampling audits on suppliers with medium-level potential risks; and joint audits together with users. Huawei's hierarchical management well secures supplier sustainability, and effectively monitors and controls risks along the supply chain.

The Huawei Green Partner (HW GP) Program aims to encourage its suppliers to systematically manage their environmental protection efforts and pay attention to green initiatives throughout product lifecycles. By controlling the use of prohibited materials from sources, coconducting energy-saving and emission-reducing innovations with suppliers, lowering carbon footprints on the supply chain, Huawei strives to establish a green supply chain, ranging from "green" design to "green" manufacturing. In 2013, Huawei upgraded the HW GP standards to GP2.0 by adding the requirements for the environment management system as well as energy and greenhouse gas management. In that year, 34 suppliers passed the HW GP certification. Huawei launched energy-efficiency and carbon emission reduction pilot projects in four suppliers of different product categories and made significant success. The four suppliers saved 2500 kWh of electricity and reduced 23,000 tons of carbon dioxide emissions in 2013. Huawei intended to launch the project in 20 suppliers in 2014.

5.2.2.4 Location Selection

Huawei publishes the name list of procurement liaison officers at Huawei's headquarters and around the globe on the company's website. According to the name list, Huawei has its procurement in over 190 countries and regions (Table 5.1), to create the globalized value chain by fully integrating quality resources around the globe. Huawei integrates its global value chain strengths with local power of innovation by working in cooperation with a due division of labor with outstanding local enterprises, to help local businesses achieve global value with their creativity. Table 5.3 lists major countries and regions where Huawei runs procurement projects outside the Chinese mainland.

TABLE 5.3 Huawei's Global Procurement Presence

Region	Countries and Areas
Asia Pacific	Australia, Pakistan, the Philippines, South Korea, Cambodia, Malaysia, Bangladesh, Nepal, Japan, Sri Lanka, Thailand, Singapore, New Zealand, India, Indonesia, Vietnam, Hong Kong, etc.
Commonwealth of Independent States	Azerbaijan, Belarus, Russia, Georgia, Kazakhstan, Kyrgyzstan, Turkmenistan, Tajikistan, Ukraine, Uzbekistan, etc.
Europe	Albania, Ireland, Austria, Bulgaria, Belgium, Iceland, Poland, Bosnia and Herzegovina, Denmark, Germany, France, Finland, the Netherlands, Czech, Croatia, Latvia, Lithuania, Luxembourg, Romania, Moldova, Norway, Portugal, Sweden, Switzerland, Serbia, Slovakia, Spain, Greece, Hungary, Italy, United Kingdom, etc.
Latin America	Argentina, Brazil, Bolivia, Panama, Ecuador, Colombia, Peru, Mexico, Venezuela, Uruguay, Chile, etc.
Middle East and North Africa	Algeria, the United Arab Emirates, Oman, Egypt, Kuwait, Qatar, Mali, Morocco, South Sudan, Senegal, Saudi Arabia, Tunisia, Turkey, Yemen, Iraq, Jordan, etc.
North America	United States, Canada, etc.
Sub-Saharan Africa	Ethiopia, Angola, Botswana, Ghana, Zimbabwe, Cameroon, Cote d'Ivoire, Kenya, Rwanda, Madagascar, Mauritius, Mozambique, Namibia, South Africa, Nigeria, Tanzania, Uganda, Zambia, etc.

Prepared by the author according to related information about Huawei.

Huawei has focused its independent production on the Chinese mainland. Shenzhen Headquarters, as the earliest one, was put into use in 2000. Located in Bantian, Longgang District, Shenzhen, the Headquarters has been Huawei's major production and operations center. Dongguan Headquarters consists of northern and southern parks, situated in the north and south of Dongguan Songshan Lake Hi-tech Industrial Development Zone. The northern park has been in use since April 2009 and Huawei's major production base outside Shenzhen in south China. The southern park is under construction and more businesses or functions of the company are expected to be relocated there. Huawei built a production base in north China in Langfang. Located on Yinhe North Road, Langfang Economic, and Technological Development Zone, the Langfang Base is the center of sales and services for Huawei's China Region. The new production base is located at a place nearer to enterprises upstream and downstream on the industrial chain and attracts more OEMs. In this way, Huawei is able to enlarge its production capacity and enhance its efficiency, while substantially reducing the costs of production. Huawei has set up joint ventures in Egypt, Russia, and Brazil. The company has succeeded in localized production via local partners in Saudi Arabia, Iran, and India. Thus, Huawei is able to better meet customer needs on overseas markets.

5.2.3 Global Marketing and Service Network

5.2.3.1 Value Objectives and Global Strategy

Huawei holds up to "Customer First." Huawei exists to serve customers, whose demands are the driving forces behind its development. The company continuously creates long-term value for customers by being responsive to their needs and requirements. Huawei measures its work against how much value it brings to customers, because it can only succeed through its customers' success. Huawei has made a complete set of strategic plans and marketing and service strategies fitting to its own development and overseas markets: market access strategy of "entering from the remote outside," national diplomacy-complying market expansion strategy, marketing strategy of "attracting customers by going on exhibitions," alliance strategy aiming "to change the global market competition via collaborative alliances," OBM strategy of hi-tech exports, and customer-centered service strategy. By implementing a variety of transformation programs, Huawei ensures prompt responses to and end-to-end management of customer requirements, further lowers its internal operating costs, and improves its business operational efficiency, thereby bolstering the company's strategic goals of globalization and sustainable development.

5.2.3.1.1 Market Access Strategy of "Entering From the Remote Outside"

Huawei began its international expansion in 1995. Initially as "a market substitute," it first entered Hong Kong with a strategy of "entering from the remote outside," before expanding into African and European and American markets. Between 1996 and 1999, Huawei began to enter Russia, Yugoslavia, Yemen, Brazil, South Africa, Ethiopia, and Laos, by way of getting in touch with local customers and participating in tenders via its local partners. Bubbles in IT industry burst in 2000. Western equipment suppliers began to withdraw from some fringe markets. Huawei then took the opportunity and gained a foothold on international markets. In this stage of full expansion, Huawei began to send a large number of its employees overseas, to establish sales offices in Saudi Arabia and Sub-Saharan Africa. Thanks to cooperation with local carriers, Huawei got soon mingled with local capital and culture. It made success in Southeast Asia, including Thailand, Singapore, and Malaysia, and in the Middle East and Africa. Huawei began to enter Europe and Americas and compete with major companies in 2001. It established four R&D centers and over 20 regional offices in Europe. By establishing partnership with reputed local agents in Western Europe, Huawei's products began to enter developed countries, including Germany, France, Spain, and Britain. In 2004, Huawei signed an agreement with the American company NTCH, to build CDMA2000 network for the latter. By then, Huawei successfully entered the United States.

5.2.3.1.2 National Diplomacy-Complying Market Expansion Strategy

Huawei has attributed its success on overseas markets to its superiorities in price, technology and market, as well as to its following the national diplomatic strategies. The company designs its marketing strategies on overseas markets in the framework of the national diplomacy. In a context of national diplomacy, Huawei has expanded its overseas markets steadfastly. In the meantime, it has contributed to China's economic and diplomatic development and won the government's support. China has in recent years adopted three major diplomatic strategies: establishing strategic partnership with powers; consolidating and promoting relationship with neighboring countries; and strengthening long-standing friendship with developing countries. Huawei has adaptively followed the national diplomatic strategies and made success in overseas markets.

China and Russia established a strategic partnership in 1996. Huawei found business opportunities in this and began to enter markets in the CIS. The Russian company Beto, Russian Telecoms and Huawei coestablished a joint venture in 1997. Huawei took this joint venture as a platform for its expansion via localization. The then President Jiang Zemin made a state visit to the United States at the invitation of the then US President Clinton in 1997. Clinton made his state visit to China in 1998. These two visits indicated that the two governments were earnest to maintain a constructive strategic partnership in the new century. In such a context, Huawei set up a research institute in Dallas and introduced an advanced management system from IBM. Huawei later established a wholly owned subsidiary FutureWei and played a full-court in the United States.

Huawei has received support from governments for its entering developing countries. Thanks to the long-standing friendship and aid programs between China and developing countries, Huawei won an opportunity of entering those countries with its product assistance and winning trust from local governments and telecom authorities. The Chinese government-presented Huawei08 was put into operation in Uzbekistan in April 1998. In August that year, Huawei's senior management paid a visit to Uzbekistan at invitation. Huawei began to enter Central Asia. Five of Huawei's mainstream products acquired market access in Uzbekistan in 2000, indicating the company's breakthrough in the Uzbekistani mobile market. In 2000, the then Vice Premier Wu Bangguo asked Mr. Ren Zhengfei to be a member of the delegation to Africa. The government helped Huawei's expansion in the African market with China's assistance programs in African countries. In 2001, Mr. Ren Zhengfei was on the delegation led by the then Vice President Hu Jintao to the Middle East, and Huawei made an advance in the Middle East market.

5.2.3.1.3 Marketing Strategy of "Attracting Customers by Going on Exhibitions"

Huawei copies its success in domestic marketing to overseas markets. The company demonstrates the best of it to customers through customer pilots, advertising, participating in telecom fairs and forums, technological exchange with customers, and inviting customers to the company sites. A localized expansion well attracts local finance and gets integrated with a local culture, so that a widely recognized brand can be created in the shortest time. Huawei made a great success when it was on Asia Pacific international telecom expo for the first time in 1994. Since then on, Huawei has been on each and every international telecom expo. The company has an annual expense of at least RMB 100 million on participating in 20-plus international exhibitions. Huawei organizes a big telecom exhibition in every new market where it starts its business. The company's booth always stands with those of international magnates of the industry, to create a visual impact at the first sight. Thus, many who have not known Huawei turn their eyes to Huawei's products and technologies. In this way, Huawei gets known to more and reshapes its brand.

5.2.3.1.4 Alliance Strategy Aiming "to Change the Global Market Competition via Collaborative Alliances"

Huawei has always attached great importance to extensive equal collaboration and strategic partnership, and made ongoing efforts on establishing mutually beneficial external collaboration in various types. Its international collaborations vary in level of cooperation, including product sales, product manufacturing, joint ventures and R&D, and sales. Huawei coestablished Cosmobic in Japan together with NEC and Panasonic, to open the market for its data communication products. Huawei-3Com joint venture in the United States helped a significant increase in sales in North America. Huawei signed an agreement of cooperation with Siemens, and the sales volumes of its data communication products increased in Europe. With international collaboration, Huawei has better understood local environments and markets, faster entered overseas markets and got into distribution channels in a fast and effective way. Also, it has gained technologies from its partners and integrated its production and distribution, to realize scale economy and lowered cost. It has been proved that collaboration and joint venture established with related enterprises, or strategic alliances, help an enterprise's expansion in overseas markets.

5.2.3.1.5 OBM Strategy of Hi-Tech Exports

Huawei chose from the very beginning of its expansion a hardest way: export of its independently developed products. With its own core technologies, Huawei has seized overseas markets with its independently developed equipment. All its hi-tech exports are the independently developed products. Huawei has long attached great significance to accumulation and protection of IPRs. Its logo, "华为" and "HUAWEI" are under special protection in over 170 Paris Convention member states and in more than 140 WTO member states. Huawei has been member to over 170 industry standard bodies and open source organizations around the globe. In extensive collaboration with international magnates in the industry, Huawei has been on its own advanced technologies, independent of those technological magnates. It is in such a condition that Huawei is able to collaborate with international tycoons on an equal footing and in a true-to-sense mutually beneficial way. In this sense, core technologies play the decisive role in Huawei's expansion on overseas markets. Huawei has upheld its own brand and independently developed products when it looks for partners and strategic investors. Huawei's success in export of independently developed and hi-tech products has to some extent enhanced the world's impression on Chinese enterprises and products.

5.2.3.1.6 Customer-Centered Service Strategy

Huawei has since the first day of its founding seen service as an important part of the company's competence. Huawei's earliest service business was in engineering installation, while after-sales maintenance was in the charge of product developers in the Central Department of Research. Huawei provided all-arranged, life-long, free-of-charge service for its customers. The company began to launch paid services in 1998, which was effectively put into operation in Technical Support Department. With expanding overseas businesses, Huawei has built its own market and service networks, providing globalized marketing services. In an overseas market, Huawei adopts relational marketing strategies that are customer retention-oriented and focus on continuous client contact and customer needs, to form an interest community with its customers. Thus, a relatively loose buyer–seller relationship has been successfully turned into a closer partnership in technological R&D, product manufacturing, and marketing. Huawei has been expanding its collaboration with carriers on various levels, and coestablished with them a future-oriented secure development model that thrives on shared success and coexistence, so as to work in cooperation with a due division of labor in a mutually beneficial way and create more value for customers around the globe.

5.2.3.2 Networks and Governance Models

With expanding overseas businesses, Huawei has built its own market and service networks, providing globalized marketing services. It is easy to acquire needed information and service on a resource-sharing platform. Huawei has an elite project management team consisting of over 760 PMPs, dedicated to providing end-to-end services for clients. The company is confident of its professional and specialized services, facilitated with and supported by the globalized IT systems, such as Global Client Request Management System (GCRMS), Engineering Project Management System (EPMS), and Spare Parts Management System (SPMS). Huawei has set up 36 training centers around the globe, fostering local technicians and vigorously promoting a localized employment in every place where the company runs businesses. The localized operations around the globe have helped Huawei better understand local markets and enabled the company to contribute to the social and economic development of countries and regions where it runs businesses. As of December 31, 2013, Huawei had over 150,000 employees in its different services around the world, about 34% in marketing, sales, delivery and service sectors.

To enable carriers to stay abreast of the trends and stay competitive in a volatile market, Huawei has set services as a priority for its strategic investment. It has established local service delivery organizations and supporting platforms around the world. In addition, Huawei has been committed to creating greater value for carriers. It has established strategic collaboration and joint innovation initiatives with carriers, increased investment in joint innovation with carriers, and constructed a number of global Centers of Expertise (COEs), such as the Service Operation Center (SOC) in Jakarta, Indonesia; the Big Data COE in Dublin, Ireland; the Global Network Evolution and Experience Center (GNEEC) in Beijing, China; and the Customer Experience Transformation Center (CETC) in Shenzhen, China. In the CEM field, the HUAWEI SmartCare CEM solution has maintained its leadership position in the industry. The scale delivery of the solution worldwide has created verifiable business value for customers. Huawei, in collaboration with the TM Forum, has led the creation of the CEM Metrics guidebook, establishing Huawei as the industry leader in standardizing CEM measurements. In Managed Services, Huawei is continuing to develop the next-generation Managed Services solutions to enhance the total value of ownership for operators. The solutions, which include Huawei's innovative strategic alignment and operations transformation model, Managed Services Unified Platform (MSUP) that supports ICT and FMC converged operations, and best-in-class OSS tools (E-iNOC), can deliver broad value including operational efficiency enhancement, together with network and service quality improvements. In the consulting and system integration field, Huawei is committed to increase the return on investment (ROI) for carriers and facilitate the smooth evolution of services and networks. Huawei will continue to strengthen investment in service solutions and related tools and platforms, develop the technical skills and working efficiency of its global service personnel, and enhance joint innovation with customers to provide them with more valuable and competitive service solutions, maximizing their ROI, and helping them transform to value-oriented operations.

As Huawei gets fully involved in deployment and operations of telecom carrier networks, engineering outsourcing is increasingly a key way of cost reduction in its overseas projects. Huawei has through its Turnkey Management Office worked in cooperation with dozens of local subcontractors. Huawei is mainly in charge of network planning, equipment supply, system integration, and operations. Subcontractors are in charge of site selection, construction, and equipment installation.

In enterprise business, Huawei gives priority two marketing and channels. Huawei Enterprise Channel business has developed steadily, supported by marketing and sales. Meanwhile, its channel policy, process, and IT support system have become significantly more mature. Also, Huawei has made efforts on finding more specialized channel partners for different products. By the end of 2013, it had more than 5000 channel partners worldwide. Huawei provides more support to its partners with respect to training and certification, Marketing Development Fund (MDF), finance, etc. and proactively conducts joint marketing activities with them. Both the quantity and quality of its partners have considerably increased. A sound and open channel ecosystem is taking shape.

Huawei Enterprise Services provides ICT services for the enterprise market, covering the entire network lifecycle. It includes technical consulting, network planning, network design, deployment, technical support, network optimization, as well as technical and presales training and certification. Huawei is dedicated to building a strong global ecosystem of service partners through certification, authorization, enablement, incentives, and all-around service support. Through this partner ecosystem it serves customers by leveraging its respective advantages. By the end of 2013, it had more than 800 certified and authorized service partners. Together with its partners Huawei delivered over 5000 projects during 2013. It has established a global ICT training and certification program, including Huawei's in-house training centers, authorized training partners, and education projects with more than 30 universities. Its training and certification program has helped develop ICT professionals around the globe. By the end of 2013, it trained more than 16,000 individuals and certified more than 3000 engineers.

4G LTE is developing rapidly. In particular, the Chinese market will lead the high-speed growth of LTE around the globe in the coming years. Huawei will improve consumer satisfaction by combining experience in hardware, applications, and services, and make success in smart terminals. In addition to consolidating the carrier channels, Huawei also actively implements a diversified channel strategy. First, it strengthens channel construction in open markets where retail accounts for a high proportion of all sales, such as China, Russia, Italy, Britain, Saudi Arabia, the Philippines, and South Africa. In addition, it has established strategic partnerships with top distributors and retailers in China, Western Europe, the Middle East, Southeast Asia, and other markets, which helped promote its large-scale sales. Its Internet marketing efforts in the Chinese e-commerce market also started to show results. In 2013, Huawei unveiled its "Make It Possible" brand proposition for the consumer business. By focusing on quality and consumer experience, the company conducted a series of branding campaigns worldwide, including sponsoring football games and clubs, such as La Liga in Spain, AC Milan in Italy, Borussia Dortmund in Germany, and Arsenal in Britain. These campaigns significantly boosted brand awareness for Huawei mobile phones.

In the meanwhile, Huawei has made brilliant achievements in service innovation, which becomes the ultimate guarantee and best booster of technological innovation. Service supply chains are set up to adapt to multiple processes of customers and fulfill customer needs. Huawei is thus able to provide value-added services for customers and discover and create market needs. With its standardized service systems and processes, Huawei serves world-class carriers with its sophisticated specialized service platforms and professional service groups. It continues to improve its service and build its competence in services. With ever-better industrialized, standardized and localized service platforms and systems worldwide, Huawei has seen its capability and brand influence well enhanced.

5.2.3.3 Location Selection

Huawei has established eight sales regions around the globe, namely China, Asia Pacific, Latin America, Europe, the CIS, the Middle East and North Africa, Sub-Saharan Africa, and North America. Customers enjoy around-the-clock service provided by 112 service organs and three Technical Access Centers (TACs) worldwide. Huawei boasts a service force of 14,300 professionals. Seventy-three percent of the company's employees outside China are local hires. Huawei is thus highly confident of its quality and effective services.

Huawei has established 11 Technical Assistance Centers (TACs) worldwide for its carrier network business by regions and service languages. See Table 5.4 for details.

Huawei has established 8 TACs worldwide for its enterprise business. See Table 5.5 for details.

5.2.4 Value Added of Huawei's Global Networks

5.2.4.1 Global Research and Development Network

IPR management remains Huawei's core strategy for open innovation and a major indicator of value-added of the company's global R&D networks. Huawei established the IPR Department in 1995. As of 2013, there were over 300 IPR-related technical experts, patent engineers, and lawyers in

TABLE 5.4 Huawei's TACs for Carrier Support

No.	TAC	Service Language	Region	Countries and Areas Covered
1	Egypt	Arabic, English, French	North Africa, Middle East, West Africa, South and East Africa	67 countries, including Egypt, Saudi Arabia, Nigeria and Angola, etc
2	Mexico	Spanish, English	Northern part of Latin America, southern part of South America	28 countries, including Mexico-Argentina, Bolivia, etc.
3	Brazil	Portuguese	Southern part of South America	Brazil
4	Romania	English	Eastern and Northern Europe, Western Europe, Central Asia and Caucasus, South Pacific	39 countries, including Romania, Germany, Turkey, Australia, etc.
5	Russia	Russian	Russia, Central Asia and Caucasus	10 countries, including Russia, Kazakhstan, etc.
6	Malaysia	English	South Pacific, Southeast Asia	24 countries, including Malaysia, Vietnam, etc.
7	China	Chinese	China	China
8	North America	English	North America	USA, Belize, Canada and Puerto Rico
9	India	English, Indian	India	India
10	Pakistan	English, Urdu	Middle East	Pakistan
11	Japan	Japanese	Japan	Japan

The authors made it based on related materials on Huawei website (http://support.huawei.com/carrier/).

copyright, trademark, and license. Huawei spends tens of millions of dollars on patent application around the world. With cross licensing, the company saves hundreds of millions of or even billions of dollars on patent royalties each year, saving significantly on cost of one-way permits and seeing wide application of its patented technologies, which generates enormous profits.

TABLE 5.5 Huawei's TACs for Enterprise Support

No.	TAC	Countries and Regions Under Service
1	China	China
2	Asia Pacific	29 countries and regions, including Malaysia, Singapore, the Philippines, Indonesia, India, Japan, Australia, etc.
3	Eastern Europe and Central Asia	29 countries and regions, including Russia, Belarus, Armenia, Turkey, Turkmenistan, etc.
4	Europe	31 countries and regions, including Germany, Italy, Switzerland, Romania, Austria and so on
5	Middle East and North Africa	11 countries and regions, including Egypt, Tunisia, Ethiopia, Saudi Arabia and South Africa, etc.
6	Sub-Saharan Africa	7 countries and regions, including South Africa, Togo, Cameroon, Kenya, etc.
7	North America	Canada and USA
8	Latin America	19 countries and regions, including Argentina, Brazil, Chile, Mexico, etc.

The authors made it based on related materials on Huawei website (https://e.huawei.com/en/service-hotline).

By the end of 2013, Huawei had joined over 170 industry standards and open source organization worldwide, including 3GPP, IETF, IEEE, ITU, BBF, ETSI, TMF, WFA, CCSA, GASM, OMA, ONF, INCITS, OpenStack and Open Daylight, and held 185 active posts. Huawei was a board member to ETSI, CCSA, OMA, OSASIS, and WFA. In 2013, Huawei submitted more than 5000 proposals to standards organizations.

Huawei has for years topped the list of invention and patent licenses published by the State Intellectual Property Office of the PRC (SIPO). According to the World Intellectual Property Organization (WIPO), in 2008, Huawei submitted 1737 international patent applications under the Patent Cooperation Treaty (PCT), ranking first for the first time. As shown in Fig. 5.9, the accumulated number of Huawei's patent applications increased from 35,773 in 2008 to 77,514 in 2013. Among them, the number of Chinese patent applications increased from 26,005 in 2008 to 44,168 in 2013; that of PCT international patent applications from 5446 to 14,555; and that of foreign patent applications from 4322 to 18,791. As of December 31, 2013, Huawei had 36,511 granted patents.

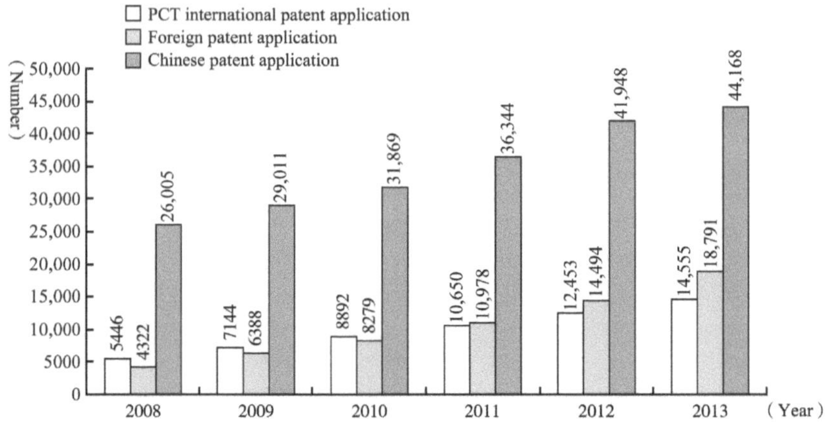

FIGURE 5.9 Cumulative numbers of Huawei's patent applications, 2008–13. Source: *Prepared by the author according to data provided in Huawei's annual reports.*

Huawei has contributed the most high-quality LTE/LTE-A standard patents since 2010. Specifically, it has demonstrated its strong capability for standards development by contributing 466 granted proposals to LTE/LTE-A core standards, achieving the global No. 1 position and constituting nearly 25% of all proposals granted globally. With its brilliant achievements in IPRs, Huawei has grown from a follower in 2G technology to a competitor for 3G technology and a leader of 4G technology. In a survey by the renowned American business media FastCompany, Top Five Innovative Corporations of 2010 were Facebook, Amazon, Apple, Google, and Huawei. Huawei has been a global model of CT innovation.

5.2.4.2 Global Production and Operation Network

Mr. Ren Zhengfei, who is the founder of Huawei, used to say, "A company is basically free from problems of management, once its ISC is in operation." With its global production and operations networks, Huawei is able to make full use of resources worldwide, speed up its production and capital turnover, lower costs and make easier management, and guarantee the global supply of its products. In addition, business and management processes are enhanced, and technological, management and service innovation boosted. Huawei's supply chain has been more competitive.

Huawei fell far behind its peers in terms of production and operations before it restructured its supply chain. The timely delivery rate was only 50% and inventory turnover at 3.6 times per year. The order-to-receipt cycle was as long as 20–25 days. Thanks to the restructured and integrated global production and operation networks, the timely delivery rate went up to 94% and inventory turnover to 9.4 times per year, while the order-to-receipt cycle was shortened to about 10 days. Accuracy of inventory and order increased from 96% to 99.5%, reaching the international average for telecom manufacturing. Huawei is now responsive-capable of zero inventory and delivery within 1 week.

5.2.4.3 *Global Marketing and Service Network*

According to Huawei, its sales increased from USD 1.4 billion in 1999 to USD 46.5 billion in 2014, up by over 30 times. Its foreign sales increased from USD 50 million to USD 28.94 billion over the same period, increasing by over 500 times. The rate of foreign sales to total sales increased from 4% to 62%, up more than 14 times (Fig. 5.10).

Huawei's sales increased from USD 1.4 billion in 1999 to USD 3.8 billion in 2014, up by 1.7 times. Its foreign sales increased from USD 53 million to USD 1.06 billion over the same period, increasing by 19 times. The rate of foreign sales to total sales increased from 4% to 28%, up by 6 times. Huawei saw its foreign sales and the rate of foreign sales to total sales keep increasing, despite a total at a relatively low level. The company made a progressive expansion on overseas markets as "a market substitute," when the domestic market was sluggish. Its "going global" strategy won initial success.

During the period of internationalization between 2004 and 2008, Huawei saw rapid and steadfast growth in its total sales, foreign sales, and rate of foreign sales to total sales. Its total sales increased from USD 5.6 billion to USD 23.3 billion, increasing by over 3 times. The foreign sales increased from USD 2.3 billion to USD 17.48 billion, up by more than 6 times. The rate of foreign sales to total sales increased from 41% to 75%, increasing by

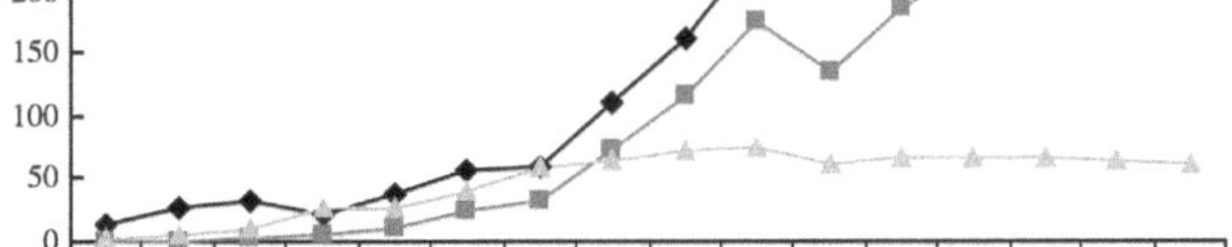

(Year)	1999	2000	2001	2001	2003	2004	2005	2006	2007	2008	2009	2010	2011	2012	2013	2014
Total sales (USD100 million)	14	27	31	21	38	56	58	110	160	233	218	280	324	354	395	465
Foreign sales (USD100 million)	0.5	1.3	3.3	5.5	10.6	23	33.6	71.5	115.2	174.8	133.7	184.6	219.8	235.7	256.2	289.4
Ratio of foreign sales to total sales (%)	4	5	11	26	28	41	58	65	72	75	61	66	68	67	65	62

FIGURE 5.10 Huawei's foreign sales and rate to total sales, 1999–2013. Source: *Prepared by the author according to data provided in Huawei's annual reports.*

34% points. Huawei made great success in the internationalization of its global marketing and service network.

Since 2009, Huawei has seen the rate of foreign sales to total sales remain 60%–70%. Its total sales and foreign sales have kept growing. The sales more than doubled from USD 21.8 billion to USD 46.5 billion. The foreign sales increased from USD 13.37 billion to USD 28.94 billion, more than double as well. Huawei has kept its momentum of growth in the postfinancial crisis times, by enlarging and consolidating its global marketing and service network for a globalized marketing service. This has contributed significantly to the success of Huawei's strategy of globalization.

5.3 HUAWEI'S GLOBAL PRODUCTION NETWORK FROM THE PERSPECTIVE OF EMBEDDEDNESS

5.3.1 Global Production Network of Communications Equipment

The world's communications equipment manufacturing is standing on a turn. Feeble in transformation, the industry has witnessed a marked slowdown since the global financial crisis, due to economic depression, plus the impacts from Internet businesses in the past decade. Major manufacturers are facing serious bottlenecks in growth and ever-fiercer competition. Cost advantage is gaining significance in competition. Thus, industry players find slowing income and reducing profits. In the meantime, as IT and CT get increasingly merged, IT software and hardware technologies are penetrating and having a profound influence on the communications equipment industry, which needs to change in development paradigm, product architecture, manufacturing mode, and industrial ecology. With those new industrial trends, it is key to build a future competence by understanding and grasping opportunities, enhancing the integrated innovation and integration of communications network and IT, promoting collaborated GPNs, and building an industrial ecosystem for independent development.

According to *2014 White Paper on Communications Equipment Industry* published by China Academy of Information and Communications Technology (CAICT) of the Ministry of Industry and Information Technology, the world's communications equipment industry has seen a low growth since 2007, which increased from USD 134.7 billion to USD 138.2 billion, with an annual growth at 0.4%. Carrier network equipment business saw a significant decline from USD 87.6 billion to USD 79.5 billion, with an annual decrease by 1.6%. Enterprise network equipment business had merely a slight annual growth at 3.7%, with a total size of USD 58.6 billion. Due to the marked slowdown of the industry, the world's major communications equipment manufacturers are falling to a new low. Half of the top six communications equipment manufacturers saw their revenues decline in 2013. Nokia's network equipment revenue declined by over 18%. ZTE and

Ericsson saw declines in revenue at 10.6% and 0.2%, respectively. ARUN lost EUR 1.3 billion, despite a 2.9% growth (calculated at fixed rate) after decline in revenue for 5 years in a row. At present, only Cisco and Huawei have been capable of continuous growth and profits. But the growth rate decreased from about 15% to 6%–8%.

Cost advantage is gaining significance in competition. The world's communications equipment industry sees profound changes in regional and business patterns. Chinese enterprises have taken more shares on the world's carrier network equipment market, from 8.5% in 2006 to 26.8% in 2012. The figure rose to 30% in 2013. It indicates an ever-higher status Chinese enterprises are in the world. In contrast, after Nortel Networks and Motorola withdrew from traditional communications equipment sector, only second-class enterprises such as Ciena and Tellabs remain on the market in North America. But in the field of Internet data communications equipment, Cisco and Juniper still hold the leading position. Cisco dominates the fields of Ethernet switch, enterprise router, and enterprise WLAN equipment, with its cutting-edge technologies, channel strength, client resources, and perfect services. Europe's market share declined from 57% to 41%, due to its weak competence in cost.

Due to a slowdown in innovation of technical infrastructure, the world's communications equipment industry has moved to cost-competitive regions. The innovation mode tends to bear characteristics of traditional industries, where technological innovation that focuses on product improvement and cost reduction is gaining importance. The capability of supply chain is taking significance in competition. The third-round large-scale restructuring has started in communications equipment industry. Manufacturers in the world has rezoned their interests, and established core links such as chips and key system software with vertical integration strategy, while outsourcing more of noncore parts such as code generating and testing. Network engineering is outsourced with turnkey. Also, network management service is provided together with sales of carrier network equipment. Enterprise ICT service is provided based upon enterprise network equipment. With the GPN-generated advantages in supply chain and cost, a player in the communications equipment industry tries to gain wider and stronger superiority in core products. IT concepts have a profound influence on the industry. A manufacturing mode based upon standard, universal, and open software and hardware architecture is on the horizon. An open platform-based ecosystem is taking shape. The pattern and operations of communications equipment industry are under influence of aforementioned trends.

5.3.2 Evolution of Huawei's Embeddedness in Europe

Europe cradled GSM and 3G and boasts world-class technologies and technological innovation and R&D. Three out of the top six communications equipment manufacturers are based in Europe, namely Ericsson, ARUN, and

Nokia. Even though Europe's market share declined from 57% to 41%, due to its weak competence in cost, European communications equipment manufacturers still hold the largest market share in the world. About 40% of the European markets are held by Nokia, about 30% by Ericsson, and about 20% by Huawei. The three top the European communications equipment market and are major rivals there. Europe is naturally a market of the most strategic importance to Huawei.

Based upon the framework of microcosmic GPN studies from the perspective of embeddedness in Chapter 3, Analytical Framework of Microcosmic GPN Studies, this section makes a detailed analysis of Huawei's embeddedness in Europe from the angles of strategy, dimension, and region (Fig. 5.11).

Huawei adopts the market access strategy of "entering from the remote outside" in Europe: Starting from lower end customers to whom Huawei's rivals had paid less attention, and applying "pressure tactic" by directing the company's limited resources on a single point, to make a breakthrough by surpassing its rivals. Strategically, Huawei's embeddedness in Europe has been pushed forward in four phases, namely exploratory embeddedness (1996–2000), strategic linkage (2001–04), strategic embeddedness (2005–10), and strategic coupling (2011 up to now). Huawei started from lower end customers in a specific market and gradually turned to large- and medium-sized enterprises. Regionally, Huawei started from less-challenging markets in the CIS and Eastern Europe to more challenging ones in Central,

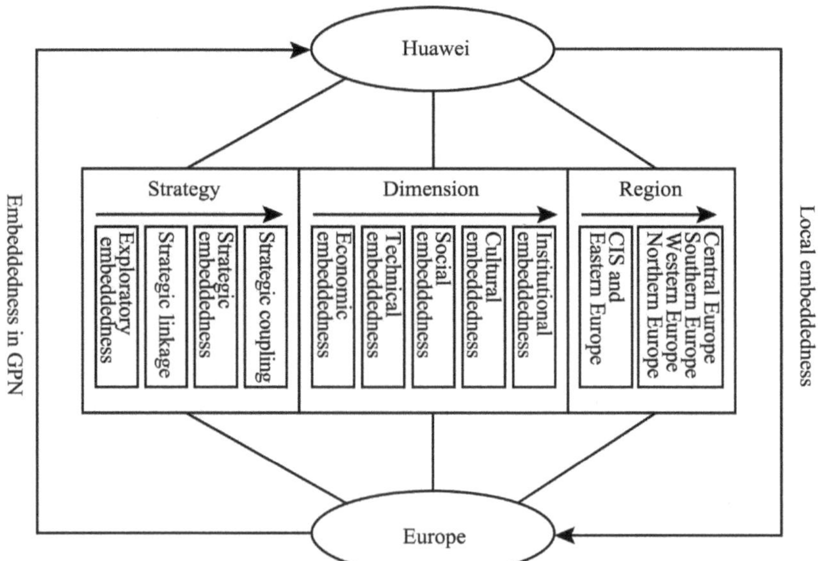

FIGURE 5.11 Analytical framework of Huawei's embeddedness in Europe. Source: *Prepared by the author.*

Southern, Western, and Northern Europe. By dimension, economic and technical embeddedness went first before social, cultural, and institutional embeddedness. Behind Huawei's strategy of "entering from the remote outside" in Europe lies its "binary learning" strategy that has boosted the company's market presence. "Binary learning" strategy has been mainly materialized by two types of strategic institutes, namely R&D centers and innovation centers. An R&D center is directly funded by Huawei itself for exploratory learning. In addition, Huawei has coestablished a number of joint innovation centers in Europe together with world-class carriers, to settle cost and technical problems for customers. That is, an innovation center is set up for exploitative learning. Huawei's "binary learning" strategy is designed "to create value for customers" in three ways, namely individualized solutions, low price, and responsiveness. Huawei has made a balance of the three and shaped its new competitive advantages, imposing acute challenges to major Western enterprises on European markets.

5.3.2.1 Exploratory Embeddedness Phase

Huawei made tentative efforts on European markets between 1996 and 2000.

Huawei started its path into Europe from Russia. It set up its Moscow Office in 1996. However, there was no order at all in the next 4 years. In 2000, Huawei made its breakthrough on the Russian market by successfully sealing contracts for Uralsvyazinform's switch project and Moscow Mobile TeleSystems's (MTS) mobile network contracts. In the same year, Huawei established an R&D center in Stockholm, Sweden, to enhance the company's core competence with Sweden's superiority in telecom resources. In this period, Huawei had no idea about how to expand its overseas markets with what organizational structure or in what way to allocate its market resources. It knew nothing about how to grasp needs of overseas markets or what types of talents it needed for an overseas expansion. The company simply tried and failed. But it never gave in. Its years of persistence was finally paid off when it made a breakthrough. In this way, Huawei accumulated precious experience in internationalization and laid a solid foundation for its internationalized operations.

5.3.2.2 Strategic Linkage Phase

Huawei tried to establish a strategic connection with European markets between 2001 and 2004.

Huawei set up the first office in Britain in 2001, indicating the company's presence in Europe and its strategic connection with European markets officially established. Huawei at first got into good partnership with world-class agents local to Europe. In 2001, Huawei's 10GSDH optical network products entered the German market. The company then cooperated with German Telecoms and made a successful breakthrough in Germany. At the end of

2002, Huawei won the contract for the Moscow–Novosibirsk long-haul optical backbone network and became one of the major telecom switch equipment suppliers on the Russian telecom market. In 2003, Huawei won the contracts for DWDM transmission backbone network of the French company LDCOM and for British Telecom's VoIP long-haul commercial network project, heralding the large-scale entry of China's high-end optical network products into telecom markets in developed European countries. Later, Huawei established a strategic partnership with multinational carriers including British Telecom, France Telecom, Telefonica, and Vodafone, expanding its presence in developed countries such as France, the Netherlands, Italy, and Spain, realizing sales of scale. A subsidiary was established in Sweden in 2003, followed by offices in Norway, Denmark, and Finland. A network covering Northern Europe was in shape. At the end of 2003, Huawei and Siemens sealed a cooperation agreement, with which Siemens would increase year by year the sales volume of Huawei's data communication products on the European market. In 2004, Huawei achieved first significant contract win in Europe valued at over USD 25 million with Dutch carrier, Telfort.

In the stage of strategic connection, Huawei's embeddedness in Europe was largely a single-way economic embeddedness, quite weak in the technical sense. It started from lower end customers in a specific market and gradually turned to large- and medium-sized enterprises. It was at the time when Huawei began to be a player in international competition, following international rules and standards and norms. It gradually became a major supplier on international telecom markets. Huawei has set up well-structured rules for its overseas operations. The rules have been drawn on regulations set by Chinese embassies or the company's experience in international and domestic markets. With its successful expansion in European markets, Huawei has further grown and seen more opportunities for an enhanced core competence. In the meanwhile, the company has created a better environment for cross-border sharing of resources and knowledge.

5.3.2.3 Strategic Embeddedness Phase

Huawei made its strategic embeddedness in Europe between 2005 and 2010.

Huawei turned to sole ownership of a foreign company in 2005, providing stronger incentive systems and solving cultural clashes in previous joint ventures. To further accelerate its pace of internationalization, Huawei adopted favorable salary and promotion polices for its overseas employees, who were guaranteed with much higher salaries and remunerations than their colleagues working back in China. Meanwhile, more authority was delegated to field offices, as what Mr. Ren Zhengfei once said in an in-company speech, "Let those who are on the forefront make a decision." During its strategic embeddedness in Europe, Huawei's single-way economic embeddedness grew into a two-way one. Based upon its economic embeddedness, Huawei took steps forward in social, cultural, and institutional embeddedness in Europe. It saw the remarkably stronger technical embeddedness as well. Local R&D and international collaboration both substantially increased.

Huawei signed in 2005 with Vodafone a Global Framework Agreement, as Vodafone's preferred telecom equipment supplier. In the same year, Huawei was selected as a preferred 21CN supplier by British Telecom, supplying BT's 21CN with multiservice network access (MSAN) components and optical transmission equipment. In 2006, Huawei Austria was established. Huawei signed with Vodafone a strategic cooperation agreement for 3G handsets in 2006, and in 2007, won Vodafone's "2007 Global Supplier Award," the only network equipment supplier to be awarded this specific accolade. Also in 2007, Huawei established a joint venture with Global Marine, providing end-to-end submarine network solutions. By the end of 2007, Huawei's core communication equipment was adopted by all the major carriers in Europe and Huawei became a partner to all the top carriers in Europe. Huawei European Research Institute was established in 2008, headquartered in Munich, Germany. In 2010, an R&D center was set up in Istanbul and the Cyber Security Evaluation Center (CSEC) established in Britain.

5.3.2.4 Strategic Coupling

Huawei has made strategic coupling in Europe since 2011.

The ICT industry was at a new starting point in 2011. Huawei intended to lay a foundation for its globalized survival, with a globalized business pattern. On the 10th anniversary of its entry into European markets, Huawei announced its intention to make Europe its second homeland and become a "European company" by investing more and getting more open-minded. The company began its strategic coupling in Europe. With ever-deepening two-way economic, technical, social, cultural, and institutional embeddedness, Huawei created in Europe "glocalized" R&D, production and operations, and marketing and service networks. Huawei and customers, suppliers, and sub-suppliers share the same business ecosystem that thrives on shared success, so that all values generated in networks are shared with customers around the globe and the industry ecosystem sees a sound and sustainable development.

In 2012, Huawei stepped up investments in Britain and established a new R&D center in Finland. Also in that year, Huawei set up local BoDs and advisory boards in France and Britain. In 2013, Huawei set up the FRCC in London to manage global financial risks and ensure that its financial operations remain efficient, secure, and standard-compliant. Huawei's European Logistics Center was put into official operation in Hungary, covering countries throughout Europe, Central Asia, the Middle East, and Africa. Huawei has completed its localization in Europe. Huawei's sales revenue in Europe reached USD 5.23 billion in 2013, an increase of 25% year-on-year. The company has at least 10% of its sales revenue put in R&D each year. It input USD 19.4 billion in R&D between 2000 and 2012. Its R&D input in Europe saw a compound annual growth rate at 28% between 2009 and 2013. Huawei had about 7700 employees in Europe in 2013, roughly about 5% of its global employment. And 80% of its European employment was local hires. In the next 5 years, Huawei was going to hire 5500 employees in Europe. Huawei has set up local BoDs and advisory boards in France and Britain, so that local senior management participates in drawing and implementing development strategies. In addition, Huawei has worked hard on local talent training. In 2013, Huawei made a total USD 3.4 billion procurement on European markets, an increase of 11% year-on-year, of components, local engineering service, and international logistics. Huawei expects a steadfast growth of procurement in Europe in future. Local procurement has boosted local employment and development, and deepened technical and innovation exchanges between enterprises, so that they all see enhanced competence.

5.3.3 Analysis of Huawei's Embeddedness in Europe by Dimension

Huawei's embeddedness in Europe has been pushed forward in four phases, namely exploratory embeddedness, strategic linkage, strategic embeddedness, and strategic coupling, all through the years of which, by dimension, it has evolved from a single-way embeddedness to a two-way one, first in economic and technical spheres and later in social, cultural, and institutional dimensions.

5.3.3.1 Economic Embeddedness

Huawei took its first important step in Europe by establishing partnership with world-class enterprises local to Europe. In backstage embeddedness, Huawei has launched procurement in many European countries. In certain high-end areas, it has established stable long-term buyer–supplier relationship with multinational magnates. In 2013, Huawei made a total USD 3.4 billion procurement on European markets, of components, local engineering service, and international logistics. In foreground embeddedness, Huawei has established strategic partnerships with all the top carriers in Europe, including Deutsche Telekom, British Telecom, France Telecom, Telefonica, and Vodafone. It has also established strategic collaboration and joint innovation initiatives with carriers, increased investment in joint innovation with carriers, and constructed a number of global COEs.

Take Huawei and Vodafone's sustainable collaboration for example. At the end of November 2005, Huawei signed with Vodafone, which is the world's largest mobile carrier, a Global Framework Agreement, as Vodafone's preferred telecom equipment supplier. The two companies announced their new collaboration on the 3GSM World Congress, Barcelona 2006 that Huawei would supply exclusive Vodafone-branded consumer 3G handsets for Vodafone across 21 countries for at least 5 years. In October 2006, Huawei and Vodafone established in Spain the first joint innovation center MIC. In October 2010, Huawei and Vodafone sealed an agreement valued at EUR 700 million or about USD 970 million. According to their strategic cooperation agreement, Huawei would provide Vodafone with all-series SingleRAN mobile communications solutions, end-to-end FTTx optical broadband and fixed-mobile convergence solutions, and engineering, construction, and other related services. Thanks to its collaboration with Vodafone, Huawei has won more opportunities of cooperation with world-class carriers. Therefore, Huawei has been able to deploy its future-oriented wireless network, and in the meantime, carriers have enhanced their competence in energy saving and emission reduction, as well as clean energy.

Huawei has stepped up investment in Europe, intending to take a firm foothold there as its second homeland. In 2011, Huawei announced its plan to double its employment in Britain to 1000 employees within 3 years. In March 2012, the company stated that it would increase its investment in Europe and enlarge its employment there. On September 11, 2012, Huawei announced in London a new investment of GBP 1.3 billion (or about RMB 13 billion) to enlarge its business in Britain. Half of the new sum would be used in investments and the rest half for procurement. Some of the sum for investment would be used to enlarge its R&D center in Ipswich Town in east England. In the coming 5 years, Huawei would invest in mobile broadband and other projects in Britain, creating 700 new jobs in the country. Huawei has set up subsidiaries in 20 European countries, including Britain, Poland, Germany, France, the Netherlands, Italy, Spain, and Sweden. The company has established six COEs for finance, marketing, and service, including the FRCC in London, Britain and the European Logistics Center in Hungary. Huawei's carrier network business has set up TACs in Romania and Russia, covering over 50 countries and regions in Europe, Central Asia, and South Pacific. Its enterprise business has established TACs in the EU, the CIS, and Turkey, covering more than 40 countries and regions in Europe and Central Asia. TACs in consumer business in Europe covered nearly 30 countries and regions.

5.3.3.2 Technical Embeddedness

Europe boasts the world-class technologies and technological innovation and R&D. With its strategy of open innovation, Huawei is able to well integrate and make good use of technological resources and talents in Europe, accelerating the pace of internationalization of its R&D. Huawei has set up 17 R&D institutes in Europe and established extensive partnerships with leading European carriers, universities and colleges, and industry organizations in fields of policy, standards, R&D, and procurement, creating value for locals and Huawei's customers around the world. Also, Huawei has established 18 innovation centers in Europe, covering all major areas of wireless access, network CT, business support system, energy, and solutions. The company has won a number of innovation awards by Vodafone and BT. Huawei won FT's Boldness in Business Award 2009 and the 2010 Corporate Use of Innovation Award by *The Economist*. Its R&D input in Europe saw a compound annual growth rate at 28% between 2009 and 2013.

Huawei entered the Russian market as early as 1997. It has established one R&D center, one training center, and dozens of service offices there. Huawei set up an R&D center in the Swede capital Stockholm in 2000 and Huawei Sweden in 2003. Later, offices were established in Norway, Denmark, and Finland, covering the entire region of Northern Europe. At present, it has one research institute in Sweden. Huawei European Research Institute was established in 2008, headquartered in Munich, Germany, with five offices based in four countries, namely Germany (Munich and Nuremberg), Italy (Milan), Britain (London), and Belgium (Brussels). Huawei started its business in Turkey in 2002 and established an R&D center in Istanbul in 2010. The business in Turkey was enlarged in 2014.

Increasingly embedded in Europe, the company aims to double its European employment in coming years. In 2010, Huawei and Option, a Belgian wireless Internet equipment manufacturer signed an agreement, to coestablish a joint innovation center in Belgium. In 2013, Huawei bought Caliopa, a Belgian company engaged in silicon photonics, and incorporated it into the R&D center in Belgium, enhancing Huawei's R&D capability in silicon photonics-based optical device solutions. In 2011, Huawei inaugurated its only Global Microwave R&D Center in Milan, Italy, for innovation of microwave technology. In 2012, Huawei set up in Helsinki, Finland, an R&D center focusing on software development for Android and Windows Phone smartphones. In 2013, Huawei set up new R&D centers in Cork City and Dublin, Ireland, focusing on HUAWEI SmartCare, the next-generation customer experience management product, so that the company is able to provide leading customer services for telecom carriers in Ireland and around the globe. Huawei established an R&D center in Ipswich Town in the east of Britain and built another one in Bristol in the southwest of the country in 2014. The two centers focus on the development of semiconductor chips and software for communications and information terminals. In 2014, Huawei inaugurated its Chip R&D Center in Sofia Antipolis in Alpes-Maritimes, France, focusing on chip design and R&D for embedded electronic equipment, to improve smartphone's camera, develop the best image signal processor and enhance the company's superiority in microelectronics and software. Huawei also set up four R&D centers for mathematics, aesthetics, home terminals, and wireless network standards. Huawei also worked on a big data R&D center in Paris, to promote the further development of technology stacks through data studies. Huawei has established 17 R&D institutes in eight European countries, namely Belgium, Finland, France, Germany, Ireland, Italy, Sweden, and Britain.

5.3.3.3 *Social Embeddedness*

As a globalized company, Huawei has not only focused on its own development, but also shouldered its social responsibility, promoting social harmony and advancement. Huawei has incorporated sustainability requirements into its business activities and built sustainability into its operations. It has continued to focus on its four sustainability strategies: bridging the digital divide, supporting stable and secure network operations, promoting environmental protection, and seeking win–win development. Its goal is to ensure harmony in the economy, environment, and society, by promoting the fulfillment of social responsibilities of the company and entire value chain.

5.3.3.3.1 Bridging the Digital Divide

Huawei has been committed to promoting equal access to the information society. Helping people with easy access to basic voice communications, Huawei focuses more on enabling broadband inclusion for all, helping people enjoy more online content and services and promoting ICT-based transformation across industries. In the meanwhile, Huawei has worked hard on nurturing ICT talent and spreading ICT knowledge, to enhance CT in every local market where the company runs its business. Huawei aims to support and build an efficient integrated system of information and logistics, and works in cooperation with its partners to bridge the digital divide and pull the world ahead. Huawei's 45 training centers around the globe provide training services in 16 languages, including English, Spanish, French, and Russian. The company has launched Telecom Seeds for the Future Program in 23 countries, including Britain, France, Germany, Italy, and Spain. By the end of 2013, the Program had been extended to benefit more than 10,000 students. As a member of the Broadband Commission, Huawei has worked together with the entire industry to increase the penetration of broadband services around the globe, so that all people, those in developing countries and regions in particular, gain convenient access to affordable high-speed broadband services. Furthermore, with its broadband services, Huawei has helped developing countries and regions enhance basic social services, such as medical service and education, facilitating to accelerate the pace in realizing the Millennium Development Goals and goals of the Post-2015 Development Agenda.

5.3.3.3.2 Supporting Stable and Secure Network Operations

Cyber security is under ever-serious threats and challenges, as ICT keeps evolving and developing. Supporting the stable and secure operations of customers' networks is its most honored commitment to its customers. Huawei has always placed stable and secure network operations over the company's commercial interests and worked together with others in unison to build a transparent network of sustainable development. Huawei provides worldwide network facilities and solutions, guaranteeing with all its strength stable and secure network operations in any conditions, even extreme ones. The company has incorporated cyber security into its technical design, construction, and deployment, building and implementing an end-to-end global cyber security assurance system. Huawei has established a multilayer cyber security evaluation process, consisting of Huawei Internal Cyber Security Lab, Britain Cyber Security Evaluation Centre (CSEC), customer evaluation teams, and third-party audit and evaluation teams. By doing so, Huawei continuously provides its customers with optimum security assurance. Huawei has published its cyber security white papers, promoting the formulation and implementation of uniform international cyber security standards, to realize the common cyber security objectives. An open, transparent, and visual framework for security problem solving facilitates the sound and sustainable development of the entire industrial chain, promotes CT innovation, and helps communications between people.

5.3.3.3.3 Promoting Environmental Protection

Huawei has energetically carried out its "Green Pipe, Green Operations, Green Partner, Green World" strategy, incorporating a green ICT idea into the entire product lifecycle and building green communications networks by ongoing innovation and enhancement of product energy efficiency. Huawei has reduced its own carbon footprints by utilizing clean energy and reducing technological and management costs. Huawei has deployed approximately 20,000 green base stations worldwide. Wind and solar energy can be leveraged to supply most of the power used by these base stations, saving 80% of fuel consumption. This helps carriers reduce carbon dioxide emissions during network expansions and lower their operational expenditures. In 2013, Huawei's Global Evaluation & Testing Center evaluated 24 products as "green products," three intelligent switches of which passed Tüv Rheinland green production evaluation. Huawei has been the world's first ICT enterprise awarded with this label. Huawei has furthermore helped and encouraged suppliers to launch energy-efficiency and emission-reducing projects. In 2013, four pilot suppliers with such a project reduced over 20,000 tons of CO_2 emissions in total. In future, Huawei will launch such projects in more suppliers, further promoting the low-carbon and environmentally friendly development of the industrial chain. Also, with its superiority and experience in energy-efficiency and environmentally friendly design, Huawei has provided and promoted green ICT solutions, promoting energy-saving and emission-reducing across industries and helping build a recycling and low-carbon society. Huawei has kept strengthening its collaboration with suppliers in green logistics. With deliveries well guaranteed, Huawei chooses the most environmentally friendly mode of transport, reducing its environmental impacts to the minimum. For instance, Huawei sent a batch of products to Duisburg, Germany, by train, the carbon emission of which was only 4.7% of that of air transport. In 2013, Huawei France launched a SmartExchange Program, encouraging consumers to trade in old mobile phones for recycling or reuse, in exchange for a discounted new smartphone. This program also created new jobs for the unemployed thus had certain social benefits.

5.3.3.3.4 Seeking Win–Win Development

Huawei aims to contribute to and create value for the sustainable development of the industrial chain, by working in cooperation with others. It gives employees access to broad platforms where they can realize their individual value. Huawei abides by business ethics, operates with integrity, and complies with applicable laws and regulations. It contributes to community development in the place where it runs business. The company keeps enhancing suppliers' capabilities and together with suppliers, leads the trend of sustainable development. Huawei stays customer-centered and makes ongoing improvements in accordance with customer needs, to enhance customer satisfaction. It saw a significant advancement year-on-year in 2013's third-party survey of customer satisfaction. Huawei's social charity activities are designed to attain four goals: bridging the digital divide, creating opportunities through education, promoting environmental initiatives, and contributing to the communities where it operates. Specifically, it contributes to local communities by supporting charity, education, environmental protection, health, and disaster relief efforts. It aims to become part of local communities, create value for them, and help them achieve prosperity and sustainability. It launched the SmartBus program in Spain and Portugal, providing ICT application training for over 10,000 students in over 100 schools; supported the Prince's Trust Foundation in Britain; supported top ICT students in Hungary through the Huawei Innovative Leaders of Tomorrow Scholarship; launched the InnoApps Challenge in Belgium to foster youth entrepreneurship; and supported the region of Van in Turkey in cooperation with the Ministry of Education, Turkcell and Turkey Education Association (TEV) for postdisaster relief.

5.3.3.4 Cultural Embeddedness

Huawei has advocated a diversified workforce. In 2009, it made adjustments to its HR structure and turned from a functional HR platform to the triple-pillar HR platform that is friendlier with employees and helps employees in their career path. The HR transformation was basically completed in 2013, building a channel for its multinational employment to realize their individual values. Huawei always attaches importance to the local development of its overseas employees. Local hires are prioritized within the framework of local laws in places it operates, to promote local employment and economic growth. In 2013, the company saw that 79% of its overseas employment was local hires and 20.7% of its overseas middle and senior management was local people. Huawei has approximately 30,000 foreign employees from more than 160 countries and regions. In 2013, Huawei saw 80% of its employment in Europe as local hires, promoting local employment and economic growth. Huawei has been a contributory corporate citizen in the places where it operates.

In addition to strictly following the laws on minimum wage, Huawei has a highly competitive compensation system. It has implemented an effective employee benefits system, which acts as an umbrella to protect its employees. Apart from mandatory insurance plans, the company offers every employee global accident insurance, critical illness insurance, life insurance, medical insurance, and business travel insurance, as well as other forms of commercial insurance. It has also implemented medical assistance mechanisms to protect its employees under special circumstances. Huawei has kept improving its global employee support system, by optimizing its management policy on overseas employee insurance and benefits. The company strictly complies with local laws and regulations. Huawei prohibits the use of forced labor. Under its antidiscrimination policy, when recruiting, training, promoting, or setting compensation for employees, it never discriminates against its employees on the basis of race, color of skin, age, gender, sexual orientation, ethnicity, disability, pregnancy, religious belief, political affiliation, society membership, or marital status. Huawei strictly prohibits the use of child labor. It shows due respect for an employee's religious belief and protects employees' privacy.

When Huawei made its entry into the smartphone market in Europe, it started from football, the favorite sport to Europeans. The company made a success there in the Europe-wide culture of football. In July 2011, Huawei made its first attempt in football events, being a sponsor to the 2011 TIM Cup (Italian Super Cup), as one of its important clients, Telecom Italia, was the founder of the football game that staged races in the Bird's Nest in Beijing, China. In April 2012, Huawei announced its sponsorship for the game between Atletico Madrid and Real Madrid. It was the first time that Huawei would be a sponsor to a sports game in Western Europe. Since then, Huawei has been a part of the European culture of football. Huawei has sponsored football games and clubs, such as Borussia Dortmund in Germany, AC Milan in Italy, Arsenal in Britain, Paris Saint-Germain in France, AFC Ajax in the Netherlands, and RSC. Anderlecht in Belgium, and established a strategic partnership with La Liga in Spain. The company has also provided wireless network services for clubs' venues that are covered with high-density WLAN networks, enabling audience to gain access to ultrafast Wi-Fi services. Huawei has chosen football clubs that are similar to it in culture and style, so that Huawei's products would get closer to potential consumers. Huawei has thus enjoyed fast-increased brand awareness in the past years, thanks to an ever-better product quality and football culture embedded marketing. Huawei has worked hard to establish a good brand image in Europe, laying a solid foundation for the sustainable development of its European business. According to a consumer survey of 32 countries conducted by the market research firm Ipsos, awareness of the Huawei brand rose from 52% in 2013 to 65% in 2014, indicating that Huawei was already known to approximately two-thirds of consumers in the world. Huawei has stood out and become one of the top three brands in terms of brand momentum. It became the first mainland Chinese company to enter Interbrand's 2014 Top 100 Best Global Brands list.

5.3.3.5 Institutional Embeddedness

Huawei intends to make Europe its second homeland and become a "European company," and it needs an in-depth institutional embeddedness in Europe. Huawei abides by ethical business practices, conforms to international conventions as well as laws of local countries, and operates with integrity. It adheres to the Huawei Business Conduct Guidelines (BCG), implements "transparent procurement" and "transparent sales," and opposes bribery and corruption. In addition, it advocates fair competition and obeys antidumping and antitrust laws and regulations defined by local countries, to create a harmonious business ecosystem. Huawei has incorporated operational compliance into its business processes. The Legal Affairs Department provides legal instructions on operations in compliance with applicable laws and regulations concerning export control, cyber security, trade competition, HR management, and antibribery and anticorruption, and identifies, evaluates, and warns of any internal or external legal risks, assisting business sectors in operations in compliance with applicable laws and regulations. Huawei protects its own IPR while respecting the rights of other IPR holders, and ensures that the company complies with international IPR regulations.

Huawei has strictly abided by all the applicable laws and regulations on export control by the United Nations, China, the United States, the EU and all other countries and regions, fulfilling its responsibility and obligations of export control and placing those obligations over the company's commercial interests. To fulfill its commitment to export control, Huawei has set up the Trade and Customs Compliance Committee and Trade Compliance Office led by Chief Legal Officer, to review and approve the company's trade compliance policy and its implementation and supervision, so that Huawei obeys in full measure applicable laws and regulations in terms of policy, organization, and process. By establishing a standard process of export inspection, Huawei is able to conduct inspections of users, end use, and hazardous factors for any export control projects, ensuring that the end-to-end Internal Control Program (ICP) is effectively put into implementation. Huawei prohibits bribery and corruption. Huawei abides by business ethics, operates with integrity, and adopts active and effective mechanism and measures for preventing and eliminating bribery and corruption. Huawei has worked out and implemented the *Regulations on Anti-Corruption and Anti-Bribery*.

Huawei's attempt at institutional embeddedness in Europe has been well incorporated into its technical strategy. In May 2000, the International Telecommunication Union (ITU) announced that China's TD-SCDMA, Europe's WCDMA, and the US' CDMA2000 would be the three international standards for 3G. Clear about its future in overseas markets, Huawei did not hesitate to set WCDMA as its R&D priority. By early 2008, Huawei had won over 100 HSPA/WCDMA contracts and established strategic partnerships with the top carriers in Europe, including Deutsche Telekom, Vodafone, British Telecom, Telefonica, and France Telecom. WCDMA finally became the mainstream worldwide in the 3G era. This had a further influence on the pattern of the 4G times. Huawei has played an active role in the mainstream international standards, to realize its worldwide expansion. By the end of 2013, Huawei had joined more than 170 industry standards and open source organizations held 185 positions in these organizations. In 2013, Huawei submitted more than 5000 proposals to standards organizations.

Governments of European countries have little intervention with operations of enterprises. Huawei had never paid an official call to any government in Europe before 2010 when the EU filed an antidumping lawsuit against Huawei. The company had not tried to communicate with governments in Europe until the EU put the case on record. The EU finally discharged the accusation of dumping against Huawei. Huawei passed the investigation of antidumping and antitrust, because of its localized business in Europe. Huawei set up the CSEC in Britain, conducting security testing on the equipment products sold on the British market, to better satisfy the British government's security requirements on communications equipment. All the testing staff in the CSEC are of British nationality with security-related qualifications issued by the British government. Under France's protectionist tendency, Huawei got in contact with French government officials, Minister of Trade, Minister of Digital Economy and officials from the Prime Minister's Office, and finally managed to get clear off Alcatel-Lucent's layoff event. By its ongoing input and in-depth embeddedness in Europe, Huawei has improved its relations with governments of European countries and attracted many talents for its own growth. It has been a development of shared success.

5.4 SUMMARY

Huawei adopted the "going global" strategy in 1996 when the domestic market was sluggish, "to live on." With the development in the two decades or so, it has solidly grown into a Fortune 500 company boasting annual sales of around RMB 240 billion. In 2014, Huawei surpassed Ericsson in terms of carrier business and sales revenue, becoming the industry leader in every sense. Huawei has become globalized in operations and investment, and particularly in a new business concept known as "glocalized" operations: As a globalized company that runs businesses in more than 170 countries and regions, Huawei has strived to build a globalized supply chain by integrating quality resources worldwide, sharing all values generated on the value chain with its customers around the globe. Huawei runs localized operations in the regions and countries where it runs businesses, to boost local employment and taxes, contributing to local development. With cooperation with excellent local enterprises based on division of labor, Huawei has fully integrated its strength in global value chain with local innovation power, helping local innovation gain global value. In the meantime, Huawei has abided by and complied with laws and regulations in the countries and regions where it operates. It has worked hard to strengthen communications with governments, media, and other external parties of interests, and strived to be a responsible corporate citizen, an enabler of ongoing innovation in the information society, and an industry contributor aiming at shared success. Seeing globalization and localization as the two sides of one coin, Huawei has successfully integrated the two and incorporated them into its business practice. It has set an example of independent innovation and global development of Chinese companies by integrating Western management science with Chinese ethos.

Based on the framework of microcosmic GPN studies proposed in Chapter 3, Analytical Framework of Microcosmic GPN Studies, this chapter analyzes, from the perspectives of value, value objectives, global strategy, networks, location selection and governance model of Huawei and its global R&D, production and operation, and marketing and service network, evaluating value-added of those networks; and details, from the perspectives of embeddedness, the global network pattern of telecommunications equipment industry, the trade structure of China's telecommunications equipment industry, and Huawei's embeddedness in Europe by economic, technical, social, cultural, and institutional dimensions. The chapter makes a comparison between GPNs of Siemens and Huawei from the perspectives of entirety, value, and embeddedness and summarizes Huawei's experience and shortcomings. The case study of Huawei's GPN in this chapter enables a better understanding of framework of microcosmic GPN studies, analyzing the driving forces and gains of GPN from a microcosmic perspective and combing the process and evolution of Chinese enterprises' embeddedness in overseas markets. It may provide a better theoretical framework for more Chinese enterprises to implement "going global" strategy by getting embedded in GPNs.

Chapter 6

A Comparison Between GPNs of Huawei and Siemens

Chapter Outline

6.1 OVERALL COMPARISON

Both Siemens and Huawei have succeeded in operating globally. As shown in Table 6.1, their global production networks are similar in value objectives, global strategy, networks, and location selection; both aim to achieve sustainable value creation and play a dominant role in producer-driven global networks. Their differences can be seen in four aspects. Huawei expands globally faster while Siemens has a higher transnationality index (TNI). Siemens focuses more on the overall global strategy "One Siemens" while Huawei gives equal importance to global expansion and local operations. Siemens's governance model is typical in Europe and the United States while Huawei's governance model is quite special. Siemens's performance in the three major regions of the world is more balanced while Huawei's performance in North America is at an average level.

Global Value Chains and Production Networks. DOI: https://doi.org/10.1016/B978-0-12-814847-1.00006-3

TABLE 6.1 Overall Comparison Between GPNs of Siemens and Huawei

Item	Siemens	Huawei
Global expansion process	Global strategy launched in 1989 and foreign sales up 2.5 times from fiscal 1990 to fiscal 2014	"Going global" strategy launched in 1996 and foreign sales up over 500 times from 1999 to 2014
TNI	77.4% in 2011	42.08% in 2011
Value objective	Achieve sustainable value creation	Achieve sustainable value creation
Global strategy	One Siemens	Glocalization
Networks	Producer-driven, including global R&D network, global manufacturing network, global supply network, and global sales and service network	Producer-driven, including global R&D network, global production and operation network, and global marketing and service network
Network governance	Follow typical European or U.S. corporate governance rules internally; adopt relational governance for core network members externally	Wholly owned by employees plus rotating CEOs; modular governance for core network members
Performance by region	Revenue in Europe Americas, and Asia-Pacific accounted for 53%, 27%, and 20% of the total, respectively for about four fiscal years	Revenue in Asia-Pacific, Americas, and Europe accounted for 51%, 13%, and 35% of the total, respectively for about four years

Prepared by the author.

6.1.1 Huawei's Faster Global Expansion Versus Siemens's Higher TNI

Siemens was founded in 1847 and launched its global strategy in 1989. Between fiscal 1990 and fiscal 2014, its foreign sales rose by 2.5 times from EUR 17.6 billion to EUR 61.1 billion and the ratio of foreign sales to total sales increased by 0.5 times from 55% to 85%. Huawei was established in 1988 and launched its "going global" strategy in 1996, which produced initial results in 1999. From 1999 to 2014, Huawei's foreign sales went up by over 500 times from USD 530 million to USD 28.94 billion and the ratio of foreign sales to total sales grew by more than 14 times from 4% to 62%. Compared with Siemens, Huawei is very fast in expanding globally.

However, Huawei is left far behind by Siemens in terms of TNI. According to UNCTAD, in 2011, the average TNI of the world's top 100 nonfinancial multinational corporations was 60.78%, and the average TNI of the top 100 nonfinancial multinational corporations in developing countries and transition economies was 40.13%. Siemens's TNI was 32.5% in 1993, kept growing after that, and reached 77.4% in 2011, making it the 30th largest company in this regard among the world's top 100 nonfinancial multinational corporations; its ranking by assets was number 17. According to the list of China's top 100 multinational corporations and TNIs released by China Enterprise Confederation and China Enterprise Directors Association, Huawei's TNI was 42.08% in 2011, slightly higher than the average TNI of the top 100 multinational corporations in developing countries (40.13%) and far lower than the average TNI of the world's top 100 multinational corporations (60.78%) and Siemens' TNI in the same year (77.4%).

TNI is calculated as the arithmetic mean of three ratios: the ratio of foreign assets to total assets, the ratio of foreign sales to total sales, and the ratio of foreign employment to total employment. There are differences between Huawei and Siemens in the three ratios. As shown in Table 6.2, the ratio of Huawei's foreign sales to total sales in 2011 was 68%, slightly lower than Siemens's 85%; the ratio of its foreign assets to total assets and that of its foreign employment to total employment in 2011 were 38% and 21%, respectively, much lower than Siemens's 79% and 68%. It suggests that Huawei still has great potential in making investments and hiring people locally in the process of carrying out its glocalization strategy.

6.1.2 Siemens Focusing on the Global Strategy "One Siemens" and Huawei Giving Equal Importance to Global Expansion and Local Operations but Both Aiming to Achieve Sustainable Value Creation

Both Siemens and Huawei aim to achieve sustainable value creation and work to create global production networks in the principle of mutual benefit and build sustainable competitive edges in a global sense. Value objectives guide their actions in network building, network governance, and location selection. In terms of global strategy, both Siemens and Huawei devise their strategies and configure their businesses and value chains from a global perspective to achieve optimal resource integration and globally consistent management based on their comparative advantages. They both consider it important to run operations in local ways and hire local people and strive to be a corporate citizen that makes outstanding contributions to local communities in every local market. They are also different in some ways. In the case of Siemens, different Sectors or Divisions formulate local strategies for specific regions/countries to support the company's overall global strategy "One Siemens." Huawei, however, attaches equal importance to global

TABLE 6.2 TNIs of Siemens and Huawei in 2011

Company	TNI	Assets (USD 100 million)			Sales (USD 100 million)			Employment (10,000)		
		Foreign Assets	Total Assets	Ratio of Foreign Assets to Total Assets	Foreign Sales	Total Sales	Ratio of Foreign Sales to Total Sales	Foreign Employment	Total Employment	Ratio of Foreign Employment to Total Employment
Siemens	77.4%	1123.6	1417.5	79%	874.2	1024.9	85%	24.4	36	68%
Huawei	42.08%	116	307	38%	220	324	68%	3	14.6	21%

UNCTAD, website of China Enterprise Confederation (www.cec-ceda.org.cn/), and Huawei's website (http://support.huawei.com/).

expansion and local operations and its "glocalization" strategy means following local rules, improving the operation and management mechanisms of its overseas subsidiaries, urging local teams to assume their local responsibilities, and creating a harmonious business environment.

6.1.3 Both Playing a Dominant Role in Producer-Driven Networks but Siemens Adopting a Governance Model of European or U.S. Style and Huawei Using a Special Model

Siemens uses the "One Siemens" framework to connect organizations within and outside the company and create global R&D, manufacturing, supply chain, and sales and service networks through various means, including direct investment (wholly owned enterprise, joint venture, M&A, etc.), trade, non-equity-based arrangements (alliance, outsourcing, cooperation agreement, contract manufacturing, etc.). Strategically, it has the ultimate ownership of global production networks and the right to control and configure them. Huawei establishes strategic partnerships with Chinese and foreign research institutes, universities, high-tech companies, multinational corporations, suppliers, distributors, and customers by way of alliance-building, substantial profit release, product compatibility, capital fusion, and stock-holding; creates global R&D, production and operation, and marketing and service networks on the basis of market management, business process reengineering (BPR), and product management; and focuses on enhancing the value added of core modules to stand on the upper part of the value chain. The two companies both have producer-driven global networks and their dominance in the networks is similar. However, core competitiveness in the field of communications equipment is more embodied in product development and software and outsourcing is a prevailing trend in this industry. Therefore, Huawei puts production, procurement, logistics, and other functions in the global production and operation network through integrated supply chain management while Siemens runs these functions in the global manufacturing network or global supply chain network.

As far as governance model is concerned, listed on both the Frankfurt Stock Exchange and the New York Stock Exchange, Siemens complies with the German Corporate Governance Code and adopts a two-tier board structure, that is, Managing Board and Supervisory Board, as well as an integrated hierarchical model featuring Corporate Functions, Sectors, Cross-sector Services, and Regional Clusters. It also observes related U.S. securities laws and regulations, and works to improve corporate governance according to the U.S. legislation, setting a great example of successfully combining typical European and U.S. governance rules. Relational governance is adopted for members of the company's external networks. In particular, the governance of tier-1 and tier-2 suppliers and production network in

Germany depends mainly on the social relations (e.g., reputation and trust) between network participants. There are also market relations between Siemens and other enterprises in the networks but modular and captive governance is rarely seen.

Huawei's internal governance model is rather rare. Huawei is a private company wholly owned by its employees. The Shareholders' Meeting comprises two shareholders: the Union and Mr. Ren Zhengfei. Representatives of shareholding employees are elected with a term of five years. Huawei's Board of Directors (BoD) and Supervisory Board operate in accordance with China's Company Law. Huawei implements the rotating CEO system under the BoD's leadership. Three Vice Chairmen take turns to act as the Rotating and Acting CEO for a tenure of 6 months. The Rotating and Acting CEO is responsible for the company's operations and crisis management during the tenure. The unique equity structure and governance model have undoubtedly contributed to Huawei's success but the strong personal character in leadership, lack of transparency in equity structure, and special decision-making body have prevented Huawei from obtaining financial resources in large-scale competition. Since it is not listed, it cannot get financing in the capital market and does not need to disclose information to the general public, which brings about the lack of transparency to some degree. This has posed some obstacles to the company's making forays into international markets, the U.S. market in particular.

6.1.4 Siemens Performing Equally Well in Three Major Regions and Huawei's Performance in the North America Market at an Average Level

Both Siemens and Huawei consider the three major economic regions of the world—Europe, Americas, and Asia-Pacific—as their main battlefield for market development. As shown in Table 6.3, between fiscal years 2011 and 2014, Siemens's revenue in Europe made up more than 50% of its total revenue, and the average level was 53%; the revenue in Americas accounted for over 25% of the total, and the average level was 28%; the ratio of revenue in Asia-Pacific to the total revenue stayed around 20%. The revenue produced in Europe, Americas, and Asia, three major economic regions of the world, all contributed to more than 20% of its total revenue so Siemens can be considered a global company in its true sense. From 2011 to 2014, the ratio of Huawei's revenue in Asia-Pacific to its total revenue was around 50%, averaging 51%; its revenue in Americas accounted for 15% or less than 15% of its total revenue, and the ratio showed a downward trend, averaging 13%; the ratio of revenue in Europe to the total revenue stayed around 35%. Huawei faces various obstacles in North America so its overall performance there is at an average level.

TABLE 6.3 Business Performance of Siemens and Huawei by Region, 2011–2014

Region	Siemens					Huawei				
	FY2011	FY2012	FY2013	FY2014	Average	2011	2012	2013	2014	Average
Europe	52%	51%	54%	54%	53%	36%	35%	35%	35%	35%
Americas	28%	29%	27%	26%	28%	15%	14%	13%	11%	13%
Asia-Pacific	20%	20%	20%	20%	20%	49%	50%	51%	53%	51%

Prepared by the author based on related data provided in the annual reports of Siemens and Huawei.

6.2 COMPARISON FROM THE PERSPECTIVE OF VALUE

Table 6.4 compares the GPNs of Siemens and Huawei from the perspective of value. Apparently, the global R&D networks of both companies are intended to continually create value for customers; they both adopt an open innovation strategy and have much in common in networks, location selection, and other aspects. However, Siemens focuses more on the development of cutting-edge technologies and the cooperation with universities and research institutes while Huawei focuses more on applied R&D and the collaboration with commercial companies and customers. Siemens's global supply chain and manufacturing networks have basically the same value objectives as those of Huawei's global production and operation network. However, Siemens considers manufacturing very important, has a large number of plants in foreign countries, and is working to increase the share of procurement from emerging markets through its Global Value Sourcing (GVS) program while Huawei outsources more of its production activities and most of its manufacturing facilities are in China. Their global marketing and service networks are also similar in terms of value objectives, governance model, etc. However, as a market leader, Siemens makes full use of its technology and brand strengths in moving from innovation differentiation to market differentiation, expanding from high-end to mid- and low-end markets, and operating in developed and developing countries at the same time. Huawei, however, leverages its low cost advantage in moving from market differentiation to innovation differentiation, spreading from mid- and low-end to high-end markets, and making forays first into developing countries and then into developed countries.

6.2.1 Global R&D Network

Both Siemens and Huawei consider innovation as the lifeblood of a company and aim to continually create value for customers. They both adopt open innovation and build global R&D networks that cross sectors and regions and involve internal and external stakeholders. They both see IPR management as a key factor of the innovation strategy and have established effective mechanisms for achieving continuous innovation. These are all crucial to the success of the two companies. In its early years, Huawei benchmarked itself against top companies such as Siemens and Ericsson and established a management and operation system up to international standards. However, as a latecomer, Huawei adopted follow-up innovation in the beginning, combining independent R&D with a take-in principle. Then it sought to meet international standards through integrated innovation, combining systematic tracking with breakthroughs, and applied a "pressure tactic" to develop core technologies. After it had gradually gathered strength in technology and market, Huawei began to make the shift from being a follower to being a leader,

TABLE 6.4 A Comparison Between GPNs of Siemens and Huawei From the Perspective of Value

Network Type	Aspect	Siemens	Huawei
Global R&D network	Value objective	To continually create value for customers	
	Global strategy	Open innovation + IPR management since fiscal year 2009	Early years: follow-up innovation, independent R&D + "take-in" principle, systematic tracking + breakthroughs
			At present: open innovation + IPR management
	Networks	CT + Sector R&D teams + external partners	2012 Laboratories + research institutes at home and abroad + external partners
	Network governance	Greater support for cutting-edge technology research	More spending on applied R&D
		More attention to the cooperation with universities and research institutes	More attention to the cooperation with commercial companies and customers
	Location selection	Set up R&D facilities in China, Japan, India, Germany, Russia, Turkey, U.S., Canada, etc.	
Global production and operation network (global manufacturing network + global supply chain network)	Value objective	To achieve global production and operations featuring fast response, high quality, low cost, flexibility, and sustainability	
	Global strategy	Supply chain management initiative (SCMI)	Integrated supply chain management
		More independent manufacturing	More outsourcing
	Networks	Global manufacturing network + global supply chain network	Global production and operation network

(*Continued*)

TABLE 6.4 (Continued)

Network Type	Aspect	Siemens	Huawei
	Network governance	Supplier management, green supply chain, global value sourcing, Siemens Production System, global manufacturing footprint, continuous quality improvement	Strategic procurement, supplier management, continuous quality improvement, green supply chain, all-inclusive risk management
	Location selection	Mainly overseas production	Mainly domestic production
		More than 285 major manufacturing sites in over 40 countries	Manufacturing sites in Egypt, Russia, Brazil, Saudi Arabia, Iran, India, etc.
Global marketing and service network	Value objective	To continually create value for customers	
	Global strategy	Develop markets through innovation differentiation and consolidate market position through market differentiation, spreading from high-end to mid- and low-end markets	Develop markets through low-cost operations and market differentiation and consolidate market position through innovation differentiation, spreading from mid- and low-end to high-end markets
	Networks	Establish a hierarchical marketing and service structure according to customer size, business and region, direct selling + distribution, direct service delivery + outsourcing	Adopt different models for different business sectors, direct selling + distribution, direct service delivery + outsourcing
	Network governance	Adopt various approaches, such as introducing cooperation, establishing joint ventures, forging alliances, etc., and stress compliance with applicable local laws and international rules and regulations in organizing marketing and service activities	
	Location selection	Operating in developed and developing countries at the same time	Operating first in developing countries and then in developed countries

Prepared by the author.

setting a great example of independent innovation and global expansion to other Chinese enterprises.

Both Siemens and Huawei have their internal global R&D networks and separate operations for the development of cutting-edge technologies and applied research and development. Given the special characteristics of the communications equipment industry, Huawei outperformed Siemens at first in the number and proportion of R&D employees but the ratio of its spending on the development of cutting-edge technologies to its total R&D spending was only 10%, which has increased to 15%—still lower than that of Siemens. Both Siemens and Huawei have R&D facilities in China, Japan, India, Germany, Russia, Turkey, the United States, Canada, and other countries. They are quite similar when selecting locations—either technology centers in developed countries or talent centers in developing countries.

Both Siemens and Huawei have long-term strategic partnerships with universities, research institutes, commercial companies, etc., forming an open innovation system. Siemens considers it very important to collaborate with universities and research institutes. It has created three models of cooperation, that is, Center of Knowledge Interchange (CKI), Ambassador Program, and business-specific cooperation, established diverse partnerships, developed a well-designed management system, and provided greater support for the development of cutting-edge technologies. Huawei considers it more important to collaborate with commercial companies and customers. It has established joint laboratories and joint ventures together with the world's top companies and drawn upon the latest research findings of the industry to increase its innovation capacity. Huawei has established 28 joint innovation centers together with customers and such centers have played an important part in continuously enhancing the strategic partnerships between them.

6.2.2 Global Production and Operation Network

Both Siemens and Huawei have fully integrated their manufacturing and supply chain networks into their value-creating networks, created a supplier management model that brings high earnings, paid close attention to overall risk management and sustainability of supply chains, developed green supply chains, and achieved global production and operation featuring fast response, high quality, low cost, and flexibility. Due to industry differences, Siemens considers independent production an important element and enhances productivity and value added by adopting the Siemens Production System (SPS) based on lean manufacturing principle and the systematic approach Global Manufacturing Footprint (GMF). Huawei, however, outsources more of its production activities and adopts integrated supply chain management to run production, procurement, logistics, and other functions in the global production and operation network and combines the global manufacturing network and global supply chain network as one.

Siemens attaches great importance to global balance in supply chain and manufacturing networks. About one-third of its purchasing is in Germany, about one-third in other parts of Europe, about one-fifth in Americas, and around one-tenth in Asia-Pacific. The linkages between purchasing and sales in Europe and Americas are also balanced. Siemens is seeking to increase the share of purchasing from emerging markets through the GVS program. It operates more than 285 major manufacturing and production sites in over 40 countries, about a half of which are in Europe, one-fourth in Americas, and one-fourth in Asia-Pacific, so its global manufacturing operations are quite balanced. Only about 8% of Huawei's employees are in the manufacturing and supply chain sector; most of its production activities are outsourced and organized at home, with only a small number of manufacturing plants in other countries such as Egypt, Russia, Brazil, Saudi Arabia, Iran, and India. Very small proportions of Huawei's assets and employees are in foreign countries, which can largely account for the fact that its TNI is far lower than that of Siemens.

6.2.3 Global Marketing and Service Network

Both Siemens and Huawei are customer-centered and committed to building their own brands on the basis of exporting high-tech products. Through cooperation, joint ventures, alliances, and other means, they combine direct selling with distribution and direct service delivery with outsourcing in creating the global marketing and service network so as to continually create value for customers. They both stress the importance of complying with applicable local laws and international rules and regulations in the organization of marketing and service activities. However, the two companies differ in the strategy of entering overseas markets due to their different competitive advantages. In the beginning, Huawei captured markets in developing countries rapidly as a market niche by starting with the remote outside, adopting low cost and market differentiation strategies, and developing a customer demand fast response mechanism and sound after-sales service. As its experience in international operations built up, Huawei began to adopt an innovation differentiation strategy in European and U.S. markets where technologies are well developed. By forging alliances with world-famous companies for years, it has achieved a dramatic growth in R&D strength and gradually made the shift from increasing market share at the cost of low profits to winning the market with competitive technology and service, thus expanding from mid- and low-end to high-end markets. Correspondingly, its marketing strategy has shifted the focus from high-end to regular customers and to all-round exchanges and information sharing with customers so as to increase the company's visibility and status in the industry. By contrast, as a market leader, Siemens first used its technology and brand advantages to open up international markets and then created marketing networks for different markets and improved its service system. Leveraging the expertise and market experience gained over the years, it has located target markets, gradually adjusted strategy to meet local needs, and cemented the ties with customers, developing markets through innovation differentiation, consolidating its market position through market differentiation, and expanding from high-end to mid- and low-end markets.

Siemens has created a hierarchical marketing and service structure according to customer size, business, and region and undertakes marketing and service activities in more than 200 countries in the world. It has account managers on regional, corporate, and global levels and runs an Executive Relationship Program (ERP) under which the Managing Board has established long-term relationships with executives from around 100 customers. Siemens's regional companies have established close local ties with small and mid-sized customers across sectors. Huawei adopts different models for different business sectors and provides customers with end-to-end services through its global resource sharing platform and elite project management teams. In the field of carrier business, Huawei has set up commercialized local service delivery organizations and platforms worldwide to collaborate with carriers, with more and more carrier network installation and maintenance services outsourced. In the field of enterprise business, Huawei seeks to build marketing and channel capacity and teams as the top priority, develops professional channel partners for different products, and has created a healthy and open partner network. In the field of consumer business, Huawei makes active efforts to promote multichannel marketing and embraces distribution, retail, and Internet marketing on the basis of consolidating operator marketing. Its ever-improving commercialized, standard, and local service platforms worldwide have eventually advanced its innovations in technology.

6.3 COMPARISON FROM THE PERSPECTIVE OF EMBEDDEDNESS

Table 6.5 compares the GPNs of Siemens and Huawei from the perspective of embeddedness. As is shown, Siemens Healthineers has experienced four phases in China, including exploratory embeddedness, strategic linkage, strategic embeddedness, and strategic coupling, as has Huawei in Europe but their paths of embeddedness are different. Siemens shifts from innovation differentiation to market differentiation, from developed to less developed regions, and from high-end to mid- and low-end markets in China while Huawei moves forward from low cost advantage and market differentiation to innovation differentiation, from mid- and low-end to high-end markets, and from less developed to developed regions in Europe. Both of them developed gradually from economic to technical, social, cultural, and institutional embeddedness and from one-way to two-way embeddedness. Siemens's backward embeddedness in China is deeper; it makes greater

TABLE 6.5 A Comparison Between GPNs of Siemens and Huawei From the Perspective of Embeddedness

Aspect	Specific	Siemens	Huawei
Strategy		Going through four phases: exploratory embeddedness, strategic linkage, strategic embeddedness, and strategic coupling, developing from economic to technical, social, cultural, and institutional embeddedness, and from one-way to two-way embeddedness	
		Shifting from innovation differentiation to market differentiation	Shifting from low cost and market differentiation to innovation differentiation
Region		Expanding from developed to less developed regions and from high-end to mid- and low-end markets	Expanding from mid- and low-end to high-end markets and from less developed to developed regions
Dimension	Economic embeddedness	Expanding from sales and services gradually to systematic and multifunctional operations, with backward embeddedness going deeper	Expanding from sales and services gradually to systematic and multifunctional operations, with forward embeddedness going deeper
	Technical embeddedness	Local R&D and production activities going from low to high value-added parts, with a bigger role in driving local industrial agglomeration and greater technology spillovers	Local R&D and production activities mainly targeted at technology-intensive and high value-added parts, with a rather small role in driving local industrial agglomeration and a focus on the introduction of technologies
	Social embeddedness	Actively fulfilling its social responsibilities, with a focus on three areas, namely environmental protection, education on science and technology, and humanitarian aid	Actively fulfilling its social responsibilities, with a focus on bridging the digital divide, supporting stable and secure network operations, promoting environmental protection, and seeking win–win development
	Cultural embeddedness	Promoting the integration of diversity and localization within the "One Siemens" framework, with more than 99% of its employees locally hired	Stressing the local growth of overseas employees, with 80% of its employees locally hired
	Institutional embeddedness	Strictly complying with local laws and regulations and ensuring global compliance, and cementing ties with local social organizations, playing a part in driving changes in China's industrial system and business model and innovations in financial management and supply chain finance	Getting deeply embedded in local systems and mechanisms to ensure global and local compliance, and focusing more on becoming embedded in local industry and complying with local rules, producing little impact on local institutional changes

Prepared by the author.

contributions to local industrial agglomeration; technology spillover effects are stronger; more of its employees are locally hired; the company plays a part in advancing reform and innovation in China's industrial system and business model. By comparison, Huawei's forward embeddedness in Europe is deeper; it focuses more on introducing technology, getting embedded in local industry, and complying with local systems and mechanisms, and has little impact on local industrial agglomeration and institutional changes.

6.3.1 Strategy

Ever since Siemens set up its representative office in China in 1982, Siemens Healthineers has experienced a process of embeddedness in China that is basically similar to that of Huawei's process of embeddedness in Europe since it launched the "going global" strategy in 1996. Both of them have gone through the four phases of exploratory embeddedness, strategic linkage, strategic embeddedness, and strategic coupling but their paths are different. Siemens Healthineers's embeddedness strategy in China is in line with China's development strategy and healthcare reform and is forward-looking to some extent, featuring a gradual transition from "footloose embedding" to deep embedding. Since it entered the strategic coupling phase in 2006, Siemens Healthineers has created R&D, manufacturing, supply chain, and sales and service networks in China featuring global–local interaction, which have reinforced Siemens's strategy in China—building a base in China, serving the Chinese market, and targeting the global market. It has achieved the transition from innovation differentiation to market differentiation and from high-end to mid- and low-end markets. China is now the second largest market for Siemens outside Germany. Huawei's market entry strategy in Europe started with the remote outside and it has made the shift from low cost advantage to market differentiation and innovation differentiation and from mid- and low-end to high-end markets. Since it entered the strategic coupling phase in 2011, Huawei has created R&D, production and operation, and marketing and service networks in Europe featuring glocalization. It intends to build a second home in Europe and grow into a "European company." Now Europe is the largest overseas market for Huawei and its revenue there is second only to that in China.

6.3.2 Region

Given the differences in target markets and competitive edges, Siemens Healthineers's path of embeddedness in China basically goes in the opposite direction to that of Huawei's path of embeddedness in Europe. Siemens Healthineers's embeddedness in China begins with coastal areas in southeast China and gradually moves to central and western areas, spreading from first- and second-tier cities to third- and fourth-tier cities and even villages and townships. Huawei's embeddedness in Europe begins with C.I.S. and Eastern Europe and then moves to Central, Southern, Western, and Northern Europe. The same low-end to high-end path is also taken in a specific country.

6.3.3 Dimension

Both Siemens Healthineers and Huawei started with economic embeddedness and then moved to technical, social, cultural, and institutional embeddedness, progressing from one-way to two-way embeddedness. As far as the division of labor is concerned, the embeddedness is embodied first in sales and services, then in local R&D, local production, and local procurement, and even in local product and brand management, spreading from simply sales and production activities to R&D centers, service centers, and regional operations headquarters. Siemens Healthineers's local sales in China begins with high-end markets and then gradually shifts to mid- and low-end markets; local production begins with labor-intensive activities and then gradually shifts to capital- and technology-intensive activities; local R&D begins with low value-added activities and the gradually shifts to high value-added activities. In the case of Huawei, local sales begins with mid- and low-end markets and then moves to high-end markets while local R&D and local production are mainly targeted at technology-intensive and high value-added activities.

In the economic sphere, both companies' embeddedness has been on deeper and deeper levels, featuring systematic and multifunctional investments, increasing investment scale, upgrading investment structure, and ever-growing economic ties with local enterprises. Huawei has subsidiaries in 20 European countries as well as six competence centers and six technical support centers. Siemens Healthineers has six operating companies, 33 offices, 60 service stations, and more than 3000 employees in China. The difference is that Siemens Healthineers has more local procurement and production activities in China, which means deeper backward embeddedness. Now 80% of the company's medical device components are procured locally in China and around 70% of its medical imaging products, 85% of its hearing aids products, and 90% of its MRI systems are exported. Huawei has strategic partnerships with leading local companies and all top carriers as well as 18 innovation centers in Europe, featuring deeper forward embeddedness.

In terms of technical embeddedness, both companies have been increasing the number of local R&D facilities and spending and boosting technical cooperation with local enterprises and research institutes. Siemens has established a Technology-to-Business Center (TTB) in Shanghai as well as a research center and a development center in China. It focuses more on the research on cutting-edge technologies and the collaboration with Chinese universities. Two of its eight CKIs in the world are in China; one is Tsinghua University and the other is Tongji University. Siemens leverages its global production network to help Chinese R&D teams go global and makes greater contributions to the concentration of R&D facilities, raised technological level, and upgraded industrial structure so technology spillovers are significant. Huawei focuses more on applied R&D and the collaboration with commercial companies and customers in Europe. Its R&D spending has been growing fast. The average compound annual growth rate of its R&D spending in Europe reached 28% between 2009 and 2013. By directly establishing R&D centers in Europe, jointly setting up innovation centers with European carriers, and executing the "binary learning" strategy, Huawei fully integrates and uses Europe's technology and human resources to build new competitive edges so technology spillovers are more prominent.

As for social embeddedness, both companies seek to be locally rooted, serve local communities, and fulfill their social responsibilities, striving to become outstanding corporate citizens. Siemens focuses on three major fields in this regard, that is, environmental protection, education on science and technology, and humanitarian aid, and has established the Employee Volunteers' Association, which is run by employees themselves. Huawei focuses on the four strategies of bridging the digital divide, supporting stable and secure network operations, promoting environmental protection, and seeking win–win development. It has established a good reputation and brand in Europe after years of hard work.

In the cultural sphere, both Siemens and Huawei are active in adapting to local culture, promoting the local application of the parent company's culture, and fostering a diverse corporate culture. Siemens seeks to integrate its corporate culture into its China business within the One Siemens framework and promote the combination of diversity and localization and the collaboration between employees with different cultural backgrounds in China. Siemens has over 30,000 employees in China who are from 26 countries and regions and more than 99% of them are Chinese. Huawei focuses more on the local development of employees and provides diverse employees with proper channels to realize their real worth. In 2013, Huawei had about 7700 employees in Europe and 80% of its employees in Europe were locally hired.

In terms of institutional embeddedness, both Siemens and Huawei consider themselves local companies, strictly observe various local laws and regulations to ensure global compliance, and cement ties with local social organizations to get deeply embedded in local systems and mechanisms. Siemens Healthineers's operations in Shanghai, Suzhou, Shenzhen, and other cities help to drive local industrial agglomeration. The "New Rural Healthcare Demonstration Center" and "green + hospital" programs launched by the company has led to institutional and business model changes in China's healthcare industry. It has also brought medical device finance leasing to China, promoting innovations and reforms in financial management and supply chain finance in China's medical system. Huawei's choice to set

WCDMA as its R&D priority in the 3G era has contributed to the global dominance of WCDMA and further influenced the development of 4G. It has established a cyber security laboratory in Britain as well as boards of directors and advisory boards in such countries as Britain and France. It seeks to improve relations with the governments of European countries mainly by complying with local systems and mechanisms.

Section III

GPN Trends in the PostCrisis Era and Conclusions

Chapter 7

GPN Trends in the Postcrisis Era

Chapter Outline

As it is noted in Chapter 1, Overview of the Research into GPNs, current GPN studies have failed to give due attention to the dynamics in the operating environment of GPNs. The "system" or "world" composed by material and economic societies constitutes the basic environment for the formation, operation, and evolution of GPNs. Now most studies are carried out on the assumption that the basic environment is relatively static and trade rules are relatively stable. The postcrisis era is marked by the mixture of recession and recovery as well as moderation and fluctuation. Instability and uncertainty is growing in the global economy. International trade disputes are intensifying, and international trade policies are full of uncertainty. Then, how does the financial crisis impact GPNs and in what direction will GPNs move in the postcrisis era? Will new changes occur to the territorial and industrial configuration of GPNs in the future? Will a new landscape emerge

Global Value Chains and Production Networks. DOI: https://doi.org/10.1016/B978-0-12-814847-1.00007-5

from the current framework? What influences will China's Belt and Road Initiative have on GPNs? This chapter explores the four driving forces behind the development of GPNs and makes dynamic analyses of the GPN trends in the postcrisis era, seeking to answer those questions.

7.1 DRIVING FORCES BEHIND THE DEVELOPMENT OF GPNS

According to traditional division of labor theories and the New Classical Trade Theory, the creation of GPNs as a new type of international production organization is caused by the endogenous evolution of division of labor; it is a direct outcome of the dilemma between the economies of division of labor and transaction costs as well as the half measure of professional decision-making by individuals. GPNs make it possible to put different stages of the production process in different countries so that the benefits of specialization can be fully exploited. Trade and investment liberalization and technology advances help to reduce institutional barriers to the division of labor worldwide considerably, thus enhancing transaction efficiency. Therefore, MNCs are active in creating and expanding GPNs. The scale, scope, and depth of the division of labor within GPNs are determined by the comparison between the marginal benefit and marginal transaction cost of the division of labor. If the former is greater than the latter, GPNs will continue to develop; if the former is lesser than the latter, GPNs will tend to shrink.

Therefore, the main driving forces behind the development of GPNs are deepening international division of labor, trade and investment liberalization, technology advances, and expansion of GPNs (Fig. 7.1). The economic efficiency achieved through the international division of labor and collaboration is the fundamental cause of GPNs' creation; trade and investment liberalization provides institutional assurance for the formation of GPNs, technology advances provide the technical assurance for the development of GPNs, and

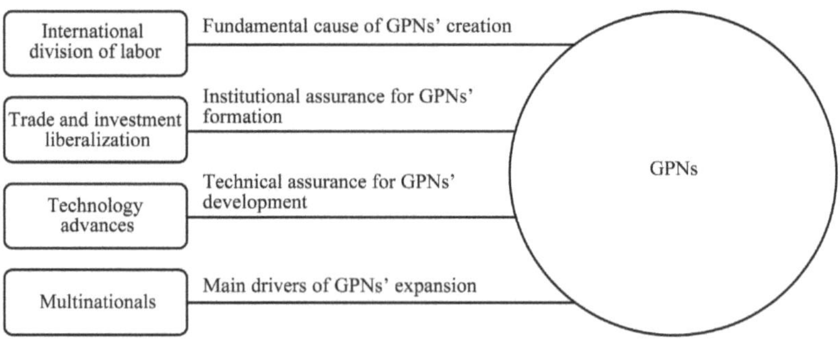

FIGURE 7.1 Driving forces behind the development of GPNs. Source: *Prepared by the author.*

multinational corporations are the players and main drivers of GPNs. How the four driving forces evolve determines the development trends of GPNs.

The issue of GPNs boils down to the division of labor since the economic efficiency achieved through international division of labor and collaboration is what GPNs are created for. The division of labor on deeper levels leads to higher levels of specialization in production, companies' dependence on trade, commercialization, diversification in economic structure, and market integration, which eventually contributes to the formation and expansion of GPNs. The expansion of GPNs then pushes the international division of labor to deeper levels so that the competitive edges of countries are consolidated and their ties cemented. In the wider context of globalization, manufacturing industries across the world are closely interconnected. Countries have all shared the "dividend of globalization" to varying degrees and are all stakeholders in GPNs. Therefore, GPNs are closely associated with the international division of labor and the competitive edges of countries.

Over the past decades, multilateral trade systems represented by the WTO and other regional, bilateral, or multilateral preferential trade agreements (PTAs) have been strongly advancing trade and investment liberalization. According to the WTO's *World Trade Report 2011*, about 300 PTAs were in force in 2010 and every WTO member was involved in 13 PTAs on average; the contents of PTAs have been changing and PTAs play an important role in the world economy. The measures to promote trade and investment liberalization enable countries to open up national markets and allow for a wide range of FDI and nonproperty rights arrangements, which thus greatly advances the globalization of markets and investments, expands market opportunities, reduces transaction costs in international trade, improves transaction efficiency, and promotes cross-border flows of production factors, leading the world economy from trade globalization to production globalization. Production globalization makes it possible to obtain cheaper labor, more expertise, and richer production inputs, thus reducing the cost of forming global production networks. Therefore, it is fair to say that the global promotion of trade and investment liberalization provides institutional assurance for the formation of GPNs.

The development of GPNs is closely related to the advances in science and technology. On the one hand, technology advances make the worldwide division of labor for production possible. On the other hand, technology advances make it easier for such factors as goods, equipment, knowledge, and people to move around the world at a faster speed and lower costs. This provides favorable technical conditions for GPNs. Over the past more than 30 years, the development of satellites, optical fiber, wireless technology, the Internet, and e-commerce have revolutionized global communications, and transportation technology has seen several major innovations, including the introduction of commercial jumbo jets, super freighters, and containers. These technologies have significantly reduced international communications costs, transportation costs, as well as coordination and transaction costs, accelerated the flow of information and commodity, and cemented the ties between regions, providing important technical support for the development of GPNs. With the wider application of information technology in various industries, the levels of process standardization and digitalization among enterprises have also been going up. As a result, the production process in the manufacturing industry has taken on some new features, such as the emergence of modular production, and the production process of products can be divided spatially. It means that the previously integrated production process can be dispersed around the world, and reconfigured according to the comparative advantages of various countries and regions. It is therefore fair to say that technology advances make it quite economical to integrate remote business activities and transport products and components worldwide for higher efficiency, and feasible to achieve production specialization and collaboration on a global scale. They provide technical assurance for the development of GPNs.

The microlevel participants of a GPN are multinational corporations (MNCs) and they assume overwhelming dominance in the network. In fact, it is their global expansion that has promoted the formation of GPNs and shaped the new landscape of international trade centering around MNCs, which is a manifestation of economic globalization on the microlevel. Economic globalization, while bringing more market opportunities to MNCs, has also involved more and more companies in cross-border competition. In response to the increasing pressure from market competition, MNCs must explore new ways of increasing efficiency, including the early expansion of their international operations to new markets and the transfer of certain production activities to reduce competition costs, or the introduction of new approaches and new ownership and contractual arrangements for international production, and the organization of new activities at new locations abroad. Thus, MNCs leverage their "ownership advantage" and "internalization advantage" to implement global strategies and reconfigure the factors of production globally. They distribute and configure different parts of the value chain according to the quality and price differences in factors between countries or regions, and combine them organically, thus forming a network dominated by MNCs and involving various participants that brings the companies new possibilities for value creation. Structurally, MNCs have gradually evolved into companies of global networks, and the traditional competition between individual enterprises is gradually replaced by the competition between these networks. MNCs have not only improved their own competitive advantages, but also produced greater and greater impact on the world economy, becoming the core organizers of international economic activities today. Therefore, MNCs are the most important driving force behind the formation and expansion of GPNs.

7.2 BASIC GPN TRENDS STAYING UNCHANGED

The basic trend of economic globalization remains unchanged in the postcrisis era (Pei, 2010). The international division of labor will undergo progressive adjustments but no disruptive changes will take place. Despite the growing restrictions on and regulation of investment, trade, and investment liberalization and investment promotion remain the focus of trade policies in the immediate term. The adjustment period of the world economy after the crisis will give rise to a new technological revolution, providing more favorable technical conditions for the cross-border flows of factors. To increase market opportunities, lower operating costs, and obtain essential resources, MNCs will continue to expand globally and move more of their R&D, production, procurement, and sales activities to emerging countries and regions. Since the basic direction of those driving forces remains unchanged, no radical changes or setbacks will occur in the development of GPNs and the MNCs-dominated GPNs will continue to expand amid adjustments and improvements.

7.2.1 No Disruptive Changes Will Appear in the International Division of Labor

After World War II, the world saw the large-scale international relocation of industries three times. Each time the relocation of industries was closely related to each country's economic development, competitive advantage change, industrial structure adjustment, etc., thus affecting the international division of labor, and getting countries closely integrated into the global division of labor system. As Zhang (2008) points out, in the tide of globalization, the major economies in the world can be put into three categories according to the division of labor, namely capital-rich and consumer countries represented by the United States, EU countries, and Japan; manufacturing countries represented by China and India; and resource-rich countries represented by countries in the Middle East, Russia, Brazil, and Australia. Some compare those three categories of countries to "head countries that produce knowledge, torso countries that produce materials, and hand-and-foot countries that supply resources." The current international division of labor has contributed to the global economic prosperity for more than two decades but it has also made the world economy imbalanced for years and caused a financial crisis so there is bound to be a rectification process of structural adjustments. However, in the postcrisis era, the growth models, competitive advantages, and industrial structures of countries will not change fundamentally, so the international division of labor will not see disruptive changes. To radically change the division of labor means that countries need to rebuild their competitive advantages. For example, developed countries with innovation advantages may develop low value-added labor-intensive industries for the purpose of creating jobs. They do not enjoy competitive edges in such industries on both national and industry levels so not just these countries themselves but all stakeholders in their global production networks will pay a huge cost.

Take the United States as an example. It has the world's most developed manufacturing industry and fastest-growing advanced manufacturing. The competitiveness of its manufacturing industry has not been affected by the financial crisis, so its reindustrialization strategy is mainly targeted at advanced industries. In 2009, the vision of revitalizing the US manufacturing was proposed, with a number of measures introduced, such as A Framework for Revitalizing American Manufacturing, American Clean Energy and Security Act, Advanced Manufacturing Partnership (AMP), and National Strategic Plan for Advanced Manufacturing. These measures mainly focus on technological innovation in advanced industries, aiming at leading the development of advanced industries. In February 2015, the Brookings Institution published a research report entitled *America's Advanced Industries: What They Are, Where They Are, and Why They Matter*. The report offers two defining criteria for an advanced industry. One is that each worker should spend more than USD 450 on R&D, or the level of R&D spending is among the top 20% of the industry. The other is that the number of STEM (science, technology, engineering, and math) degree holders in the industry is larger the national average, or makes up 21% of the industry's total. Any industry that meets both criteria is a so-called advanced industry. According to the report, America has 50 advanced industries, including 35 in advanced manufacturing, three in advanced energy and 12 in advanced services. Between 1980 and 2013, the output of America's US advanced industries grew at an average annual rate of 5.4%, around 30% higher than the US economic growth rate. Advanced industries create USD 2.7 trillion value added each year, accounting for 17% of America's GDP. As of 2013, the 50 advanced industries hired 12.3 million workers, including 80% of the country's engineers, engaged in 90% of private sector research and development, held about 85% of US patents, and contributed to around 60% of US exports.

After the crisis, both employment and output of US advanced industries rose sharply and the number of employees grew significantly across all sectors, particularly in the service sector. From 2010 to 2013, the United States witnessed about one million new jobs created in advanced industries, whose employment and output growth rates were, respectively, 1.9 times and 2.3 times higher than those in other sectors. Advanced industries offered jobs, directly or indirectly, to more than 39 million people, about one-fourth of the country's employed population. It is worth noting that 35, or 70%, of the 50 industries mentioned in the report are advanced manufacturing industries (Brookings Institution, 2015). Apparently, advanced manufacturing is a major component of US advanced industries. Advanced services contributed to the employment boom after the recession, creating about 65% of the new jobs.

The United States will not change its service-oriented economic structure or its long-standing "consumer-oriented" economic model, and the offshoring of jobs in its manufacturing and service sectors has not ceased. Reports about American companies moving production lines from low-wage countries back to the United States began in 2009. Consulting firms such as Boston Consulting Group (BCG) and A.T. Kearney also joined the discussion about the return of American manufacturing. But the discussion was usually focused on one-way flows. In fact, in the same period, a large number of companies were expanding their outsourcing business. Considering that American multinationals have more than 25,000 foreign subsidiaries and more than 36 million employees worldwide, the number of backflow cases is clearly not at the same level of magnitude. The Peterson Institute for International Economics (PIIE) carried out a study based on data about the global operations of most US companies. It argues that while some companies have changed their previous decision to outsource, there is no evidence that the backflow of manufacturing is a general trend and the global supply chain is still thriving and is unlikely to change in the short term. However, such a situation should not be seen as a threat to the US economy or to American workers, whereas continued outsourcing enhances the competitiveness of the US manufacturing sector and benefits American workers and consumers (Oldenski, 2015).

The PIIE defined the outsourced products of American manufacturers as goods that are produced overseas and returned to the United States for sale to check whether the backflow of American manufacturing has become a prevailing trend. It used official data from all US multinationals to explore whether the outsourcing business of these companies had undergone a dramatic reversal. Fig. 7.2 shows the imports of US multinationals between

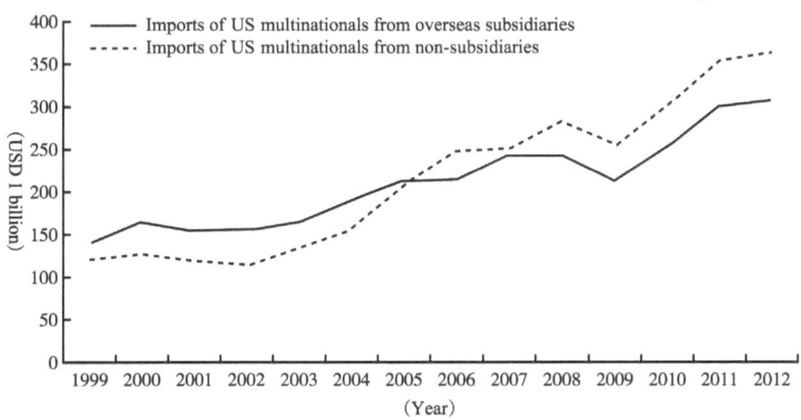

FIGURE 7.2 Imports of US multinationals, 1999–2012. Source: *US Bureau of Economic Analysis (BEA), https://www.bea.gov/.*

1999 and 2012. The solid line indicates the imports of US multinationals from overseas subsidiaries while the dotted line indicates the imports of US multinationals from nonsubsidiaries, which are usually from overseas OEM production or trade. The solid line is a good indicator of the outsourcing business because those goods are produced overseas and shipped back to the United States for final assembly or sale. As of 2012, this type of outsourcing business had basically maintained an upward trend, except for a fall in 2009, which is only a cyclical response to the postcrisis recession.

Fig. 7.3 shows the made-in-America exports of US multinationals from 1999 to 2011. Apparently, while outsourcing business was growing, made-in-America exports were also rising. Thus, while the backflow of companies is not a widespread phenomenon, it does not detract from the competitiveness of American manufacturing. In fact, US multinationals will continue to carry out outsourcing business, but they will become stronger in promoting production at home and US exports.

Apparently, in the postcrisis era, countries are committed to the development of new international industrial structure and the establishment of a new international division of labor system, but this is based on relatively stable competitive advantages. In the context of economic globalization, no country can build an industrial system alone; all need to cooperate with other countries to draw upon each other's advantages and develop a full range of industries. Therefore, the change and improvement of international division of labor is unavoidable in the postcrisis era, but the international division of labor system will not collapse or be completely built. Instead, gradual adjustments will be made according to the new global economic climate and the competitive advantages of countries. Moreover, the new international

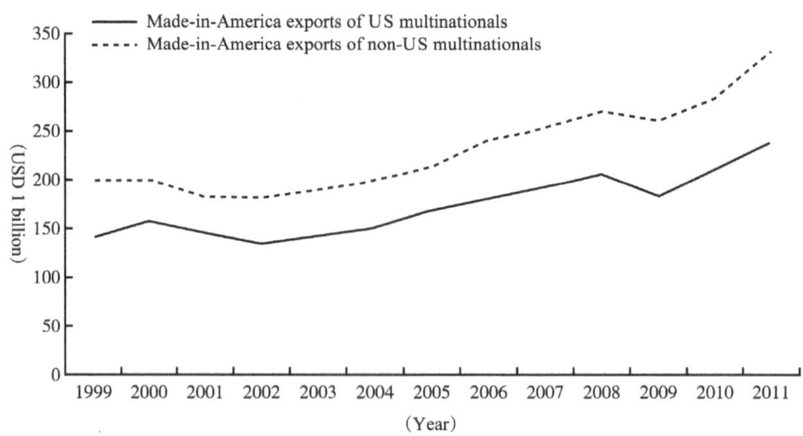

FIGURE 7.3 Made-in-America exports of US multinationals, 1999–2011. Source: *US Bureau of Economic Analysis (BEA).*

division of labor, cooperation, and trade systems will usher in a new period of development after restructuring and become the driving force behind the development of GPNs, trade, and investment.

7.2.2 Trade and Investment Liberalization Will Remain the Focus of Trade Policy in the Immediate Term

The global promotion of trade and investment liberalization provides institutional assurance for the formation of GPNs. Despite the growing restrictive measures, administrative procedures, and risks of investment protectionism in the last few years, investment liberalization, promotion, and facilitation remain the focus of trade policies in the immediate term. The integration of and interaction between transnational investment and trade will continue. There is still a solid institutional foundation for the development of GPNs.

7.2.2.1 IIAs: Rising Total Number and Number of Regional Agreements

According to the World Investment Report (WIR) issued by the United Nations Conference on Trade and Development (UNCTAD), as new investment treaties and related arbitrations have grown in number in recent years, the scope and scale of international investment system have expanded somewhat and a systematic evolution is emerging, that is, a better system that accommodates the rights and obligations of both states and investors is about to take form. In 2014, 31 new international investment agreements (IIAs) were concluded. As of the end of 2014, the total number of IIAs had reached 3271, including 2926 bilateral investment treaties (BITs) and 345 other IIAs (mainly free trade agreements, economic partnership agreements, and regional agreements that cover investment provisions). On the whole, BITs have been decreasing in number annually but more and more countries are conducting negotiations on IIAs on regional and subregional levels. For example, around 90 countries are involved in the ongoing negotiations about five agreements, including the Trans-Pacific Partnership (TPP), the Transatlantic Trade and Investment Partnership (TTIP), and the Regional Comprehensive Economic Partnership (RCEP).

The annual number of new BITs has been on the decline in recent years but due attention should be given to some facts. First, the total number of IIAs is still increasing. Second, the size and complexity of IIAs is difficult for both governments and investors to deal with. Yet they are not sufficient to cover all possible bilateral investment relations. All parties concerned are intensifying their research into the future direction of the IIA regime and its development implications. Third, while bilateral agreements still make up the majority, regional trade agreements are strengthening, which has also brought systemic changes to the IIA system. In addition, consideration of sustainability has become an increasingly important factor in influencing IIAs and in the formulation of international investment policies by governments. The current IIA system is at a crossroads and there is an urgent need to reform it so that it can better adapt to today's policy challenges. The 2015 WIR presents a number of policy options for such reform, covering different aspects of reform (substantive provisions of IIAs, and investment dispute resolution) and the different dimensions of reform-oriented policy-making (national, bilateral, regional, and multilateral). In general, these policy options on reform address the standard elements covered by IIAs, and match the typical provisions found in IIAs so that IIAs can be used to promote sustainable and inclusive growth and to identify the most effective means of safeguarding regulatory authority, while providing investment protection and facilitation.

7.2.2.2 National Investment Policies: Liberalization and Intensifying Regulation

Despite the rise in new regulatory and restrictive measures in recent years, there has not been a significant increase in trade protectionism so far. Many countries continue to adopt loose and promotion policies for foreign investment in various sectors to stimulate economic growth. They have also introduced many FDI outflow restrictions and return incentives for foreign investors. In the meantime, the interaction between FDI policy and industrial policy has been strengthened. According to the WIR issued by the UNCTAD, in 2014, 37 countries and economies adopted at least 63 measures affecting foreign investment. Of these measures, 47 are related to investment liberalization, and facilitation, nine are to the imposition of new restrictions or regulations on investment, and the remaining seven are neutral. Some countries have introduced or amended their investment laws or guidelines to give new investment incentives or facilitate the process of investment. Some countries have loosened restrictions on foreign ownership or opened up new business activities (e.g., in infrastructure and services) to foreign investment. The new investment restrictions or regulations primarily concern national security considerations and strategic sectors (e.g., transportation, energy and defense). In 2013, 59 countries and economies implemented 87 FDI-related measures, 61 of which were more favorable and 23 were about FDI regulation and restriction. The share of liberalization and promotion measures in 2014 increased markedly to 84% from 73% in 2013. While it is still too early to say that the trend of tightening policy controls in recent years has been reversed, countries in general continue to pursue investment liberalization and promotion policies.

7.2.3 Technology Advances Will Continue to Drive the Flow of Production Factors Across Countries

Scientific and technological innovation has always been an important driving force behind profound changes in the ways of work and life, and technological advancement provides important technical support for the development of GPNs. With the advent of the 21st century, a new scientific and technological revolution and industrial changes are emerging; the world's scientific and technological innovation are presenting new development trends and characteristics. The interdisciplinary fusion is speeding up, new disciplines are emerging, and the frontiers of science are being pushed back. Major breakthroughs are being or expected to be made in basic sciences such as structure of matter, evolution of the universe, origin of life, and nature of consciousness. The wide penetration of information technology, biotechnology, new materials, and new energy has led to the mass technological revolution characterized by greenness, intelligence, and ubiquity in almost all areas. The famous British magazine *The Economist* noted in 2012 that manufacturing digitization would lead the Third Industrial Revolution. Thanks to new materials, new processes, intelligent software, smart robots, network collaboration, producer services, etc., manufacturing will be digitized, and the mode of production in the manufacturing sector will be thus changed (The Economist, 2012). Jeremy Rifkin (2012) projected in his book *The Third Industrial Revolution* that a new economy, built on the combination of the Internet and new energy, was coming. Transition to renewable energy, decentralized production of micropower plants, storage of intermittent energy using hydrogen and other storage technologies, distribution through the Energy Internet, and shift of traditional transportation to plug-in and fuel cell powered vehicles constitute the five pillars of the Third Industrial Revolution, which will give rise to a new paradigm of economics. Germany's National Academy of Science and Engineering and Siemens put forward the idea of Industry 4.0 in 2011. They believe that the previous three industrial revolutions resulted from mechanization, electrification, and information transformation of production, and the promotion and application of cyber-physical systems (CPSs) in the manufacturing sector now is triggering the Fourth Industrial Revolution with digital manufacturing as its core (Fig. 7.4). Industry 4.0 will integrate real and virtual production; enable flexible manufacturing with the help of various information technologies such as digital engineering and digital manufacturing, based on big data, the Internet, and people; and achieve horizontal integration between enterprises through the value chain and value network and vertical integration through enterprises' internal networked manufacturing systems that can be restructured flexibly so as to meet customization needs in a cost-efficient way (ACATECH, 2013).

Be it the Third Industrial Revolution or Industry 4.0, the new scientific and technological revolution and industrial changes are about the high-level

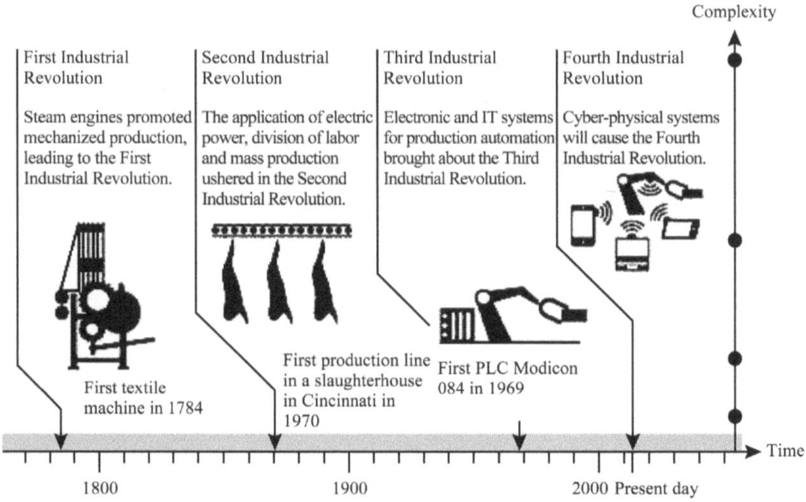

FIGURE 7.4 History of industrial revolutions. Source: *Siemens, "Industry 4.0—Future of Manufacturing".*

integration of industrialization and informatization that advances the trend towards digital, networked, and intelligent manufacturing. Information technology enables information sharing among enterprises in different parts of the value chain. They can quickly find and dynamically change their partners in the world. The advantageous resources of enterprises can be integrated to achieve decentralized production in the aspects of research and development, manufacturing, service, and logistics, and promote the transformation from centralized production to networked collaborative production. Universal connectivity and pervasive computing will be ubiquitous. Virtualization technology, 3D printing, industrial Internet, big data, and other technologies will reshape the manufacturing technology system. For example, 3D printing will add new materials, digital technology, and intelligent technology to products so that their functions are changed radically. With the strong support of the Internet, IoT, cloud computing, and big data, manufacturers, producer service providers, and users can interact on open, shared online platforms, and small-scale customized production will gradually replace production lines. CPS-based intelligent plants will become the main form of future manufacturing, with more and more repetitive work and jobs that require general skills done by intelligent equipment. With the focus of the value chain shifted from production to R&D, design, marketing and service, the manufacturing sector will change from being production-oriented to being service-oriented. Crowdsourcing, collaborative design across regions, mass customization, precision supply chain, etc. are building new competitive advantages of enterprises. Life cycle management, total integration and general contracting, Internet finance, e-commerce, etc. are gathering speed in restructuring the value chain.

The deep integration of the new-generation information technology and manufacturing will effect profound changes in the mode of manufacturing, the way of organizing production and the landscape of the industry. The organizational model in the new industrial transformation of the world is the networked, flat structure, and the network of small and medium-sized enterprises plays a role together with international business giants. This is what Jeremy Rifkin called the "flat world." Every building is a power plant, and can sustain production using the energy generated on its own. Everyone can be a producer and have their own company, which is also true with logistics. With the help of information technology, orders and user needs can be communicated to producers in time, and producers can also display and sell their products on the Internet, which cuts almost all unnecessary costs. Some large enterprises may disappear, but most large enterprises will become aggregators that no longer engage in large-scale production of goods, but provide logistics platforms, trading platforms, and services for small enterprises. In the future, manufacturing will be revolutionized. Competitive markets will increasingly give way to cooperative networks. The traditional centralized operations will gradually be replaced by networked, decentralized enhanced operations. The traditional hierarchical economic and political power will yield to flat power organized in nodes. In brief, the new scientific and technological revolution will advance the global manufacturing sector, and GPNs will also be better developed.

7.2.4 Multinationals Will Continue to Expand International Production

MNCs are the microlevel participants of GPNs and the main driving force behind the formation and expansion of GPNs. The main motivation for MNCs to create GPNs is to increase market opportunities, reduce operating costs, and gain access to essential resources. From the perspective of resources, after the financial crisis countries have adjusted industrial and investment policies to further protect their strategic resources, so MNCs have to increase investment in host countries to obtain essential resources. From a cost point of view, the financial crisis may have changed the cost advantage of countries, but will not eliminate them. The differences between countries in the cost of production factors remain and MNCs still have the incentive to organize and optimize international production for the purpose of reducing costs. From a market point of view, the United States, European countries, Japan, and other developed countries are not seeing smooth economic recovery or any rebound in consumer demand, while emerging economies will generate huge consumer demand, so MNCs will focus more on emerging markets. To gain larger shares in emerging markets in the context of intensifying competition, MNCs must adopt a sound localization strategy; further increase investment in the host country, closely connect procurement, manufacturing, and sales activities based on local R&D and production; and make a correct and rapid response to customer needs. Therefore, MNCs will continue to expand their GPNs in the postcrisis era.

According to the WIR published by UNCTAD, international production indicators, including overseas sales, output, exports, employees, and assets of MNCs, saw increases during the period 2009–14, all above the average during the period 2005–07 before the financial crisis. The number of employees in the foreign subsidiaries of MNCs was about 69 million in 2011, the majority of whom were in developing countries and economies in transition; there were 18.2 million in China alone. The sales and output of foreign subsidiaries of MNCs reached USD 28 trillion and USD 7 trillion, and the output contributed to about one-tenth of the global GDP. Their exports amounted to about USD 7.4 trillion, accounting for about one-third of global total. According to a survey, international production of the world's 100 largest MNCs is on the rise, and their overseas sales, employees, and assets are growing faster than those at home. In 2014, foreign affiliates of MNCs had 75 million employees, over USD 36 trillion sales, about USD 7.8 trillion output, and about USD 7.8 trillion exports. Thus, market-seeking, cost-seeking, and strategic asset-seeking MNCs will all continue to expand international production, and GPNs will be developed further in the postcrisis era.

7.3 NEW FEATURES OF GPNS

Although the basic GPN trends stay unchanged in the postcrisis era, there are still some major changes in the driving forces behind the development of GPNs. For example, the new round of industrial upgrading and relocation of industries will lead to major adjustments in the global industrial competition landscape. The increase in regional trade agreements will bring about the creation of multiple regional manufacturing centers. In particular, China's Belt and Road Initiative will exert significant impact on GPNs; new protectionist policies and trade barriers will increase trade frictions; the new scientific and technological revolution will make "Internet + intelligent manufacturing" a prevailing trend. MNCs will be deeply embedded in local economic and social networks and the changes in driving forces will give GPNs some new features.

7.3.1 Global Industrial Landscape Facing Major Adjustments

After World War II, the world witnessed large-scale industry relocation across countries and regions three times. The nature of relocation each time might be different but the pattern was the same. The rising costs in transferring countries and the international expansion of markets are the internal cause of industry relocation while the cost advantage and huge market of receiving countries are the fundamental driver of industry relocation. After reaching a balance and developing to a certain extent, the division of labor system will become unbalanced and experience a new round of adjustments and improvements. The international division of labor system is always in the cycle of "balance, development, imbalance, adjustment, and rebalancing." The imbalance in the international division of labor is one of the root causes of the global financial crisis that broke out in 2008. After the global financial crisis, developed countries have adopted the reindustrialization strategy one after another to boost manufacturing innovation, build new competitive edges of the manufacturing industry, and accelerate the creation of a new global trade and investment landscape. Some developing countries are also speeding up planning, and actively participating in the new international division of labor by taking in industries and capital from other countries, to gain more opportunities in international markets and an upper hand in the new competition. The financial crisis has changed the economic environment and competitive advantages of all countries. The new round of economic rebalancing will promote industrial upgrading and industry relocation. The industry chains of emerging industries will gradually take shape. The global industrial and capital reconfiguration will facilitate adjustments and improvements in the international division of labor, and affect the landscape of GPNs. Since China is considered a "factory of the world," its manufacturing industry faces severe challenges from both developed countries and other developing countries. Therefore, China should adopt a global vision and focus on building its manufacturing competitiveness so as to translate challenges into opportunities and gain the high ground in the new round of manufacturing competition.

In the field of advanced manufacturing, the United States, European countries, Japan, and other developed countries remain the global leaders in innovation and R&D. In the postcrisis era, major developed countries all consider technological innovation as a fundamental means to reshape their competitive advantages and walk out of the economic crisis. On the one hand, they continue to increase strategic investments in technological innovation; on the other hand, they focus on the development of high technologies, including climate change, energy, life and health, environmental protection, information, nanotechnology, and other technologies. Moreover, they are committed to speeding up the commercial application of high-tech achievements to lead the future development and gain the high ground in the new round of scientific and technological competition. The US government issued the *Strategy for American Innovation* in 2009, 2011, and 2015. The Strategy stresses that technological innovation is the basis for economic growth and competitiveness enhancement in the future. The 2015 edition of the Strategy highlighted nine strategic areas, including advanced manufacturing, precision medicine, BRAIN Initiative, advanced vehicles, smart cities, clean energy and energy-saving technology, educational technology, space exploration and new frontiers in computing, as well as six key elements, including investing in the building blocks of innovation and delivering innovative government with and for the people. It also proposed strategic initiatives to ensure that the United States would continue to lead the global innovation economy and develop future industries, and help the United States overcome the difficulties encountered in economic and social development. In November 2011, the European Commission announced the European 2020 strategy for the development of the European Union. The strategy is the EU's agenda for scientific research and innovation for the next decade, and specifies the priorities and targets of the EU in the next decade: developing an intelligent economy based on knowledge and innovation, enhancing efficiency of energy use to improve competitiveness and achieve sustainable development, and raising employment levels to boost social cohesion. The strategy attaches great importance to innovation. Building innovative alliances is defined to be the top of seven flagship initiatives, and the other six initiatives are all related to innovation. Fig. 7.5 shows the rankings of countries by innovation index between 2006 and 2016. According to the *Global Competitiveness Report* published by the World Economic Forum, starting from 2006, Switzerland, Japan, Germany, and the United States are largely among the top five by innovation index; Britain's ranking fell slightly in previous years and rebounded back to the top 10 in 2012; the rankings of France and South Korea started to decline slowly in 2008 and are now around No. 20; China's

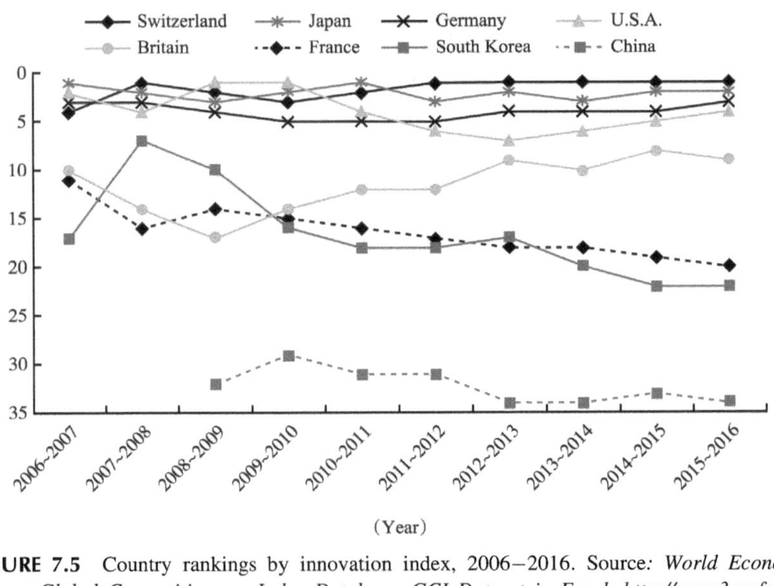

FIGURE 7.5 Country rankings by innovation index, 2006–2016. Source*: World Economic Forum, Global Competitiveness Index Database. GCI Dataset in Excel, http://www3.weforum.org/docs/gcr/2015-2016/GCI_Dataset_2006-2015.xlsx.*

ranking has stayed in the range of No. 30–35 since 2008. Apparently, the financial crisis does not have significant impact on the innovation and R&D capacity of the United States, European countries, Japan, and other developed countries have a major impact, and the developed countries; they still have marked competitive advantages in terms of innovation system.

On July 24, 2015, McKinsey Global Institute published its research paper *The China Effect on Global Innovation*. It divided innovations in all industries into four types, namely scientific research innovation, engineering technology innovation, customer-centered innovation, and efficiency-driven innovation. Fig. 7.6 lists the ratio of Chinese enterprises' contribution to the total revenue of the global industry to Chinese economy's contribution to the global GDP, which was 12% in 2013. Figures about US companies were calculated in the same way. In terms of customer-centered innovation, Chinese companies can commercialize such innovations rapidly thanks to the enormous consumer market in China and successful examples in this regard include Baidu, Alibaba, and Haier. As for efficiency-driven innovation, China has consolidated its status as a global manufacturing center thanks to its widespread, all-inclusive manufacturing ecosystem. In the field of engineering technology innovation, the Chinese government has been striving to

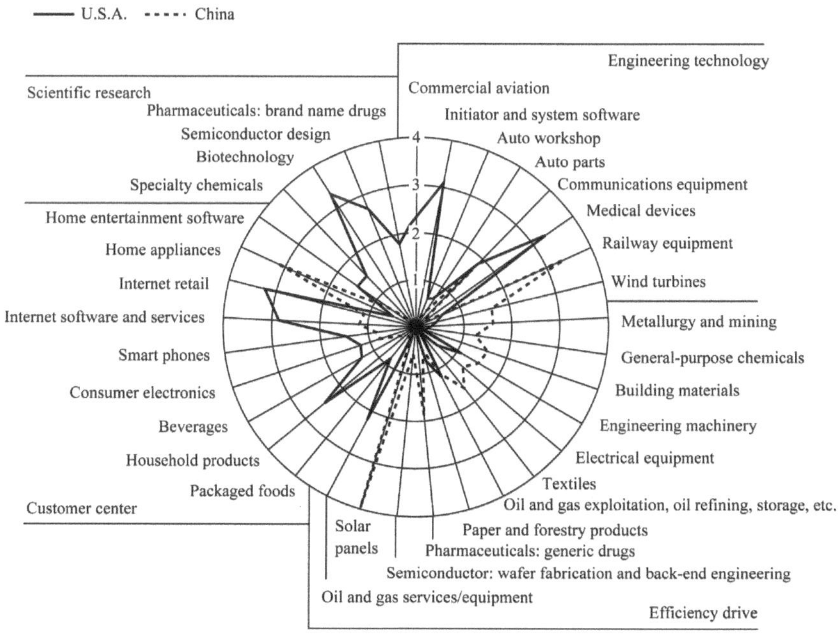

FIGURE 7.6 Comparison of industry performance between China and the United States. Source*: McKinsey Global Institute,* The China Effect on Global Innovation. *https://www.mckinsey.com/~/media/McKinsey/Featured%20Insights/Innovation/Gauging%20the%20strength%20of%20Chinese%20innovation/MGI%20China%20Effect_Full%20report_October_2015.ashx.*

promote enterprises' faster learning but there are imbalances between different fields. Regarding scientific research innovation, China's research and innovation system now has some problems such as the quality of research results being disproportional to the heavy investment so China is not a global leader in this regard. Over the past three decades, China's economic growth has mainly relied on a model of innovation sponge, that is, catching up with leading countries by introducing and adapting a large number of advanced technologies and best practices. The contribution of innovation to economic growth has declined in recent years, coupled with fast-aging population and rising debt levels. All these have posed brand new challenges to China's economic development. It is pressing for China to make a fast shift from innovation sponge to innovation leadership, developing more groundbreaking innovations to compete in the global market. Innovation is essential to the sustainable growth of the Chinese economy in the long run.

After the financial crisis, developed countries reflected on the deindustrialization policy of the 1980s–90s and successively put forward the reindustrialization strategy, stressing the importance of revitalizing manufacturing and consolidating the real economy. Manufacturing has once again become a highlight of global economic competition. The United States has unveiled such plans as Advanced Manufacturing Partnership and National Network for Manufacturing Innovation. Germany has introduced the Industry 4.0 strategy. France has adopted the New Industrial France strategy. Japan stressed the importance of developing robots in the 2014 White Paper on Manufacturing Industries. Britain has published the Manufacturing 2050 vision. The reindustrialization strategy will help developed countries to develop strategic emerging industries and advanced manufacturing. They can use new technologies to upgrade the industries and gain a competitive edge in the new industrial revolution. The strategy will also encourage domestic companies to move their overseas plants back and attract investment from foreign companies, thus reducing imports and creating more jobs. Eventually, growth will be mainly driven by export and the manufacturing sector.

At present, manufacturing is showing the sign of returning to the developed countries. GE, for example, has decided to relocate its water heater production line in China, and some of its high-end refrigerator production in Mexico back to the United States. GM has announced the plan to move the production of next-generation Cadillac SRX SUV from Mexico back to the United States. Caterpillar has opened a new factory in the United States to produce excavators. Apple has set up plants in the United States to make computers. Panasonic will move its vertical washing machine and microwave oven lines from China back to Japan. Sharp plans to produce more models of LCD televisions and refrigerators in Japan. TDK will also move some of its electronic component production from China to Akita, Japan. BCG has released a series of research reports, suggesting that the manufacturing backflow is an important trend and is expected to grow significantly. According to a report published by A.T. Kearney in 2014, more than 300 US companies in 2014 chose to return, compared with 201 in 2013, 104 in 2012, 64 in 2011, and only 16 in 2010. Chinese companies have increased their investment in developed countries. According to a report published in May 2015 by Rhodium Group, a New York research firm, Chinese companies' new projects and acquisitions in the United States were worth USD 46 billion from 2000 to 2014. In 2013, KEER chose to open its first overseas factory in the United States, which was also the first manufacturing plant established by a Chinese textile company in the country. In 2014, the Sun Fiber Company established a polyester plant in Richburg, South Carolina. A $100 million precision copper tube project invested by the Golden Dragon Group in the United States has been completed and put into production. Lenovo has decided to relocate some of its commercial computer production lines to Japan, and move some of its PC production lines to its factories in Sommerda, Germany, and will start making PCs in the United States.

In terms of mid- and low-end manufacturing, in the context of rising manufacturing costs and population aging, China will say farewell to the "cheap production" era, and its competitiveness in many mid- and low-end labor-intensive, technology-intensive industries or production activities will continue to be weakened. Vietnam, India, and some other Southeast Asian countries will begin developing mid- and low-end manufacturing leveraging their comparative advantages in resources and labor, bringing in labor-intensive manufacturing industries at lower costs. Some MNCs invest directly in emerging countries, setting up factories there, while others are considering relocating factories in China to other emerging countries. Some companies have adopted the "China + 1" strategy, which means opening a factory in another country for market testing and backup. It is also necessary to note that excessive debts make it hard for developed countries to sustain consumption growth and financing. Market consumption is difficult to recover quickly after the crisis. China's traditional export markets are sluggish, and the effort to stimulate domestic demand has yet to produce results. Therefore, China's status as a global manufacturing base and an exporter of various manufactured goods will be impacted.

The most direct factor in the relocation of China's mid- and low-end manufacturing is the rise in manufacturing costs. On the one hand, China's wage costs have been on the rise. According to a BCG report in 2011, the wages and benefits of Chinese workers were growing at an annual rate of 15%–20%; the hourly wage in China's manufacturing sector was 52 cents in 2000 and reached 4.41 dollars in 2015. It does not have any advantages anymore compared with Southeast Asian countries, India, and other countries. On the other hand, international transportation costs have been on the rise. Increasing oil prices and the reduction in ship and container capacity have pushed up logistics costs. Companies tend to organize production activities at locations closer to the market. Eastern European countries, Mexico, and other countries are thus showing greater competitive advantages for geographical reasons. In 2014, BCG published a research report entitled *The Shifting Economics of Global Manufacturing*. The report constructs a manufacturing cost index based on four influencing factors, that is, manufacturing wages, labor productivity, energy costs, and exchange rate, and analyzes the top 25 leading export economies in the world. The index of the United States is considered the benchmark level, that is, 100, and China's manufacturing cost index is 96. Between 2004 and 2014, the estimate of China's manufacturing cost advantage over the United States fell from 14% to 4%. Ten years ago, the average productivity-adjusted wage in China's manufacturing sector was about 4.35 dollars per hour and it was 17.54 dollars in the United States. Today, the average wage level in China has nearly tripled, reaching 12.47 dollars per hour, while the level in the United States has increased by only 27% to 22.32 dollars. From 2004 to 2014, industrial electricity cost in China rose by 66%, and natural gas cost soared 138%. The sharp increases in labor and energy costs weakened China's competitiveness. Since 2004, Mexican workers' wages have risen by only 67%, and Mexico's electricity and natural gas costs are competitive. Therefore, Mexico's productivity-adjusted labor costs are now 5% lower than China's, and it has the biggest increase in manufacturing costs among all of the 25 economies. Mexico and the United States are becoming rising stars of the global manufacturing community, thanks to their low-wage growth, continued productivity gains, stable exchange rates, and a huge energy cost advantage.

Rising costs have led to the withdrawal of foreign factories and foreign orders from China. For example, Microsoft plans to shut down Nokia's Dongguan plant, and some of its equipment has been transferred to Hanoi, capital city of Vietnam. Famous companies such as Nike, Uniqlo, Samsung, Funai, and Foxconn have opened new plants in Southeast Asia and India. Adidas has closed its only directly run plant in Suzhou. In 2010, Vietnam overtook China as the world's largest producer of Nike shoes marking MNCs' search for new low-cost manufacturing bases in Southeast Asia, India, Eastern Europe, Mexico, and other destinations. For example, foreign investment in Mexican factories has been growing again in some of China's dominant industries. During the period 2006–13, Mexico's exports of electronic goods increased by more than twice to USD 78 billion. Asian companies such as Sharp, Sony, and Samsung contribute to one-third of Mexico's electronics manufacturing investment, compared with about 8% a decade ago. Due to the falling number of overseas orders, a large number of low-cost factories in China have been shut down or moved abroad. Some Chinese companies have even transferred a lot of domestic orders overseas, or invested more overseas to build factories. Midea has set up production bases in Vietnam, Egypt, Brazil, Argentina, and India. Gree has opened factories in Brazil, Pakistan, Vietnam, the United States, and other countries. With more facilities developed overseas, more than half of Gree's air conditioners are expected to be produced in foreign countries. Foxconn is Mexico's second largest exporter, following GM. The Foxconn plant in San Jerónimo, Chihuahua, Mexico has 5500 workers and exports 8 million PCs a day. The plant is now in a massive expansion. It can be predicted that the declining share of Chinese manufacturing in the mid- and low-end export market will be a long-term trend.

7.3.2 Regional Trade Agreements Playing an Increasingly Bigger Role

After the financial crisis, the economy slumped rapidly in many countries, protectionism gathered strength worldwide, and the WTO negotiations for promoting free trade were blocked. As a result, countries began to emphasize regional openness and sector coordination in regional cooperation, accelerating the pace of economic integration. According to the WIR published by UNCTAD, the conclusion of traditional bilateral investment agreements has lost momentum in recent years and policy has shifted towards the formulation of regional agreements, most of which are free trade agreements (FTAs). Bilateral agreements still make up the majority, but regional agreements are increasingly important in the economic sense. Such a shift would consolidate and unify investment rules, demonstrating that countries have taken the first step towards multilateralism. The introduction of new agreements does not require the elimination of old treaties and the old treaties cannot replace new agreements, either. While regional trade agreements will make the IIA network more complex and prone to overlapping and inconsistency, this reflects the need for broader policy considerations, and will lead to systemic changes in the international investment regime, thus bringing opportunities for establishing a more coherent, balanced, and effective international investment regime conducive to development. In the postcrisis era, regional economic integration is likely to usher in a new climax, leading to the further development of two-way trade and investment in the region. For China, on the one hand, regional economic integration in Asia will enhance China's position in Asian production networks; on the other hand, the development of economic integration in other regions will promote the creation of regional manufacturing bases, such as Mexico, Eastern Europe, and other countries and regions mentioned earlier. This will affect the global export of Chinese manufacturers.

For example, the EU has expanded its regional cooperation through the Union for the Mediterranean. The East African Community (EAC), the Common Market for Eastern and Southern Africa (COMESA), and the Southern African Development Community (SADC) have decided to strengthen cooperation. Mexico has formally signed FTAs with five countries in Central America. The China-ASEAN Free Trade Area established on January 1, 2010 is a highlight of regional integration in Asia amid the crisis. In the past more than 10 years, there has been substantial progress in regional economic cooperation in Asia. ASEAN + 1 (China-ASEAN, Japan-ASEAN, and South Korea-ASEAN FTAs), and ASEAN + 3 (ASEAN plus China, Japan, and South Korea) cooperation mechanisms have been put into operation. The ASEAN + 6 (ASEAN plus China, Japan, South Korea, India, Australia, and New Zealand) mechanism is also in the making. The ASEAN + 6 is officially known as the RCEP, which covers about half of the world's population. The countries involved contribute to one-third of the global GDP. This is the latest product of regional economic integration. ASEAN's FTAs with the six countries will be integrated and improved, and a free trade area with a trade liberalization rate of up to 95% will be created. According to the study by Lin Guijun et al. in 2011, the share of internal trade in Asia has surpassed that of Asia's trade with the EU since 2004, and the interdependence of Asian plants is the highest among the world's top three groups of plants (Asian, European, and North American plants), which suggests a higher level of integration in Asian production networks. In the postcrisis era, with the gradual reduction of trade barriers among the countries in Asia, economic integration in the region and the division of labor within Asian production networks will go deeper, and China will further integrate into the Asian production networks, especially into East Asian production networks.

For many years, ASEAN has been playing a leading role in regional economic cooperation in Asia, but this is the result of competition among major East Asian countries (China, Japan, and South Korea). ASEAN's capacity has been increasingly questioned for its limited economic strength. China has overtaken Japan as Asia's largest economy, and China is the largest trading partner or export destination for most of the other Asian countries. After the international financial crisis, China has become the main driving force of Asian economic growth (Park and Shin, 2009). Asian economies are highly dependent on the activities of Asian factories, while Asian factories have the highest degree of dependence on China, with Japan ranking third in this regard. China is increasingly becoming a core platform in Asian and even global production networks (Lin and Deng, 2011). Therefore, an Asian economic cooperation landscape with China as its core is emerging. At present, all countries consider industrial restructuring and upgrading as a major move to develop their own economies. If China can effectively make use of the opportunity presented by the shift in the division of labor in Asian production networks, improve its status in the division of labor and trade in such networks, and adjust its industrial structure with an opener mind, it will greatly promote the optimization and upgrading of the entire industrial chain.

7.3.3 BRI Helping to Boost the Transformation of Globalization

Despite the growing protectionism in various forms, the tide of economic globalization is irreversible. Modern means of production, GPNs, and information technology have brought most countries in the world closer. The world today is seeking new reform mechanisms for economic globalization and reform requires new thinking and new models. At this historical juncture, China as the world's second largest economy has proposed the Belt and Road Initiative that involves 64 countries, including 43 in Asia, 16 in Central and Eastern Europe, the C.I.S., and one in Africa. The area of these countries accounts for 62% of the earth's total land area and most of them have not very high levels of economic development. They have strong desire for increasing connectivity and trade exchanges with the rest of the world. There are no fixed standards. Countries can choose the proper cooperation models according to their own development levels and needs. The initiative offers arrangements that have vague boundaries and involve more cooperation elements, including the Asian Infrastructure Investment Bank (AIIB) and Silk Road Fund, free trade areas, and infrastructure connectivity. Existing mechanisms such as the Shanghai Cooperation Organization (SCO), ASEAN Plus China (10 + 1), Asia-Pacific Economic Cooperation (APEC), Asia-Europe Meeting (ASEM), Asia Cooperation Dialogue (ACD), Conference on Interaction and Confidence-Building Measures in Asia (CICA), China-Arab States Cooperation Forum (CASCF), China-Gulf Cooperation Council Strategic Dialogue, Greater Mekong Subregion (GMS) Economic Cooperation, and Central Asia Regional Economic Cooperation (CAREC) will be made full use of to attract more countries and regions to participate in the Belt and Road Initiative so that all countries can work together to build a community with shared future featuring mutual benefit, coexistence, and common prosperity based on diversity.

The Belt and Road Initiative provides an institutional framework for the transformation of globalization, including the basic objective, basic principles, and basic philosophy. The initiative is intended to build a path of peace, prosperity, openness, innovation, and culture. This objective is in line with the basic direction of world and human development. The basic principles of the initiative include not interfering in other countries' internal affairs, not exporting social system and development model, not imposing anything on other countries, not adopting geopolitical maneuvering, and not forming any cliques that undermine stability. This sends out two important messages. One is to promote open, nondiscriminatory cooperation; the other is to respect different choices of different paths. Therefore, the cooperation advocated in this initiative is based on equality and for achieving mutual benefit and shared prosperity and peace. The Belt and Road are to be jointly built through consultation to meet the interests of all. The basic approach is to promote pragmatic cooperation. Plans will be formulated and discussed to reach consensus and plan implementation will be well monitored. The priority areas of cooperation include policy coordination, facilities connectivity, unimpeded trade, financial integration, and people-to-people bonds. This is highly consistent with the ultimate goal of globalization, which is to achieve integration in every aspect. When connectivity in every aspect reaches a high level, the transformation towards globalization is completed and a better state of integrated development is achieved. In addition, the initiative helps to reinforce the economic circle in Asia with China as the center. This will contribute to the creation of various GPNs and the faster implementation of regional economic integration strategy.

7.3.4 New Trade Barriers Growing in Number

In the postcrisis era, countries have introduced many investment control and restrictive measures, in part because of increased emphasis on the protection of strategic industries and specific industries, resources, and national security. Emerging economies pay more attention to environmental and social protection, while the least developed countries are filling blanks in their regulatory frameworks. High unemployment rates have caused a concern that FDI outflows may lead to job export and the weakening of the domestic industrial base. To stabilize foreign exchange rate and improve the balance of payments, countries have taken many measures to control FDI outflows and provide incentives for foreign investors. At the same time, there have been strengthened interactions between FDI policy and industrial policy. The measures introduced on the grounds of industrial policy include adjustments to the access policy for foreign investors (e.g., in agriculture and pharmaceuticals). More regulatory policies have been formulated for the extractive industries, including requirements on the extent of state ownership, requisition or withdrawal of funds, and increasing corporate income tax rate, royalties, and the cost of renegotiation of contracts. Governments have adjusted FDI policies to better balance the relationship between investment liberalization and the regulation of some core industries.

The introduction of various investment control policies, while facilitating the creation of domestic production networks, has also increased trade protectionism to some extent. The unclear direction of the global economy, unclear trade demands of countries, and the diverse international trade measures have led to marked uncertainty in trade policies in postcrisis era (Huang, 2012). Under the WTO framework, it becomes increasingly difficult to employ traditional tariff and nontariff barriers as protectionist measures. Some countries, especially developed countries, use hidden, vague, and uncertain policies and measures to set more invisible trade barriers, such as the implementation of trade remedies that can only be used under special circumstances, and the introduction of intellectual property, technical and low-carbon trade barriers in some areas to protect domestic markets and safeguard national interests. Some countries have taken advantage of flaws in the WTO's exception clause to introduce discriminatory protectionist measures. For example, in the China–US tire case of June 2009, the United States adopted special safeguard measures against China for the first time, which produced significant negative effects. In addition, the United States also uses legislation to introduce discriminatory trade protectionist policies, such as Article 727 of the *Omnibus Appropriations Act of 2009*, passed by the House of Representatives. There is also abuse of discriminatory trade protectionist measures through special empowerment, such as the "Super 301" clause in Section 301 of the *Omnibus Trade and Competitiveness Act*. The United States also employs political and diplomatic means to push forward discriminatory trade protectionist measures. For example, it requires India, China, Turkey, South Korea, and other countries to reduce oil imports from Iran to the satisfaction of the US side to be exempted from US sanctions.

It is foreseeable that China's economic and trade frictions with developed and developing countries will be more intense in the future, which will adversely affect China's export of manufactured goods. It should also be noted that the importance of corporate social responsibility standards and their increasing impact on global supply chains have posed a greater challenge to Chinese manufacturing. Such standards have complex social and environmental requirements, and MNCs often require suppliers in different industries to comply with. Compliance with such standards provides opportunities for companies and also poses serious challenges to them, particularly to small suppliers in developing countries. Such international standards tend to be above China's regulations and prevailing market practices. In some cases, Chinese rules may be in conflict with the standards conflict, thus undermining the competitiveness of Chinese companies.

7.3.5 "Internet + Intelligent Manufacturing" Becoming the Prevailing Trend

After the 2008 global financial crisis, the diffusion of the new-generation information technology and its integration with industries has provoked the international community's heated discussion about the Third Industrial Revolution, Energy Internet, Industrial Internet, Industry 4.0, and a series of other development ideas and models. The deep integration of the new-generation information technology with manufacturing is effecting industrial changes of far-reaching influence, leading to the creation of new modes of production, forms of business, new business models, and new sources of economic growth. Countries are boosting scientific and technological innovation to enable new breakthroughs in 3D printing, mobile Internet, cloud computing, big data, bioengineering, new energy, new materials, and other fields. CPS-based intelligent equipment and intelligent plants are leading the change of manufacturing approach. Crowdsourcing, collaborative design, mass customization, precision supply chain, full life cycle management, e-commerce, etc. are reshaping the industrial value chains. Wearable smart products, smart appliances, smart cars, and other smart devices keep expanding fields of manufacturing. It is foreseeable that with the deep integration of information technology and manufacturing, the combination of the Internet and intelligent manufacturing has broad prospects and unlimited potential. It is an irresistible trend that is producing strategic and overall impact on the economic and social development of all countries.

"Internet + intelligent manufacturing" means to integrate the innovation of the Internet with manufacturing; promote the application of such technologies as cloud computing, IoT intelligent industrial robots, and additive manufacturing in the production process; promote the intelligent upgrading of production equipment; process transformation and basic data sharing; and effectively support the intelligent transformation of the manufacturing industry, creating an intelligent manufacturing ecosystem based on the Internet, featuring openness, sharing, and collaboration. The extension of the Internet to the field of industrial production and service gives rise to new models such as crowdsourcing design and customization, which will promote the real-time interaction between producers and consumers, and the transformation from mass production to mass customization. The boundaries between Internet companies and manufacturers, and between producers and service providers are becoming increasingly blurred; the focus of production has been shifted from traditional product manufacturing to providing customers with rich products, services and even total solutions. This has helped traditional manufacturers to transform themselves into service providers that operate across sectors. The Internet is closely integrated with IoT and service network, connecting products, machines, resources, and people organically, promoting the integration of the physical world and the digital world, and advancing digital, networked, and intelligent operations throughout the whole life cycle of products and the whole manufacturing process. When the world's industrial systems are integrated with advanced computing, analysis, sensing technology and the Internet, an open and globalized Industrial Internet will be formed (GE, 2015). The Industrial Internet, coupled with software and big data analysis, will reshape global industries and stimulate productivity.

The new scientific and technological revolution and industrial changes converge with China's endeavor to rapidly change its growth model. The Chinese government has worked out a series of plans to make its manufacturing sector competitive, including Made in China 2025 action plan and Internet Plus action plan. The Internet Plus action plan is about giving full play to China's strength in the size and application of the Internet. Eleven initiatives including Internet plus entrepreneurship and innovation, and Internet plus collaborative manufacturing, have been designed to promote application of the Internet from the consumer field to the production area, accelerate industrial development, and enhance the innovation capacity of various industries, creating new engines of economic and social development (State Council of China, 2015).

7.3.6 Multinationals to Be Embedded Further Locally

The sluggish consumption in developed countries has drawn the attention of MNCs to emerging markets. To adapt to local development, MNCs have embedded their R&D, production, purchasing, marketing, and management activities into the local economic and social environment, so as to respond to the customer needs correctly and quickly. Strengthening local R&D can help to attract more local talents and companies, better understand customer needs, and acquire related knowledge and technology faster, thereby shortening the innovation cycle and improving customer satisfaction. Strengthening local sourcing and production can reduce the costs of production, procurement, and transportation, and help deliver products to the market faster. Improving local sales and after-sales service will also draw more attention from customers and help improve customer satisfaction and awareness.

Due to the importance of China's huge consumer market, a large number of MNCs have implemented a "deep embedding" strategy in China. MNCs' embeddedness in China's economic and social networks has become an engine of China's economic development. It has contributed to the sound and fast growth of the Chinese economy and also exerted greater impact on local Chinese enterprises. The competition between MNCs and local enterprises used to be in different markets. For example, MNCs focus on first- and second-tier cities while local enterprises take root in third- and fourth-tier cities; MNCs focus on high-end markets while local enterprises focus on mid- and low-end markets. In other words, the cooperation between MNCs and local enterprises is greater than competition. MNCs help local SMEs to upgrade production technology, get embedded in global production networks, and enter into international markets. After the financial crisis, however, MNCs come to expand to third- and fourth-tier cities as well as mid- and low-end markets, while local enterprises aim at high-end markets and overseas markets. There will be more intense competition between MNCs and Chinese enterprises. Take the medical device industry as an example. According to statistics, the sales revenue of China's medical device industry has been growing at an average rate of over 20% in recent years, but its share in high-end medical device market is less than one-third. The introduction of China's new healthcare reform plan in April 2009 has brought more opportunities for MNCs, especially the huge demand for new equipment and equipment upgrading at second-grade county hospitals. The three giants in high-end medical device market—GE, Philips, and Siemens are also starting to target the primary healthcare market (Luo, 2012). GE plans to change the ratio of high-end business to community-level business in the Chinese market from 8:2 to 5:5 in 3–5 years. Early in 2006, Siemens launched the SMART program to design products that are simple to use, easy to maintain, affordable, durable, and put to the market in a timely manner. This has helped the company enter into China's basic healthcare market. MNCs' advantages in capital, technology, channels, and influence will break the monopoly of Chinese companies in primary healthcare; the industry may even be reshuffled.

7.4 TOPOLOGICAL STRUCTURE OF GPNS IN THE POSTCRISIS ERA

In brief, the changes in the four driving forces, that is, international division of labor, trade and investment policies, science and technology, and MNCs, have led to some new features of global production networks, including multiple centers, more decentralized operations, closer ties, and more balanced development. The topological structures of GPNs before and after the crisis are shown in Fig. 7.7.

The topological structure shows that GPNs are developing in a direction towards multiple centers and decentralization, as embodied in consumption, research and development, manufacturing, and resource supply. In the consumer sector, the rapid growth of emerging economies creates huge consumer demand, while the market in developed countries will shrink and the center of global market will be shifted to emerging economies. GPNs will see the change in the centers of consumption from US, European, and Japanese markets to both developed countries and emerging markets. In terms of research and development, the production-oriented countries will greatly increase their R&D capacity by making heavier investments in this regard. While the developed countries are still competitive in innovation and

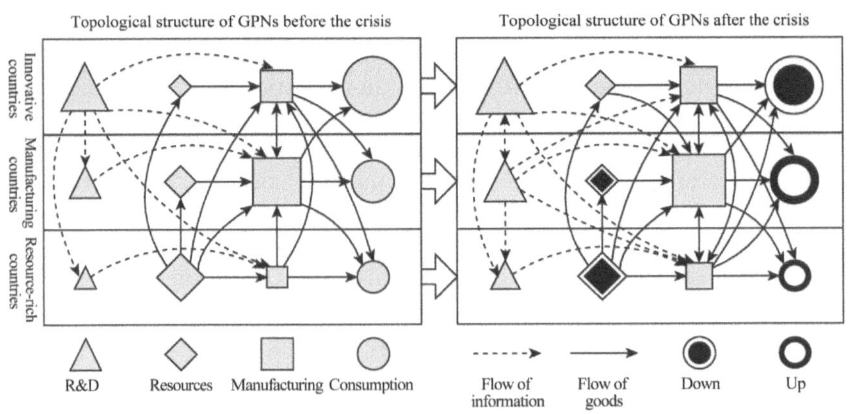

FIGURE 7.7 Topological structure of GPNs before and after the crisis. Source*: Prepared by the author.*

research, their leading edge will be narrowed. The distribution of global R&D centers at the national level will be more decentralized. In the manufacturing sector, developed countries will continue to lead advanced manufacturing. Coupled with the backflow of manufacturing, they will intensify efforts to promote local manufacturing and export. Production-oriented countries will catch up rapidly through industrial upgrading, while consolidating the advantages of traditional manufacturing and industrial chain, adopting industry rotation and regional rotation, and using domestic consumption to offset the decline in export. However, many low-end industries and production activities will be shifted to resource-rich and primary production countries, leading to the expansion of the latter's local production and more geographically dispersed international production. The same trend can be seen in the resource segment. Given the increase in local manufacturing in developed countries and the rising prices of productive resources, the constraints on strategic resources in developing countries, and the advances in technology for new energy development, will lead to increased availability of local resources so that global resource supply will become more decentralized.

Multiple centers and decentralization will not bring countries back to the era of closed-door development. On the contrary, global economic integration will deepen and GPNs will make countries more closely connected and more balanced developed. For example, the global knowledge and technology flows will not only be from innovative countries to production-oriented countries; there will be two-way flows and they will accelerate the transfer of technology to resource-rich and primary production-oriented countries. The Energy Internet will allow dispersed renewable energy sources to flow freely across national borders. Every region and every building will become nodes of a global, borderless green power network for the better sharing and distribution of energy. From the point of view of intraproduct international division of labor, production-oriented countries will engage in more production of components or processes with a high technology content, and have closer collaboration with innovative countries and resource-rich countries on product processes. There will be more frequent two-way flows of products. As the development of GPNs brings about the general improvement of national production capacity, and technology advances will greatly remove regional barriers, every country is likely to supply goods to the world, so all parts of the networks including research and development, production, resource supply, and market consumption will be more balanced. GPNs will enable each participant to better share the fruits of globalization.

More decentralized GPNs means that the elements of GPNs are also more decentralized. According to the GPN framework proposed by Henderson et al. (2002), value, power, and embeddedness are the three elements of a global production network. From the point of view of value, there will be more participants in technology transfer and knowledge diffusion in postcrisis era and it will be on higher levels and larger scales and in shorter cycles. Value creation, enhancement, or capture will be more decentralized in GPNs. From the perspective of power, developing countries will make outstanding contributions to world economic growth in the postcrisis era, with their international status greatly enhanced, and G20 will replace G8 to play a greater role in international affairs. GPNs will consolidate the cooperation between the North and the South and strengthen South–South cooperation. The rise of socialized production will give rise to a large number of SMEs and even individual entrepreneurs. Business operations in GPNs will be more decentralized and organizational structure will be flatter. Therefore, the distribution of power between developed and developing countries, as well as among enterprises, institutions, and collectives, and even between lead firms and secondary suppliers in GPNs will be more equitable and balanced. From the perspective of territorial embeddedness, with the enhancement of regional trade agreements and the evolution of the Third Industrial Revolution, the formation of intercontinental economic and political alliances will gather speed, GPNs will move towards continentalization, the importance of continental production networks will increase, and a number of regional manufacturing centers will emerge. From the point of view of network embeddedness, amid more dispersed, tighter network connections, the key nodes and links of GPNs will be redundant, which means that backup nodes that are not economically optimal are allowed. This helps to prevent natural disasters or political turmoil as key links, such as Japan's earthquake and nuclear leakage, Thailand's floods, and the Arab Spring movement, from exerting enormous impact on GPNs, thus making them more secure and reliable.

Chapter 8

Conclusions and Research Prospects

Chapter Outline

8.1 CONCLUSIONS

This book builds on microcosmic GPN studies and combines theoretical exploration with case studies. It presents an overview of the research into GPNs, examines the theoretical basis of microcosmic GPN studies, offers the analytical framework of such studies, discusses the Siemens and Huawei cases, and describes GPN trends in the postcrisis era and their impact on China. This research can help make up the deficiencies in previous GPN studies in terms of theoretical framework, research content, attention to new developments, etc. The following are the main conclusions drawn from the research.

8.1.1 Value and Embeddedness Are the Core Elements of Microcosmic Global Production Network Studies

According to the GPN framework proposed by Henderson et al. (2002), value, power, and embeddedness are the three elements of a global

Global Value Chains and Production Networks. DOI: https://doi.org/10.1016/B978-0-12-814847-1.00008-7

production network (GPN). On the microlevel, GPNs are in essence global value networks and the agents in networks with the same value objectives form an ecosystem of symbiosis. Value is a driving force behind the evolution of GPNs. Therefore, value stands at the center of microcosmic GPN studies. Embeddedness connects multinational corporations (MNCs) closely with local economy and contributes to their interactions and common development, so it also plays an important role in microcosmic GPN studies. Power is largely influenced by firms' contribution to network value and how they are embedded in the networks so it is not desirable to consider power a key subject of microcosmic GPN studies. Therefore, value and embeddedness are the most important elements in the microlevel research into GPNs. The analytical framework of microcosmic studies proposed in this book comprises one framework from the perspective of value and another from the perspective of embeddedness. In addition, how each element of the framework is studied is explained, making microcosmic GPN studies more operable.

8.1.2 Chinese Enterprises Should Make Full Use of Global Production Networks to Go Global

As the international division of labor deepens and GPNs develop further, the competition between individual firms has been gradually replaced by that between consortiums of firms. The competitive edges of countries are embodied no longer in industries, but in businesses or one part or several parts of the product value chain. In any market of the world, Chinese companies face fierce competition from MNCs. To succeed in going global, they need to draw upon the successful experience of the world's leading companies and make full use of GPNs to enhance their international competitiveness. The improvement of companies' capabilities will also help build the country's competitive edges.

From a global perspective, Chinese companies should be fully aware of how important value is in GPNs and shared value objectives are the cornerstone of network organizations' formation and survival. Value creation determines the position and power of companies in GPNs. Chinese companies should aim to achieve sustainable value creation, formulate corresponding global strategies, and leverages the location advantages of different countries and regions and pool global resources and technologies to build highly efficient global R&D, manufacturing, supply chain, and sales and service networks for the better configuration of factors worldwide. The networks can be used to configure business across regions and industries and integrate value-creating activities globally, turning the competition in price to the competition in value and fighting alone into working together. With the continuous increase of company value, competitive edges in a global sense can be established and sustainable profit growth can be achieved.

From a local perspective, Chinese companies should be fully aware of how important embeddedness is to corporate and local development. Embeddedness is far greater than localization in terms of both connotation and implications. Embeddedness stresses the interactions between MNCs and local communities as well as the full integration with local environment in economic, social, technical, institutional, and other dimensions. Therefore, embeddedness plays a key role in MNCs' global and local economic linkages. When entering an overseas market, Chinese companies should devise the proper embeddedness strategy and path according to local policies, developing from exploratory embeddedness to strategic linkage, strategic embeddedness, and strategic coupling, and from economic to technical, social, cultural, and institutional embeddedness. As embeddedness reaches deeper levels and fulfills more functions, Chinese companies will eventually be rooted in local economic and social networks, achieving win–win interactions with local industries. One-way embeddedness will gradually evolve into two-way embeddedness. Companies' embeddedness also helps the local economy to be better integrated into GPNs while they improve the configuration of factors worldwide to enhance their competitiveness.

8.1.3 Global Production Networks Will Move Toward a Landscape Characterized by Multiple Centers, More Scattered Operations, Closer Ties, and More Balanced Development

The main driving forces behind the creation of GPNs are deepening international division of labor, trade and investment liberalization, technology advances, and expansion of GPNs. How the four driving forces evolve determines the development trends of GPNs. The basic trend of economic globalization remains unchanged in the postcrisis era. The international division of labor will undergo progressive adjustments but no disruptive changes will take place. Despite the growing restrictions on and regulation of investment, trade and investment liberalization and investment promotion remain the focus of trade policies in the immediate term. The adjustment period of the world economy after the crisis will give rise to a new technological revolution, providing more favorable technical conditions for the cross-border flows of factors. To increase market opportunities, lower operating costs, and obtain essential resources, MNCs will continue to expand globally and move more of their R&D, production, procurement and sales activities to emerging countries and regions. Since the basic direction of those driving forces remains unchanged, no radical changes or setbacks will occur in the development of GPNs and the MNCs-dominated GPNs will continue to expand amid adjustments and improvements.

However, Chinese manufacturing will face competition in both high-end and low-end markets since the new round of industrial upgrading and industry relocation in the postcrisis era will push forward adjustments in the international division of labor. The increase in regional trade agreements will lead to the creation of multiple regional manufacturing centers and influence China's export but China's status in Asian production networks will continue to be raised. The Third Industrial Revolution will make production organizations more decentralized, flatter, and more socialized. Digital manufacturing means more challenges than opportunities to Chinese manufacturers. MNCs' deep embeddedness in local economic and social networks has greater impact on Chinese companies, while promoting China's industrial system. The changes in the four driving forces, that is, international division of labor, trade and investment policies, science and technology, and MNCs, have led to some new features of GPNs, including multiple centers, more scattered operations, closer ties, and more balanced development, which have significant impact on the Chinese economy.

8.2 RESEARCH PROSPECTS

8.2.1 To Improve the Theories and Research Methods Regarding Microcosmic Global Production Network Studies

The theoretical research into GPNs on the microlevel is still at the exploratory stage. Improvements are yet to be made in the integration and expansion of the theoretical system, the discussion about the relationships between various elements and the influence, as well as the integration of the Managerial School and Manchester School, macrolevel and microlevel studies, and the studies on value, embeddedness, and power. More specific research methods should be introduced. There should be specific research methods, reference models, and supporting tools for every element so that a set of methodological tools is developed for microcosmic GPN studies.

8.2.2 To Expand the Empirical and Case Studies of Global Production Networks on the Microlevel

In terms of case study, the GPN framework proposed in this book can be used to compare the GPNs of MNCs with different home countries and in different business sectors. In particular, comparisons should be made between the GPNs of MNCs in developing countries, global small and medium-sized enterprises, and emerging industries such as new energy. Comparative studies can also be conducted on the embeddedness of the same company in different countries and regions or the embeddedness of different companies in the same country or region. The interactions between companies of newly industrialized countries should be analyzed. In this way, the results of microcosmic GPN studies will be enriched. In addition, masses of corporate data should be gathered and related theoretical models should be used to analyze them empirically, thus providing a solid foundation for the application of the microlevel GPN framework.

8.2.3 To Analyze the Development of Global Production Networks Dynamically

On the dynamic analysis front, a discrete choice model or continuous selection model should be built, and new influencing factors, new country/region classification, new functions, and new network linkage categories should be introduced to make dynamic, systematic analyses of GPNs' future development trends in different industries.

Bibliography

ACATECH, 2013. Recommendations for implementing the strategic initiative INDUSTRIE 4.0. Final report of the Industrie 4.0 Working Group. Available from: https://en.acatech.de/944/?tx_ttnews%5Btt_news%5D = 916&cHash = 49e3bb346e0888ce34ccf543280b08e6.

Andersson, U., 1996. Subsidiary embeddedness and control in the multinational corporation. Int. Bus. Rev. 5, 487–508.

Andersson, U., Forsgren, M., Pedersen, T., 2001. Subsidiary performance in multinational corporations: the importance of technology embeddedness. Int. Bus. Rev. 10 (1), 3–23.

Andersson, U., Forsgren, M., Holm, U., 2002. The strategic impact of external networks: subsidiary performance and competence development in the multinational corporation. Strat. Manag. J. 23 (11), 979.

Andersson, U., Bjorkman, I., Forsgren, M., 2005. Managing subsidiary knowledge creation: the effect of control mechanisms on subsidiary local embeddedness. Int. Bus. Rev. 14 (6), 521–538.

Antras, P., 2003. Firms, contracts, and trade structure. Quart. J. Econ. 11, 1375–1418.

Antras, P., 2005. Incomplete contracts and the product cycle. Am. Econ. Rev. 95, 1054–1073.

Antras, P., Helpman, E., 2004. Global sourcing. J. Polit. Econ. 6, 552–580.

Antras, P., Helpman, E., 2006. Contractual Frictions and Global Sourcing. National Bureau of Economic Research.

Arndt, S.W., 1997b. Globalization and the open economy. N. Am. J. Econ. Fin. 8 (1), 71–79.

Bair, J., 2005. Global capitalism and commodity chains: looking back, going forward. Comp. Change 9 (2), 153–180.

Bair, J., Gereffi, G., 2003. Upgrading, uneven development, and jobs in the North American apparel industry. Glob. Netw. 3 (2), 143–169.

Baldwin, R.E., Nicoud, F.R., 2004. The Impact of Trade on Intra-Industry Reallocations and Aggregate Industry Productivity: A Comment, Working Paper.

BCG, 2014. The Shifting Economics of Global Manufacturing: How Cost Competitiveness Is Changing Worldwide. Available from: https://www.bcg.com/publications/2014/lean-manufacturing-globalization-shifting-economics-global-manufacturing.aspx.

Bernard, A.B., Eaton, J., Jensen, J.B., Kortum, S., 2003. Plants and productivity in international trade. Am. Econ. Rev. 93 (4), 1268–1292.

Bernard, A.B., Redding, S.J., Schott, P.K., 2007. Comparative advantage and heterogeneous firms. Rev. Econ. Stud. 74 (1), 31–66.

Brandenburger, A.J., Nalebuff, Barry J., 1996. Co-Opetition: A Revolution Mindset That Combines Competition and Cooperation: The Game Theory Strategy That's Changing the Game of Business. Publishing Group Inc., New York, NY.

Brander, J.A., Krugman, P., 1983. A 'reciprocal dumping' model of international trade. J. Int. Econ. 15 (3–4), 313–321.

Brookings Institution, 2015. America's Advanced Industries: What They Are, Where They Are, and Why They Matter. Available from: https://www.brookings.edu/research/americas-advanced-industries-what-they-are-where-they-are-and-why-they-matter/.

Bu, G.Q., 2007. Impact of GPN governance model on processing trade upgrading. Product. Res. 21.

Buckley, P.J. (Ed.), 2009. What is International Business? (Z.X. Zhao, W.H. Wang, Trans.). University of International Business and Economics Press, Beijing.

Buckley, P.J., Casson, M.C., 1976. The Future of the Multinational Enterprise. Macmillan, London.

Buckley, P.J., Casson, M.C., 1998. Analysing foreign market entry strategies: extending the internalization approach. J. Int. Bus. Stud. 29 (3), 539–561.

Callon, M., 1986. The sociology of an actor-network: the case of the electric vehicle. In: Callon, M., Law, J., Rip, A. (Eds.), Mapping the Dynamics of Science and Technology: Sociology of Science in the Real World. Palgrave Macmillan, UK, pp. 19–34.

Chai, Y., 2003. Latest developments in international investment and international production. Qiushi 23.

Chen, J.H., 2010. A Study on Industrial Agglomeration in China's Development Areas: From the Perspective of MNC Embeddedness. People's Publishing House, Beijing.

Chi, G.H., 2008. EVA Management and Performance Evaluation System. Science Press, Beijing.

China Academy of Information and Communications Technology, May 2014. White Paper on the Communications Equipment Industry. <http://www.miit.gov.cn/>.

China Enterprise Confederation and China Enterprise Directors Association, 2011. China's Top 100 MNCs and TNI. <http://www.cec-ceda.org.cn/>.

Christaller, W., 1933. In: Chang, Z.W., Wang, Z.X. (Eds.), Central Places in Southern Germany. The Commercial Press, Beijing.

Coe, N., et al., 2004. Globalizing regional development: a global production networks perspective. Trans. Inst. Br. Geogr. 29 (4), 468–484.

Collis, D.J., Montgomery, C.A., 1995. Competing on resources: strategy in the 1990s. Harv. Bus. Rev. 73.

Collis, D.J. and Montgomery, C.A., 2005.Corporate Strategy: A Resource-Based Approach (Y. G. Wang, Y.H. Yang, Trans.). Dongbei University of Finance & Economics Press, Dalian.

Deardorff, A.V., 2001. Fragmentation across cones. Fragmentation: New Production Patterns in the World Economy. Oxford University Press, Oxford and New York, pp. 35–51.

Depner, H., Bathlt, H., 2005. Exporting the German model: the establishment of a new automobile industry cluster in Shanghai. Econ. Geogr. 81, 53–81.

Dicken, P., Thrift, N., 1992. The organization of production and the production of organization: why business enterprises matter in the study of geographical industrialization. Trans. Inst. Br. Geogr. 17, 279–291.

Dicken, P., Kelly, P., Olds, K., Yeung, H., 2001. Chains and networks, territories and scales: towards an analytical framework for the global economy. Glob. Netw. 1 (2), 89–112.

Dunning, J.H., 1977. Trade, Location of Economic Activity and the MNE: A Search for an Eclectic Approach. Palgrave Macmillan, UK.

Dunning, J.H., 1979. Explaining changing patterns of international production: in defence of the eclectic theory. Oxf. Bull. Econ. Stat. 11.

Dunning, J.H., 1993. The Theory of Transnational Corporations. Routledge, London.

Dunning, J.H., 1998. Location and the multinational enterprise: a neglected factor? J. Int. Bus. Stud. 29.

Ernst, D., 2003a. How sustainable are benefits from global production networks? Malaysia's upgrading prospects in the electronics industry. East-West Center Working Papers. p. 57.

Ernst, D., 2003b. Pathways to innovation in the global network economy: Asian upgrading strategies in the electrics industry. East-West Center Working Papers. p. 58.

Ernst, D., Kim, L., 2002. Global production networks, knowledge diffusion, and local capability formation. Res. Pol. 31, 1417–1429.

Falvey, R.E., 1981. Commercial policy and intra-industry trade. J. Int. Econ. 11 (4), 495–511.

Fan, Y., 2007. 'New' new trade theory and progress. Int. Econ. Trade Res. 12.

Feenstra, R.C., Hanson, G.H., 1996. Globalization, outsourcing, and wage inequality. Am. Econ. Rev. 86 (2), 240–245.

Gao, B.Y., Liu, W.D., Norcliffe, G., Du, C., 2011. Influences of international trade barriers on global production networks: taking China–Canada Trade in bicycles as an example. Acta Geogr. Sin. 4.

GE, 2015. Industrial Internet: Pushing the Boundaries of Minds and Machines. China Machine Press, Beijing.

Gereffi, G., 1995. Global production systems and third world development. In: Stallings, B. (Ed.), Global Change, Regional Response. Cambridge University Press, New York.

Gereffi, G., 1999a. International trade and industrial upgrading in the apparel commodity chain. J. Int. Econ. 48 (1), 37–70.

Gereffi, G., 1999b. A Commodity Chains Framework for Analysing Global Industries. Mimeo, Department of Sociology.

Gereffi, G., Korzeniewicz, M. (Eds.), 1994. Commodity Chains and Global Capitalism. Praeger, Westport.

Gereffi, G., Memedovic, O., 2003. The Global Apparel Value Chain: What Prospects for Upgrading by Developing Countries. <http://www.Unido.org>.

Gereffi, G., Humphrey, J., Kaplinsky, R., Sturgeon, T., 2001. Globalization, value chains and development. IDS Bull. 32 (3).

Gereffi, G., Humphrey, J., Sturgeon, T., 2005. The governance of global value chains. Rev. Int. Pol. Econ. 12 (1), 78–104.

gkzhan.com, June 25, 2014. China's Manufacturing Costs to Reach U.S. Level and Local Companies Busy Investing Overseas. Available from: https://www.gkzhan.com/news/Detail/44486.html.

Granovetter, M., 1985. Economic action and social structure: the problem of embeddedness. Am. J. Soc. 91 (3), 481–510.

Halinen, A., Tomroos, J.A., 1998. The role of embeddedness in the evolution of business networks. Scand. J. Manag. 14 (3), 187–205.

Han, W., 2010. A study on value-driven corporate network changes. J. Bus. Econ. 11.

Han, L.P., 2011. Building a EVA-centered corporate value management system. J. Changsha Telecommun. Technol. Voc. Coll. 4.

Heckscher, E., Ohlin, B., 1991. In: Flam, H., Flanders, M.J. (Eds.), Heckscher–Ohlin Trade Theory. MIT Press, Cambridge, MA.

Helpman, E., 2006. Trade, FDI and the organization of firms. J. Econ. Lit. 44, 589–630.

Helpman, E., Krugman, P., 1985. Market Structure and Foreign Trade. MIT Press, Cambridge, MA.

Helpman, E., Melitz, M.J., Yeaple, S.R., 2004. Export versus FDI with heterogeneous firms. Am. Econ. Rev. 94 (1), 300–316.

Helpman, E., M.J. Melitz, Y. Rubinstein, 2007. Estimating trade flows: trading partners and trading volumes. NBER Working Paper. <http://www.nber.org/papers/w12927>.

Henderson, J., Dicken, P., Hess, M., Coe, N., Yeung, H., 2002. Global production networks and the analysis of economic development. Rev. Int. Pol. Econ. 9, 436–464.

Hess, M., Coe, N., 2006. Making connections: global production networks, standards, and embeddedness in the mobile-telecommunications industry. Environ. Plan. A 38 (7), 1205–1227.

Hess, M., Rodrigue, J.-P., 2006. Transportation and the geographical and functional integration of global production network. Growth Change 37 (4), 510–525.

Hess, M., Yeung, H., 2006. Whither global production networks in economic geography? past, present and future. Environ. Plan. A 38, 1193–1204.

Hochtberger, K., Zademach, H.M., Grimes, S., 2003. Aspiring Affiliates, Global Project Networks and Local Embeddedness: Evidence from Bangalore, India. Centre for Innovation and Structural Change.

Huang, L., 2012. On features of trade protectionist policies in the post-crisis era. China Bus. Trade 22.

Huawei, 2017. Huawei Investment & Holding Co., Ltd. 2017 Annual Report. Available from: https://www.huawei.com/en/.

Hummels, D., 1999. Toward a geography of trade costs. GTAP Working Papers. Center for Global Trade Analysis, Department of Agricultural Economics, Purdue University.

Humphrey, J., Schmitz, H., 2002. How does insertion in global value chains affect upgrading in industrial clusters? Reg. Stud. 36 (9), 1017–1027.

Hymer, S., 1960. The International Operations of International Firms: A Study of Direct Investment (Unpublished Ph.D. Dissertation). Massachusetts Institute of Technology, Cambridge, MA.

Isard, W., 1956. Location and Space-dconomy, 33. MIT Press, Cambridge, MA, p. 363 (4).

Jiang, Z.H., Li, J., October 2011. Huawei's International Model of Independent Innovation, Seventh Academic Conference on Science and Technology Policy and Management, Nanjing.

Jones, R.W., Kierzkowski, H., 1990. The role of services in production and international trade: a theoretical framework. The Political Economy of International Trade: Essays in Honor of Robert E. Baldwin. Blackwell, Oxford and Cambridge, Mass, pp. 31–48.

Jones, R.W., Kierzkowski, H., 2001. A framework for fragmentation. Fragmentation: New Production Patterns in the World Economy. Oxford University Press, Oxford and New York, pp. 17–34.

Kathandaraman, Wilson, D.T., 2001. The future of competition-value-creating networks. Ind. Market. Manag. 30.

Kindleberger, C.P., 1969. American business abroad. Int. Exec. 11 (2), 11–12.

Kogut, B., 1985a. Designing global strategies: profiting from operational flexibility. Sloan Manag. Rev. 27, 27–39.

Kogut, B., 1985b. Designing global strategies: comparative and competitive value-added chains. Sloan Manag. Rev. 26, 75–98.

Krugman, P.R., 1991. Increasing returns and economic geography. J. Pol. Econ. 99.

Krugman, P.R., Venables, A., 1995. Globalization and the inequality of nations. Quart. J. Econ. 110, 135–156.

Latour, B., 1987. Science in Action: How to Follow Scientists and Engineers Through Society. Open University Press, Milton Keynes, England.

Latour, B., 1999. On recalling ANT. Soc. Rev. 47, 15–25.

Law, J., 1987. Technology and heterogeneous engineering: the case of portuguese expansion. In: Bijker, W., Hughes, T., Pinch, T. (Eds.), The Social Construction of Technological Systems. MIT Press, Cambridge, MA, pp. 111–134.

Law, J., 1999. After ANT: complexity, naming and topology. Soc. Rev. 47 (1), 1–14.

Law, J., Hassard, J. (Eds.), 1999. Actor Network Theory and After. Blackwell, Oxford.

Levy, D.L., 2005. Offshoring in the new global political economy. J. Manag. Stud. 42 (3), 685–693.

Li, F., Lin, H.C., Liu, Y., 2010. Building of China's advanced manufacturing model oriented towards global value network and its dynamic evolution: a case study of Huawei. Econ. Manag. J. 12.

Li, J., Ding, Y., 2007. An empirical study on agglomeration and vertical disintegration in the manufacturing sector. Technol. Econ. 4.

Li, J., Ning, Y.M., 2011. Emergence of global production networks: a research framework for local development based on globalization. Shanghai Econ. Rev. 9.

Li, J., Ning, Y.M., Wang, M.F., 2008. Computer industry's global production network and its development in the Chinese mainland. Acta Geogr. Sin. 4.

Li, C.D., 2010. literature review on 'New' new trade theory. World Econ. Pap. 1.

Lin, G.J., Deng, S.Z., 2011. Asian factories and correlation analysis. World Econ. Pol. 11.

Liu, C.S., 2008. Creation of Global Production Networks and China's Strategic Choices. China Renmin University Press, Beijing.

Liu, C.S., 2011. China's Path of Opening Up in the Context of Global Production Networks. China Business Update. p. 2.

Liu, D.X., et al., 2006. Global Production Networks and Processing Trade Upgrading. Economic Science Press, Beijing.

Liu, J.L., 2011. Internal support of Huawei's international expansion. China Econ. Trade Herald 4.

Liu, W., Dicken, P., 2006. Transnational corporations and 'obligated embeddedness': foreign direct investment in China's automobile industry. Environ. Plan. A 38 (4), 168–182.

Liu, Z.G., Yin, Y.M., 2006. Relationship between economic geography and economics from a historical perspective. Econ. Geogr. 3.

Losch, 1944. The Economics of Location: A Pioneer Book in the Relations Between Economic Goods and Geography. Yale University Press, New Haven, pp. 234–236.

Lu, F., 2004. Intra-product specialization. China Econ. Quart. 4.

Luo, D., 2012. GE Healthcare in Rural China. 21st Century Business Review, (13), 72–73.

Manova, K., 2008. Credit Constraints, Heterogeneous Firms, and International Trade. NBER Working Paper. <http://www.nber.org/papers/w14531>.

Marshall, A., 1890. In: Lian, Y.J. (Ed.), Principles of Economics. Huaxia Publishing House, Beijing.

Marx, K., 1867. In: Guo, D.L., Wang, Y.N. (Eds.), Das Kapital. Shanghai Joint Publishing, Shanghai.

McKinsey Global Institute, 2015. The China Effect on Global Innovation. Available from: https://www.mckinsey.com/~/media/McKinsey/Featured%20Insights/Innovation/Gauging%20the%20strength%20of%20Chinese%20innovation/MGI%20China%20Effect_Full%20report_October_2015.ashx.

Melitz, M.J., 2003. The impact of trade on intra-industry reallocation and aggregate industry productivity. Econometrica. 71, 1695–1725.

Melitz, M.J., Ottaviano, G., 2008. Market size, trade and productivity. Rev. Econ. Stud. 75 (1), 295–316.

Moses, L.N., 1958. Location and the theory of production. Quart. J. Econ. 72 (2), 259–272.

Nohria, N., Garcia-Pont, C., 1991. Global strategic linkage and industry structure. Strat. Manag. J. 12, 105–124.

Nohria, N., Ghoshal, S., 1994. Differentiated fit and shared values: alternatives for managing headquarters-subsidiary relations. Strat. Manag. J. 15 (6), 491–502.

Oldenski, L., 2015. Reshoring by U.S. Firms: What Do the Data Say? (SIFL Institute, Trans.), PIIE Policy Briefs. Available from: https://piie.com/sites/default/files/publications/pb/pb15-14.pdf.

Palpacuer, F., Parisotto, A., 2003. Global production and local jobs: can global enterprise networks be used as levers for local development? Glob. Netw. 3 (2), 97–120.

Park, D., Shin, K., 2009. Can trade with the people's republic of china be an engine of growth for developing asia? Asian Dev. Rev 27-1 (1), 160–181.

Parthasarathy, B., Aoyama, Y., 2006. From software services to R&D services: local entrepreneurship in the software industry in Bangalore, India. Environ. Plan. A 38, 1269–1285.

Pei, C.H., 2010. Economic globalization in the post-crisis era: new features and new situations. Int. Econ. Rev. 4.

Polanyi, K., 1944. The Great Transformation: The Political and Economic Origins of Our Time. Beacon Press.

Porter, M., 1980. In: Chen, X.Y. (Ed.), Competitive Strategy. Huaxia Publishing House, Beijing.

Porter, M., 1985. In: Chen, X.Y. (Ed.), Competitive Advantage. Huaxia Publishing House, Beijing.

Prahalad, C.K., G. Hamel, 1990. The core competencies of the corporation. Harvard Bus. Rev.

Pu, H.L., Zhang, J., 2008. Dynamic optimization of GPNs and industrial structure: an empirical study of the automobiles industry. Sci. Technol. Manag. Res. 10.

Qiu, G.D., Chen, J.H., 2010. A study on MNCs' embeddedness in China's coastal areas. Res. Fin. Econ. Issues 9.

Ricardo, D., 1817. In: Zhou, J. (Ed.), On the Principles of Political Economy and Taxation. Huaxia Publishing House, Beijing.

Rifkin, J., 2012. The Third Industrial Revolution: How Lateral Power is Transforming Energy, the Economy, and the World (T.W. Zhang, Y.N. Sun, Trans.). China CITIC Press, Beijing.

Roper, S., Grimes, S., 2005. Wireless valley, silicon wadi and digital Island—Helsiniki, Tel Aviv and Dublin and the ICT global production network. Gen. Inform. 36, 297–313.

Sally, W., 2006. The embeddedness of global production networks: the impact of crisis in Fiji's garment export sector. Environ. Plan. A 38 (7), 1249–1267.

Sanyal, K.K., Jones, R.W., 1982. The theory of trade in middle products. Am. Econ. Rev. 72 (1), 16–31.

Schmitz, H. (Ed.), 2004. Local Enterprises in the Global Economy. Edward Elgar, Cheltenham.

Siemens A.G., Annual Report 2014. Available from: https://www.siemens.com/global/en/home.html.

Slywotzky, A.J., 2002. The Profit Zone: How Strategic Business Design Will Lead You to Tomorrow's Profits (X.D. Ling, Trans.). China CITIC Press, Beijing.

Smith, A., 1776. In: Guo, D.L., Wang, Y.N. (Eds.), An Inquiry into the Nature and Causes of the Wealth of Nations. The Commercial Press, Beijing.

Smith, A., Rainnie, A., Dunford, M., Hardy, J., Hudson, R., Sadler, D., 2002. Networks of value, commodities and regions: reworking divisions of labour in macro-regional economies. Prog. Hum. Geogr. 26, 41–63.

Smith, D.M., 1971. Industrial Location: An Economic Geographkai Analysis. John Wiley& Sons, New York, NY.

State Council of China, July 4, 2015. Guidelines on Actively Promoting the Internet Plus Action Plan. Available from: http://www.gov.cn/zhengce/content/2015-07/04/content_10002.htm.

Sturgeon, T., 2002a. How do we define value chains and production networks? IDS Bull. 32 (3), 9–18.

Sturgeon, T., 2002b. Modular production networks: a new American model of industrial organization. Ind. Corp. Change 11 (3), 451–496.

Su, G.F., Liu, D.X., Bu, G.Q., 2005. Analysis of GPN governance mechanism. On Econ. Prob. 2.

Tang, H.Y., Zhang, H.Q., 2009. China's status in the new international division of labor system: analysis from the value chain perspective. J. Int. Trade 2.

Tang, Y.C., 2014. Huawei: growing invisible wings for strategic development. Petrol. Petrochem. Mater. Procur. 3.

Taudes, A., Trcka, M., Lukanwicz, M., 2002. Organizational learning in production networks. J. Econ. Behav. Org. 47, 141–163.

The Economist, 2012. The Third Industrial Revolution (Xing, H,F., Trans.) Shanghai Enterprise (6), 54–56.

UNCTAD, World Investment Report.

Vernon, R., 1966. International investment and international trade in the product cycle. Quart. J. Econ. 80 (2), 190–207.

Von Thunen, J.H., 1826. Der Isolierte Staat. Beziehung auf Landwirtschaft und National? Konomie.

Wang, Z.X., 2013. Huawei: open innovation makes it a top hi-tech MNC of the world. Globalization 3.

Weber, A., 1909. Ueber den Standort der Industrien. 1. Reine Theorie des Standorts. Mohr.

WEF, The Global Competitiveness Report.

Wen, H., Yang, Y.R., Hou, J.J., 2007. A study on embeddedness and FDI-driven industry clusters: taking Pudong's IC industry cluster as an example. Econ. Geogr. 5.

Whitel M.C., 2003. Inward investment, firm embeddedness and place: an assessment of Ireland's multinational software sector. CISC Working Paper.

Wild, J.J., Wild, K.L., Han, J., 2009. International Business, fourth ed. (Y. Chen, Trans.). Peking University Press, Beijing.

WTO, World Trade Report.

www.askci.com, 2012. The Development of China's Medical Device Industry (2011–2012).

Yang, X.K., 1997. Contemporary Economics and Chinese Economy. China Social Sciences Press, Beijing.

Yang, X.K., 1998. Principles of Economics. China Social Sciences Press, Beijing.

Yang, X.K., Huang, Y.G., 1999. Specialization and Economic Organization (Y.G. Zhang, Trans.). Economic Science Press, Beijing.

Yang, X.K., Zhang, Y.S., 2003. New Classical Economics and Inframarginal Analysis (Revision). Social Sciences Academic Press, Beijing.

Yang, Y.R., Xia, Z.J., 2005. Local embeddedness of GPNs and evolution of local institutions: taking greater Dongguan as an example. City Plan. 3.

Ye, Q.X., 2008. Local Embeddedness of MNCs: Theory, Empirical Study and Policy Options. Zhejiang University Press, Hangzhou.

Yeaple, S.R., 2005. A simple model of firm heterogeneity, international trade, and wages. J. Econ. 65, 1–20.

Yeung, H.W.-C., 2009. Regional development and the competitive dynamics of global production networks: an east asian perspective. Reg. Stud. 43 (3), 325–351.

Yi, H., Su, N., Li, Y.M., 2011. On supply chain collaboration in companies' transnational operations. Bus. Econ. 7.

Young, A.A., 1928. Increasing Returns and Economic Progress (G.L. Jia, Trans.). Comparative Economic & Social Systems, p. 2.

Zhang, M.N., 2008. China's economic growth in the post-crisis era in the context of unbalanced division of labor. Reform 12.

Zhang, Z.B., 2010. Global production networks: development process and momentum. Econ. Res. Guide 30.

Zhao, P., 2004. Embeddedness and industry cluster competitiveness: theoretical research and analytical framework. Southeast Acad. Res. 6.

Zhao, L.F., Gu, Q.L., 2010. A study on the relationship between wage levels and industrial upgrading in global industrial networks. Soc. Sci. Front 8.

Zhao, Z.X., Lv, Z. (Eds.), 2009a. International Trade Theories and Policies. Peking University Press, Beijing.

Zhao, Z.X., Lv, Z., 2009b. Review of studies on influencing factors of companies' export: from a perspective of trade models with heterogeneous firms. J. Int. Trade 9.

Zhou, X., Cheng, L.R., 2011. A study on MNCs' value network operation: concepts and portfolio models. Stat. Decis. 10.

Zhu, T., Sun, Q.J., 2008. Theoretical analysis of MNCs' global production model: from an industry perspective. Mod. Econ. Sci. 1.

Zhuo, X., 2011. A comparative analysis of global value chain governance models. J. Shanghai Bus. Sch. 2.

Zukin, S., DiMaggio, P., 1990. Structures of Capital: The Social Organization of the Economy. Cambridge University Press, Cambridge, MA.

Zysman, J., Doherty, E., Schwartz, A., 1997. A tales from the 'global' economy: cross-national production networks and the reorganization of the European economy. Struct. Changes Econ. Dyn. 8, 45–85.

Index

Note: Page numbers followed by "*f*" and "*t*" refer to figures and tables, respectively.

H

I

U

V

W

FSC
www.fsc.org
MIX
Papier | Fördert
gute Waldnutzung
FSC® C083411

Kolibri 360 GmbH
Ferdinand-Jühlke-Straße 7
99095 Erfurt, Deutschland
produktsicherheit@kolibri360.de